Non-Chemical Methods for Disinfestation of Stored Products

This reference book covers traditional and recent technologies in the disinfestation of stored foods. Storage of durables is challenging and demands scientific management protocols. Chemical fumigants have been used for decades, and several reports detail their impact on the environment, in addition to other concerns. This book focuses on the use of non-chemical methods of disinfestation in durables like cereals, pulses, oilseeds, millets, and spices. It discusses important disinfestation techniques like biological control, extreme-temperature-based approaches, insect-repellent packaging, and nano-based techniques. The book also covers the detection of pests and some chemical methods of disinfestation. It is an important reference for professionals, researchers, industry personnel, and post-graduate students in the field of food and grain storage.

Key Features:

- Focuses on the non-chemical methods for grain disinfestation
- Reviews the rising need for a shift to non-chemical methods of disinfestation of stored foods
- Includes a blend of topics on entomology, food engineering, food design, and regulations
- Discusses conventional methods of disinfestation as well as chemical approaches
- Covers topics like extreme temperature-based approaches, dielectric heating, insect traps, and use of inert materials

Prof. M. Loganathan is the Dean of Academics at the National Institute of Food Technology, Entrepreneurship and Management – Thanjavur (NIFTEM-T), an Institute of National Importance under the Ministry of Food Processing Industries, Government of India.

Dr. Jeyan A. Moses is Assistant Professor and In-charge of the Computational Modeling and Nanoscale Processing Unit at the National Institute of Food Technology, Entrepreneurship and Management – Thanjavur (NIFTEM-T).

Non-Chemical Methods for Disinfestation of Stored Products

Edited by
M. Loganathan and Jeyan A. Moses

CRC Press
Taylor & Francis Group
Boca Raton London New York

CRC Press is an imprint of the
Taylor & Francis Group, an **informa** business

Designed cover image: Shutterstock

First edition published 2025
by CRC Press
2385 NW Executive Center Drive, Suite 320, Boca Raton FL 33431

and by CRC Press
4 Park Square, Milton Park, Abingdon, Oxon, OX14 4RN

CRC Press is an imprint of Taylor & Francis Group, LLC

ISBN: 9781032976310 (hbk)
ISBN: 9781041004370 (pbk)
ISBN: 9781003609834 (ebk)

DOI: 10.1201/9781003609834

Typeset in Times
by KnowledgeWorks Global Ltd.

Contents

List of Contributors

S. Anandakumar
National Institute of Food Technology, Entrepreneurship and Management – Thanjavur
Thanjavur, Tamil Nadu, India

R. Anbarasan
National Institute of Food Technology, Entrepreneurship and Management – Thanjavur
Thanjavur, Tamil Nadu, India

Nairiane Bilhalva
Universidade do
Vale do Rio dos Sinos, Brazil

Indumathi Chokkanathan
National Institute of Food Technology, Entrepreneurship and Management
and
SRM Institute of Science and Technology
Thanjavur, Tamil Nadu, India

Monisha Chandran
National Institute of Food Technology, Entrepreneurship and Management – Thanjavur
Thanjavur, Tamil Nadu, India

V. Chandrasekar
National Institute of Food Technology, Entrepreneurship and Management – Thanjavur
Thanjavur, Tamil Nadu, India

V. Chelladurai
Department of Agricultural Engineering
Bannari Amman Institute of Technology
Sathyamangalam, Erode, Tamil Nadu, India

Paulo Carteri Coradi
Universidade do
Vale do Rio dos Sinos, Brazil

Aldrey Nathália Ribeiro Corrêa
Universidade do
Vale do Rio dos Sinos, Brazil

Sankarganesh E.
Bihar Agricultural University
Bhagalpur, Bihar, India

Cristiano Dietrich Ferreira
Mestrado Profissional em Nutrição e Alimentos e Pesquisador no Instituto Tecnológico em Alimentos para a Saúde
Itt Nutrifor Unidade Acadêmica de Pesquisa e Pós-Graduação
Universidade do
Vale do Rio dos Sinos, Brazil

L. Gopianand
Insect Taxonomy Laboratory, Department of Entomology, Faculty of Agriculture
Annamalai University
Annamalainagar, Tamil Nadu, India

Digvir S. Jayas
University of Lethbridge
Lethbridge, Canada

Fuji Jian
Department of Biosystems Engineering
University of Manitoba
Winnipeg, Manitoba, Canada

C. Kathirvelu
Stored Product Research Laboratory, Department of Entomology, Faculty of Agriculture
Annamalai University
Annamalainagar, Tamil Nadu, India

L. Kavitha
National Institute of Food Technology, Entrepreneurship and Management – Thanjavur
Thanjavur, Tamil Nadu, India

S. Ashwin Kumar
National Institute of Food Technology, Entrepreneurship and Management – Thanjavur
Thanjavur, Tamil Nadu, India

M. Loganathan
National Institute of Food Technology, Entrepreneurship and Management – Thanjavur
Thanjavur, Tamil Nadu, India

B. O. Madhu
National Institute of Food Technology, Entrepreneurship and Management – Thanjavur
Thanjavur, Tamil Nadu, India

R. Mahendran
National Institute of Food Technology, Entrepreneurship and Management – Thanjavur
Thanjavur, Tamil Nadu, India

Irengbam Barun Mangan
Kelappaji College of Agricultural, Engineering and Food Technology
Tavanur, Kerala, India

A. Mariadoss
National Institute of Plant Health Management
Rajendranagar, Hyderabad, Telangana State, India

R. Meenatchi
National Institute of Food Technology, Entrepreneurship and Management – Thanjavur
Thanjavur, Tamil Nadu, India

R. Jagan Mohan
National Institute of Food Technology, Entrepreneurship and Management – Thanjavur
Thanjavur, Tamil Nadu, India

Sarma Mohan
Department of Agricultural Entomology
Tamil Nadu Agricultural University
Coimbatore 641 003, India

Jeyan A. Moses
National Institute of Food Technology, Entrepreneurship and Management – Thanjavur
Thanjavur, Tamil Nadu, India

N. Muthukrishnan
Stored Product Research Laboratory, Department of Entomology, Faculty of Agriculture
Annamalai University
Annamalainagar, Tamil Nadu, India

V. Eyarkai Nambi
National Institute of Food Technology, Entrepreneurship and Management – Thanjavur
Thanjavur, Tamil Nadu, India

Aditi Negi
National Institute of Food Technology, Entrepreneurship and Management – Thanjavur
Thanjavur, Tamil Nadu, India

Ricardo Tadeu Paraginski
Universidade do
Vale do Rio dos Sinos, Brazil

Vinass Jamali P.
National Institute of Food Technology, Entrepreneurship and Management – Thanjavur
Thanjavur, Tamil Nadu, India

P. Paulin Patricia
National Institute of Food Technology, Entrepreneurship and Management – Thanjavur
Thanjavur, Tamil Nadu, India

D. Praveenkumar
Department of Agricultural Engineering
Bannari Amman Institute of Technology
Sathyamangalam, Erode, Tamil Nadu, India

Larissa Alves Rodrigues
Universidade do
Vale do Rio dos Sinos, Brazil

Vismaya K. Sachithanandhan
National Institute of Food Technology, Entrepreneurship and Management – Thanjavur
Thanjavur, Tamil Nadu, India

P. Santhoshkumar
National Institute of Food Technology, Entrepreneurship and Management – Thanjavur
Thanjavur, Tamil Nadu, India

Kannan Senguttuvan
Department of Agricultural Entomology
Tamil Nadu Agricultural University
Coimbatore, India

T. Senthilkumar
University of Prince Edward Island
Charlottetown, PEI, Canada

S. Shanmugasundaram
National Institute of Food Technology, Entrepreneurship and Management – Thanjavur
Thanjavur, Tamil Nadu, India

Anantika Sharma
National Institute of Food Technology, Entrepreneurship and Management – Thanjavur
Thanjavur, Tamil Nadu, India

C. B. Singh
Lethbridge College
Lethbridge, Alberta, Canada

Mariappan Suganthy
Department of Agricultural Entomology
Tamil Nadu Agricultural University
Coimbatore 641 003, India

C. Alice R. P. Sujeetha
National Institute of Plant Health Management
Rajendranagar, Hyderabad, Telangana State, India

Ashok Suraj
National Institute of Food Technology, Entrepreneurship and Management – Thanjavur
Thanjavur, Tamil Nadu, India

M. D. Tejeshwini
National Institute of Food Technology, Entrepreneurship and Management – Thanjavur
Thanjavur, Tamil Nadu, India

J. Thendral
National Institute of Food Technology, Entrepreneurship and Management – Thanjavur
Thanjavur, Tamil Nadu, India

Newiton da Silva Timm
Universidade do
Vale do Rio dos Sinos, Brazil

Valmor Ziegler
Universidade do
Vale do Rio dos Sinos, Brazil

1 Grain Storage Ecosystem

Fuji Jian
Department of Biosystems Engineering,
University of Manitoba, Winnipeg, Canada

Digvir S. Jayas
President's Office, University of Lethbridge, Lethbridge, Canada

1.1 INTRODUCTION

The concept of a grain storage ecosystem was developed more than 70 years ago, and the ecosystem approach is widely applied in grain storage management now. This practice follows the general trend of less application of chemicals and more application of physical methods, monitoring tools, and new techniques to reduce the resistance of bioagents to pesticides and fungicides, less chemical residue, less pollution, less energy cost, and sustainability for the grain industry. Customers are now more sensitive to air pollution, chemical residues, and fresh taste and color of processed foods. The concerns over its best monetary value have also increased. These trends urge the grain industry to focus its investment and efforts on efficient and reliable grain quality management techniques. Climate change makes grain storage more challenging and requires more energy consumption for grain drying, handling, and storage. This challenge and change need the application of a thorough ecosystem approach coupled with new techniques for the development of sustainable methods to maintain grain quality. Any method in the development procedure should consider grain storage as a whole ecosystem in a set of agricultural ecosystems. Therefore, it is needed to have a detailed description of the grain storage ecosystem, including its boundary, major factors, distinctive properties and development, diversity, complex energy flow network and interaction among factors of a grain storage ecosystem, and general recommendations and guidelines for grain ecosystem management.

1.2 GRAIN STORAGE ECOSYSTEM

The term ecosystem is arbitrarily defined as a combination of the biotic community (or called biotic or living factors) and its nonliving environment (or called abiotic or nonliving factors) interacting with each other as a functioning system of complementary relationships, including transfer and circulation of energy and materials (Dunkel, Barak, and Harein 1982). The biotic community comprises a group of populations of plants and animals in the system. Population denotes a group of individuals of a single or multiple species living together in the system. Grain (cereal grains, oil seeds, and pulses) can be stored in any type of structure, including containers (different sizes, materials, and shapes of bags, sacks, cribs, bottles, boxes, pails, drums, and buckets) and building structures (different sizes, materials, and shapes of bins, silos, underground pits, and warehouses). Any bulk of stored grain in any kind of storage structure is a human-made and solar-powered ecosystem that is composed of dormant grain or serves as an energy source and habitat for many biotic communities, including fungi, bacteria, insects, mites, and occasionally rodents and birds. The solar-powered system means that the stored grain is originally produced by photosynthesis, and the physical and chemical status of the stored grain is also influenced by its ambient environment, which is affected by solar radiation and any form of energy, such as wind, which is originally powered by the sun.

DOI: 10.1201/9781003609834-1

For a well-developed stored grain ecosystem, there are usually three well-represented communities: the fungal community, the insect community, and the mammal and avian communities. The fungal community includes those that can develop in an oxygen environment and those that can only develop in a reduced or no oxygen atmosphere. The insect community includes predators, parasitoids, and scavengers, as well as those insects which actually cause loss of stored grain. A well-managed grain storage ecosystem has no mammal communities. Birds, rodents, and voles are common members of mammal and avian communities. Squirrels and shrews are other pest species that may appear in a storage structure. Domestic mammals and humans are the intended end-users of the system. Squirrels, shrews, domestic mammals, and humans can be counted as members of the mammal and avian communities. The population of each community member generally has a dynamic trend, and at the same time as the dynamic population, a succession of different species occurs. Jian et al. (2018) summarized the process of the succession of insects in stored grain ecosystems. They also described the development of hotspots and demonstrated the succession of insect and fungal communities in hotspots.

Grain storage is a way or process by which harvested products are kept for future use. During storage, the stored biomaterials are ecologically unstable due to the activity of biotic factors and the interaction between biotic and abiotic factors. This activity usually results in deterioration of the stored biomaterials. The objective of grain storage managers is to preserve the biomaterials for future consumption by humans or domesticated animals or for future planting. To meet these objectives, managers intervene to prevent or reduce the interactions among biotic and abiotic factors so the energy and nutrient flow through the consumers and decomposers can be delayed. If the stored grain is well managed, the population of the fungal, insect, mammal, and avian communities is minimized. Therefore, stored grain in granaries can be considered a human-made ecosystem with a relatively simple structure, a non-regenerating food energy supply, and minimized biotic communities.

1.2.1 Boundary of Grain Storage Ecosystem and Interaction among Ecosystems

A grain storage ecosystem is usually considered an isolated system because it is physically separated from other systems by the walls of storage structures and the low level of interaction with other ecosystems (Figure 1.1). Even though the physical boundary, such as bin walls, is usually defined as the boundary of the storage ecosystem, a grain storage ecosystem is an open system in which energy and materials constantly enter and leave the storage structure. A bag of grain, a grain

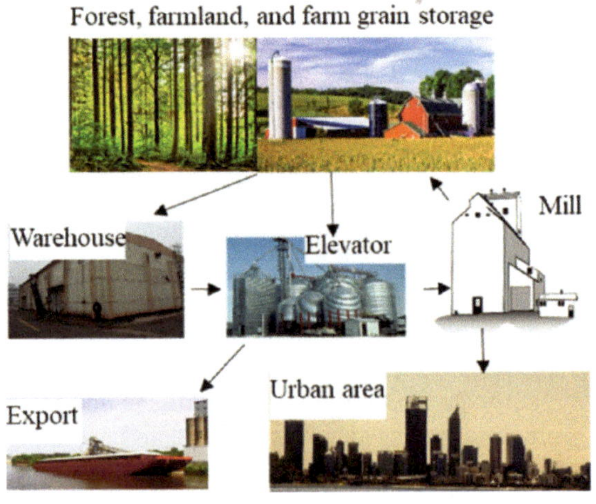

FIGURE 1.1 A set of ecosystems showing the main logistic chains of grain and its processed products from producers to consumers.

bin, a grain elevator, or a region where there are many granaries can all be considered as a whole or part of the ecosystem if they interact with each other (Figure 1.1). Insects can migrate between inside and outside of the grain bin. In the U.S.A., the natural habitat of *Rhyzopertha dominica* (F.) is probably the forest ecosystem. Insects can migrate from the forest ecosystem to the stored grain bins (the grain storage ecosystem). The movement of this insect from desert to storage to urban ecosystems has been documented (Dunkel, Barak, and Harein 1982). Another example of the open ecosystem is the global distribution of insects and fungi. Even with customs inspection, insects move in small amounts of stored products brought by travelers. When one or few reproductive females arrive in a mated condition in a new territory, it may establish a colony. With this grain in transit, insects, genetic materials, and fungi move. For example, the fungi *Aspergillus flavus* and *Aspergillus parasiticus*, primarily found in nuts and maize, are now located around the world on many products.

Another example of the open grain storage ecosystem is that the ambient environment is usually counted as an abiotic factor of the grain storage ecosystem, even though the bin walls are usually counted as the boundary of the system. The boundary of a grain storage ecosystem is continually influenced by the ambient environments. For example, bin walls have a continuous energy exchange with the ambient air and receive solar radiation, resulting in a flow of energy across the boundary. Thus, the grain storage ecosystem can be a set of systems, a system within a system connected by the transportation systems, field or forest systems, and the systems of commerce and food distribution chains (Figure 1.1). The grain storage ecosystem is usually connected by the transportation systems, the systems of commerce, and international food aid distribution systems.

Compared with the interaction of biotic and abiotic factors inside the systems, the grain storage systems have a low level of interaction with other systems because the interactions with other systems are mostly managed by human activities. For example, good preparation and management of grain before storage can narrow down the initial infestation of insects and fungi (such as the infestation of field and storage fungi) by decreasing grain temperature and moisture content using forced air ventilation of grain.

1.2.2 Major Factors of the Stored Grain Ecosystem and Their Interaction

The parameters influencing stored grain ecosystems include nonliving (abiotic) factors, living factors, and grain management operations. Nonliving factors include time, weather, bin structure, and the physical environment of the stored grain. Time is one of the main factors influencing the grain ecosystem because any actions, such as quality change and insect multiplication, need time. Weather includes temperature, relative humidity, wind speed and direction, solar radiation, and precipitation. Bin structure includes the shape, size, and wall materials of the bin and its geographical location. The physical environment includes temperature, grain moisture content, dockage and foreign materials in grain, air relative humidity, intergranular gaseous composition, and chemicals. The gaseous composition includes oxygen (O_2), nitrogen (N_2), carbon dioxide (CO_2), and odor. Chemicals include pesticides, fungicides, and chemical components of the grain, including carbohydrates, fat acidity, protein, and water. Living (biotic) factors include the living grain, consumers (insects, mites, fungi, and occasionally birds and rodents), and decomposers (fungi). Insect, mites, and fungal fragments might be consumed by other insects, mites, or fungi (Figure 1.2). Therefore, these insects, mites, and fungi can also be the secondary producers in the ecosystem. Decomposers include fungi, bacteria, and other microorganisms that break down the biological material into small chemical compounds (Figure 1.2). These biotic and abiotic factors interact with each other at any level of energy and nutrition flow in the system (Figure 1.2). Other factors of the nonliving environment that slightly influence the ecosystem are light, radiation, chemicals, and pressure. Grain management operations include any management actions modifying the physical, chemical, and biological environments, such as aeration and drying.

The factors mentioned above can also be categorized as physical, chemical, and biological environments of the storage ecosystem (Figure 1.3). These environments interact with each other,

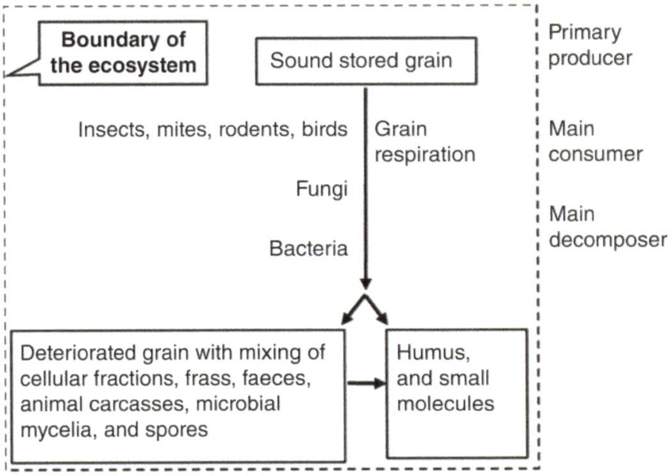

FIGURE 1.2 Nutrient flow in a stored grain ecosystem due to consumption and decomposition of the stored grain. The deteriorated grain, with the mixing of other fractions, can be further downgraded to humus and small molecules by the consumers and decomposers. (Reprinted with permission from Fuji Jian and Jayas [2012].)

which affects the grain storage ecosystem. The grain will significantly change its quality and quantity when all those environments favor the reduction of quality and quantity of the stored grain. Figure 1.3 also indicates that controlling the physical environment is critical because both chemical and biological environments are influenced by the physical environment, even though both chemical and biological environments can change the physical environment sometimes. For example, the multiplication of insects and fungi can change physical, chemical, and biological environments. However, this change is usually slow and in a small area, while the change of physical environment is in a large area and usually the entire storage ecosystem.

During the grain storage period, the reduction of grain quality and quantity is usually caused by one factor at the very beginning. After the reduction of grain quality and quantity is initiated, multiple chain actions occur that further reduce grain quality and quantity. For example, insects can infest the newly binned grain. The insects can come from outside of the bin through migration. Some insects might already infest the bin before the grain is binned because most insects only need

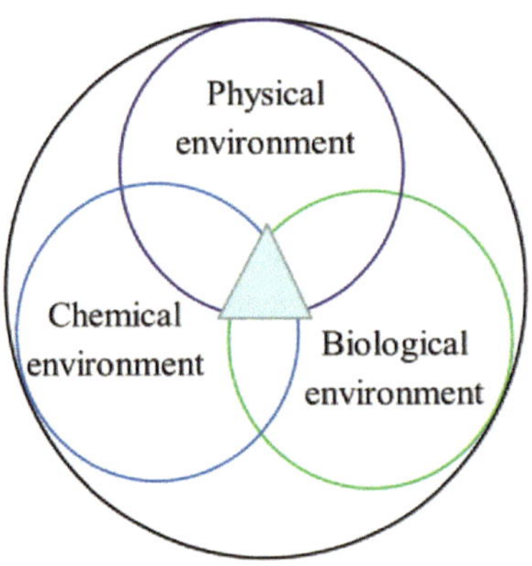

FIGURE 1.3 Interaction of the storage grain environments. The triangle indicates the condition favoring the reduction of the stored grain quantity and quality.

one or a few grain kernels to complete their life cycle, and they can feed on the residue of the grain if the bin is not thoroughly cleaned before grain loading. The grain might be already infested before it is loaded into the bin because the insects can come from the transportation and handling equipment. Insects can also fly into the grain (Jian 2019). One of the major factors that mostly initiates the reduction of grain quality and quantity is mold infection. Most mold can multiply on grain when the temperature is higher than 5°C in grain with higher than 0.7 water activity. However, mold multiplication is very slow at these low temperatures and increases with an increase in one or both the temperature and water activity. There is a significant increase in mold development and multiplication after insect infestation because respiration of the insects and grain can produce heat and water, which favor the development of microflora (Sinha, Liscombe, and Wallace 1962; Sinha, Wallace, and Chebib 1969). Mold multiplication and development will further increase the moisture content and temperature of the grain because mold multiplication will produce more water and heat. After heavy mold infection, grain quality and quantity are significantly reduced (Seitz, Sauer, and Mohr 1982; Nithya et al. 2011; Sun et al. 2014).

The reasons for the chain actions are as follows: (1) the initiation of the reduction of grain quality and quantity gradually or rapidly changes the physical, chemical, and biological environments of the ecosystem; (2) those changes in the storage ecosystem usually favor the downgrade of the grain quality and quantity, such as the multiplication and development of mold and insects because they multiply fast at warm and wet conditions; and (3) chain action further and quickly downgrades the grain quality because most of the chemical reaction and biological activity produce heat and water, which favor the multiplication and development of insects, mites, and mold. For example, insect infestation can damage grain kernels, and damaged grain has a higher chemical reaction. Those infestations and chemical reactions produce water and heat, which result in the multiplication of mold. Multiplication of mold produces more water and heat, which further encourages the multiplication of mold.

The most important factors influencing grain quality are the temperature and moisture content because any biological activity (e.g., multiplication and development of insects and mold) and chemical reaction require heat and water. For example, most stored product insects cannot complete development when the temperature is lower than 7–17.6°C, depending on insect species. Therefore, controlling grain temperature and moisture content is the basic requirement for maintaining grain quality during the management of stored grain ecosystems.

1.2.3 Distinctive Properties and Development of Stored Grain Ecosystems

The main difference between the grain storage ecosystem and other natural ecosystems is that the storage ecosystem has a limited biomass, a short time span (limited storage time), and an open system that is made by man and managed (interrupted) by human beings (human intervention). So, diversities of animal species are artificially selected and reduced by human management to maintain stored grain quality with minimum loss of the stored grain. There are diversities of stored grain/product ecosystems, e.g., some systems are simple, i.e., cereal grain, oilseed, and legume seeds stored in small containers (baskets, boxes, gourds, bags, small bins), and others are bags packed on pallets in warehouses to create a controlled atmosphere or large containers for keeping the materials. The most modern grain/product system is a grain or bulk in a bin (silo) or warehouse equipped with handling, monitoring, and ventilation systems and only stores one single variety of a crop type. Different types of grain might be stored in bags and piled in one warehouse or storage structure. These systems require energy or petroleum-based input for proper management of the systems, such as pesticide application, drying, and aeration.

For a well-managed stored grain ecosystem, the system stays at its initial stage. Its biotic communities might only have grain, and fungal spores and bacteria are at their dormancy stage. The grain storage management practice is to maintain this system at its initial stage as long as possible. The diversity of the system will increase along with the development of the ecosystem. For example, infestation of insects and fungi will increase the abundance of the system. After the ecosystem is well developed and reaches its mature stage, the abundance of the system is rich because it will be full of fungal and insect communities. The energy used for ecosystem development mostly comes from the stored grain.

The composition of insect species changes as the grain deteriorates, and dominance by certain species is associated with the grain quality. Arbogast and Mullen (1988) and Fuji Jian and Jayas (2012) reviewed the insect succession in stored grain ecosystems, which demonstrated the diversity of the insect species and infestation level in the grain storage ecosystem. Even though a grain storage ecosystem is much simpler than a field ecosystem, a grain storage ecosystem has its complexity and diversity. The complexity includes the complex interactions of the factors and complex network of energy flows.

1.2.3.1 Diversity of Stored Grain Ecosystems

The diversity of stored grain ecosystems is caused by different sizes and shapes of storage structures, different types and varieties of stored grain, different moisture contents and percentages of dockage and foreign materials. This grain can be infested by different insects, mites, and fungi. The diversity is different at different geographical locations because other local ecosystems and climates at the location influence the stored grain ecosystem. Abiotic factors such as temperature and relative humidity regulate insect and fungal communities. Therefore, abiotic factors mostly influence the abundance of biotic factors. This abundance could also influence the diversity due to insect migration and fungal diffusion among different ecosystems. There are lots of studies on insect population dynamics and abundance at different temperatures (Aspaly et al. 2007; Jian et al. 2018; Tripathi et al. 2021), grain moisture content (Arbogast 1976; Armitage, Cook, and Baxter 2002), and diet and food sources (White et al. 1995; Ahmad, Walter, and Raghu 2012; Athanassiou, Kavallieratos, and Boukouvala 2016). Biotic factors also influence insect abundance and diversity through competition. White and Sinha (1980) found high bacterial counts were associated with high FAV (fat acidity value) levels; *Tarsonemus* numbers were positively related to *Aspergillus*; *Alternaria* and seed germination were negatively related to FAV, bacteria, and grain damage; and that the number of insects was related to the presence of *Aspergillus* and negatively related to the presence of bacteria. Seed germination and *Alternaria* infection often decrease rapidly, presumably because of infection by fungi of the *Aspergillus glaucus* group. The combined action of *R. dominica* and *Aspergillus* sp. enhanced seed damage and increased grain moisture content, thus promoting bacterial growth, which in turn inhibited insect and mold growth. Fat acidity values increased with time unless seed damage and bacterial infection were extensive (White and Sinha 1980). At a local scale, the diversity, distribution, and dominance of these stored grain insects are determined by the prevailing local climate and available crop genotypes (Sinha and Watters 1985).

Grain trade and personnel traveling make most stored products, insects, and fungi cosmopolitan. Grain is always stored at certain recommended temperatures and moisture contents regardless of the country. Thus, the diversity of stored grain ecosystems becomes similar in different countries. However, diversity in each stored grain ecosystem increases because it might have more than a few insect species and fungi multiply in the stored grain ecosystem. More than 500 insect species are reported to be associated with grain (Hagstrum and Subramaniyam 2009) and 150 species of insects and mites are listed as grain and its product pests, but only 10–15 species occur frequently, and most are beetles. The diversity of the ecosystem indicates a more occasional infestation of uncommon insects. Harish et al. (2018) surveyed six districts of Hyderabad Karnataka region, India and found 25 species on pulses, and different numbers of insects were found in different districts and different pulses. Naveena, Subramanya, and Setty (2015) surveyed the tribal community living in the forest areas of Biligirirangana Hills in southern Karnataka, India (about 540 km^2) and found 13 species of insects infesting stored grains. *Sitophilus oryzae* was the most widely occurring and abundant species, while *Corcyra cephalonica* was the least abundant and less frequently occurring species. Species richness was influenced by the number of food grain types stored by the tribal community. They concluded that biotic and abiotic factors (e.g., temperature and relative humidity) did not influence the diversity and distribution of stored grain insects, but anthropogenic factors like food grain sharing habitat among the Soligas area and accessibility to the towns and cities from their settlements influenced the diversity of the insects. They concluded that the grain movement pattern is a driving force for the insect distribution in the tribal areas. However, most researchers consider major environmental factors like temperature, RH, and food sources to regulate the diversity and distribution of organisms (Sinha and Watters 1985; Savopoulou-Soultani et al. 2012).

1.2.3.2 Complex Energy Flow Network and Interaction among Factors

Even though there are few specific studies focusing on the complexity of the energy flow network and interaction among factors, almost all published studies demonstrated the complexity of the energy flow network and interactions among factors in a stored grain ecosystem. For example, when a safe storage condition of a new crop is studied, several storage temperatures and moisture contents should be considered because fungi will have different multiplication rates under different temperatures and moisture contents. The interaction between the temperature and moisture content influences the germination of the crop. The crop type and its physical and chemical conditions should also be considered. Another example is aeration – a common practice to reduce or control insects. When aeration is studied, the following factors should be considered: insect species and stages of the insects, the cold hardiness of the insect species and their life stages, the minimum development temperature and moisture content, ambient temperature and relative humidity and their change during the aeration period, and the bin structure and configuration. Fuji Jian and Jayas (2012) provided a simple interaction map of those interactions (Figure 1.4). Compared with an agricultural ecosystem, a grain storage ecosystem might be simpler. However, full modeling of

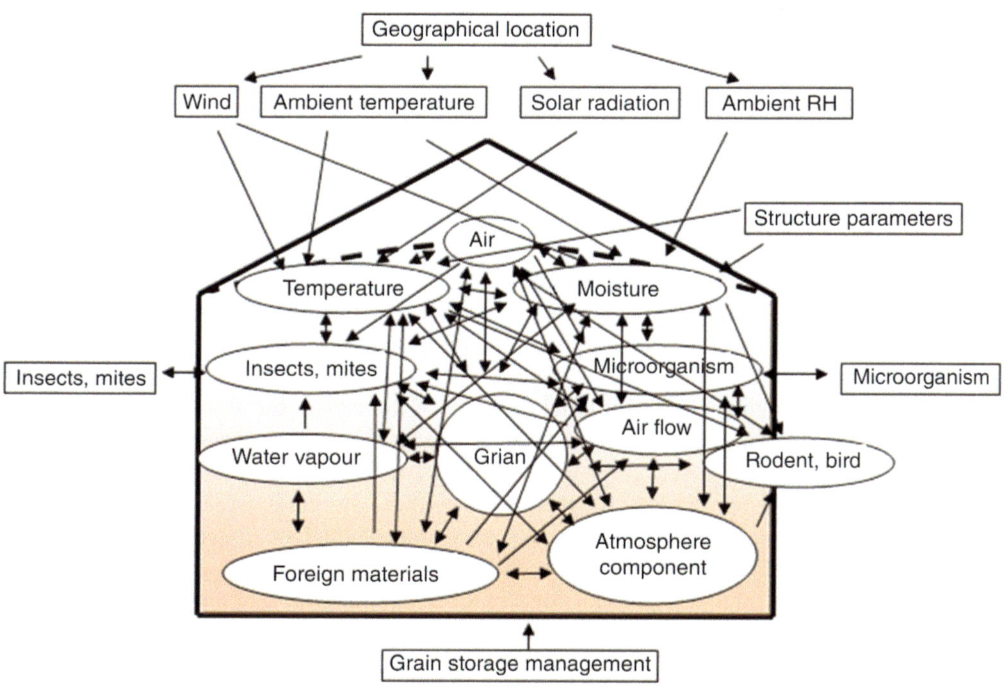

FIGURE 1.4 Factors in a stored grain ecosystem and the interactions among factors in non-aerated and unsealed bulk grain. (Reprinted with permission from Fuji Jian and Jayas [2012].)

the storage grain ecosystem is not achieved. The most complex model published up to now is the model that can predict temperature, moisture content, and convection current in a bin. Some of those models have been combined with insect movement (Mani et al. 2001) or aeration (Arthur and Flinn 2000).

Interactions also occur among grain storage ecosystems and other ecosystems, e.g., agricultural field systems. The harvested grain should be quickly dried to its recommended safe storage moisture contents. After drying, it should be loaded into bins, and aeration should be conducted if the binned grain temperature is warm (e.g., higher than 13°C), and natural air drying should be conducted if the grain moisture content is higher than the recommended safe moisture content. During this process, the grain storage ecosystem is naturally associated with the agricultural field system. The infesting insects and fungi in the field might also be loaded into the bin. The grain storage ecosystem is also influenced by the transportation and grain loading practices, which are influenced by the local weather and ecological systems. During the grain storage period, the stored grain ecosystem will be influenced by other ecosystems associated with the grain bin (Lawrence and Maier 2011a, 2011b; Lawrence et al. 2012; Jian et al. 2015). For example, the natural habitat of *R. dominica* (F.) is probably the forest ecosystem. The insects fly and migrate among the ecosystems. Insects can move into or out of the bins. Insect migration from the southern U.S.A. to Canada during summer influences the population and diversity of insects in Canada.

The energy network of a grain storage ecosystem is always in the direction of becoming more forms due to consumption by animals and decomposition by microflora. The plants harvest solar energy and produce grain seeds. The stored grain seeds at the initial stage of the stored grain ecosystem only have the form of grain kernels. After the stored grain is infested by the fungal communities consumed by insect, mammal, and avian communities, the stored grain becomes deteriorated grain with a mixing of cellular fractions, frass, feces, animal carcasses, microbial mycelia, and spores. Along with this energy flow, CO_2, heat, and water are produced through the respiration of the grain itself, insects, and fungi. Because the CO_2 and heat gradually diffuse out of the stored grain bin, the total energy is gradually decreased with the dry mass decrease of the stored grain. The target during grain storage is to limit dry mass loss to less than 0.5% because more than 0.5% of dry mass loss will result in a significant quality drop, such as germination decrease and FAV increase if this dry mass loss is caused by mold multiplication. After the ecosystem loses its energy, more energy forms (such as energy in the forms of insect and mold bodies, rotten grain kernels, gas, and heat) would be developed.

1.3 ECOSYSTEM MANAGEMENT

The best method to manage the stored grain ecosystem is to use the ecosystem approach because any modification of the system through grain storage management will influence the factors of the grain storage ecosystem. The ecosystem approach to stored grain is "a strategy for the integrated management of stored grain that promotes conservation and sustainable methods to manage stored grain to deliver high-quality grain and products at the end of the storage period" (Jian and Jayas 2012). The application of the ecosystem approach is to use the concept of an ecosystem to understand and manage the stored grain ecosystem thoroughly. In an ecosystem, all factors are intertwined and linked to economic and optimized use of the stored ecosystem without damaging the ecosystems; therefore, those interactions among factors should be considered when short- or long-term goals are set. The main goal of the application of the ecosystem approach is to achieve an environmentally friendly and sustainable solution for stored grain management.

To manage the stored grain ecosystem, understanding the ecosystem and application of ecosystem approach are the keys (Jian and Jayas 2012). Fuji Jian and Jayas (2012) reviewed the progress in understanding the stored grain ecosystem. It is noticed that the ecosystem approach has been used in research and development widely now, and most recommendations and guidelines for grain storage management are developed by using the concept of an ecosystem approach.

1.3.1 General Recommendation and Guideline for Grain Ecosystem Management

The general recommendations and guidelines for safely storing grain are to clean bins, put grain in dry, cool grain, and sample for insects. Due to the complex interaction of living and nonliving factors in a stored grain ecosystem, grain storage management should focus on implementing preventive management rather than solving storage problems post-occurrence. These recommendations and guidelines are developed based on the concept of the ecosystem approach and should be used for the preventive management of stored grain.

"Clean bins" means thoroughly cleaning bins and grain before the grain is binned. The cleaning could be conducted during harvesting, handling, or after drying. The main idea of this recommendation is to limit any source of mold, insects, dockage, foreign materials, rodents, birds, and any sources that might influence grain quality and/or provide the source for the multiplication of any biotic factors.

"Put grain in dry" means the grain should be kept at the recommended safe stored grain moisture content. The safe storage time for almost all main grain types has been developed (Mills and Sinha 1979; Nithya et al. 2011; Jian 2022). The safe storage time can be predicted by using the moisture content of the stored grain (Jian 2022). Charts and tables of safe storage times for different crops are available (Figure 1.5). The main idea of this recommendation is that stored fungi cannot multiply and develop slowly at lower than 70% relative humidity. Keeping grain in a dry condition is also an effective management tool for mite control (Stejskal et al. 2019). Even though the safe storage time is developed by evaluating the time of 20% of germination reduction from the initial germination,

Grain Temp. °C	Grain Moisture Content (%)											
	6	7	8	9	10	11	12	13	14	15	16	17
5	>300					<300	150-250	100-240	80-105	20-50	10-30	8-15
10	>300				<300	130-210	130-200	100-200	60-90	10-40	10-30	5-15
15	>300			<300	130-200	100-180	60-100	40-80	20-70	10-30	5-8	5-8
20	>300	<300	<300	100-150	70-120	40-60	30-60	20-40	8-15	5-8	3-5	
25	>300	150-240	100-180	70-105	30-70	20-40	15-30	8-15	5-8	3-5		
27	<300	130-210	70-150	50-90	20-50	8-15	8-15	5-8	3-5			
30	150-210	90-130	40-80	30-70	8-15	5-8	5-8	3-5				

FIGURE 1.5 Safe storage time (days) of canola. (This chart is developed based on a 20% loss in germination from its initial value for three tested high-oil cultivars reported by Sun et al. [2014].)

germination loss during the safe storage time study was mainly influenced by the water activity (relative humidity) at a given temperature (Fuji Jian, Vignesh, and Carlin 2022). High water activity is one of the main reasons causing grain quality and quantity loss. Therefore, "put grain in dry" is the basic requirement from the view of ecosystem approach.

"Cool grain" indicates the grain should be kept at 5–13°C. This temperature range is the guideline and target temperature of aeration (Arthur et al. 2008). It is developed by using the knowledge of the biology of insects and mold (such as their development at different temperatures and moisture contents), a physical property of the stored grain (such as grain temperature gradients and moisture migration in convection currents), and economic feasibility of grain storage practices. Even though microflora can develop and multiply at lower than the recommended temperature, they develop slowly and might stop their development and multiplication when at the recommended safe storage moisture content and the recommended temperature. The recommended temperature range mainly targets insect control because insects cannot complete their life cycle when the temperature is lower than 5–13°C (Stejskal et al. 2019).

"Sample for insects" indicates that samples are required only for detecting insect infestations and no need for detecting mold infestations if the grain is stored under the recommended safe storage moisture contents. This recommendation is due to the fact that (1) insects can develop and multiply in grain with lower than the recommended safe storage moisture contents; (2) insects can migrate into the bin by flying and walking; (3) insects can infest empty bins by hiding in refuge sites, and most of them only need one-grain kernel; and (4) insects can infest the grain before the grain is binned.

From the view of the ecosystem approach, the general recommendations and guidelines for safely storing grain should be practiced as a whole. Emphasizing one at the expense of another will result in the negligence of the grain storage ecosystem as a whole system in a set of ecosystems. This negligence might result in high energy consumption and the multiplication of insects, mites, and mold, which cause quality and quantity loss of the stored grain.

1.3.2 Stored Grain Management

Any practice of ecosystem management should be treated as a component part of the stored grain ecosystem and align with the general recommendations and guidelines. "A component part" indicates that any practices of stored grain management should not contradict others and should support each other. It is the best new practice that can provide a synergy effect on "clean bins, put grain in dry, cool grain, and sample for insects." For example, application guidelines for any pesticide and fungicide should be developed based on this general recommendation and guideline. There are times, however, when grain should be at temperatures above this range, i.e., during fumigation, grain temperature should be 25°C, but in most cases, the basic requirement of general recommendations and guidelines for grain ecosystem management should be followed. It should not increase its percentage of dockage and broken grain, decrease its bulk and kernel density, increase grain temperature and moisture content, and increase the chance of insect and fungi infestation.

The main management method of stored grain includes a prediction of the safe storage time of the grain, and prevention and control of grain quality and quantity loss. Prediction of safe storage time of the grain has uncertainty due to the variation of the following factors: carry-over from previous storage, indeterminate harvest conditions, different harvesting and handling methods, changeable ambient weather conditions during storage, and different storage structure and size. Due to those uncertainties, the safe storage time of the grain is estimated by only considering the grain temperature and moisture content.

The most used methods of prevention of grain quality and quantity loss are grain drying, cleaning, cooling (aeration), heating, monitoring, controlled or modified atmospheres, grain turning and coring, and physical methods used to eliminate insects, microflora, and rodent harborages. The insect control method includes physical and chemical methods. The physical methods include heat and cold treatment, removal of spoiling/spoiled grain, controlled (modified) atmosphere storage, hermetic storage, and chemical (such as fumigants) application. Some methods like grain drying, cooling (aeration), modified atmospheres, and grain turning are both prevention and control methods. Therefore, these methods should be considered as the primary measure in the entire storage period. The grain managers and farmers should always use the prevention methods and only apply the control methods (such as pesticide application) as a supplementary and remedial measure under urgent situations, such as when the insect number is higher than the economic threshold.

1.3.3 Stored Grain Management Using New Technology

Many new technologies have been introduced into grain storage management at different times. The most used techniques now are to monitor and predict the condition of the stored grain such as measuring grain temperature, moisture content, and CO_2 concentration; detecting insect activity and density; and automatically controlling grain drying and aeration. These techniques can be coupled with the internet (Internet of Things), so remote monitoring and control of the stored grain is possible, and this technique has been commercialized. These introduced emerging technologies extend beyond the traditional binary relationship and offer greater scope for scientific and intelligent theoretical analysis of grain storage practices. Artificial intelligence (AI) and machine vision have been used in grain storage management, such as insect species classification, grain temperature and moisture mode analysis, and data-driven approach to stored grain ecosystem. Digital twin technology will make the following management techniques possible: (1) tracking grain in the supply chain from producer to consumer; (2) simulating and mapping the dynamic condition of stored grain in storage structures; (3) supporting assessments and early warnings; and (4) analyzing, summarizing, and optimizing management operations.

The introduced technologies provide new opportunities for better grain storage management. However, these new techniques require (1) scientifically efficient, multidisciplinary, interdisciplinary, and intelligent studies; (2) a deep understanding of the grain storage ecosystem; and (3) sustainable approaches which include no pesticide and fungicide, no pollution, no insect and fungi resistance, low cost and energy input, and high quality of stored grain. Given the diversity and complexity of the stored grain ecosystems, innovation in multidisciplinary and interdisciplinary coupling theory is crucial. Therefore, the ecosystem approach of stored grain ecosystem should be used as the basic guideline, and it is imperative to explore mathematical models to simulate the real dynamic ecosystem of stored grain in any type of stored structure.

1.4 CONCLUSION

Stored grain bulk is an open ecosystem in a set of agricultural ecosystems and includes the combination of the biotic community and its nonliving environment interacting with each other as a functioning system of complementary relationships, including the transfer and circulation of energy and materials. The main difference between the grain storage ecosystem and other natural ecosystems is that the storage ecosystem has a limited biomass, a short time span (limited storage time), and an open system that is made by man and managed (interrupted) by human beings (human intervention). So, diversities of animal species are artificially selected and reduced by human management to maintain stored grain quality with minimum loss of the stored grain/products. Even though a grain storage ecosystem is much simpler than a field ecosystem, a grain storage ecosystem has its complexity and diversity. The complexity includes the complexity of interactions of factors and a complex network of energy flows. The best method to manage the stored grain ecosystem is to use the ecosystem approach because any modification of the system through grain storage management will influence the factors of the grain storage ecosystem. The general recommendation and guideline for safely storing grain are "clean bins, put grain in dry, cool grain, and sample for insects." Any practices of the stored grain ecosystem should be treated as a component part of the stored ecosystem and align with the general recommendations and guidelines as much as possible. Even though new technologies provide new opportunities for better grain storage management, one or all of the following ecosystem approach methods should be applied when a new technique is used: (1) multidisciplinary, interdisciplinary, and intelligent studies; (2) deep understanding of grain storage ecosystem; (3) sustainable approaches; and (4) mathematical model development to simulate the real dynamic ecosystem of stored grain in any stored structures.

REFERENCES

Ahmad, F., G. H. Walter, and S. Raghu. 2012. "Comparative Performance of *Tribolium castaneum* (Herbst) (Coleoptera: Tenebrionidae) across Populations, Resource Types and Structural Forms of Those Resources." *Journal of Stored Products Research* 48 (January): 73–80. doi:10.1016/j.jspr.2011.09.005.

Arbogast, R. T. 1976. "Population Parameters for *Oryzaephilus surinamensis* and *O. mercator*: 1 Effect of Relative Humidity 2." *Environmental Entomology* 5 (4): 738–742. doi:10.1093/ee/5.4.738.

Arbogast, R. T., and M. A. Mullen. 1988. "Insect Succession in a Stored-Corn Ecosystem in Southeast Georgia." *Annals of the Entomological Society of America* 81 (6): 899–912. doi:10.1093/aesa/81.6.899.

Armitage, D. M., D. A. Cook, and D. E. Baxter. 2002. "Farm-Scale Experiments to Compare Infestation and Quality Changes in Malting Barley Stored at Three Moisture Contents." *Journal of the Institute of Brewing* 108 (2): 178–186. doi:10.1002/j.2050-0416.2002.tb00538.x.

Arthur, F. H., and P. W. Flinn. 2000. "Aeration Management for Stored Hard Red Winter Wheat: Simulated Impact on Rusty Grain Beetle (Coleoptera: Cucujidae) Populations." *Journal of Economic Entomology* 93 (4): 1364–1372. doi:10.1603/0022-0493-93.4.1364.

Arthur, F. H., Y. Yang, L. T. Wilson, and T. J. Siebenmorgen. 2008. "Feasibility of Automatic Aeration for Insect Pest Management for Rice Stored in East Texas." *Applied Engineering in Agriculture* 24 (3): 345–350. doi:10.13031/2013.24489.

Aspaly, G., V. Stejskal, S. Pekár, and J. Hubert. 2007. "Temperature-Dependent Population Growth of Three Species of Stored Product Mites (Acari: Acaridida)." *Experimental and Applied Acarology* 42 (1): 37–46. doi:10.1007/s10493-007-9074-1.

Athanassiou, C. G., N. G. Kavallieratos, and M. C. Boukouvala. 2016. "Population Growth of the Khapra Beetle, *Trogoderma granarium* Everts (Coleoptera: Dermestidae) on Different Commodities." *Journal of Stored Products Research* 69 (October): 72–77. doi:10.1016/j.jspr.2016.05.001.

Dunkel, F. V., A. V. Barak, and P. K. Harein. 1982. "Geographical Distribution of *Cynaeus angustus* (LeConte) (Coleoptera: Tenebrionidae) and Its Association with Stored Products." *Journal of Biogeography* 9 (4): 345. doi:10.2307/2844721.

Hagstrum, D. W., and B. Subramaniyam. 2009. *Stored-Product Insect Resource*, 518. Elsevier Inc.

Harish, G., A. Naganagoud, A. G. Sreenivas, Somashekar, S. Hiregoudar, and B. Kisan. 2018. "Diversity and Distribution of Stored Grain Insect Pests of Pulses and Their Natural Enemies." *Indian Journal of Agricultural Research*, 52 (2): 157–161.

Jian, F. 2019. "Influences of Stored Product Insect Movements on Integrated Pest Management Decisions." *Insects* 10 (4): 100. doi:10.3390/insects10040100.

Jian, F. 2022. "A General Model to Predict Germination and Safe Storage Time of Crop Seeds." *Journal of Stored Products Research* 99 (December): 102041. doi:10.1016/j.jspr.2022.102041.

Jian, F., V. Chelladurai, D. S. Jayas, and N. D. G. White. 2015. "Three-Dimensional Transient Heat, Mass, and Momentum Transfer Model to Predict Conditions of Canola Stored Inside Silo Bags under Canadian Prairie Conditions: Part II. Model of Canola Bulk Temperature and Moisture Content." *Transactions of the ASABE* (August): 1135–1144. doi:10.13031/trans.58.11053.

Jian, F., and D. S. Jayas. 2012. "The Ecosystem Approach to Grain Storage." *Agricultural Research* 1 (2): 148–156. doi:10.1007/s40003-012-0017-7.

Jian, F., D. S. Jayas, P. G. Fields, N. D. G. White, H. Zhang, and P. Tang. 2018. "Demography of Rusty Grain Beetle in Stored Bulk Wheat: Part I, Population Dynamics at Different Temperatures and Grain Bulk Sizes." *Environmental Entomology* 47 (2): 244–255. doi:10.1093/ee/nvy019.

Jian, F., R. Vignesh, and L. Carlin. 2022. "Dry Mass Loss of Different Moisture Content Canola at Different Temperatures." *Canadian Biosystems Engineering* 62 (1): 3.01–3.10. doi:10.7451/CBE.2020.62.3.1.

Lawrence, J., and D. E. Maier. 2011a. "Development and Validation of a Model to Predict Air Temperatures and Humidities in the Headspace of Partially Filled Stored Grain Silos." *Transactions of the ASABE* 54 (5): 1809–1817. doi:10.13031/2013.39820.

Lawrence, J., and D. E. Maier. 2011b. "Three-Dimensional Airflow Distribution in a Maize Silo with Peaked, Levelled and Cored Grain Mass Configurations." *Biosystems Engineering* 110 (3): 321–329. doi:10.1016/j.biosystemseng.2011.09.005.

Lawrence, J., D. E. Maier, J. Hardin, and C. L. Jones. 2012. "Development and Validation of a Headspace Model for a Stored Grain Silo Filled to Its Eave." *Journal of Stored Products Research* 49 (April): 176–183. doi:10.1016/j.jspr.2012.02.002.

Mani, S., W. E. Muir, D. S. Jayas, and N. D. G. White. 2001. "Computer Modelling of Insect-Induced Hot Spots in Stored Wheat." *Canadian Biosystems Engineering* 43 (1): 7–14.

Mills, J. T., and R. N. Sinha. 1979. "Safe Storage Periods for Farm-Stored Rapeseed Based on Mycological and Biochemical Assessment." *Postharvest Pathology and Mycotoxins* 70 (6): 541–547.

Naveena, N. L., S. Subramanya, and S. Setty. 2015. "Diversity and Distribution of Stored Grain Insects among the Soliga Tribal Settlements of Biligirirangana Hills, Karnataka, India." *Journal of Stored Products Research* 62 (May): 84–92. doi:10.1016/j.jspr.2015.04.002.

Nithya, U., V. Chelladurai, D. S. Jayas, and N. D. G. White. 2011. "Safe Storage Guidelines for Durum Wheat." *Journal of Stored Products Research* 47 (4): 328–333. doi:10.1016/j.jspr.2011.05.005.

Savopoulou-Soultani, M., N. T. Papadopoulos, P. Milonas, and P. Moyal. 2012. "Abiotic Factors and Insect Abundance." *Psyche: A Journal of Entomology* 2012: 1–2. doi:10.1155/2012/167420.

Seitz, L. M., D. B. Sauer, and H. E. Mohr. 1982. "Storage of High-Moisture Corn: Fungal Growth and Dry Matter Loss." *Cereal Chemistry* 59 (2): 100–105.

Sinha, R. N., E. A. R. Liscombe, and H. A. H. Wallace. 1962. "Infestation of Mites, Insects and Microorganisms in a Large Wheat Bulk after Prolonged Storage." *The Canadian Entomologist* 94 (5): 542–555. doi:10.4039/Ent94542-5.

Sinha, R. N., H. A. H. Wallace, and F. S. Chebib. 1969. "Principal-Component Analysis of Interrelations among Fungi, Mites, and Insects in Grain Bulk Ecosystems." *Ecology* 50 (4): 536–547. doi:10.2307/1936244.

Sinha, R. N., and F. L. Watters. 1985. *Insect Pests of Flour Mills, Grain Elevators, and Feed Mills and Their Control.* Agriculture Canada.

Stejskal, V., T. Vendl, Z. Li, and R. Aulicky. 2019. "Minimal Thermal Requirements for Development and Activity of Stored Product and Food Industry Pests (Acari, Coleoptera, Lepidoptera, Psocoptera, Diptera and Blattodea): A Review." *Insects* 10 (5): 149. doi:10.3390/insects10050149.

Sun, K., F. Jian, D. S. Jayas, and N. D. G. White. 2014. "Quality Changes in High and Low Oil Content Canola during Storage: Part I – Safe Storage Time under Constant Temperatures." *Journal of Stored Products Research* 59 (October): 320–327. doi:10.1016/j.jspr.2014.05.008.

Tripathi, R., D. S. Jayas, F. Jian, and V. A. D. Hervet. 2021. "Population Dynamics of Red Flour Beetle (*Tribolium castaneum* (Herbst)) under Different Temperatures and Patch Sizes." *Journal of Stored Products Research* 94 (December): 101865. doi:10.1016/j.jspr.2021.101865.

White, N. D. G., C. J. Demianyk, H. Kawamoto, and R. N. Sinha. 1995. "Population Growth of *Cryptolestes ferrugineus* and *C. pusillus* (Coleoptera: Cucujidae) Alone, or in Competition in Stored Wheat or Maize at Different Temperatures." *Bulletin of Entomological Research* 85 (3): 425–429. doi:10.1017/S0007485300036154.

White, N. D. G., and R. N. Sinha. 1980. "Principal Component Analysis of Interrelations in Stored-wheat Ecosystems Infested with Multiple Species of Insects." *Population Ecology* 22 (1): 33–50. doi:10.1007/BF02513534.

2 Insects Attacking Stored Products

P. Paulin Patricia, R. Meenatchi, and M. Loganathan
National Institute of Food Technology Entrepreneurship and Management, Thanjavur, Tamil Nadu, India

2.1 INTRODUCTION

Insects that destroy or cause damage to stored grains are generally considered as stored product insects. More than hundreds of species of insects associated with stored products, including beetles, weevils and moths, have indeed evolved over millions of years ago. The postharvest losses of total food grains in India are more than 10%. This loss might be due to improper storage structures, storage methods and cross-contamination by insects and microbes. The World Bank reported that 12–16 MMT (million metric tons) of food grains are lost every year in India. Coleoptera (larva and adult) and Lepidoptera (larva) order of insects cause severe damage to stored products. The loss occurred not only due to damage caused by insects, insect wastes and webbing, leading to moisture migration, fungal activity and deterioration of grains. This chapter deals with the stored product insects commonly encountered during storage.

2.2 CLASSIFICATION OF STORED PESTS

The stored product insects are generally classified as follows:

- *Primary pest*: The insects that cause damage to undamaged or whole grains are called primary pests.
- *Secondary pest*: The insects that damage broken grains and milled and already infested products are called secondary pests. Based on the feeding characteristics of insects, they are grouped into internal feeders and external feeders.
- *Internal feeders*: These insects feed only the internal portion of the kernel. It is very difficult to detect until the emergence of an adult from the kernel. It spent both the larval and pupal stages inside the grains. Thus, internal feeders cause severe damage to stored products (Table 2.1).
- *External feeders*: They only feed on the surface of the grain, damaged grains, milled and final products (Table 2.1).

2.2.1 Internal Feeders

2.2.1.1 Weevils (*Sitophilus oryzae, S. zeamais* and *S. granarius*)

2.2.1.1.1 Identification

- Weevils are characterized by their distinct beak or snout.
- Adult weevils typically have a size of about 3 mm.
- Weevils have elbowed antennae, and these antennae are slightly clubbed in shape.
- Weevil coloration can range from reddish brown chocolate to almost brown, indicating that they come in various shades within this color spectrum.

 DOI: 10.1201/9781003609834-2

TABLE 2.1
Internal and External Feeders of Stored Products

Common Name	Scientific Name	Family and Order	Pest Destructive Stage	Infested Domain
		A. Internal Feeders		
Rice weevil (or) black weevil	*Sitophilus oryzae* (L.)	Curculionidae, Coleoptera	Grubs and adults	All cereals
Maize weevil	*Sitophilus zeamais* (Motsch.)	Curculionidae, Coleoptera	Larvae and adult	Raw or processed cereals such as wheat, oats, barley, sorghum, rye and buckwheat
Granary weevil	*Sitophilus granarius* (L.)	Curculionidae, Coleoptera	Grubs and adults	All small grains (notably wheat, corn, rye, oats, barley, sorghum)
Lesser grain borer	*Rhyzopertha dominica* (F.)	Bostrichidae, Coleoptera	Grubs and adults	All cereals and some pulses
Pulse beetle	*Callosobruchus maculatus* (F.) *C. chinensis* L.	Bruchidae, Coleoptera	Grubs and adults	All pulses
Cigarette beetle	*Lasioderma serricorne* (F.)	Anobiidae, Coleoptera	Grubs and adults	Tobacco products, cocoa, dried cassava, black and red pepper, ginger, turmeric, dried fruits, chili powder, spices etc.
Drug store beetle	*Stegobium paniceum* (L.)	Anobiidae, Coleoptera	Grubs	Turmeric, coriander and ginger
Tamarind beetle or groundnut bruchid	*Caryedon serratus* (Oliv.)	Bruchidae, Coleoptera	Grubs and adults	Tamarind seeds, groundnut
Angoumois grain moth	*Sitotroga cerealella* (Oliv.)	Gelechiidae, Lepidoptera	Larvae	All cereals
		B. External Feeders		
Khapra beetle	*Trogoderma granarium* (Everts)	Dermestidae, Coleoptera	Grub and adults	Cereal products such as wheat, barley, oats, rye, maize, rice, flour, malt and noodles
Red flour beetle	*Tribolium castaneum* (Herbst.) *T. confusum*	Tenebrionidae, Coleoptera	Grubs and adults	All milled products
Warehouse moth (or) almond moth	*Ephestia cautella* (Walker)	Phycitidae, Lepidoptera	Larvae	Wheat, rice, maize, jowar, groundnuts and spices
Indian meal moth	*Plodia interpunctella* (Hubner)	Phycitidae, Lepidoptera	Larvae	Broken grains and milled cereal products such as flour, flakes, pasta and spices
Rice moth	*Corcyra cephalonica* (Staint.)	Galleriidae, Lepidoptera	Larvae	Rice, sorghum, maize, gram and groundnut

(Continued)

TABLE 2.1 *(Continued)*
Internal and External Feeders of Stored Products

Common Name	Scientific Name	Family and Order	Pest Destructive Stage	Infested Domain
		II. Secondary Pest		
Saw toothed beetle	*Oryzaephilus surinamensis* (Linn.)	Silvanidae, Coleoptera	Grubs and adults	Rice, wheat, maize and cereal products, oil seeds and dry fruits
Flat grain beetle	*Cryptolestes ferrugineus* (Stephens)	Cucujidae, Coleoptera	Grubs and adults	All cereals and some pulses (preferably milled products)
Long-headed flour beetle	*Latheticus oryzae* (Water)	Tenebrionidae, Coleoptera	Grubs and adults	Cereals, maize, sorghum, oatmeal, pasta, dried cassava and tea
Grain lice	*Liposcelis divinatorius* (Muller), *L. transvaalensis* (Enderlein)	Liposcelidae, Psocoptera	Larvae and adult	All food grains

- Weevils generally have hind wings, except for a species called *Sitophilus granarius.*
- Among the three species mentioned, *Sitophilus zeamais* is the largest one.

2.2.1.1.2 Life History and Habits

The female may lay translucent white eggs on the grain surface. An adult lays up to 150–250 eggs during its life span. Once the eggs hatch, larvae emerge and feed on the endosperm of the grain for about 20 days. After the larval stage, the pupal stage typically lasts for four to six days. When development is complete, the adult comes out by leaving a circular hole in the surface of the kernel. The life span of an adult is usually three to five months; during this period, they breed and lay eggs. The whole process from egg to adult stage is completed in 26 days at a temperature of 30°C and 75% R.H. They usually hibernate or remain less active during the winter season. These conditions are conducive to their rapid development. The germination power of the grain is lost because adult feeds on starchy endosperm, and larvae feed only on the germ (Singh et al. 2017).

2.2.1.1.3 Nature of Damage

Weevils are known for hollowing out grains and reducing them to powdery form. This kind of feeding behavior causes significant damage to grains. *Sitophilus oryzae* and *S. zeamais* are field carry-over insects that will start to attack grains from the field. Adult weevils are responsible for cutting circular holes in grains. These holes are typically an indication of weevil infestation in stored grain products.

2.2.1.1.4 Geographical Range

S. oryzae is native to India and spreads later throughout the world.

2.2.1.2 Lesser Grain Borer (*Rhyzopertha dominica*)

2.2.1.2.1 Identification

- Small cylindrical shape with a length of 3 mm and width of 1 mm.
- Its color ranges from brown to black, with a large three-segmented antenna.
- There is a prominent constriction between prothorax and elytra.
- One of the most important characteristic features of the lesser grain borer is that its head is deflexed downward to such an extent that it is almost hidden when viewed from above. This gives them a hunched appearance.

2.2.1.2.2 Life History and Habits

The adult may lay a single or cluster of eggs on a grain surface. Approximately a female can lay 550 eggs in her lifetime. The grubs normally enter the kernel, grow and pupate after the third instar. It takes about 25 days to complete one life cycle and feeds on the starchy endosperm portion of the grains. An ideal climatic condition for lesser grain borer is 34°C with R.H. 70%, and the moisture content of seed remains 14% (Edde 2012). This insect prefers dark, dingy stores. Infestation in bagged storage is not normally confined to the periphery.

2.2.1.2.3 Nature of Damage

The life span of adults is longer; they voraciously feed on the internal portion of the kernels and reduce them to mere shells. The irregular holes in the package and small piles of messy waste flour or webbing in the package are common signs of infestation (Biosci et al. 2015).

2.2.1.2.4 Geographical Range

It originated from the Indian subcontinent and distributed throughout the tropics of North America, South America and Australia.

2.2.1.3 Pulse Beetle (*Callosobruchus maculatus* and *C. chinensis*)

2.2.1.3.1 Identification

- It is 3–4.5 mm in length, colored brown with gray and black spots on the abdomen, which is heart-shaped.
- The male beetle abdomen is smaller than the female.
- The antenna is pectinate (male) and serrate (female).

2.2.1.3.2 Life History and Habits

The adult lays a single egg, which is glued on the surface of the grains. The fresh eggs are translucent in color and turn opaque once hatched. The eggs hatch within three to five days. A female adult can have the ability to produce 80–100 eggs. The larva is fleshy and creamy white in color with black-colored mouth. The complete life cycle takes 28–35 days at 30°C and 70% R.H. The pulse beetle breeds more in the months of July and August.

2.2.1.3.3 Nature of Damage

The larva feeds on the endosperm portion of the grains. Adults make exit holes to come out of the grain (Kumari 2020). The life span of adults is short, and they will not feed on grains.

2.2.1.3.4 Geographical Range

It is a cosmopolitan pest that originated in Africa. Later, it abundantly distributed to tropical and subtropical regions of Asia.

2.2.1.4 Cigarette Beetle (*Lasioderma serricorne*)

2.2.1.4.1 Identification

- It is oval-shaped, brown-colored, with 2.0–2.5 mm long.
- The elytra of the beetles are smooth and have very short hairs.
- The antennae are relatively short, approximately half the length of the body. The fourth to tenth segments of the antennae have a serrated or saw-like appearance.
- They tuck their head under the protective shield-like structure called the pronotum when they feel disturbed.
- The larvae of these beetles are white and have a scarab beetle-like shape.

2.2.1.4.2 Life History and Habits

Cigarette beetles can lay around 100–110 eggs loosely on the grain surface. Once hatched, the larvae often eat their eggshells. At first, the larva is quite active and will move around on or bore into the commodity. They will move deeply into a loosely packed commodity than into one that is tightly packed, generally four to six larval generations per year. The larval period lasts for four to five weeks (Edde 2019). The pupal stage takes eight to ten days. Adults are active fliers, flying freely in the evening and at night. Adults live for two to four weeks and do not feed.

2.2.1.4.3 Nature of Damage

The larvae are very active and make cylindrical holes into the commodity.

2.2.1.4.4 Geographical Range

It is found worldwide and prefers to survive in tropical and subtropical climatic zones than in temperate regions.

2.2.1.5 Drug Store Beetle (*Stegobium paniceum*)

2.2.1.5.1 Identification

- It is 2.25–3.5 mm long, cylindrical in shape and reddish brown in color.
- The elytra are surrounded by fine hairs longitudinally.

- Similar to *L. serricorne* but can be distinguished by its antennae (not serrated).
- The elytra are smooth in appearance.

2.2.1.5.2 Life History and Habits

The biology of this insect is similar to that of *Lasioderma*, except that *L. serricorne* performs better at higher temperatures and grows more rapidly than *S. paniceum*.

2.2.1.5.3 Nature of Damage

Grubs damage the product by boring cylindrical holes.

2.2.1.5.4 Geographical Range

It is distributed worldwide and found abundantly in temperate regions.

2.2.1.6 Tamarind Beetle (*Caryedon serratus*)

2.2.1.6.1 Identification

- The fresh egg is oval, translucent and milky white in color, which turns into grayish white.
- The adult is a small and brownish-gray-colored beetle that lays an egg singly on the grain surface.
- It has a long serrated antenna, and the abdomen is not fully covered by elytra.
- The hind femur can be enlarged, which helps to jump and run.
- The female beetle is larger and broader than the male (Behera et al. 2016).

2.2.1.6.2 Life History and Habits

Eggs are laid on the pods or shells. Once hatched, larvae penetrate through the shell into the kernel. It feeds on the kernel and develops into an adult. It takes 42 days to complete one life cycle at 25°C and 75% R.H. The grubs have the habit of migrating from place to place, especially during the last instar and pupate, even outside storage containers or outside the pods or bags.

2.2.1.6.3 Nature of Damage

Larvae damage the groundnut and tamarind by making circular holes in the shell or pod (Sharma et al. 2017). This insect infestation is identified by visualizing the larvae emergence hole and the presence of cocoons outside the pods.

2.2.1.6.4 Geographical Range

It is commonly found in India, Egypt, Thailand, Vietnam and Australia.

2.2.1.7 Grain Moth (*Sitotroga cerealella*)

2.2.1.7.1 Identification

- They are typically small and grayish brown in color.
- The wings can expand up to 10–14 mm, and the rear edges of the wings are fringed.
- Adults fly during nighttime and are attracted to light.
- It leaves tiny specks when they come in contact with windows and walls.

2.2.1.7.2 Life History and Habits

An adult lays 100 eggs one by one on the surface of the grains. Depending upon climatic conditions, the incubation period deviates from 4 to 30 days. On hatching, the larva enters into the internal portion of the grain and starts feeding. It is brownish gray in color. It bears true legs (three pairs) and prolegs (four pairs) with 5 mm long. Larva undergoes four instars after about six days each at a temperature of 30°C and 76% R.H. At a pupal stage, it may hibernate during the winter season.

Matured larva spins a cocoon to pass the pupal stage inside the grain. Afterward, the adult is pushed out of the cap, leaving a neat circular emergence hole with a typical flap door. The only stage noticed in granaries is adult, which is short-lived.

2.2.1.7.3 Nature of Damage

The larva drills into the endosperm portion of the grain, pupate and emerge through the circular hole. The infestation is confined to a 30 cm depth from the top surface.

2.2.1.7.4 Geographical Range

It prefers to survive in mild to moderate climatic conditions.

2.2.2 External Feeders

2.2.2.1 Khapra Beetle (*Trogoderma granarium*)

2.2.2.1.1 Identification

- The adult is oval, and the size ranges from 1.5 to 3 mm and is reddish brown in color. There is no distinct separation of body parts. The abdomen is relatively larger. The entire body is covered with fine hairs.
- The sexual dimorphism of females is well-developed and comparatively larger than males.
- The larvae are typically straw-colored with distinct hairy bands on each segment.

2.2.2.1.2 Life History and Habits

The life cycle of the khapra beetle completes in 25 days, with four to five generations per year. Breeding takes place from April to October and hibernates in the larval stage from November to March. On starvation, larvae have the capability to reduce their size, hide in cracks or crevices and become inactive. It can survive without food for one year. But adults are harmless, do not eat, never fly and have short life spans.

2.2.2.1.3 Nature of Damage

Larvae generally feed on the germ portion of grains. The presence of skins of larvae, frass and debris indicates the infestation of grains. Damage is observed on the peripheral layer of the bag and bulk storage.

2.2.2.1.4 Geographical Range

It originated in Punjab and spread to many parts of the Middle East, Asia, Europe and Africa. It has the capability to sustain in extremely dry climates.

2.2.2.2 Red Flour Beetle (*Tribolium castaneum, T. confusum*)

2.2.2.2.1 Identification

- It is flat, oblong in shape, ranges from 3 to 4 mm and is colored reddish brown.
- The eyes are completely notched for *Tribolium confusum* in, whereas in *Tribolium castaneum*, the eyes *are* notched partially.
- *T. castaneum* has a clear, three-segmented clubbed antenna.
- *T. confusum* is found in imported food grains and in hilly regions.
- The larvae are yellowish-white, cylindrical, with pale brown heads covered with fine hairs.

2.2.2.2.2 Life History and Habits

Females can lay from 150 to 300 eggs in their life span (five to eight months). About 50% of eggs are laid in the first four to five weeks (Abdel-Hady et al. 2021). Eggs are white, translucent, sticky in nature and take 5–12 days to hatch. Larvae feed on broken grains and milled products. Adults are active during the rainy season. It takes three to five weeks to complete one life cycle at the temperature of 25–27°C and 60–70% R.H. An adult can survive for more than a year and rapidly multiplies during the rainy season. Males remain fertile for the entire life span, while females can lay eggs for up to one year.

2.2.2.2.3 Nature of Damage

Red flour beetle is considered a primary insect to flour and milled products and a secondary insect to grains. It prefers to feed on the germ portion. It constructs tunnels when moved through flour. It also releases gaseous quinines (producing an acid odor) during a high infestation, which turns the flour unfit for consumption (Villaverde, Juárez, and Mijailovsky 2007).

2.2.2.2.4 Geographical Range

It is found worldwide, especially in tropical and subtropical regions.

2.2.2.3 Warehouse Moth (*Ephestia cautella*)

2.2.2.3.1 Identification

- The moth's length is 13 mm; the forewings are broader toward the apex and may expand to 12 mm.
- The larval head is dark brown in color.
- It makes cocoon coverings with food particles and sticky secretion to pupate.
- It is nocturnal in nature and hides in dark places during day time.

2.2.2.3.2 Life History and Habits

Adult lays whitish eggs in cracks and crevices and incubates for three to four days. The larval period lasts for 30–40 days, during which it feeds on the germ portion of grains and spins a silken cocoon. Pupation occurs inside the cocoon and lasts for 10–12 days. Once matured, the adult comes out from the cocoon. The average life span of an adult is approximately 14 days. At 28–30°C with R.H. above 70%, it takes 25 days to develop from the egg to the adult stage.

2.2.2.3.3 Nature of Damage

Larva feeds on only the germ portion of the grains, which significantly affects the nutritional value and grain quality. During bulk infestation, damage caused by larvae is limited to the peripheral region. The insect can create webbing as a part of their feeding and pupation process. This leads to clogging in mills, and processing and handling of grains becomes challenging.

2.2.2.3.4 Geographical Range

It is commonly found in non-arid regions of tropical and subtropical zones.

2.2.2.4 Indian Meal Moth (*Plodia interpunctella*)

2.2.2.4.1 Identification

- The wingspan of adults is 15 and 8 mm long at rest.
- The forewing is dual colored with silver white and reddish brown.
- It has long hind wings with silky fringes.
- The antenna is two-thirds of body size when at rest.

2.2.2.4.2 Life History and Habits

The female moth may lay 100–365 eggs and incubates for 2–14 days. These eggs are grayish white, spherical and less than 1 mm in diameter. Emerged larvae start feeding immediately and have the capability to enter packaging material through cracks and holes (Johnson 2013). Matured larvae can grow up to 13 mm long. Larva lasts for about four to five weeks and pupal period takes ten days. It takes 26 days to develop fully at 20°C and 70% R.H.

2.2.2.4.3 Nature of Damage

The feeding and webbing behavior of larvae causes intense damage to the grains. Adults fly from infested grain to other commodities, which results in cross-contamination.

2.2.2.4.4 Geographical Range

It basically originated in Asia and adopted a wide range of climatic conditions.

2.2.2.5 Rice Moth (*Corcyra cephalonica*)

2.2.2.5.1 Identification

- Adult gray colored, 12 mm length at rest with 15 mm wingspan.
- Projected tuft of scale present in the head.
- Dark veins present in the forewings and hind wings are translucent.

2.2.2.5.2 Life History and Habits

A small, oval, whitish egg incubates for a period of three to five days. Emerged larva actively feeds on the broken grains. Fully matured larva can grow up to 12 mm in size and form dense webbing. Larvae proceed to pupate in 10–15 days. Pupation occurs for a period of ten days. It has a short life span of 7–14 days. It takes 28–35 days to complete one life cycle at 23–32°C and 75% R.H. It is nocturnal in nature, and commonly, infestation occurs in the peripheral region.

2.2.2.5.3 Nature of Damage

It attacks especially rice and is commonly present in flour mills. Larva causes severe damage to the grains (Menge et al. 2019). The rice quality is reduced by larval debris, feeding behavior and webbing.

2.2.2.5.4 Geographical Range

It spreads in Europe, Asia, North America and Africa. Also, it is commonly found in India and Pakistan.

2.2.3 Secondary Pest

2.2.3.1 Saw Toothed Beetle (*Oryzaephilus surinamensis*)

2.2.3.1.1 Identification

- Adults are dark brown in color with a 2.5 mm length. Six saw teeth-like projections on each side of the thorax.
- Males can be differentiated from females with the presence of a tooth on the femur of the hind leg.
- The abdomen is completely covered by the elytra.

2.2.3.1.2 Life History and Habits

It lays about 375 eggs freely or in crevices. Eggs hatch in 3–5 days, and the larval period lasts for 12–15 days. The emerged larvae are flattened and light yellow in color. Larva builds a silky web spun with viscous secretion and food particles to pupate. It takes three weeks to complete one life cycle at 30°C and 75% of R.H. Adults may live for three years with eight generations. It generally survives for more than seven months. Adults are winged but rarely fly and tend to migrate from commodity (Trematerra et al. 2012). So, it is difficult to eradicate.

2.2.3.1.3 Nature of Damage

The grains become rough and uneven due to severe damage caused by adults and larvae. Insect debris and secretions generate unpleasant odors in stored grains.

2.2.3.1.4 Geographical Range

Saw toothed beetle prefers to survive in warmer, temperate and tropical regions.

2.2.3.2 Flat Grain Beetle (*Cryptolestes ferrugineus*)

2.2.3.2.1 Identification

- It is flat, reddish brown in color, and size ranges from 1.5 to 2 mm long. So, it is considered the smallest one among stored pests.
- A long, thin, hair-like antenna is located at the head.
- A fine longitudinal line is present on both sides of the neck. Larva is white and grows up to 3 mm long.

2.2.3.2.2 Life History and Habits

Eggs are laid freely in flour, grain and cracks. It incubates for a period of five days, and yellowish-white larvae emerge with 0.7 mm length. The average lifetime of an adult is between 6 and 12 months. From egg to adult, it takes six weeks, but in summer it may take nine weeks. The life cycle completes in six weeks at 28°C optimum temperature and 70% R.H. It may increase to nine weeks during the summer season. Flat grain beetles are more tolerant to cold climates.

2.2.3.2.3 Nature of Damage

Both larvae and adults feed on broken kernels and milled products. High infestation and enormous feeding behavior of larvae lead to heat generation and an increase in moisture content. This results in mold growth, and stored grains become unfit for consumption.

2.2.3.2.4 Geographical Range

It is distributed all around the world.

2.2.3.3 Long-Headed Flour Beetle (*Latheticus oryzae*)

2.2.3.3.1 Identification

- Adult 2–3 mm long with light brown in color.
- It appears like *Tribolium* spp. But the proportion of the head is longer than body.
- It has a short antenna and five clubbed apical segments.

2.2.3.3.2 Life History and Habits

The life cycle consists of four stages: egg, larva, pupa and adult. Around 400 white, smooth and cylindrical eggs are laid randomly in grains and crevices of the stored bag. The larvae are creamy white and can grow to about 4–5 mm in length. They feed on the food products, creating tunnels and causing damage during this stage. Long-headed flour beetles are typically scavengers and feed on broken grains, fragments and food residues. They can be transported through infested food products, making it essential to practice proper food storage and sanitation to prevent infestations.

2.2.3.3.3 Nature of Damage

The larvae and adults feed on milled products only. So, it is considered a secondary pest.

2.2.3.3.4 Geographical Range

It is distributed all over the world and is commonly found in South and Southeast Asia.

2.2.3.4 Grain lice/Psocids (*Liposcelis divinatorius*)

2.2.3.4.1 Identification

- Psocids are indeed tiny insects, yellowish-white in color and are quite active.
- They have long, filiform antennae, which is a distinguishing feature of their appearance.
- They tend to have a more elongated or oblong body shape (Baz 2004).

2.2.3.4.2 *Life History and Habits*

About 7–60 eggs are laid, which hatch into nymphs (like adults but smaller in size, i.e., metamorphosis is incomplete). There are few instars. They are more commonly present in coastal areas or during the rainy season. They cannot survive when R.H. is less than 60% (Phillips and Throne 2010). They usually present along with other insects and mites.

2.2.3.4.3 *Nature of Damage*

Psocids are scavengers during heavy infestations; they primarily feed on organic matter like insect fragments, broken grains and other decaying materials. They are not typically known to cause significant damage to storage structures, but their presence indicates unhygienic storage conditions.

2.2.3.4.4 *Geographical Range*

It originated from tropical and subtropical regions.

2.3 INSECTS IN OIL SEED CROPS, TREE NUTS AND DRIED FRUITS

These insects are polyphagous in nature, cause physical damage, contaminate, generate unpleasant odor and reduce the quality of the commodity. Infestation occurs either in the field or during storage. This significantly affects the economic value of high-valued commodities such as cocoa, coffee beans, dates, raisins, apricots, figs, copra, cashews and other nuts. The family, species and infesting commodity of some common insects infesting oil seed crops, tree nuts and dried fruits are given in Table 2.2.

TABLE 2.2
Insect Species in Oil Seed Crops, Tree Nuts and Dried Fruits

Common Name	Scientific Name	Family and Order	Infested Domain
Coffee bean weevil	*Araecerus fasciculatus*	Anthribidae, Coleoptera	Cocoa, coffee bean, corn, nut megs and peanuts
Groundnut seed beetle	*Caryedon serratus*	Chrysomelidae, Coleoptera	Maize, groundnuts, Brazil nuts and spices
Red-legged ham beetle	*Necrobia rufipes*	Cleridae, Coleoptera	Copra, bone meal, dried figs and palm nut kernels
Larger cabinet beetle	*Trogoderma inclusum LeConte*	Dermestidae, Coleoptera	Dried fruits and nuts
The hairy fungus beetle	*Typhaea stercorea* (L.)	Mycetophagidae, Coleoptera	Barley, oats, bran, rice, flour, bread and dried fruit
Sap beetle	*Carpophilus hemipterus*	Nitidulidae, Coleoptera	Dried fruits, cocoa and oil cake
The merchant grain beetle	*Oryzaephilus mercator* (Fauvel)	Silvanidae, Coleoptera	Dried fruits, oil seeds and oil cake
The foreign grain beetle	*Ahasverus advena* (Waltl)	Silvanidae, Coleoptera	Blue berries, strawberries, pineapples, melons and corn
Carob moth	*Spectrobates ceratoniae*	Pyralidae, Lepidoptera	Dates, figs and citrus fruits
The raisin moth	*Ephestia figulilella*	Pyralidae, Lepidoptera	Raisins, cottonseed cake, cacao beans and cashew kernels
The dried fruit mite	*Carpoglyphus lactis*	Carpoglyphidae, Acari	Dried fruits such as apricots, raisins and figs

2.3.1 Coffee Bean Weevil (*Araecerus fasciculatus*)

2.3.1.1 Identification

- It is dome-shaped and relatively larger in size when compared with other stored pests.
- Normally, an adult measures 3–5 mm long.
- It is dark brown-colored with prothorax, and elytra appear mottled.
- The antennae are slender in nature. Particularly in male weevils, it is larger than the body, with three large segments clubbed together at the end (Navarro and Navarro 2018). The elytra don't cover the last segment of the abdomen.

2.3.1.2 Life History and Habits

The adult lays around 50 eggs, incubates for six days and pupates in seven days. It takes 46–66 days to complete one life cycle at 28°C and 70% R.H. Insects multiply quickly at high moisture content and degrade at low humidity. At 60% R.H., the average life span of an adult is 17 weeks. Lowering the relative humidity further significantly affects the longevity of adults. Temperature below 22°C is not considered a favorable condition for the eggs to develop into adults.

2.3.1.3 Nature of Damage

It is a notorious pest that can cause significant damage to high-value commodities such as cocoa and coffee beans. It can cause infestation both in the field and during storage. The larvae initially feed on the coffee cherries, then tunnel into the seeds, pupate and finally, adults emerge by boring holes through the seed. The degree of infestation totally depends on the type and quality of the commodity, storage condition and contamination level.

2.3.1.4 Geographical Range

It is commonly found in cocoa, coffee beans, corn, nutmegs, peanuts and dried fruits. It originated in India in 1775 but was later distributed worldwide in tropical and subtropical regions during trading.

2.3.2 Groundnut Seed Beetle (*Caryedon serratus*)

2.3.2.1 Identification

- Adult ranges 3–5 mm long. It is reddish brown in color with scattered, dark, irregular spots on the elytra.
- The prothorax is trapezoidal with soft, fine, dense gray hairs and broader hind legs.
- The antenna is serrate, 11 segmented and larger in males.
- In females, the pygidium is projected beyond the elytra (Deshmukh et al. 2022).
- The larvae are whitish with brown heads.

2.3.2.2 Life History and Habits

The adult lays about 105 eggs at 27°C and 70% R.H. Incubates for a period of 2–5 days and pupates within 10–22 days (Sakhare, Mutkule, and Kharade 2018). At 45°C and 70% R.H. adults survived for three to four days. The survival rate of adults totally depends on the atmospheric condition. At lower temperatures (27–30 °C) and higher relative humidity (70–90%), its life span increases (21 days). Moreover, the life span of females is more than males.

2.3.2.3 Nature of Damage

It infests both peanuts (*Arachis hypogaea* Linnaeus) and tamarind (*Tamarindus indica* Linnaeus). It is the most important noxious stored pest that causes 20% damage to ground nuts. The heat and moisture generated by beetles accelerate mold growth, which results in the development of aflatoxin in groundnuts.

2.3.2.4 Geographical Range

It is found in North-Eastern and West Africa, Europe, West Indies, South and Central America (Dobie et al. 1985) and Northern Asia (excluding China).

2.3.3 Red-Legged Ham Beetle (*Necrobia rufipes*)

2.3.3.1 Identification

- The average length of an adult is 4.5 mm long.
- The head, elytra and thorax are green or bluish in color with a dark blue abdomen.
- Legs and antennae are reddish brown in color (Hasan et al. 2020).
- Antennae are 11 segmented and clubbed at the end.
- Females are differentiated from males with the presence of fine hairs on the elytra.

2.3.3.2 Life History and Habits

Eggs are oval, smooth and shiny. An adult can lay approximately 2000 eggs in its lifetime, depending on the type of food and atmospheric conditions. Eggs are laid on the surface of the food and in cracks or cervices. Hatching occurs in four to eight days, and young larvae initially prefer to feed on eggs rather than food source. Larva develops in 35–130 days. An optimum temperature required for insect multiplication is 30–34°C. It takes a minimum of 45 days to develop from egg to adult. Adults are very active and wander on the surface of the food.

2.3.3.3 Nature of Damage

Adult only feeds on the surface of the food commodity. It infests high-protein food sources such as dried coconut meat or copra, dried fish, dried egg, cheese, bone meal, processed meat, palm nut kernel and cashews.

2.3.3.4 Geographical Range

It is distributed throughout the world; it is also considered a well-known pest in India, China, Brazil and the United States.

2.3.4 Larger Cabinet Beetle (*Trogoderma inclusum* LeConte)

2.3.4.1 Identification

- Adults are 2.1–2.4 mm long and 1.1–1.3 mm wide.
- It is brown to black in color, with darker or lighter color patches (Hasan et al. 2020).
- The dorsal region is covered with coarse, golden brown and white hairs.
- Antennae are 11 segmented, and 3 segments broadened and clubbed at the end.
- It consists of distinct notch in the inner margin of the eyes, distinguishing it from other *Trogoderma* species.
- Larvae are indeed cream-colored and have tufts of hair and spines on their bodies.

2.3.4.2 Life History and Habits

An adult can lay about 45 eggs and will hatch in 12 days. Larvae live for five months and undergo molting every two weeks. It takes six months to complete one life cycle. Adult starts to mate and lay eggs after one month.

2.3.4.3 Nature of Damage

The larvae are considered a diverse feeder that feeds on a wide range of food materials such as pet foods, grains, cereal products, dried fruits and nuts (Hinton 1945).

2.3.4.4 Geographical Range

It is native to the United States and France and later spreads to the United Kingdom, Japan, Poland, Italy and Spain (Castañé et al. 2020).

2.3.5 The Hairy Fungus Beetle (*Typhaea stercorea* (L.))

2.3.5.1 Identification and Life History

- Adults are oval, oblong and light brown in color with a length of 3 mm.
- It resembles a drugstore beetle but is smaller in size.
- The elytra are covered with parallel lines of coarse hairs.
- It takes nearly 21–33 days to develop from the egg to the adult stage.
- Adults can fly, and larvae move easily and faster.

2.3.5.2 Nature of Damage

It generally infests peanut, tobacco, moldy cereals and hay. It is a scavenger that also feeds on the mold and fungi on the high-moisture grains. Its presence indicates poor storage conditions and fungal attacks in stored commodities. It is found in temperate and tropical regions.

2.3.6 Sap Beetles (*Carpophilus hemipterus* and *Carpophilus dimidiatus*)

2.3.6.1 Identification

- Adults are 3–5 mm long, oval, brown to black in color, with yellowish brown patches on the elytra.
- They have short antennae with clubbed ends. Elytra are often shortened, exposing two or three abdominal segments.
- The yellow patches are not present in the *Carpophilus dimidiatus*, which differentiates it from *Carpophilus hemipterus*.

2.3.6.2 Life History and Habits

Eggs are milky white and sausage-shaped; females can lay approximately 1000 eggs singly. Hatching occurs in 3 to 5 days, and the larval development period lasts for 6–14 days. After pupation (3–8 days), adults emerge in 4–16 days. The adult flies actively and normally lives for three months; in rare cases, it may survive up to one year. Due to the short generation time, nearly three generations can occur in a year.

2.3.6.3 Nature of Damage

The dried fruit beetle (*C. hemipterus*) *and* dried sap beetle (*C. dimidiatus*) are found both in the plantation field and in-store. It has a wide range of feeding behaviors and generally feeds on flowers, fruits, sap, fungi and ripening or decomposing plant tissues. These beetles are present in blueberries, strawberries, pineapples, melons, corn and dried fruits.

2.3.6.4 Geographical Range

It is distributed worldwide.

2.3.7 The Merchant Grain Beetle (*ryzaephilus mercator* Fauv.)

2.3.7.1 Identification and Life History

The morphological features of *Oryzaephilus mercator* are identical to *Oryzaephilus surinamensis*. *O. mercator* is distinguished by laying two-thirds of eggs laid by *O. surinamensis*. It prefers to infest products with higher oil content and feeds on seed-borne fungi (Sinha 1965). It has a narrow temple with larger eyes and is known to fly. The *O. surinamensis* prefers to infest grains with higher carbohydrates and feeds on eggs and dead moths (Lecato 1973). It has a triangular-shaped head with a wider temple.

2.3.7.2 Nature of Damage

It is commonly present in products with higher oil content: peanut, cashew, coconut, bran and sunflower seeds. The larvae and adults cause damage rapidly, and the infestation is not identified immediately. The metabolic behavior of insects increases grain moisture content, which results in mold growth.

2.3.7.3 Geographical Range

It is found globally, but intolerant to temperate regions.

2.3.8 The Foreign Grain Beetle (*Ahasverus advena* (Waltl))

2.3.8.1 Identification

- Adults are 2 mm long, small, reddish brown in color. The young larvae are white and matured are grayish.
- They have slender antennae with enlarged last three segments. Elytra are often shortened, exposing two or three abdominal segments.

2.3.8.2 Life History and Habits

Female can lay single or clustered eggs of about 100-300 eggs and hatch in 4 to 5 days. Larva takes 9 to 15 days to develop fully, pupation period 2 days and adult emerges in 3 to 5 days. The life span of matted beetle is less as compared to unmated. Insect will not prefer to breed at relative humidity less than 65%.

2.3.8.3 Nature of Damage

It preferably feeds embryo, molds, dead insects and damaged foods. These beetles are commonly present in grains, oil seeds, cereal products, corn and dried fruits.

2.3.8.4 Geographical Range

It is found worldwide.

2.3.9 Carob Moth (Spectrobates (Ectomyelois) Ceratoniae (Zeller))

2.3.9.1 Identification

- The adult moth is dark gray in color and measures 0.8–1.0 cm long with a wingspan of 20 mm.
- The forewings and hind wings are pale brown and white in color.
- Matured larvae are whitish pink in color with a browny head capsule.

2.3.9.2 Life History and Habits

It is a fruit moth, and an adult can lay up to 200 eggs. After hatching, the larvae prefer to enter the damaged fruits through cracks and feed only on the pulpy portion of the fruit until it matures. Depending on climatic conditions carob moth takes 32–43 days to develop fully. The life span of the adult moth is five to ten days.

2.3.9.3 Nature of Damage

It is polyphagous in nature that infests carbs, dates, pomegranates, citrus fruits, almonds, walnuts, pistachio and figs in the field and also attacks a variety of nuts during storage. Crop infestation results in premature yellowing of fruits.

2.3.9.4 Geographical Range

It is distributed in Mediterranean regions of America, Australia and South Africa.

2.3.10 The Raisin Moth (*Ephestia figulilella* Gregson (*Cadra figulilella*))

2.3.10.1 Identification

- Adult raisin moth is 1 cm long, with a wingspan of 12–20 mm.
- It is grayish brown in color with a reddish brown head.
- Their body is surrounded horizontally by purple color dots.
- Matured larvae measure 10 mm long.

2.3.10.2 Life History and Habits

Adult laid eggs in the range of 350–690, which are small and creamy white in color. The matured egg turns into a greenish-gray color. Eggs hatch in 3–6 days, and larvae mature in 32 days (Perring, El-Shafie, and Wakil 2015). Based on temperature and humidity, larvae undergo four to five generations per year. The average life span of an adult female is 16 days, and a male is 11 days. Overall, it takes 43 days to complete full development at 28.3°C.

2.3.10.3 Nature of Damage

The matured larvae infest the freshly harvested dates, figs, grapes, carobs, mulberries and dried raisins, cocoa beans and cashew kernels.

2.3.10.4 Geographical Range

The favorable climatic condition is high temperature with low relative humidity. It spreads throughout the world, especially in the Middle East, North Africa, Middle Asia and some parts of America and Australia.

2.3.11 The Dried Fruit Mite (*Carpoglyphus lactis* (Linnaeus))

2.3.11.1 Life History and Habits

The duration of the oviposition period is four to six days. At 18°C and 65% R.H., the adult laid 41 eggs, and the maximum life span of an adult was 68 days in this condition. It takes only 14 days to develop completely. *Carpoglyphus lactis* has a nature of penetrating into polypropylene and aluminum foil packaging material to attack stored dry fruits. Mites can trigger allergic reactions and act as vectors for mycotoxin-producing fungi (Hubert et al. 2011). Hence, proper storage and handling practices prevent contamination and retain food quality.

2.3.11.2 Nature of Damage

Fruit mites are the most important species that commonly attack dried apricots. It feeds on saccharides present in dried fruits such as dates, raisins, apricots, figs, plums and old honeycombs, leading to fungal growth, which results in off-odor development. These species significantly affect the economic value of the dried fruits. *C. lactis* recorded the highest infestation of 15,534 mites per kg of dried apricot (Çobanoğlu 2009).

2.3.11.3 Geographical Range

It is distributed in Mediterranean basins, especially in apricot producing regions Malatya and Elazig of Turkey.

2.4 CONCLUSION

The duration of each stage in the life cycle may vary significantly among different stored pests. There are many factors, such as temperature, humidity, food source, food availability and crop contamination, which will influence the growth and development of insects. Understanding the life cycle of specific insects is important for adopting effective prevention and management techniques as it helps in determining the most appropriate timing for intervention measures and also to follow the IPM.

REFERENCES

Baz, A. 2004. *Encyclopedia of Entomology. Encyclopedia of Entomology.* Dordrecht: Kluwer Academic Publishers. doi:10.1007/0-306-48380-7.

Behera, P., A. Mohanty, D. Patra, and D. S. Karn. 2016. "Life Cycle of Ground Nut Seed Beetle (*Caryedon serratus*)." *Indian Journal of Oilseed Research* 5 (6): 4–6.

Biosci, I. J., M. Z. Majeed, T. Mehmood, M. Javed, F. Sellami, M. A. Riaz, and M. Afzal. 2015. "Biology and Management of Stored Products' Insect Pest *Rhyzopertha dominica* (Fab.) (Coleoptera: Bostrichidae)." *International Journal of Biosciences (IJB)* 7 (5): 78–93.

Castañé, C., N. Agustí, P. del Estal, and J. Riudavets. 2020. "Survey of *Trogoderma* spp. in Spanish Mills and Warehouses." *Journal of Stored Products Research* 88 (September): 101661. doi:10.1016/j.jspr.2020.101661.

Çobanoğlu, S. 2009. "Mite Population Density Analysis of Stored Dried Apricots in Turkey." *International Journal of Acarology* 35 (1): 67–75. doi:10.1080/01647950902884645.

Dobie, P., C. P. Haines, R. J. Hodges, and P. F. Prevett. 1985. "Insects and Arachnids of Tropical Stored Products." In *Their Biology and Identification*, 271. Berks, UK: Tropical Development and Research Institute (TDRI) Slough.

Edde, P. A. 2012. "A Review of the Biology and Control of *Rhyzopertha dominica* (F.) the Lesser Grain Borer." *Journal of Stored Products Research* 48 (January): 1–18. doi:10.1016/j.jspr.2011.08.007.

Edde, P. A. 2019. "Biology, Ecology, and Control of *Lasioderma Serricorne* (F.) (Coleoptera: Anobiidae): A Review." *Journal of Economic Entomology* 112 (3): 1011–1031. doi:10.1093/jee/toy428.

Hinton, H. E. 1945. "A Monograph of the Beetles Associated with Stored Products." *Annals of the Entomological Society of America* 38 (4): 443. doi: 10.1093/aesa/38.4.602

Hubert, J., T. Erban, M. Nesvorna, and V. Stejskal. 2011. "Emerging Risk of Infestation and Contamination of Dried Fruits by Mites in the Czech Republic." *Food Additives & Contaminants: Part A* 28 (9): 1129–1135. doi:10.1080/19440049.2011.584911.

Kumari, S., S. S. Yadav, K. Rolania, and S. Dhanda. 2019. "The biology of pulse beetle, C. chinensis on stored mungbean." *Journal of Entomology and Zoology Studies* 8 (3): 1200–1203.

Lecato, G. L. 1973. "Sawtoothed Grain Beetle:1 Population Growth on Peanuts Stimulated by Eating Eggs or Adults of the Indian Meal Moth2,4." *Annals of the Entomological Society of America* 66 (6): 1365–1365. doi:10.1093/aesa/66.6.1365.

Menge, A. K., S. K. Mehendale K. V. Naik, and P. Raut. 2016. "Biology of rice moth, Corcyra cephalonica (Stainton) on groundnut variety Konkan Tapora." *Advances in Life Sciences*, 5 (21): 9733–9735.

Navarro, S., and H. Navarro. 2018. "Insect Pest Management of Oilseed Crops, Tree Nuts and Dried Fruits." In *Recent Advances in Stored Product Protection*, 99–141. Berlin, Heidelberg: Springer Berlin Heidelberg. doi:10.1007/978-3-662-56125-6_6.

Perring, T. M., H. A. F. El-Shafie, and W. Wakil. 2015. "Carob Moth, Lesser Date Moth, and Raisin Moth." In *Sustainable Pest Management in Date Palm: Current Status and Emerging Challenges*, 109–167. Cham: Springer International Publishing. doi:10.1007/978-3-319-24397-9_6.

Phillips, T. W., and J. E. Throne. 2010. "Biorational Approaches to Managing Stored-Product Insects." *Annual Review of Entomology* 55 (1): 375–397.

Sakhare, V. M., D. S. Mutkule, and V. G. Kharade. 2018. "Biology of Groundnut Bruchid, *Caryedon serratus* (Olivier) on Different Groundnut Varieties." *Journal of Entomology and Zoology Studies* 6 (4): 1577–1580.

Sharma, A., A. Sharma, S. K. Khinchi, and K. C. Kumawat. 2017. "Studies on Biology of *Caryedon serratus* (Olivier) on Groundnut *Arachis hypogaea* (L.)." *Journal of Entomology and Zoology Studies* 5 (3): 582–597.

Singh, S., D. K. Sharma, S. Bhatia, and A. Singh. 2017. "Effect of Various Plant Powders on Rice Weevil in Stored Wheat." *Journal of Environmental Biology* 38 (03): 501–508. doi:10.22438/jeb/38/3/MS-164.

Sinha, R. N. 1965. "Development of *Cryptolestes ferrugineus* (Stephens) and *Oryzaephilus mercator* (Fauvel) on Seedvelril Fungi 1." *Entomologia Experimentalis et Applicata* 8 (4): 309–313. doi:10.1111/j.1570-7458.1965.tb00867.x.

Trematerra, P., J. E. Throne, M. Fernandez, and R. Knox. 2012. "Insect and Mite Pests of Durum Wheat." In *Durum Wheat Chemistry and Technology*, 2nd edition. St. Paul, MN: AACC International, 73–83.

Villaverde, M. L., M. P. Juárez, and S. Mijailovsky. 2007. "Detection of *Tribolium castaneum* (Herbst) Volatile Defensive Secretions by Solid Phase Microextraction–Capillary Gas Chromatography (SPME-CGC)." *Journal of Stored Products Research* 43 (4): 540–545. doi:10.1016/j.jspr.2007.03.003.

3 Detection of Stored Product Pests

T. Senthilkumar
University of Prince Edward Island, Charlottetown, PEI, Canada

V. Chelladurai
Department of Agricultural Engineering, Bannari Amman Institute of Technology, Sathyamangalam, Erode, Tamil Nadu, India

C. B. Singh
Advanced Post-Harvest Technology Centre, Lethbridge College, Lethbridge, AB, Canada

3.1 INTRODUCTION

In 2021, global primary crop production reached 9.5 billion tonnes, valued at USD 2.8 trillion, with value addition bringing the total to USD 3.8 trillion (FAO 2023). Cereal grains represent nearly one-third of this production, with cereal trade reaching 508 million tonnes in the same year (FAO 2023). The agriculture sector employed 873 million people worldwide in 2021. The FAO predicts that for the 2024/2025 season, the total cereal supply will be 2846 million tonnes, while demand will reach 2851 million tonnes, indicating a higher demand than supply (FAO 2024). As of 2021, an estimated 735 million people are undernourished (FAO 2023).

Approximately 30 percent of grains are lost post-harvest due to improper storage, insect, and other pest infestations (Kumar and Kalita 2017). Proper storage of food grains from farm to consumer is crucial to feeding the growing global population. Stored product pests, including insects, mites, and fungi, infest and damage stored grains, posing significant challenges to food security and safety. Early detection and identification of stored product insects enable producers and processors to take timely corrective actions, preventing quality degradation and economic losses. The type of insects infesting stored grains varies with climate and storage practices. Insects not only reduce the quantity of stored grains but also degrade their quality, affecting market value. Infestation hotspots foster the growth of fungi and mites, leading to further deterioration. Major insects like *Rhyzopertha dominica* (lesser grain borer), *Sitophilus granarius* (grain weevil), and *Tribolium castaneum* (red flour beetle) are primary pests causing significant damage, while others are secondary pests (Cai et al. 2022). Insects generally do not multiply when grain temperature and moisture content are low; however, hotspots within grain masses can occur, allowing insect proliferation.

Insect-damaged grains exhibit reduced germination, undesirable odors, decreased mass, and the presence of insect fragments, leading to quality issues. Many countries have strict regulations regarding insect presence in stored grains. For instance, Canada has zero tolerance for live insects in stored grains, while the USA allows up to two live weevils per kilogram of grain; exceeding this limit classifies the grain as infested. Early detection and accurate estimation of insect populations allow for timely corrective actions, such as insecticide application and fumigation, minimizing excessive pesticide use and reducing food safety risks. Storage practices vary globally, with North America and Europe predominantly using bins, while other regions use bags stored in concrete or steel-roofed buildings. Detection methods also vary, with vacuum-operated sampling probes

DOI: 10.1201/9781003609834-3

used for bins in Canada and the USA, and smaller probes for bag storage to prevent grain spillage and pest attraction. Existing insect detection methods are often subjective and less accurate (Neethirajan et al. 2007). There is a need for fast and accurate detection techniques. Researchers are working toward achieving this goal.

This chapter discusses the importance of stored grain, the negative impacts of insect presence, and conventional insect detection methods like visual inspection, traps, and probes, highlighting their advantages and disadvantages. It also provides an overview of emerging detection techniques, including carbon dioxide sensors, volatile organic compound (VOC)-based detection, bioacoustics, thermal imaging, NIR hyperspectral imaging, and soft X-ray imaging, and their current usage in grain storage. Conventional methods offer rough estimates of insect populations and are often time-consuming and labor-intensive. Emerging techniques provide more accurate insect quantification, with some advanced systems capable of detecting infestation stages, aiding grain storage managers in formulating effective insect control plans.

3.2 CONVENTIONAL METHODS

Insect monitoring and detection in grains is crucial to identify the species and make corrective actions like pesticide and fumigation applications and crucial for when to stop the corrective actions. The modern-day grain structures are equipped with temperature, moisture sensors, and fan control systems to maintain the grains at the desired low temperature and moisture content throughout the storage period; however, some localized spots can occur inside the storage structures for the insects to develop and multiply, so a frequent insect monitoring and detection should take place to store the grains without any quality and quantity loss. There are three categories of conventional insect monitoring and capture, as shown in Figure 3.1.

3.2.1 Visual Inspection

Visual inspection is a very critical and common practice to check for insect presence in storage structures; however, it can only be a qualitative assessment but can't estimate insect infestation. The inspectors, in addition to checking the presence of live insects, should also look for insect-damaged kernels, dead insects, and insect fragments to validate the infestation in the grains. Another drawback of visual inspection is that the inspectors mostly check the presence of insects only in the daytime for their own safety climbing up the grain bins and the absence of a lighting system in the grain bins; however, the insects are more active at night compared to daytime (Toews and Nansen 2012). So, grain inspectors should rely on other trapping and sensing technologies in addition to visual inspection to detect the presence of insects in grain bins.

3.2.2 Traps

Traps are used in grain storage structures for insect management, and the main aim of placing traps is to identify the insect species, monitor and estimate the insect population over the storage period, and to take corrective actions based on the insect population. The grain storage managers/professionals should have insect interpretation skills and knowledge about different pesticides and fumigants and their dosage levels based on insect density. There are four common types of traps being used in grain storage facilities: (1) light traps, (2) aerial traps, (3) surface traps, and (4) bulk traps (Figure 3.1).

3.2.2.1 Light Traps

Insects in general get attracted to ultraviolet, green, and blue lights (Kammar et al. 2020). The commercial light traps commonly use ultraviolet light to attract the insects; the attracted insects are either captured or killed by the electrocuting grid placed around the trap. The light traps are

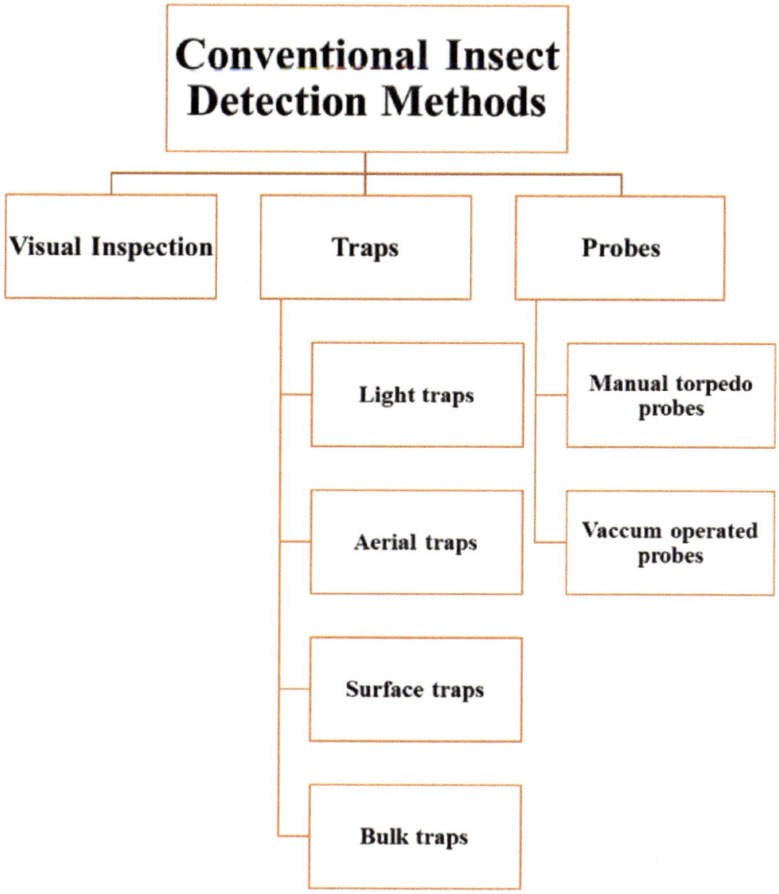

FIGURE 3.1 Conventional insect detection methods.

commonly used in food processing facilities, bakeries, and restaurants to control adult flying insects; they can be mounted to the walls or suspended from the roofs (Figure 3.2).

There are many advantages of using light traps: (1) easy to install due to its portable design, (2) sustainable solution; no chemical residues, and (3) the insect species can be easily identified even after being killed by electrocuting. The biggest disadvantages of the light traps are (1) only effective to certain insect species, (2) mostly adult stages are attracted, and (3) the temperature, humidity, and moisture content of stored grains play a role in insects attracted to the traps (Shimoda and Honda 2013). It is very important for grain managers to know about the effectiveness of light traps and to make sure the traps are cleaned and bulbs replaced at regular intervals.

3.2.2.2 Aerial Traps

Aerial traps are used to attract adult flying insects by using pheromones or food materials. The attracted insects were captured by a sticky surface or by an escape-proof collector placed alongside the trap. Like light traps, aerial traps are wall-mounted or suspended from the roof and sometimes placed on warehouse shelves. The common types of aerial traps are bucket traps, funnel traps, and sticky traps (Figure 3.3).

Funnel traps and bucket traps are reusable, sturdy, and long-lasting and are very effective with a pheromone or food bait. Funnel traps and bucket traps are designed with an insect collection reservoir with an insecticide strip or with a liquid to make the insect stay inside the reservoir. The funnel

FIGURE 3.2 Light trap suspended to the roof inside a storage facility. (Image Provided by Dr. Partho Dang.)

traps are used in the outdoor environment to monitor lesser grain borer (Leos-Martinez et al. 1987). The traps are designed to have small openings or holes to avoid dust accumulation. The modern-day sticky traps are reusable by removing the sticky surface with a steel leveler or knife and reapplication fresh TANGLEFOOT® TANGLE-TRAP® sticky coatings.

3.2.2.3 Surface Traps

Surface or pitfall traps are used to capture crawling insects. The surface traps are placed on horizontal surfaces, and they can capture mainly adult species and occasionally other insect developmental stages. The surface trap uses both pheromone and food attractants as bait to lure the insects. Pitfall traps have been used since the 1960s to detect stored grain insects (Loschiavo and Atkinson 1967), and pitfall traps capture efficiency depending on the grain temperature, moisture content, and trap position (White and Loschiavo 1986; Fargo et al. 1989). A study by Reed et al. (1991) found that pitfall traps were very effective during the early storage period by detecting *R. dominica* (F.). The pitfall traps are also effective in capturing other important stored grain insects *Sitophilus granarius*,

(a) (b) (c)

FIGURE 3.3 Different types of aerial traps: (a) funnel, (b) bucket, and (c) sticky traps.

FIGURE 3.4 Pitfall traps. (martinlishman.com.)

Oryzaephilus surinamensis, and *T. castaneum* (Aulicky et al. 2016). The pitfall traps were able to capture 25 insects per trap placed inside the storage structures during the month of September, and it decreased to 22 and 9 in November and January, respectively (Reed et al. 1991). The pitfall traps are placed 5–6 m from each other inside the storage structures for effective trapping (Figure 3.4).

3.2.2.4 Bulk Grain Traps

Bulk grain traps are modified pitfall traps designed to use in bulk storage structures like concrete silos and steel bins. The trap is placed inside the grain mass for a longer period, and insects are captured in the non-escape chamber. The pheromone or food attractant baits are commonly not used in bulk grain traps. The bulk traps are also called probe traps; they are designed in such a way as to capture the insects in the probe, excluding the grains, and the traps are placed just below the top surface because insects are mostly found in this area (Flinn et al. 2010). There are different sizes of bulk grain traps (Figure 3.5). The reduced-length probe trap was found to be more effective

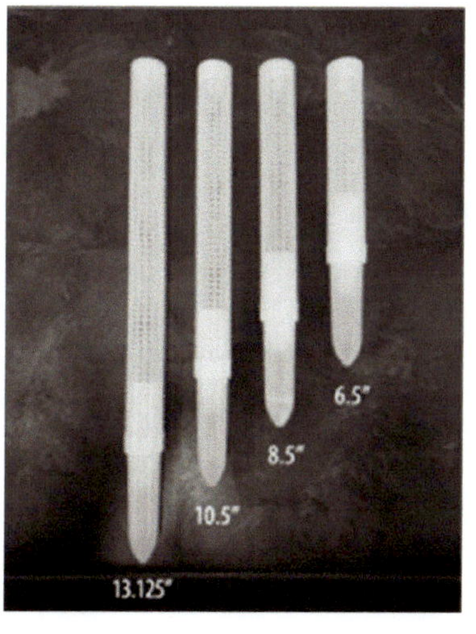

FIGURE 3.5 Bulk grain traps with different lengths (Konemann 2021).

than the standard 13.125 in., probe trap. The study used three different sizes of probe traps, 6.5, 8.5, 10.5 in., and the standard 13.125 in., to compare the effectiveness; the results showed that all the smaller size probe traps are significantly better than standard size probe traps, and the 8.5 in. probe trap captured more insects than the other smaller size probe traps (Konemann 2021).

3.2.3 Probes

3.2.3.1 Manual Torpedo Probe

A torpedo-style probe or tier probe can be used to collect grain samples from grain storage structures, where placing traps is difficult. The probes are of different lengths, and samples can be collected at different depths, as shown in Figure 3.6.

The probes need to be carried to the top of the grain bin or grain bags to get the sample for checking the presence of insects in the grains. The probe should be (1) inserted completely into the grain bins, then (2) the probe gates need to be opened for sufficient time to collect the grains (3) the gates should be closed, and (4) the probes are removed with the collected grains. The collected grain samples should be sieved using a 2-mm aperture sieve for wheat and barley, and a 0.85-mm aperture sieve for canola, and the sieve should be shacked thoroughly to separate the insects from the grains. To get certified by the grain export agency, the 1000 g of grain samples collected using the grain probe from the grain bins were placed in the Berlese funnel. The Berlese funnel is an insect extraction equipment consisting of a light bulb suspended above a funnel with grain samples, a water jar placed under the funnel will keep insects from escaping. The light and heat from the bulb make the insects move out of the grains and get collected in the jar for insect interpretation.

In addition to the sieve and the Berlese funnel to capture insects from the grains collected using the probes, the Carter dockage tester is used to collect the insects from the grains. The insects are either collected at the aspirator pan or the pan for the smallest sorting (Canadian Grain Commission (CGC) 2021).

3.2.3.2 Vacuum-Operated Probes

The vacuum-operated probes are mainly used to collect grain samples from the containers reaching the export elevators from the grain bins (Figure 3.7). The Canadian Grain Commission recommends eight different sampling places per load for a single container and ten sampling locations for a partitioned truck (Canadian Grain Commission (CGC) 2021). The collected grain samples were

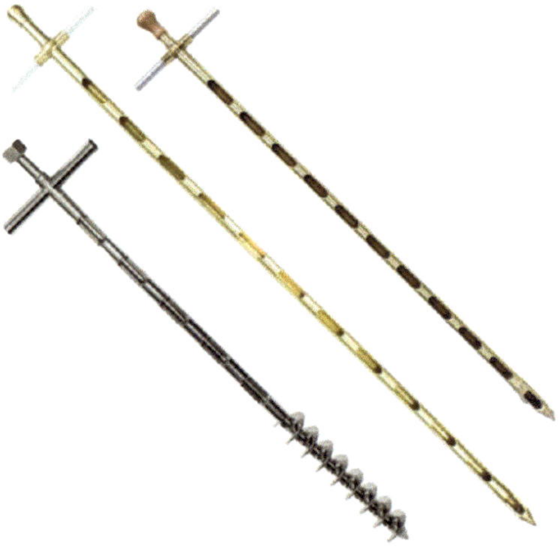

FIGURE 3.6 Grain probes with multiple tires.

FIGURE 3.7 Vacuum-operated grain probes. (Canadian Grain Commission (CGC) 2021.)

sieved using a Berlese funnel or Carter dockage tester to capture the insects and further species interpretation.

3.3 EMERGING METHODS

There are numerous conventional insect detection methods like visual inspection, traps, and probes being used at grain storage structures to monitor stored grain insects; however, there are many drawbacks: they are subjective, inaccurate, and cannot provide the real insect population estimate (Neethirajan et al. 2007). With these limitations in conventional methods, researchers all over the world are developing new technologies to detect insects in stored grains; some techniques are already being deployed in storage structures, and some are in the implementation process. This section will cover the following new emerging techniques: (1) VOCs-based detection methods, (2) carbon dioxide sensors, (3) bioacoustic methods, (4) insectors, and (5) image-based techniques.

3.3.1 Volatile Organic Compounds and Biosensors

Insects release VOCs to communicate with other insects; the identified VOC can be used to build biosensors to detect insect presence in stored grains. The headspace samples from the container stored with wheat and *T. castaneum* (red flour beetle) were extracted using an automatic headspace sampler and analyzed using GC-MS to identify the VOC. The *T. castaneum* produced methyl 1,4 benzoquinone, ethyl 1,4 benzoquinone, and 1 tridecene, and the VOC concentration increased with an increase in insect density (Senthilkumar et al. 2012). A similar study by Lim et al. (2021) identified the volatile compounds in the headspace of wheat samples with *R. dominica*, and *S. granarius* in the headspace. The *R. dominica* produced acetone and ethanol, and *S. granarius* produced dominicalure-1 and dominicalure-2 and 3-hydroxy-2-butanone. The VOC's 4,8-dimethyl-4, 8-decadien-10-olide, and 3-dodecen-11-olide were produced by *Cryptolestes ferrugineus* (rusty grain beetle) (Wong et al. 1983). The insects at extreme hot and cold exposure produced higher VOC compared to insects at optimum temperatures, so the VOC method can be used as a validation method to detect whether the treatments killed the insects or not (Senthilkumar et al. 2012). The development of sensors based on the VOC is yet to be fully explored. A study used the E-nose equipped with 12 metal oxide sensors to detect *T. castaneum* and *C. ferrugineus* insects in wheat samples based on the captured headspace volatiles, and the E-nose was able to detect samples infested with *T. castaneum* at low moisture content and higher insect density. However, the E-nose failed to detect *C. ferruginous*-infested samples.

3.3.2 Carbon Dioxide Sensors

An increase in carbon dioxide (CO_2) in stored grain structures is an indication of the presence of insects, fungi, and mites. The CO_2 concentration above 500 ppm indicates an increase in respiration or pest activity in grains. The CO_2 sensor can be used as an early indicator of spoilage of the grains caused by insects, fungi, or mites; however, we cannot specifically say whether it is because of insects, fungi, or mites; we still need to rely on probe sampling to detect the presence of insects. The moisture sensors and temperature sensors are predominantly deployed in grain storage structures to detect spoilage in grains. Recently, CO_2 sensors have been added to the grain storage management package in addition to other sensors. Recently, based on a study conducted at Lethbridge College to identify a suitable place to install the CO_2 sensors in the grain bin, a Calgary-based grain management company added CO_2 sensors to their portfolio. The study conducted at Lethbridge College recommended using CO_2 sensors at both the headspace and plenum to detect the elevated CO_2 levels in the grain bins effectively and to make proper rectifying measures (Senthilkumar, Panigrahi, and Singh 2024). The CO_2 sensors are an effective tool, and more grain storage managers have started adopting the technology to monitor the grain condition, and when combined with probe sampling can be an effective combination to detect the presence of insects in stored grains.

3.3.3 Acoustic and Bioacoustics Methods

Insects, when they are feeding and moving on the grains, create acoustic signatures or sounds, which can be recorded using acoustic sensors. Researchers started exploring the acoustic detection of insects first in the 1900s, and only after the 1980s the interest in acoustic methods started increasing because of the availability of economical sensors and data processing systems (Mankin et al. 2011). The acoustic sensing system typically constitutes a piezoelectric sensor, an acoustic amplifier, and a data acquisition and processing system (Eliopoulos et al. 2015; Eliopoulos, Potamitis and Kontodimas 2016). The piezoelectric sensor converts the sound waves into voltage variations, and the amplifier attached to the sensor improves the signal strength. The converted and amplified signals were further analyzed to detect the unique acoustic signatures produced by the insects (Eliopoulos et al. 2015). The bioacoustics method was used to detect *Sitophilus oryzae* (the rice weevil), *R. dominica* (the lesser grain borer), *Tribolium confusum* (the confused flour beetle), *O. surinamensis* (the saw-toothed grain beetle), *C. ferrugineus*, *Trogoderma granarium Everts* (the khapra beetle), and *Lasioderma serricorne* (the cigarette beetle) in wheat mass. The bioacoustics device can detect insects accurately (74–100 percent) when there are one or two adult insects per kilogram, however, the lower insect levels like one adult insect per 10 kg and one adult insect per 2 kg were detected with less accuracy. The bioacoustics method has the potential to detect insects in stored grain mass, and with more research they can be deployed in grain bins, which will be chemical free and more economical for the farmers.

3.3.4 Electronic Insect Detectors

Similar to conventional insect probes, electronic insect detectors have been developed to quantify stored product insects in grain bulks, such as those found in grain silos. These electronic insect detectors facilitate real-time quantification of insect populations and detection of insect types without any physical sampling, a common necessity in conventional insect traps. These electronic detectors combine the functionalities of pitfall insect probe traps and infrared sensor systems (Figure 3.8). They feature a perforated outer cover with multiple openings through which insects can crawl into the probe. Advanced electronic insect detectors, such as the Insector® (OPI Systems Inc., Calgary, Canada), are suspended inside grain bins similar to moisture and temperature sensor cables.

The Insector® is equipped with two perpendicular infrared beams. When insects fall into the pitfall trap through the perforations in the Insector®'s wall, they interrupt the infrared light beams. The amount of light blocked is automatically recorded by the control system. The size and shape of the insect determine the extent of the light blockage, allowing the system to identify the species of insect and count the number of insects that have fallen into the trap. Additionally, the Insector® measures the temperature at the location and the time when the insect fell into the trap. This data is displayed on the control system to which the Insector® units are connected. The number of Insector®

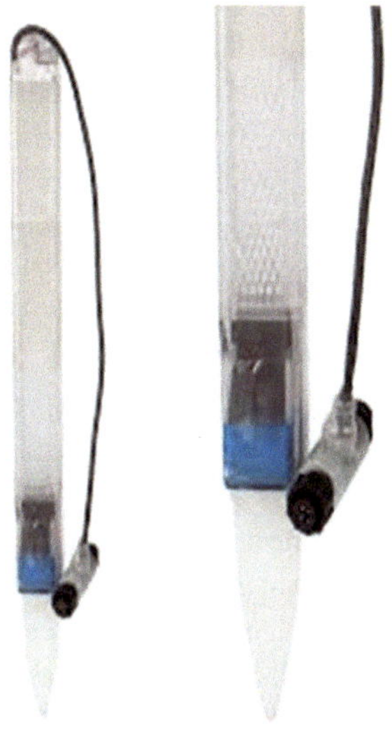

(a)

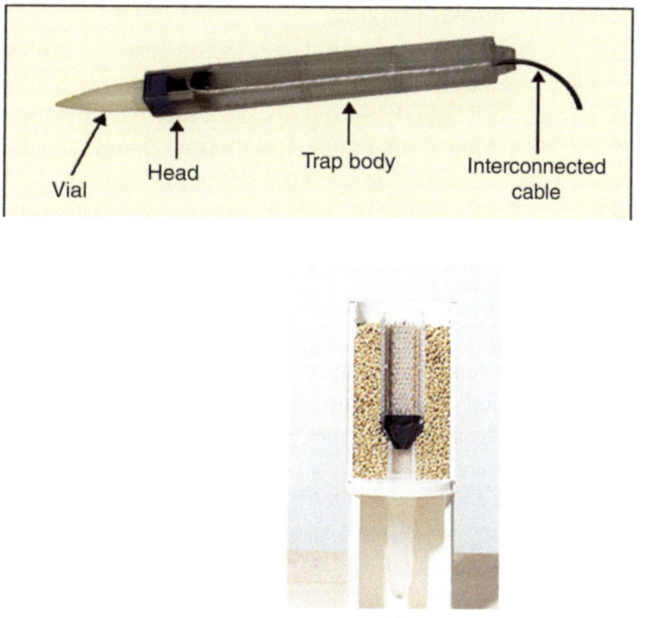

(b)

FIGURE 3.8 (a) Insector®; (b) cross-sectional view of Insector®.

traps required per grain bin depends on factors such as grain type, storage location, storage structures, and weather conditions. This comprehensive data (temperature, insect count, and species) enables grain storage managers to make informed decisions about corrective actions needed to protect stored grain from quantitative and qualitative losses.

3.3.5 Imaging Techniques

Image-based systems are rapid, non-chemical, and waste-free (non-destructive) methods, where the tested products can be reused if the samples are found to be insect-free. There are different types of imaging systems: (1) thermal imaging, (2) NIR hyperspectral imaging, and (3) X-ray imaging techniques were deployed based on the requirements, as shown in Figure 3.9.

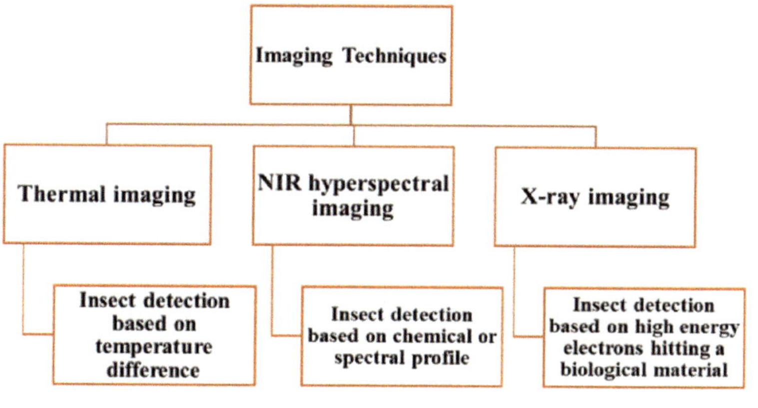

FIGURE 3.9 Imaging techniques to detect stored grain insects.

3.3.5.1 Thermal Imaging

Thermal imaging techniques detect the presence of insects based on the difference in temperature profile caused by applied external heat or cold to the biological materials. The temperature difference is not visible to the naked eye, so a thermal imaging system was used to capture the temporal profile of biological materials. The thermal imaging system consists of a thermal camera that can capture the two-dimensional temperature profile of the biological product, a thermal unit to heat the sample or a refrigerator to cool the sample, an image capture and processing system (Figure 3.10) (Manickavasagan, Jayas, and White 2008; Manickavasagan et al. 2010; Chelladurai, Jayas, and White 2010) used the thermal imaging system with applied cooling to detect wheat kernels internally infested with six different developmental stages of *C. ferrugineus* (four larvae, one pupa, and one adult stage). The internally infested kernels were placed in the refrigerator at 5°C for 60 seconds and placed at room temperature for 20 seconds before subjecting to thermal imaging. The processed image provided a higher temperature for the larvae stage because more respiration produced more heat. The thermal images were able to differentiate between normal and insect-infested kernels with higher classification accuracy of above 80 percent; however, the thermal camera provided lower classification accuracy when differentiating between six insect developmental stages.

The presence of insects in grain bins increases the temperature, and localized hot spots can occur. Manickavasagan et al. (2006) placed artificial heat sources at 30, 40, 50, and 60°C at four different locations in a 1.5 m diameter and 1.5 m length bin filled with barely and acquired thermal images of the wall side and top side up to 48 hours. The results showed that the temperature on top of the silo was significantly higher than the atmospheric temperature because of the presence of hot spots in silos. The acquired temperature is reduced with an increase in the moisture content of grains and an increase in wind velocity. The thermal imaging system can be deployed in grain storage structures to detect hot spots produced by insects, and the same thermal camera can be used to detect insect-infested kernels when samples are obtained using probes.

3.3.5.2 NIR Hyperspectral Imaging System

NIR hyperspectral imaging system is also referred to as chemical imaging because of its potential to provide a chemical profile of biological materials. The hyperspectral images are three-dimensional (one spectral and two spatial dimensions), so it is a combination of spectroscopy and digital imaging. The NIR hyperspectral imaging system comprised of a NIR hyperspectral camera, a lighting source to emit light in the wavelength region between 900 and 2500 nm, a sample holder or a motion device to carry the sample and a data acquisition and processing unit (Figure 3.11).

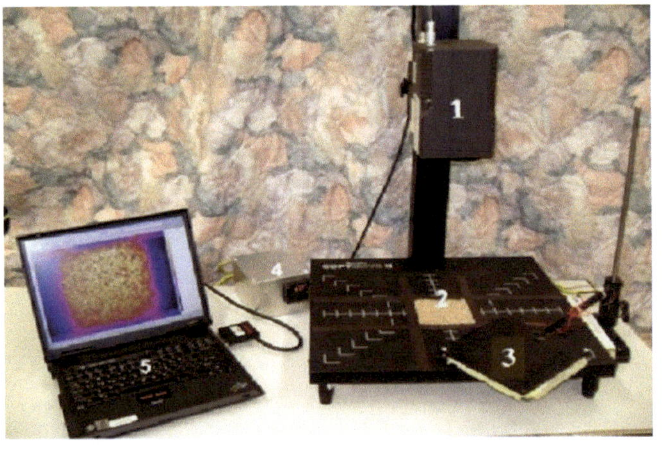

FIGURE 3.10 Thermal imaging system (1. thermal camera, 2. wheat sample, 3. heating element, 4. PID temperature controller, and 5. data processing system).

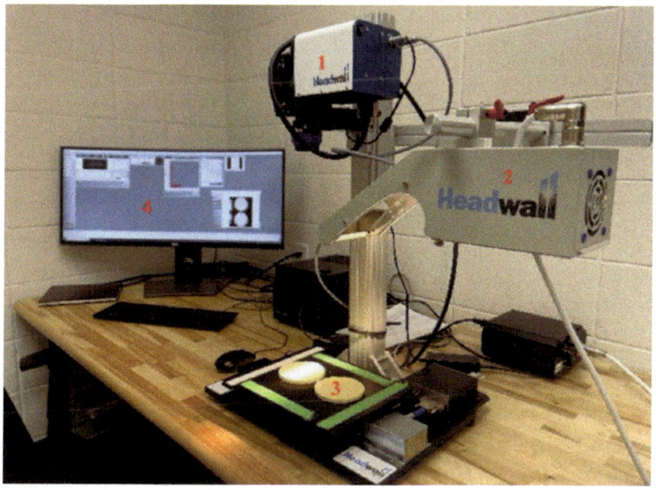

FIGURE 3.11 NIR hyperspectral imaging system (1. NIR camera, 2. lighting system, 3. sample holder, 4. data acquisition and processing system).

NIR hyperspectral imaging system in the wavelength region of 1000–1600 nm with 10 nm spectral resolution to detect six developmental stages (four larvae, one pupa, and one adult stage) of *Callosobruchus maculatus* (the cowpea weevil) in mung bean. The 300 control mung beans and 300 insect-infested mung beans from every developmental stage were subjected to NIR hyperspectral imaging. A total of 2100 individual mung bean images were acquired and processed into two-dimensional reflectance intensities. The statistical and histogram features from the two-dimensional images were extracted and subjected to statistical classifiers. The statistical classifiers were able to detect the control and infested mung beans with a classification accuracy of more than 80 percent. The pupae and adult stages were easily differentiated from the control compared to the four larvae stages. Another research study used NIR hyperspectral camera in the wavelength region between 960 and 1700 nm by following a similar procedure used to differentiate *C. maculatus*-infested soybeans from control soybeans. The statistical classifiers detected the control and insect-infested soybeans with a classification accuracy of 90 and 94 percent, respectively (Chelladurai et al. 2014).

The penetration depth on NIR hyperspectral imaging is only a few millimeters, so the NIR system cannot be deployed at the grain bin level. The NIR hyperspectral imaging system can replace the Berlese funnel and the Carter dockage tester because the Berlese and Carter dockage tester cannot detect the different developmental stages of insects accurately.

3.3.5.3 Soft X-Ray Imaging

Soft X-ray imaging operates in the wavelength region between 0.1 and 10 nm and energies between 0.12 and 12 keV. The soft X-rays have higher penetrating power compared to NIR hyperspectral imaging, so they were predominantly used to detect insects present inside the kernels. The soft X-ray imaging system has (1) X-ray source, (2). lixi fluoroscope and data acquisition and a processing system. A soft X-ray system was used to detect the different developmental stages of *R. dominica*. The soft X-ray images were acquired, and textural, histogram, and shape moments from the acquired images were extracted. The statistical classifiers and Back Propagation Neural Networks (BPNN) differentiated the control wheat kernels from infested wheat kernels with a classification accuracy of 93 and 98 percent, respectively. The soft X-ray system proved to be effective in detecting all developmental stages of *R. dominica* with a higher classification accuracy (Karunakaran, Jayas, and White 2004). Another study by Chelladurai et al. (2014) used soft X-ray imaging to detect different developmental stages of *C. maculatus* in soybeans. A soft X-ray imaging system operated at 16.1 kV and 150 μA was used to acquire images. The textural, histogram, and GLCM features from the images were extracted and used as inputs for statistical classifiers. The healthy and infested kernels were able to differentiate with the highest classification accuracy of more than 90 percent.

The soft X-ray imaging system used in the above two studies is a lab-level system. A line scan-type imaging system is needed to match the industry speed of operation. The soft X-ray system, like the NIR hyperspectral system, cannot be directly deployed at grain storage structures; they can replace the Berlese funnel and the Carter dockage tester.

3.4 CONCLUSION

In a modern-day grain storage structure, the storage managers and farmers rely on the moisture and temperature cables to access the storage conditions of the grains, and they occasionally climb up to check the grain conditions manually. When there were any elevated temperatures and moisture content, probes were used to get samples and checked for the presence of insects manually. The grain bins, when needed to be certified insect-free, then the grain samples were sent to certification agencies to check the presence of insects using a Berlese funnel. CO_2 sensors have started coming into the grain management portfolios to detect the presence of insects or other pests in the grain storage structures. Modern techniques like bioacoustics and VOCs-based biosensors are very promising because they can provide an estimate of the insect population. Thermal cameras can detect hot spots, but we cannot identify the insect species. The NIR hyperspectral imaging and X-ray imaging can be used to replace the Berlese funnel but cannot be deployed at grain storage structures because of its penetration capacity inside the grain bin. So, modern-day grain managers are relying on conventional probing techniques and the Berlese funnel to detect the presence of insects, which will soon be replaced by probing and imaging techniques. More research is needed to deploy any biosensor or bioacoustics-type insect monitoring devices in grain storage structures.

REFERENCES

Aulicky, R., V. Stejskal, Z. KuCerova, and P. Trematerra. 2016. "Trapping of Internal and External Feeding Stored Grain Beetle Pests with Two Types of Pitfall Traps: A Two-Year Field Study." *Plant Protection Science* 52 (1): 45–53. doi:10.17221/30/2015-PPS.

Cai, L., S. Macfadyen, B. Hua, H. Zhang, W. Xu, and Y. Ren. 2022. "Identification of Biomarker Volatile Organic Compounds Released by Three Stored-Grain Insect Pests in Wheat." *Molecules* 27 (6): 1963. doi:10.3390/molecules27061963.

Canadian Grain Commission (CGC). 2021. "Pneumatic Truck Probe Sampling." https://www.grainscanada.gc.ca/en/grain-quality/sampling-grain/guide-taking-representative-sample/02-truck-probe-sampling.html

Chelladurai, V., D. S. Jayas, and N. D. G. White. 2010. "Thermal Imaging for Detecting Fungal Infection in Stored Wheat." *Journal of Stored Products Research* 46 (3): 174–179. doi:10.1016/j.jspr.2010.04.002.

Chelladurai, V., K. Karuppiah, D. S. Jayas, P. G. Fields, and N. D. G. White. 2014. "Detection of *Callosobruchus maculatus* (F.) Infestation in Soybean Using Soft X-Ray and NIR Hyperspectral Imaging Techniques." *Journal of Stored Products Research* 57 (April): 43–48. doi:10.1016/j.jspr.2013.12.005.

Eliopoulos, P. A., I. Potamitis, D. C. Kontodimas, and E. G. Givropoulou. 2015. "Detection of Adult Beetles Inside the Stored Wheat Mass Based on Their Acoustic Emissions." *Journal of Economic Entomology* 108 (6): 2808–2814. doi:10.1093/jee/tov231.

Eliopoulos, P.A., I. Potamitis and D. C. Kontodimas. 2016. "Estimation of Population Density of Stored Grain Pests via Bioacoustic Detection." *Crop Protection* 85: 71–78.

FAO. 2023. "World Food and Agriculture – Statistical Yearbook 2023."

FAO. 2024. "Crop Prospects and Food Situation." Triannual Global Report, No. 2, July 2024. https://openknowledge.fao.org/handle/20.500.14283/cd1503en

Fargo, W. S., D. Epperly, G. W. Cuperus, B. C. Clary, and R. Noyes. 1989. "Effect of Temperature and Duration of Trapping on Four Stored Grain Insect Species." *Journal of Economic Entomology* 82 (3): 970–973. doi:10.1093/jee/82.3.970.

Flinn, P. W., D. W. Hagstrum, C. Reed, and T. W. Phillips. 2010. "Insect Population Dynamics in Commercial Grain Elevators." *Journal of Stored Products Research* 46 (1): 43–47.

Kammar, V., A. T. Rani, K. P. Kumar, and A. K. Chakravarthy. 2020. "Light Trap: A Dynamic Tool for Data Analysis, Documenting, and Monitoring Insect Populations and Diversity." In *Innovative pest Management Approaches for the 21st Century: Harnessing Automated Unmanned Technologies*, pp.137–163.

Karunakaran, C., D. S. Jayas, and N. D. G. White. 2004. "Detection of Internal Wheat Seed Infestation by *Rhyzopertha dominica* Using X-Ray Imaging." *Journal of Stored Products Research* 40 (5): 507–516. doi:10.1016/j.jspr.2003.06.003.

Konemann, C. E. 2021. "WB II Probe Trap: Reduced Length and Efficacy on Trap Catches." https://www.trece.com/in-the-news/wb-ii-probe-trap-reduced-length-and-efficacy-on-trap-catches/

Kumar, D., and P. Kalita. 2017. "Reducing Postharvest Losses during Storage of Grain Crops to Strengthen Food Security in Developing Countries." *Foods* 6 (1): 8. doi:10.3390/foods6010008.

Leos-Martinez, J., T. A. Granovsky, H. J. Williams, S. B. Vinson and W. E. Burkholder. 1987. "Pheromonal Trapping Methods for Lesser Grain Borer, *Rhyzopertha dominica* (Coleoptera: Bostrichidae)." *Environmental Entomology*, 16 (3): 747–751.

Lim, C. M., M. Carey, P. N. Williams, and A. Koidis. 2021. "Rapid Classification of Commercial Teas According to Their Origin and Type Using Elemental Content with X-Ray Fluorescence (XRF) Spectroscopy." *Current Research in Food Science* 4 (February). Elsevier Ltd: 45–52. doi:10.1016/j.crfs.2021.02.002.

Loschiavo, S. R., and J. M. Atkinson. 1967. "A Trap for the Detection and Recovery of Insects in Stored Grain." *The Canadian Entomologist* 99 (11): 1160–1163. doi:10.4039/Ent991160-11.

Manickavasagan, A., D. S. Jayas, N. D. G. White, and F. Jian. 2006. "Thermal Imaging of a Stored Grain Silo to Detect a Hot Spot." *Applied Engineering in Agriculture* 22 (6): 891–897. doi:10.13031/2013.22243.

Manickavasagan, A., D. S. Jayas, and N. D. G. White. 2008. "Thermal Imaging to Detect Infestation by *Cryptolestes ferrugineus* Inside Wheat Kernels." *Journal of Stored Products Research* 44 (2): 186–192. doi:10.1016/j.jspr.2007.10.006.

Manickavasagan, A., D. S. Jayas, N. D. G. White, and J. Paliwal. 2010. "Wheat Class Identification Using Thermal Imaging." *Food and Bioprocess Technology* 3 (3): 450–460. doi:10.1007/s11947-008-0110-x.

Mankin, R. W., D. W. Hagstrum, M. T. Smith, A. L. Roda, and M. T. K. Kairo. 2011. "Perspective and Promise: A Century of Insect Acoustic Detection and Monitoring." *American Entomologist* 57 (1): 30–44. doi:10.1093/ae/57.1.30.

Neethirajan, S., C. Karunakaran, D. S. Jayas, and N. D. G. White. 2007. "Detection Techniques for Stored-Product Insects in Grain." *Food Control* 18 (2): 157–162. doi:10.1016/j.foodcont.2005.09.008.

Reed, C. R., V. F. Wright, T. W. Mize, J. R. Pedersen, and B. J. Evans. 1991. "Pitfall Traps and Grain Samples as Indicators of Insects in Farm-Stored Wheat." *Journal of Economic Entomology* 84 (4): 1381–1387. doi:10.1093/jee/84.4.1381.

Senthilkumar, T., D. S. Jayas, N. D. G. White, M. S. Freund, C. Shafai and D. J. Thomson. 2012. "Characterization of Volatile Organic Compounds Released by Granivorous Insects in Stored Wheat." *Journal of Stored Products Research* 48: 91–96.

Senthilkumar, T., S. S. Panigrahi and C. B. Singh. 2024. "Continuous Carbon Dioxide Monitoring in a Commercial Bin Filled with Canola." In: Jayas, D. S., C. B. Singh and F. Jian (eds.), *Proceedings of the 12th International Conference on Controlled Atmosphere and Fumigation in Stored Products (CAF2024)*, CAF Permanent Committee Secretariat, Winnipeg, Canada. https://caf.phtcenter.com/wp-content/uploads/2024/11/CAF2024-Proceedings.pdf

Shimoda, M., and K.-i. Honda. 2013. "Insect Reactions to Light and Its Applications to Pest Management." *Applied Entomology and Zoology* 48 (4): 413–421. doi:10.1007/s13355-013-0219-x.

Toews, M. D., and C. Nansen. 2012. "Trapping and Interpreting Captures of Stored Grain Insects." In *Stored Product Protection*, 243. https://www.atchison.k-state.edu/docs/crops/stored_product_protection.pdf#page=250

White, N. D. G., and S. R. Loschiavo. 1986. "Effects of Insect Density, Trap Depth, and Attractants on the Capture of *Tribolium castaneum* (Coleoptera: Tenebrionidae) and *Cryptolestes ferrugineus* (Coleoptera: Cucujidae) in Stored Wheat." *Journal of Economic Entomology* 79 (4): 1111–1117. doi:10.1093/jee/79.4.1111.

Wong, J. W., V. Verigin, A. C. Oehlschlager, J. H. Borden, H. D. Pierce, A. M. Pierce, and L. Chong. 1983. "Isolation and Identification of Two Macrolide Pheromones from the Frass of *Cryptolestes ferrugineus* (Coleoptera: Cucujidae)." *Journal of Chemical Ecology* 9 (4): 451–474. doi:10.1007/BF00990219.

4 Chemical Methods of Disinfestation

Cristiano Dietrich Ferreira and Valmor Ziegler
Technological Institute in Food for Health, University of Vale do Rio dos Sinos, São Leopoldo, Brazil

Ricardo Tadeu Paraginski
Federal Institute of Education, Science and Technology Farroupilha, Campus Santo Augusto, Farroupilha, Brazil

4.1 INTRODUCTION

Integrated Grain Pest Management is an association of techniques that aims to reduce the incidence and severity of pest attacks on stored grains, providing knowledge of the situation of the grains and the storage unit. One of the main tools is the identification of the species and populations of pests occurring, in this case, using products registered for each species, respecting the dosage, care in application and shortage. Grain storage around the world is done with different levels of technology, from rustic clay containers with no control over storage conditions to large, automated concrete or metal structures. The characteristics common to all storage systems are susceptibility to attack by insect pests and storage fungi, which, associated with mites and bacteria, are mainly responsible for the loss of quality. Pests that affect stored grain may originate in the field or survive in warehouse structures between harvests. In order to be successful in pest control, regardless of the method, it is necessary to reduce the initial inoculum of these pests. The main hiding places for pests are corridors, walkways, tunnels, elevators and hoppers. Therefore, these places need to be cleaned carefully, as it is not always possible for the active ingredients to reach these places.

The pesticides used to control pests in warehouses are molecules or synthetic chemicals similar to those used in the field; however, the management used in warehouses is different. The search for cleaner post-harvest management has encouraged the development and registration of new molecules designed to control pests in stored products. Therefore, the molecules have been used for a long time, which leads to cases of pest resistance. The two main application methods are fumigation and liquid application. Purging is a curative technique for eliminating infesting insects in stored grains and seeds through the use of chemical products that release vapors and gases, also called fumigation, being applied in hermetic systems, where the gas released and/or introduced into it must remain in adequate concentration and time to be lethal to insects, depending on the molecule to be used.

The application carried out directly on the units' storage structures aims to eliminate initial outbreaks of pests, as well as the migration of insects from the outside to the inside of the structures. Spraying structures present better results in the absence of grains, that is, before receiving the grains, but can be carried out with stored raw material.

4.2 MANAGEMENT AND CLEANING OF STORAGE UNITS TO INCREASE THE EFFICIENCY OF CHEMICAL CONTROL

Quality management strategies focused on reducing losses have been adopted in several grain storage units. HACCP (Hazard Analysis and Critical Control Points) is used to identify, manage and

DOI: 10.1201/9781003609834-4

mitigate food safety and quality risks. HACCP can be used at all stages of the food chain, from primary production to final consumption, constituting an important part of food safety programs in storage units. The principles of HACCP are to perform a hazard analysis, determine critical control points (CCPs), establish critical limits for each CCP, establish a system to monitor PCC control and establish corrective action when monitoring indicates that a specific CCP is not under control; Establish verification procedures; Establish documentation relating to all procedures and records relevant to HACCP principles; and its application.

Grain storage around the world is carried out with different levels of technology, from rustic clay containers with no control over storage conditions to large, automated concrete or metal structures. Common characteristics of all storage systems are susceptibility to attack by insect pests and storage fungi, which, associated with mites and bacteria, are mainly responsible for loss of quality. Pests that affect stored grains can originate in the field or survive in warehouse structures between harvests. In order to be successful in pest control, regardless of the method, it is necessary to reduce the initial inoculum of these pests. The main hiding places for pests are corridors, walkways, tunnels, elevators and hoppers. Therefore, these places need to be cleaned carefully, as it is not always possible for the active ingredients to reach these places.

After the grains have been completely removed from the warehouse, maintenance of the structures begins, which also involves cleaning them. Simple measures such as sweeping floors and walls and removing grain residue and dust prevent the proliferation of insects and fungi, which can cause reinfestation of storage units. It should be noted that washing with water is a suggested practice for optimal cleaning of the units. With the units free of dirt and food fragments, the first chemical control is carried out, generally through thermoforming and/or spraying with insecticides to eliminate insects present on walls and equipment.

As mentioned previously, cleaning grain reduces the entry of insects into silos and warehouses, as dirt generally carries insects and fungal spores. The pre-cleaning stage carried out prior to drying and storage, is carried out by machines that use sieving and ventilation to remove impurities very efficiently, leaving the grain mass with foreign matter and impurities values of around 1–5%. Briefly, the main objective of cleaning grains and structures is to reduce pest reinfestation, as it reduces the initial inoculum but also minimizes the survival of resistant insect strains within the structures.

4.3 CHEMICAL METHODS

4.3.1 Pest Control by Chemical Purging or Fumigation

Purging is a curative technique to eliminate infesting insects in stored grains and seeds through the use of chemical products that release vapors and gases, therefore also called fumigation; being applied in hermetic systems, where the gas released and/or introduced, it must remain in adequate concentration and time to be lethal to insects, depending on the molecule to be used (Marco Aurélio G Pimentel et al. 2007; Paul et al. 2020). The main molecules used and studied are detailed below.

4.3.1.1 Aluminum Phosphide and Magnesium Phosphide

Phosphine is the most used fumigant in the world for controlling insects in stored grains and is normally used in aluminum phosphide or magnesium phosphide tablets, with records of its first commercial application around 1950 (Paul et al. 2020). The tablets, in contact with humid air (above 40% RH) at a temperature of 15°C or more, promote the release of phosphine gas (Arora, Stanley, and Srivastava 2021). The time for lethal action in the different life stages (egg, larva, pupa and adults) of insects will depend on the target species, the maintenance of concentration in the environment to be purged, relative air humidity and temperature and can vary from 4 to 25 days (Nayak et al. 2020).

The minimum concentration of phosphine gases in the environment to be purged must be maintained above 400 ppm for at least 120 hours (Lorini et al. 2015). Brabec et al. (2021) report that to eliminate insects resistant to *Rhyzopertha dominica*, a constant concentration of 700 ppm is necessary for 96 hours. A study with *Sitophilus oryzae*, *R. dominica* and *Tribolium castaneum* demonstrated that a concentration of 500–600 ppm, for seven days, was efficient for the elimination of these insects in the different phases of life in the storage of 5 Tons of wheat in piles, with a temperature of 25–27°C and relative humidity of 38–45% (Arora, Stanley, and Srivastava 2021).

In addition to the application of tablets to release the gas, when they come into contact with humidity in the environment, the gas can also be injected into the environment to be purged through cylinders or gas generators. It is important to highlight that maintaining these concentrations constitutes a relevant operational challenge in large silos and warehouses due to deficiencies and imperfections in the sealing of these environments, which often make it impossible to reach phosphine concentrations that represent lethality for insects (Chidemo, Musundire, and Mashavakure 2023). The mode of action of phosphine has been studied by some researchers, which is generally characterized as a reducing agent capable of inhibiting insect respiration, and the mechanisms that cause the inhibition of this respiration can be varied:

a. Phosphine acts on the mitochondrial electron transport chain, causing the generation of highly reactive oxyradicals, which react with vital enzymes, leading to insect lethality (Chaudhry and Price 1992).
b. Phosphine can act to reduce the enzymatic activity of cytochrome oxidase, NADH dehydrogenase and succinate dehydrogenase, reducing the respiratory rate and, consequently, ATP synthesis (Dua and Gill 2004).
c. Phosphine inhibits the mechanism of action of dihydrolipoamide dehydrogenase, which is the main respiratory enzyme of insects (Schlipalius et al. 2012).
d. Another study revealed that phosphine reduces the respiratory rate by inhibiting the antioxidant enzyme catalase and its coding gene (Liu et al. 2015).

Fumigation of stored products is applied with two objectives that have slightly different characteristics between them:

a. *For import or export*: It is carried out in the form of quarantine, with the aim of ensuring that the product does not take any insects associated with it to other regions. These are more rigorous fumigations, with the aim of eliminating 100% of the pest population. To achieve this, this operation is carried out in chambers and/or environments built for this purpose, which allow for tightness, control of temperature, relative humidity, fumigant concentration and aeration after the end of the process (Pimentel et al. 2007).
b. *Control in silos, warehouses and sheds*: It is carried out under real product storage conditions, which can vary substantially in terms of storage capacity, storage method and construction standard, among others. Operation is necessary to prevent degradation (quantitative and qualitative) of the product, extending its useful life (Pimentel et al. 2007).

The homogeneity of phosphine gas, in the concentrations already mentioned, throughout the environment to be purged is a challenge, which is why some management techniques with phosphine recirculation have been studied. One of them consists of forced recirculation, where ducts capture the gas in the upper part of the silo and are ventilated again in the lower part of the silo in a closed system. Another technique is called passive recirculation, where movement is produced by the effect of temperature and solar radiation on an external piping system, called a thermosiphon (Plumier and Maier 2018).

Recirculation has been shown to be effective in eliminating insects, including resistant insects. An important point is to identify the most appropriate recirculation rate that can reconcile the lowest silo leakage rates with reaching the necessary phosphine concentrations in the most vulnerable areas of the silo. In the study by Plumier and Maier (2018), a recirculation rate of 37.5% proved to be efficient in this balance between leakage and reaching the necessary concentrations in vulnerable areas of the silo; however, these same authors report that this rate can vary and adapt to the size and sealing conditions of each silo. Furthermore, the study by Plumier and Maier (2018) indicates that thermosiphon recirculation becomes more viable when compared to forced recirculation, as in general, the lowest possible flow rates for regulation in fans are still well above the ideal flow rate identified with the use of the thermosyphon.

Computer technologies have also been applied to assist in the management and control of phosphine distribution in storage environments. The study by Agrafioti et al. (2020) evaluated the use of a computational fluid dynamics method, using phosphine sensors associated with the OpenFOAM software, to monitor the distribution of phosphine and the control of susceptible and resistant insects. The findings of this study showed that recirculation monitored by technology improves the space-time distribution of the gas, making it possible to eliminate even resistant insects, which reduces the multiplier effects of resistance, which is in line with "precision fumigation."

Another important factor to be considered during phosphine purging is the amount that is absorbed by the grain, as well as the appropriate conditions to promote desorption and release to ensure operational safety in the storage environment and minimize toxicological impacts on the grain. There is a consensus, and in some countries, some standards, that the concentration of phosphine must be below 0.3 ppm in purged grain handling environments, as inhalation of phosphine gas presents serious risks of morbidity and mortality (Sudakin 2005). In this context, aeration conditions after opening the purge become a critical point and difficult to standardize since factors such as species, grain moisture and airspeed during aeration affect the desorption speed.

A study carried out by Plumier et al. (2020) in a silo loaded with 80 tons of wheat demonstrates that the phosphine desorption rate is 10 times greater than the sorption rate and the 24-hour aeration time would be sufficient in the environmental context of Australia, to reduce the phosphine levels below 0.3 ppm. It should also be noted that the better the silo's tightness, the greater the absorption of phosphine by the grains, which requires longer periods of ventilation. Due to the high variability of environmental and structural conditions in storage environments, the constant measurement of phosphine concentration in the environments after opening the purge is recommended since the release is not linear or homogeneous, which may lead to inaccurate diagnoses of the real concentrations of phosphine environmental.

The measurement of the residual phosphine content present in the grains still involves a series of interferences, ranging from the use of the appropriate extraction and quantification methodology as well as the understanding of the chemical reactions involved in the absorption of phosphine. Although without many exhaustive studies and investigations, the evidence shows that phosphine can be practically completely desorbed from the grains in prolonged aeration (reaching 0.0067 mg/kg of wheat grain), as there does not seem to be enough chemical interactions for it to be retained inside the grain (Khanchi et al. 2010).

4.3.1.1.1 Phosphine Resistance

Phosphine resistance occurs in numerous insect species and is a genetically controlled and heritable trait that allows carriers to survive a dose of phosphine that would normally be lethal. In several grain storage locations around the world, the incidence of resistant insect populations has been studied and recorded, as can be observed for *R. dominica* (Rajendran 1999; Beckel, Lorini, and Lázzari 2004; Opit et al. 2012; Collins et al. 2017; Afful et al. 2018), to *S. oryzae* (Holloway et al. 2016;

Nguyen et al. 2016), to *Sitophilus zeamais* (Pimentel et al. 2009) to *T. castaneum* (Gautam, Opit, and Hosoda 2016; Cato et al. 2017; Nayak et al. 2017) for *Lasioderma serricorne* (Sağlam, Edde, and Phillips 2015), among others (Nayak et al. 2020).

The mechanism by which insects acquire resistance to phosphine is being revealed and appears to be centered on two enzymes. The enzyme dihydrolipoyl dehydrogenase (DLD) generates reactive oxygen species (ROS) as a byproduct of its normal role in aerobic respiration (Tahara et al. 2007), while the FADS enzyme generates unsaturated fatty acids, which are targets of ROS. Exposure to phosphine enhances the production of ROS, which causes oxidative damage to fatty acids in cell membranes. However, the key point of resistance is related to populations of homozygous insects that tend to have cell membranes that are more resistant to the action of ROS; in addition, homozygous populations tend to produce less ROS. The association between the low production of ROS and the greater resistance of cell membranes to the ROS produced results in populations that are extremely resistant to phosphine (Nayak et al. 2020).

Other studies demonstrate the enhancement of cuticular protein genes in phosphine-resistant insect populations, as the cuticle is a natural barrier for insects against the invasion of external agents, such as pesticides. Insects reduce pesticide penetration by altering the structure and thickness of their cuticles, increasing their resistance to phosphine (Guo, Guo, and Liu 2022; Tang et al. 2023). In this context, the concern about the spread of insects with resistant genes to other regions becomes relevant and critical in the agricultural scenario.

As already discussed in this chapter, there are numerous factors that interfere with an effective purge, mainly due to the operational difficulties of maintaining adequate levels and times of phosphine so that the purge is capable of completely eliminating the insect population present in the purged environment. This set of factors gives insects the opportunity to begin adapting to these conditions through different mechanisms already mentioned. In this context, Nayak et al. (2020) report the importance of an integrated approach to the topic through monitoring and genetic mapping of phosphine-resistant species, fumigation protocols and strategic use of alternative fumigants, as well as associated with other management techniques that recommend the best practices for storage grain.

4.3.1.2 Sulfur Fluoride

Sulfur Fluoride (SF) is one of the most promising fumigants, after phosphine, but still far from reaching the commercial and cost-benefit potential that phosphine presents. The active ingredient consists of interrupting the glycolysis pathway in insects, depriving them of metabolic energy. Furthermore, evidence indicates that SF acts by inhibiting the main respiratory enzymes in addition to generating ROS oxidation of lipids and proteins. In theory, the mode of action of SF differs from that of phosphine in some mechanisms, which is positive, as it limits the chances of cross-resistance between these two fumigants (Opit et al. 2016).

Commercialization and use on a large scale are currently only permitted in some countries, such as the United States, Australia, Japan, Italy and Great Britain, but are still not authorized in several others (Rajendran and Sriranjini 2007). In addition to its possibility of being applied as an alternative to phosphine, in some cases it is used interspersed with phosphine, aiming to reduce phosphine-resistant insect species. It stands out for its ability to penetrate difficult-to-access areas, is non-flammable and does not react with electronic and structural materials.

Some limiting points for its use include the fluoride residue left in grains, which can be from 0.9 to 2.2 ppm, and because it is inert and stable, its concentration remains constant under environmental conditions (Rajendran and Sriranjini 2007), in addition to being considered less effective in controlling insect eggs (Athanassiou et al. 2012). In post-embryonic stages, for *T. castaneum*, a concentration of 31.25 mg/L showed promise (Jagadeesan et al. 2015).

In this context, SF can be an alternative, mainly for structural fumigation, due to its better penetrability in structures; however, the generation of waste and ineffectiveness in insect eggs constitute relevant obstacles that prevent the growth of its use on a large-scale commercial (Paul et al. 2020).

4.3.1.3 Methyl Bromide

Methyl bromide stands out as an effective fumigant for most different species of insects that attack stored grains. It is a colorless gas, highly toxic, non-flammable, non-corrosive, fast-acting and highly penetrable, in addition to having a low operating cost (Arora, Stanley, and Srivastava 2021). However, several adverse effects on human health and the environment have been reported, being considered a neurotoxicant that causes respiratory, cardiovascular, reproductive and developmental toxicity in animals (Park et al. 2020).

Due to the toxic health effects and destructive capacity of the ozone layer already reported by methyl bromide, its use was banned from 2005 onward in developed countries and from 2015 onward in developing countries, its use being limited only in some quarantines and pre-boarding treatments, being completely banned by the European Union for all purposes in 2010 (Paul et al. 2020). From this perspective, in recent years, a series of studies have been carried out with the intention of replacing methyl bromide and also being an alternative to the use of phosphine, as can be seen in the following items.

4.3.1.4 Nitric Oxide

Nitric oxide is one of the potential substitutes for more traditional fumigants. Its efficiency depends on the amount of oxygen in the environment, as it reacts quickly with O_2 to form nitrogen dioxide. Hermetic conditions appear to be more efficient, in addition to other factors such as the type of agricultural product, the height and volume of the grain mass and the degree of contamination (Cao et al. 2019). A greater number of studies and use on real scales is observed in other agricultural products, such as apples, strawberries, lettuce and other fruits and with a broader objective of quality control of these products, which involve mites and microorganisms (Granella et al. 2022).

Like other fumigants, the protocol for using nitric oxide is quite rigorous, considering that nitric oxide is quite reactive with O_2 and generates relevant amounts of NO_2, thus ensuring strict control of the O_2 concentration during the fumigation treatment must be observed to avoid excess NO_2 in the grains. Therefore, in grains, O_2 concentrations should not be greater than 0.1% (Liu, Yang, and Masuda 2017), which represents a suitable condition for the formation of up to 0.1% of NO_2, which is proved to be efficient in combating storage fungi, however, for effective insect control, this concentration still needs to be studied in more depth.

A study carried out by Liu (2013) evaluated the insecticidal capacity of nitric oxide in *Tribolium confusum* and *S. oryzae*, at different life stages, fumigated between 0.1 and 2.0% nitric oxide under oxygen levels ≤50 ppm in 1.9 L glass bottles, with a temperature between 2 and 25°C. These researchers found that the control of larvae, pupae and adults of *T. confusum* was verified after 24 hours at a concentration of 0.5% nitric oxide at 20°C, while complete mortality of eggs was verified after 24 hours with 0.5% nitric oxide at 10°C. To control *S. oryzae*, this was carried out with 1% nitric oxide in 24 and 48 hours, respectively, for adults and eggs, at 25°C. The results of this study show that nitric oxide can be a potential insecticide for use on stored grains, but different full-scale studies need to be carried out.

4.3.1.5 Ethyl Format

Ethyl formate (EF) is constituted as an ethyl ester of formic acid, which breaks down into ethanol and formic acid when exposed to air. Commercially it is sold as a non-flammable mixture of EF and liquid carbon dioxide with registrations in some countries such as Australia, New Zealand, Indonesia and Israel (Paul et al. 2020). The active ingredient of EF is based on the inhibition of the cytochrome oxidase enzyme in the electron transport chain and metabolic acidosis, in addition to reacting with insect body fluids to produce formic acid, which causes neuroexcitatory effects in insects (Song and Scharf 2008).

Because it is used as a food additive in some countries, it is considered safe for fumigation without leaving toxic residues. Its main studied uses are associated with other fumigants, with the aim of eliminating resistant populations of insects. The mixture of EF with CO_2 represented 92% mortality in all stages of *S. oryzae* in just 1 hour of exposure to 193 g/m^3, and the pupal stage was more tolerant (Damcevski et al. 2010). In another study, it was found that a concentration of EF at 70 g/m^3 for 24 hours can control all stages of *Tribolium* and *Rhyzopertha*, which attack stored grains (Ryan and De Lima 2014). Furthermore, the study carried out by Smit et al. (2020) demonstrated that a minimum concentration of 50 g/m^3 of EF for 1 hour was necessary to obtain 100% mortality of *Macchiademus diplopterus*. The synergistic effect of EF with methyl salicylate on *Callosobruchus chinensis* was evaluated by Chiluwal et al. (2020), and concluded that the synergistic effect can significantly reduce the dose needed to control different life stages. The large-scale application of EF still requires a lot of caution and studies; however, it may be promising for resistant populations and can also be associated with other insecticidal substances, with a positive difference in presenting low toxicity.

4.3.2 Pest Control by Liquid Application

4.3.2.1 Products Used

Pesticides are molecules or synthetic chemicals used to control insects, and currently, there are several groups and molecules launched every year; however, for use in controlling insects and storage pests, the number of these is small and also investments aimed at new. Among the various mechanisms of action to control stored grain pests, we can highlight acetylcholinesterase enzyme inhibitors, acetylcholine agonists and sodium channel modulators (Table 4.1).

Acetylcholinesterase enzyme inhibitors, which have synaptic transmission, bind to the acetylcholinesterase enzyme, preventing it from binding to acetylcholine, and as a result, there is an accumulation of acetylcholine in the synapse, causing hyperexcitability of the central nervous system due to the continuous and uncontrolled transmission of nerve impulses, and causes muscle paralysis, preventing breathing and causing death, with the products acephate, chlorpyrifos, dimethoate, diazinone, malathion, methamidophos, methamidophos, profenofós and terbufós standing.

Acetylcholine agonists affect the enzyme acetylcholinesterase, which under normal conditions binds to acetylcholine, whereas acetylcholine agonists simulate the effect of acetylcholine, constantly binding to acetylcholinesterase. Thus, it maintains the long-lasting activation of acetylcholine, causing hyperexcitability of the central nervous system due to the continuous and uncontrolled transmission of nerve impulses, resulting in tremors, convulsions and, occasionally, death due to the absence of oxygen in the brain. In this group, we can highlight neonicotinoids (acetamiprid, imidacloprid, thiamethoxam), nicotine and spinosins (Spinosad and Spinetoram).

Sodium channel modulators act to deregulate the entry and exit of sodium (Na) and potassium (K) in the nerve cell. Under normal conditions, sodium enters the cell to trigger nerve impulse transmission. In the case of the use of insecticides with this mechanism of action, there is a permanent entry of sodium into the cell, causing repetitive impulses and leading to exhaustion and death. In this group, pyrethroids and pyrethrins stand out (alphamethrin, betacypermetry, betacyfluthrin, bifenthrin, cyfluthrin, cypermethrin, deltamethrin, esfenvalerate, fenvalerate, fenpropatrin).

The Integrated Pest Management of Stored Grains provides knowledge of the situation of the grains and the storage unit and one of the main tools is to identify the species and populations of pests occurring, in which case using the products registered for each species, respecting dosage, care in application and also grace period. It is noteworthy that currently some pests already have resistance to some insecticides, and this knowledge needs to be taken into consideration when using this practice.

In this context, the liquid application can be carried out preventively and/or correctively to control stored grain pests; however, the products used for both purposes are the same, with approval for the use of pesticides to control pests following each country's own regulations.

TABLE 4.1
Different Insecticides for Controlling Stored Grain Pests

Species	Product	Title	Authors
Sitophilus oryzae	Deltamethrin, bifenthrin, thiamethoxam, spinosad and abamectin	Combined effects of contact insecticides and 50°C temperature on *S. oryzae* (L.) (Coleoptera: Curculionidae) in wheat grain	Pražić Golić, Andrić, and Kljajić (2016)
Sitophilus granarius, S. oryzae and Sitophilus zeamais	Spinosad and spinetoram	Effectiveness of spinosad and spinetoram against three *Sitophilus* species: Influence of wheat endosperm vitreousness	Andrić et al. (2019)
S. granarius and S. zeamais	Deltamethrin and spinosad	Deltamethrin- and spinosad-mediated survival, activity and avoidance of the grain weevils *S. granarius* and *S. zeamais*	Vélez et al. (2017)
Rhyzopertha dominica	Pyrethroid and organophosphate	Control failure and insecticides resistance in populations of *R. dominica* (Coleoptera: Bostrichidae) from Colombia	Ortega et al. (2021)
Mite pests (ácaros)	Organophosphate	A review of alternatives to organophosphorus compounds for the control of storage mites	Collins et al. (2017)
R. dominica, S. oryzae, Tribolium castaneum, Oryzaephilus surinamensis	Spinosad	Effectiveness of spinosad as a grain protectant against resistant beetle and psocid pests of stored grain in Australia	Nayak, Daglish, and Byrne (2005)
O. surinamensis and S. granarius	Pyrethroid	Initial efficacy of pyrethroids, inert dusts, their low-dose combinations and low temperature on *O. surinamensis* and *S. granarius*	Karimzadeh, Salehpoor, and Saber (2021)
Molecular residue	Chlorpyrifos-methyl, pirimiphos-methyl and permethrin	Stability of organophosphate and pyrethroid pesticides on wheat in storage	Afridi, Parveen, and Masud (2001)
Trogoderma granarium	Pyrethroid and insect growth regulators	Evaluation of pyrethroid insecticides and insect growth regulators applied to different surfaces for control of *T. granarium* (Coleoptera: Dermestidae) the Khapra Beetle	Arthur et al. (2018)
Tribolium confusum	Beta cyfluthrin, alpha cypermethrin and deltamethrin	Residual toxicity of beta cyfluthrin, alpha cypermethrin and deltamethrin against *T. confusum* Jacquelin du Val (Coleoptera: Tenebrionidae) on stored wheat	Athanassiou et al. (2004)
Cryptolestes ferrugineus, O. surinamensis, T. castaneum, T. confusum, Dermestes maculatus, Liposcelis bostrychophila and *Liposcelis paeta*	Beta-Cyfluthrin and Imidacloprid and Beta-Cyfluthrin Alone	Efficacy of a Combination of Beta-Cyfluthrin and Imidacloprid and Beta-Cyfluthrin Alone for Control of Stored-Product Insects on Concrete	Athanassiou et al. (2013)
R. dominica, S. oryzae and *T. confusum*	Spinetoram and spinosad	Evaluation of spinetoram and spinosad for control of *Prostephanus truncatus, R. dominica, S. oryzae*, and *T. confusum* on stored grains under laboratory tests	Athanassiou and Kavallieratos (2014)
T. confusum	Spinetoram, imidacloprid, thiamethoxam and chlorantraniliprole	Comparison of spinetoram, imidacloprid, thiamethoxam and chlorantraniliprole against life stages of *T. confusum* Jacquelin du Val (Coleoptera: Tenebrionidae) on concrete	Saglam, Athanassiou, and Vassilakos (2013)

(Continued)

TABLE 4.1 (Continued)
Different Insecticides for Controlling Stored Grain Pests

Species	Product	Title	Authors
S. granarius, S. oryzae, R. dominica, T. confusum, O. surinamensis, P. truncatus and Liposcelis entomophila	Pirimiphos-methyl	Comparison of two pirimiphos-methyl formulations against major stored-product insect species	Rumbos, Dutton, and Athanassiou (2013)
P. truncatus	Thiamethoxam and alpha-cypermethrin	Immediate and delayed mortality of the larger grain borer, *P. truncatus* (Horn), on different surfaces treated with thiamethoxam and alpha-cypermethrin	Doganay et al. (2018)
S. granarius, S. oryzae, O. surinamensis, T. castaneum and *C. ferrugineus*, as well as adults and larvae of *T. confusum*	Spinetoram and thiamethoxam	Insecticidal effect of spinetoram and thiamethoxam applied alone or in combination for the control of major stored-product beetle species	Rumbos, Dutton, and Athanassiou (2018)
P. truncatus, R. dominica, S. oryzae, S. granarius, O. surinamensis, T. confusum and *C. ferrugineus*	Thiamethoxam	Insecticidal effect of thiamethoxam against seven stored-product beetle species	Tsaganou, Vassilakos, and Athanassiou (2021)
S. oryzae, O. surinamensis, T. confusum and *P. truncatus*	Cyphenothrin and prallethrin	Insecticidal efficacy of a binary combination of cyphenothrin and prallethrin, applied as surface treatment against four major stored- product insects	Karanika et al. (2019)
Ephestia kuehniella	Cypermethrin and dichlorvos	Individual and combined effects of contact insecticides and high temperatures on Mediterranean flour moth, *E. kuehniella* Zeller (Lepidoptera: Pyralidae)	Javanshir et al. (2022)
S. oryzae, R. dominica and *T. confusum*	Spinetoram	Persistence and efficacy of spinetoram against three major stored grain beetle on wheat	Vassilakos, Athanassiou, and Tsiropoulos (2015)
R. dominica	14 chiral amides	Synthesis, insecticide activity, and phytotoxicity of novel chiral amides	Aguiar et al. (2019)
R. dominica	Deltamethrin	Prevalence of resistance to deltamethrin in *R. dominica* (F.) in eastern Australia	Daglish and Nayak (2018)
R. dominica	Spinosad and spinetoram	Residual efficacy of spinosad and spinetoram on traditional and new improved rice varieties on the mortality of *R. dominica* (F.) (Coleoptera: Bostrychidae)	Dissanayaka, Sammani, and Wijayaratne (2020)
S. granarius, R. dominica, and *T. confusum*	Pirimiphos-methyl	Efficacy of two formulations of pirimiphos-methyl as surface treatment against *S. granarius, R. dominica,* and *T. confusum*	Rumbos, Dutton, and Athanassiou (2014)

4.3.2.2 Direct Application in Storage Structures

The application carried out directly on the units' storage structures aims to eliminate initial outbreaks of pests, as well as the migration of insects from outside to the inside of the structures. Spraying structures present better results in the absence of grains, that is, before receiving the grains, but it can be carried out with stored raw material. This form of application is one of those recommended in integrated pest management, mainly because the application does not generate large volumes of insecticide affecting the raw materials.

The toxicological class of these products is extremely important since residues from these products can come into contact with the grains due to cargo movement. However, in this type of application, priority should be given to chemical products with a residual effect, ranging from 7 to 21 days, in order to eliminate all stages of insect development (eggs, larvae, pupae and adults). We can consider that pests are more easily controlled in the adult stage, followed by the larvae, pupa and egg stages; therefore, it is necessary that at the time of egg hatching and adult emergence, there is still product residue to carry out control.

As mentioned previously, it is necessary that warehouse structures are clean, minimizing the areas of refuge for insects at the time of application, whether in the application prior to receipt or after dispatch of the grains.

4.3.2.3 Direct Application to Grains

The application of pesticides via liquid directly to the grains must take into account the time that these grains will remain stored, and application is not recommended for storage periods of less than three months. Although the pesticides used in this type of pest control are capable of controlling pests immediately after application, they are more often used as preventive treatments. For this reason, the grains must be clean and dry, as this promotes better coating of the grains with pesticides.

Liquid spraying and fumigation are two control methods used alone or in combination. Liquid spraying obtains better results when applied in the absence of adult insects, as the dose and volume of spray generally need to be greater. That is why phosphine purge was previously used to reduce insect population levels. However, in bulk warehouses and vertical silos, purging is difficult to carry out due to the large volume of grains and phosphine gas leak points. Therefore, pests must be combatted preventively, with the application of insecticides on the conveyor belts and, if possible, equipped with pallets and/or tippers to turn the grains and standardize the mixture with insecticide during filling.

The chemical method seeks preventive control of pests, as they find optimal conditions for their development in the stored grains, and in this method, insecticides with a residual effect are applied, which aim to create an unfavorable environment for the development of pests. Preventive treatment consists of applying liquid insecticides to the grains on the conveyor belt when loading the warehouse and homogenizing them so that the entire grain receives insecticide, and this insecticide will protect the grain against attack by pests that will try to settle in the grain mass.

Spraying must be carried out with stabilized grains; that is, treatment must not be carried out with the hot grain mass immediately after it has left the dryer, as hot grains present a series of inconveniences for the treatment, which can result in ineffectiveness. Therefore, it is advisable to let the grains cool for some time and then spray them with insecticides and proceed with proper storage.

For application, a spray bar must be properly installed on the conveyor belt, in the tunnel or on the walkway, with three or five nozzles, distributed in such a way that all the grain receives the insecticide, and tippers must also be placed on the conveyor belt to ensure that the grains are mixed as they pass under the spray bar.

As mentioned previously, spray nozzles must be installed in places where there is grain movement, whether conveyor belts or other types of movers. But easily accessible locations are generally preferred, such as between the unloading pit and the elevator, or even after the cleaning machines.

The number of nozzles to be installed takes into account the capacity of the mover, with 1 sprinkler nozzle being recommended for every 35 tons of grain per hour. The height of the nozzle in the installation must be 40–50 cm from the conveyor belt, with a distance of 45–50 cm between one nozzle and another. For the entire grain mass to be reached, the placement of mixers and side protectors is essential.

For pest control to be successful, temperature and humidity conditions must not be favorable to the development of pests, as the greater the number of generations, the more easily insects and fungi acquire resistance to active ingredients. Another factor to be considered is the degradation of the active ingredient at high temperatures, which causes a reduction in control effectiveness and grace period.

A complementary practice of direct application is nebulization, which is the practice of applying insecticides using microparticles that are released into a stream of smoke produced by equipment that burns mineral oil, producing and releasing a jet of smoke into the environment. The low-density smoke carries the insecticide microparticles to higher points, where spraying is difficult, and can help control flying insects, such as moths, and the insecticide dosage is calculated depending on the volume of the internal space of the structure that will be occupied by smoke.

4.3.2.4 Pest Resistance

The most commonly used pest control method currently is the use of chemical insecticides; however, this method has been experiencing restrictions in use as problems arise regarding pest resistance to products used in stored grains. The selection of resistant insects is an example of the evolution of species and demonstrates how they can survive and change physiologically due to the great selection pressure of chemical products.

Likewise, it is necessary to adopt a rigorous monitoring system for pests, temperature and humidity of the grain mass; knowledge of the feeding habits of each pest constitutes an important element to define the management to be implemented in the mass of grain. Grains. Resistance of stored grain pests to insecticides is increasing worldwide, and some examples date back to 1991, such as the resistance of *Plutella xylostella* (L.) to the microbial insecticide *Bacillus thuringiensis* (Tabashnik, Finson, and Johnson 1991) and resistance has now been recorded to growth-regulating insecticides from different orders, such as reinforce the importance of evaluating resistance as a rational basis for chemical control of stored grain pests, indicating that studies involving behavioral changes in target species are important to elucidate some points about resistance, and according to Georghiou (1983), the behavior of a pest on an insecticide-treated surface is a biological factor that can influence selection for resistance.

Resistance mechanisms can be developed by reducing the penetration of the insecticide through the insect cuticle (1), by detoxification or metabolization of the insecticide by enzymes (2), by reducing sensitivity at the site of action of the insecticide by the nervous system (3) and (4) behavioral changes caused by chemicals (Narahashi, 1983; Oppenoorth, 1985; Soderlund and Bloomquist, 1990). In this context, integrated management practices need to be adopted to avoid the development of resistance mechanisms, which can make management in warehouses difficult. Furthermore, there are some biological agents, that is, natural enemies of pests that can be used to avoid this problem.

According to Beckel, Lorini, and Lázzari (2004), the management of insect resistance in stored products is particularly important due to the small number of active ingredients indicated for pest control, as well as the problems of toxic residues in products resulting from the increase in doses necessary to achieve satisfactory control. In this context, deltamethrin and pirimiphos-methyl are considered two of the main insecticides used in grain storage and several resistant populations have already been identified (Lorini and Galley 1996; Lorini and Galley 1999; Beckel, Lorini, and Lázzari 2004; Lorini et al. 2015). Some cases of resistance are *Rhyzopertha dominica* to the pyrethroid insecticide deltamethrin (Lorini and Galley 1996, 1999) and its resistance to the insecticides pirimiphos-methyl, chlorpyrifos-methyl and permethrin. Work carried out by Baena-Moreno et al. (2019) sought to evaluate the resistance of *T. castaneum, Oryzaephilus surinamensis, R. dominica* and *S. oryzae* in different grain storage units in Brazil to the insecticides deltamethrin and pirimiphos-methyl, to assist in integrated pest management actions. Stored grains and the results indicated resistance of adults of *O. surinamensis* with a resistance factor greater than 14 times, *T. castaneum and S. oryzae* with 2.5 times and *R. dominica* with 41 times, to the insecticide deltamethrin, and when In the case of the insecticide pirimiphos-methyl, no studies on resistance or susceptible populations were found to allow comparisons to be made with *T. castaneum* population.

Chemical pest control has been studied for a long time, mainly focusing on the products deltamethrin, bifenthrin, thiamethoxam, spinosad, beta cyfluthrin and pirimiphos-methyl for the control of species of the genus *Sitophilus, R. dominica, T. castaneum, O. surinamensis* and *Cryptolestes ferrugineus.* These insects mentioned above are pests that predominantly attack cereals and flour, foods that are more susceptible when compared to protein and oilseed seeds; therefore, a greater concentration of studies is justified.

Scientific studies using the active ingredients mentioned above have sought to meet what is being adopted in warehouses. These active ingredients used recurrently and without rotation, can intensify pest resistance. Therefore, studies with dosage, application conditions and rotation of active ingredients are necessary. Another point of research that has been gaining ground is the association of chemical methods and alternative methods used mainly to overcome pest resistance.

4.4 CONCLUSION

The use of pesticides to control pests in warehouses presents distinct challenges compared to field management. The pursuit of cleaner post-harvest management has driven the development of new molecules to reduce pest resistance, a recurring problem due to the prolonged use of the same chemicals. The main application techniques are fumigation and liquid application. Fumigation, commonly performed with phosphine, is effective but faces operational challenges due to the need to maintain adequate gas concentrations in large and often poorly sealed environments. Phosphine resistance, present in various insect species, is a growing concern. Liquid application directly to the grains requires consideration of storage time and is more effective as a preventive measure. Spraying storage structures is recommended before receiving the grains, minimizing the use of large volumes of insecticides and reducing the risk of toxic residues. Despite the effectiveness of chemical insecticides, pest resistance remains a significant issue, exemplifying the evolution of species under the selective pressure of chemical products. Therefore, the integration of management methods, along with the development of new molecules, is crucial for sustainable pest control in warehouses.

REFERENCES

Afful, E., B. Elliott, M. K. Nayak, and T. W. Phillips. 2018. "Phosphine Resistance in North American Field Populations of the Lesser Grain Borer, *Rhyzopertha dominica* (Coleoptera: Bostrichidae)." *Journal of Economic Entomology* 111 (1): 463–469. doi:10.1093/jee/tox284.

Afridi, I. A. K., Z. Parveen, and S. Z. Masud. 2001. "Stability of Organophosphate and Pyrethroid Pesticides on Wheat in Storage." *Journal of Stored Products Research* 37 (2): 199–204. doi:10.1016/S0022-474X(00)00020-5.

Agrafioti, P., E. Kaloudis, S. Bantas, V. Sotiroudas, and C. G. Athanassiou. 2020. "Modeling the Distribution of Phosphine and Insect Mortality in Cylindrical Grain Silos with Computational Fluid Dynamics: Validation with Field Trials." *Computers and Electronics in Agriculture* 173 (June): 105383. doi:10.1016/j.compag.2020.105383.

Aguiar, A. R., E. S. Alvarenga, E. M. P. Silva, E. S. Farias, and M. C. Picanço. 2019. "Synthesis, Insecticidal Activity, and Phytotoxicity of Novel Chiral Amides." *Pest Management Science* 75 (6): 1689–1696. doi:10.1002/ps.5289.

Andrić, G., P. Kljajić, M. P. Golić, S. Trdan, T. Bohinc, and M. B. Solarov. 2019. "Effectiveness of Spinosad and Spinetoram against Three Sitophilus Species: Influence of Wheat Endosperm Vitreousness." *Journal of Stored Products Research* 83 (September): 209–217. doi:10.1016/j.jspr.2019.07.001.

Arora, S., J. Stanley, and C. Srivastava. 2021. "Temporal Dynamics of Phosphine Fumigation against Insect Pests in Wheat Storage." *Crop Protection* 144 (June): 105602. doi:10.1016/j.cropro.2021.105602.

Arthur, F. H., M. N. Ghimire, S. W. Myers, and T. W. Phillips. 2018. "Evaluation of Pyrethroid Insecticides and Insect Growth Regulators Applied to Different Surfaces for Control of *Trogoderma granarium* (Coleoptera: Dermestidae) the Khapra Beetle." *Journal of Economic Entomology* 111 (2): 612–619. doi:10.1093/jee/toy040.

Athanassiou, C. G., and N. G. Kavallieratos. 2014. "Evaluation of Spinetoram and Spinosad for Control of *Prostephanus truncatus, Rhyzopertha dominica, Sitophilus oryzae*, and *Tribolium confusum* on Stored Grains under Laboratory Tests." *Journal of Pest Science* 87 (3): 469–483. doi:10.1007/s10340-014-0563-9.

Athanassiou, C. G., N. G. Kavallieratos, F. H. Arthur, and J. E. Throne. 2013. "Efficacy of a Combination of Beta-Cyfluthrin and Imidacloprid and Beta-Cyfluthrin Alone for Control of Stored-Product Insects on Concrete." *Journal of Economic Entomology* 106 (2): 1064–1070. doi:10.1603/EC12406.

Athanassiou, C. G., N. G. Kavallieratos, B. J. Vayias, C. B. Dimizas, A. S. Papagregoriou, and C. T. Buchelos. 2004. "Residual Toxicity of Beta Cyfluthrin, Alpha Cypermethrin and Deltamethrin against *Tribolium confusum* Jacquelin Du Val (Coleoptera: Tenebrionidae) on Stored Wheat." *Applied Entomology and Zoology* 39 (2): 195–202. doi:10.1303/aez.2004.195.

Athanassiou, C. G., T. W. Phillips, M. J. Aikins, M. M. Hasan, and J. E. Throne. 2012. "Effectiveness of Sulfuryl Fluoride for Control of Different Life Stages of Stored-Product Psocids (Psocoptera)." *Journal of Economic Entomology* 105 (1): 282–287. doi:10.1603/EC11209.

Baena-Moreno, F. M., M. Rodríguez-Galán, F. Vega, B. Alonso-Fariñas, L. F. Vilches Arenas, and B. Navarrete. 2019. "Carbon Capture and Utilization Technologies: A Literature Review and Recent Advances." *Energy Sources, Part A: Recovery, Utilization, and Environmental Effects* 41 (12): 1403–1433.

Beckel, H., I. Lorini, and S. M. N. Lázzari. 2004. "Comportamento de Adultos de Diferentes Raças de *Rhyzopertha dominica* (Fabricius) (Coleoptera, Bostrichidae) Em Superfície Tratada Com Deltamethrin." *Revista Brasileira de Entomologia* 48 (1): 115–118. doi:10.1590/s0085-56262004000100019.

Brabec, D., W. Morrison, J. Campbell, F. Arthur, A. Bruce, and K. Yeater. 2021. "Evaluation of Dosimeter Tubes for Monitoring Phosphine Fumigations." *Journal of Stored Products Research* 91 (March): 101762. doi:10.1016/j.jspr.2021.101762.

Cao, Y., D. Guan, W. Huang, J. Yang, Y. Cao, and Y. Qiao. 2019. "Pedestrian Detection with Unsupervised Multispectral Feature Learning Using Deep Neural Networks." *Information Fusion* 46 (March): 206–217. doi:10.1016/j.inffus.2018.06.005.

Cato, A. J., B. Elliott, M. K. Nayak, and T. W. Phillips. 2017. "Geographic Variation in Phosphine Resistance among North American Populations of the Red Flour Beetle (Coleoptera: Tenebrionidae)." *Journal of Economic Entomology* 110 (3): 1359–1365. doi:10.1093/jee/tox091.

Chaudhry, M. Q., and N. R. Price. 1992. "Comparison of the Oxidant Damage Induced by Phosphine and the Uptake and Tracheal Exchange of 32P-Radiolabelled Phosphine in the Susceptible and Resistant Strains of *Rhyzopertha dominica* (F.) (Coleoptera: Bostrychidae)." *Pesticide Biochemistry and Physiology* 42 (2): 167–179. doi:10.1016/0048-3575(92)90063-6.

Chidemo, S. C., R. Musundire, and N. Mashavakure. 2023. "Higher Dosage of Phosphine Is Required to Control Resistant Strains of Pests in Outdoor Grain Storage Systems: Evidence from Zimbabwe." *Journal of Stored Products Research* 100 (January): 102046. doi:10.1016/j.jspr.2022.102046.

Chiluwal, K., B. H. Lee, T. H. Kwon, J. Kim, S. D. Bae, G. H. Roh, Y. Ren, B. Li, and C. G. Park. 2020. "Synergistic Effect of Fumigation with Ethyl Formate and Methyl Salicylate on Mortality of Life Stages of Adzuki Bean Beetle, *Callosobruchus chinensis* (L.)." *Journal of Asia-Pacific Entomology* 23 (2): 483–491. doi:10.1016/j.aspen.2020.03.010.

Collins, P. J., M. G. Falk, M. K. Nayak, R. N. Emery, and J. C. Holloway. 2017. "Monitoring Resistance to Phosphine in the Lesser Grain Borer, *Rhyzopertha dominica*, in Australia: A National Analysis of Trends, Storage Types and Geography in Relation to Resistance Detections." *Journal of Stored Products Research* 70 (January): 25–36. doi:10.1016/j.jspr.2016.10.006.

Daglish, G. J., and M. K. Nayak. 2018. "Prevalence of Resistance to Deltamethrin in *Rhyzopertha dominica* (F.) in Eastern Australia." *Journal of Stored Products Research* 78 (September): 45–49. doi:10.1016/j.jspr.2018.06.003.

Damcevski, K. A., G. Dojchinov, J. D. Woodman, and V. S. Haritos. 2010. "Efficacy of Vaporised Ethyl Formate/Carbon Dioxide Formulation against Stored-Grain Insects: Effect of Fumigant Concentration, Exposure Time and Two Grain Temperatures." *Pest Management Science: Formerly Pesticide Science* 66 (4): 432–438.

Dissanayaka, D. M. S. K., A. M. P. Sammani, and L. K. W. Wijayaratne. 2020. "Residual Efficacy of Spinosad and Spinetoram on Traditional and New Improved Rice Varieties on the Mortality of *Rhyzopertha dominica* (F.) (Coleoptera: Bostrychidae)." *Journal of Stored Products Research* 88 (September): 101643. doi:10.1016/j.jspr.2020.101643.

Doganay, I., P. Agrafioti, A. A. Isikber, O. Saglam, and C. G. Athanassiou. 2018. "Immediate and Delayed Mortality of the Larger Grain Borer, *Prostephanus truncatus* (Horn), on Different Surfaces Treated with Thiamethoxam and Alpha-Cypermethrin." *Journal of Stored Products Research* 76 (March): 1–6. doi:10.1016/j.jspr.2017.10.003.

Dua, R., and K. D. Gill. 2004. "Effect of Aluminium Phosphide Exposure on Kinetic Properties of Cytochrome Oxidase and Mitochondrial Energy Metabolism in Rat Brain." *Biochimica et Biophysica Acta (BBA) – General Subjects* 1674 (1): 4–11. doi:10.1016/j.bbagen.2004.05.003.

Gautam, S. G., G. P. Opit, and E. Hosoda. 2016. "Phosphine Resistance in Adult and Immature Life Stages of *Tribolium castaneum* (Coleoptera: Tenebrionidae) and *Plodia interpunctella* (Lepidoptera: Pyralidae) Populations in California." *Journal of Economic Entomology* 109 (6): 2525–2533. doi:10.1093/jee/tow221.

Georghiou, G. P. 1983. "Management of Resistance in Arthropods." In *Pest Resistance to Pesticides*, 769–792. Boston, MA: Springer US. doi:10.1007/978-1-4684-4466-7_32.

Granella, S. J., T. R. Bechlin, D. Christ, and S. R. M. Coelho. 2022. "A Potential Role of Nitric Oxide in Postharvest Pest Control: A Review." *Journal of the Saudi Society of Agricultural Sciences* 21 (7): 452–459. doi:10.1016/j.jssas.2021.12.002.

Guo, P.-L., Z.-Q. Guo, and X.-D. Liu. 2022. "Cuticular Protein Genes Involve Heat Acclimation of Insect Larvae under Global Warming." *Insect Molecular Biology* 31 (4): 519–532. doi:10.1111/imb.12777.

Holloway, J. C., M. G. Falk, R. N. Emery, P. J. Collins, and M. K. Nayak. 2016. "Resistance to Phosphine in *Sitophilus oryzae* in Australia: A National Analysis of Trends and Frequencies over Time and Geographical Spread." *Journal of Stored Products Research* 69 (October): 129–137. doi:10.1016/j.jspr.2016.07.004.

Jagadeesan, R., M. K. Nayak, H. Pavic, K. Chandra, and P. J. Collins. 2015. "Susceptibility to Sulfuryl Fluoride and Lack of Cross-Resistance to Phosphine in Developmental Stages of the Red Flour Beetle, *Tribolium castaneum* (Coleoptera: Tenebrionidae)." *Pest Management Science* 71 (10): 1379–1386. doi:10.1002/ps.3940.

Javanshir, M., R. Karimzadeh, M. J. Hejazi, and H. Shayestehmehr. 2022. "Individual and Combined Effects of Contact Insecticides and High Temperatures on Mediterranean Flour Moth, *Ephestia kuehniella* Zeller (Lepidoptera: Pyralidae)." *Journal of the Science of Food and Agriculture* 102 (5): 2165–2171. doi:10.1002/jsfa.11558.

Karanika, C., C. I. Rumbos, P. Agrafioti, and C. G. Athanassiou. 2019. "Insecticidal Efficacy of a Binary Combination of Cyphenothrin and Prallethrin, Applied as Surface Treatment against Four Major Stored-Product Insects." *Journal of Stored Products Research* 80 (January): 41–49. doi:10.1016/j.jspr.2018.10.008.

Karimzadeh, R., M. Salehpoor, and M. Saber. 2021. "Initial Efficacy of Pyrethroids, Inert Dusts, Their Low-Dose Combinations and Low Temperature on *Oryzaephilus surinamensis* and *Sitophilus granarius*." *Journal of Stored Products Research* 91 (March): 101780. doi:10.1016/j.jspr.2021.101780.

Khanchi, A. R., M. Moein, A. Mansoubi, M. H. Hosseini, F. Bani, and M. K. Mahani. 2010. "Evaluation of the Factors Affecting the Phosphine Residue in the Fumigation–Aeration Process of Stored Wheat Using Orthogonal Array Design." *Food Analytical Methods* 3 (3): 169–174. doi:10.1007/s12161-009-9114-z.

Liu, Y.-B. 2013. "Nitric Oxide as a Potent Fumigant for Postharvest Pest Control." *Journal of Economic Entomology* 106 (6): 2267–2274. doi:10.1603/EC13249.

Liu, T., L. Li, F. Zhang, and Y. Wang. 2015. "Transcriptional Inhibition of the Catalase Gene in Phosphine-Induced Oxidative Stress in *Drosophila melanogaster*." *Pesticide Biochemistry and Physiology* 124 (October): 1–7. doi:10.1016/j.pestbp.2015.05.005.

Liu, Y.-B., X. Yang, and T. Masuda. 2017. "Procedures of Laboratory Fumigation for Pest Control with Nitric Oxide Gas." *Journal of Visualized Experiments*, no. 129 (November). doi:10.3791/56309.

Lorini, I., and D. J. Galley. 1996. "Changes in Resistance Status of *Rhyzopertha dominica* (F.) (Coleoptera: Bostrychidae), a Pest of Stored Grain in Brazil, with and without Deltamethrin Selection." *Resistant Pest Management Newsletter* 8: 12–14.

Lorini, I., and D. J. Galley. 1999. "Deltamethrin Resistance in *Rhyzopertha dominica* (F.) (Coleoptera: Bostrichidae), a Pest of Stored Grain in Brazil." *Journal of Stored Products Research* 35 (1): 37–45. doi:10.1016/S0022-474X(98)00028-9.

Lorini, I., F. C. Krzyzanowski, J. D. B. França-Neto, A. A. Henning, and F. A. Henning. 2015. *Manejo Integrado de Pragas de Grãos e Sementes Armazenadas*. Empresa Brasileira de Pesquisa Agropecuária.

Narahashi, T. 1983. "Resistance to Insecticides due to Reduced Sensitivity of the Nervous System." In *Pest Resistance to Pesticides*, 333–352. Boston, MA: Springer US.

Nayak, M. K., G. J. Daglish, and V. S. Byrne. 2005. "Effectiveness of Spinosad as a Grain Protectant against Resistant Beetle and Psocid Pests of Stored Grain in Australia." *Journal of Stored Products Research* 41 (4): 455–467. doi:10.1016/j.jspr.2004.07.002.

Nayak, M. K., G. J. Daglish, T. W. Phillips, and P. R. Ebert. 2020. "Resistance to the Fumigant Phosphine and Its Management in Insect Pests of Stored Products: A Global Perspective." *Annual Review of Entomology* 65: 333–350. doi:10.1146/annurev-ento-011019-025047.

Nayak, M. K., M. G. Falk, R. N. Emery, P. J. Collins, and J. C. Holloway. 2017. "An Analysis of Trends, Frequencies and Factors Influencing the Development of Resistance to Phosphine in the Red Flour Beetle, *Tribolium castaneum* (Herbst) in Australia." *Journal of Stored Products Research* 72 (May): 35–48. doi:10.1016/j.jspr.2017.03.004.

Nguyen, T. T., P. J. Collins, T. M. Duong, D. I. Schlipalius, and P. R. Ebert. 2016. "Genetic Conservation of Phosphine Resistance in the Rice Weevil *Sitophilus oryzae* (L.)." *Journal of Heredity* 107 (3): 228–237. doi:10.1093/jhered/esw001.

Opit, G. P., T. W. Phillips, M. J. Aikins, and M. M. Hasan. 2012. "Phosphine Resistance in *Tribolium castaneum* and *Rhyzopertha dominica* From Stored Wheat in Oklahoma." *Journal of Economic Entomology* 105 (4): 1107–1114. doi:10.1603/EC12064.

Opit, G. P., E. Thoms, T. W. Phillips, and M. E. Payton. 2016. "Effectiveness of Sulfuryl Fluoride Fumigation for the Control of Phosphine-Resistant Grain Insects Infesting Stored Wheat." *Journal of Economic Entomology* 109 (2): 930–941. doi:10.1093/jee/tov395.

Oppenoorth, F. J., Kerkut, G. A., and Gilbert, L. I. 1985. "Biochemistry and Genetics of Insecticide Resistance." In *Comprehensive Insect Physiology, Biochemistry and Pharmacology: Insect Control* (Vol. 12), 731–773. Oxford: Pergamon Press.

Ortega, D. S., T. Bacca, A. P. Nascimento Silva, N. A. Canal, and K. Haddi. 2021. "Control Failure and Insecticides Resistance in Populations of *Rhyzopertha dominica* (Coleoptera: Bostrichidae) from Colombia." *Journal of Stored Products Research* 92 (May): 101802. doi:10.1016/j.jspr.2021.101802.

Park, M.-G., J. Choi, Y.-S. Hong, C. G. Park, B.-G. Kim, S.-Y. Lee, H.-J. Lim, H.-ho Mo, E. Lim, and W. Cha. 2020. "Negative Effect of Methyl Bromide Fumigation Work on the Central Nervous System." Edited by Amitava Mukherjee. *PLOS ONE* 15 (8): e0236694. doi:10.1371/journal.pone.0236694.

Paul, A., M. Radhakrishnan, S. Anandakumar, S. Shanmugasundaram, and C. Anandharamakrishnan. 2020. "Disinfestation Techniques for Major Cereals: A Status Report." *Comprehensive Reviews in Food Science and Food Safety* 19 (3): 1125–1155. doi:10.1111/1541-4337.12555.

Pimentel, M. A. G., L. R. D'A. Faroni, R. N. C. Guedes, A. H. Sousa, and M. R. Tótola. 2009. "Phosphine Resistance in Brazilian Populations of *Sitophilus zeamais* Motschulsky (Coleoptera: Curculionidae)." *Journal of Stored Products Research* 45 (1): 71–74. doi:10.1016/j.jspr.2008.09.001.

Pimentel, M. A. G., L. R. D. A. Faroni, M. R. Tótola, and R. N. C. Guedes. 2007. "Phosphine Resistance, Respiration Rate and Fitness Consequences in Stored-product Insects." *Pest Management Science* 63 (9): 876–881. doi:10.1002/ps.1416.

Plumier, B. M., and D. E. Maier. 2018. "Sensitivity Analysis of a Fumigant Movement and Loss Model for Bulk Stored Grain to Predict Effects of Environmental Conditions and Operational Variables on Fumigation Efficacy." *Journal of Stored Products Research* 78 (September): 18–26. doi:10.1016/j.jspr.2018.05.012.

Plumier, B. M., M. Schramm, Y. Ren, and D. E. Maier. 2020. "Modeling Post-Fumigation Desorption of Phosphine in Bulk Stored Grain." *Journal of Stored Products Research* 85 (January): 101548. doi:10.1016/j.jspr.2019.101548.

Pražić Golić, M., G. Andrić, and P. Kljajić. 2016. "Combined Effects of Contact Insecticides and 50°C Temperature on *Sitophilus oryzae* (L.) (Coleoptera: Curculionidae) in Wheat Grain." *Journal of Stored Products Research* 69 (October): 245–251. doi:10.1016/j.jspr.2016.09.004.

Rajendran, S. 1999. "Resistência à Fosfina Em Pragas de Insetos de Grãos Armazenados Na Índia." In *In Proceedings of the 7th International Working Conference on Stored Product Protection*, 635–641.

Rajendran, S., and V-rao Sriranjini. 2007. "Use of Fumigation for Managing Grain Quality." *Stewart Postharvest Review* 3 (6): 1–8. doi:10.2212/spr.2007.6:9.

Rumbos, C. I., A. C. Dutton, and C. G. Athanassiou. 2013. "Comparison of Two Pirimiphos-Methyl Formulations against Major Stored-Product Insect Species." *Journal of Stored Products Research* 55 (October): 106–115. doi:10.1016/j.jspr.2013.08.003.

Rumbos, C. I., A. C. Dutton, and C. G. Athanassiou. 2014. "Efficacy of Two Formulations of Pirimiphos-Methyl as Surface Treatment against *Sitophilus granarius*, *Rhyzopertha dominica*, and *Tribolium confusum*." *Journal of Pest Science* 87 (3): 507–519. doi:10.1007/s10340-014-0599-x.

Rumbos, C. I., A. C. Dutton, and C. G. Athanassiou. 2018. "Insecticidal Effect of Spinetoram and Thiamethoxam Applied Alone or in Combination for the Control of Major Stored-Product Beetle Species." *Journal of Stored Products Research* 75 (January): 56–63. doi:10.1016/j.jspr.2017.10.004.

Ryan, R. F., and C. P. F. De Lima. 2014. "Ethyl Formate Fumigation an Overview Update." In *In Proceedings of the 11th International Working Conference on Stored Product Protection*, 24–28.

Saglam, O., C. G. Athanassiou, and T. N. Vassilakos. 2013. "Comparison of Spinetoram, Imidacloprid, Thiamethoxam and Chlorantraniliprole against Life Stages of *Tribolium confusum* Jacquelin Du Val (Coleoptera: Tenebrionidae) on Concrete." *Crop Protection* 53 (November): 85–95. doi:10.1016/j.cropro.2013.05.007.

Sağlam, Özgür, P. A. Edde, and T. W. Phillips. 2015. "Resistance of *Lasioderma serricorne* (Coleoptera: Anobiidae) to Fumigation with Phosphine." *Journal of Economic Entomology* 108 (5): 2489–2495. doi:10.1093/jee/tov193.

Schlipalius, D. I., N. Valmas, A. G. Tuck, R. Jagadeesan, L. Ma, R. Kaur, A. Goldinger, et al. 2012. "A Core Metabolic Enzyme Mediates Resistance to Phosphine Gas." *Science* 338 (6108): 807–810. doi:10.1126/science.1224951.

Smit, R., M. M. Jooste, M. F. Addison, and S. A. Johnson. 2020. "Ethyl Formate Fumigation: Its Effect on Stone and Pome Fruit Quality, and Grain Chinch Bug (*Macchiademus diplopterus*) Mortality." *Scientia Horticulturae* 261 (February): 108845. doi:10.1016/j.scienta.2019.108845.

Soderlund, D. M., and J. R. Bloomquist. 1990. "Molecular Mechanisms of Insecticide Resistance." In *Pesticide Resistance in Arthropods*, 58–96. Boston, MA: Springer US.

Song, C., and M. E. Scharf. 2008. "Formic Acid: A Neurologically Active, Hydrolyzed Metabolite of Insecticidal Formate Esters." *Pesticide Biochemistry and Physiology* 92 (2): 77–82. doi:10.1016/j.pestbp.2008.06.005.

Sudakin, D. L. 2005. "Occupational Exposure to Aluminium Phosphide and Phosphine Gas? A Suspected Case Report and Review of the Literature." *Human & Experimental Toxicology* 24 (1): 27–33. doi:10.1191/0960327105ht496oa.

Tabashnik, B. E., N. Finson, and M. W. Johnson. 1991. "Managing Resistance to *Bacillus thuringiensis*: Lessons from the Diamondback Moth (Lepidoptera: Plutellidae)." *Journal of Economic Entomology* 84 (1): 49–55. doi:10.1093/jee/84.1.49.

Tahara, E. B., M. H. Barros, G. A. Oliveira, L. E. S. Netto, and A. J. Kowaltowski. 2007. "Dihydrolipoyl Dehydrogenase as a Source of Reactive Oxygen Species Inhibited by Caloric Restriction and Involved in *Saccharomyces cerevisiae* Aging." *The FASEB Journal* 21 (1): 274–283. doi:10.1096/fj.06-6686com.

Tang, P-A., H-Y. Hu, W-W. Du, F-J. Jian, and E-H. Chen. 2023. "Identification of Cuticular Protein Genes and Analysis of Their Roles in Phosphine Resistance of the Rusty Grain Beetle *Cryptolestes ferrugineus*." *Pesticide Biochemistry and Physiology* 194 (August): 105491. doi:10.1016/j.pestbp.2023.105491.

Tsaganou, F. K., T. N. Vassilakos, and C. G. Athanassiou. 2021. "Insecticidal Effect of Thiamethoxam against Seven Stored-Product Beetle Species." *Journal of Stored Products Research* 93 (September): 101843. doi:10.1016/j.jspr.2021.101843.

Vassilakos, T. N., C. G. Athanassiou, and N. G. Tsiropoulos. 2015. "Persistence and Efficacy of Spinetoram against Three Major Stored Grain Beetle on Wheat." *Crop Protection* 69 (March): 44–51. doi:10.1016/j.cropro.2014.08.010.

Vélez, M., W. F. Barbosa, J. Quintero, M. Chediak, and R. N. C. Guedes. 2017. "Deltamethrin- and Spinosad-Mediated Survival, Activity and Avoidance of the Grain Weevils *Sitophilus granarius* and *S. zeamais*." *Journal of Stored Products Research* 74 (December): 56–65. doi:10.1016/j.jspr.2017.10.002.

5 Challenges and Issues Associated with Chemical Methods of Fumigation

A. Mariadoss and Alice R. P. Sujeetha
National Institute of Plant Health Management, Rajendranagar, Hyderabad, India

Sankarganesh E
Bihar Agricultural University, Bhagalpur, Bihar, India

5.1 STORED PRODUCTS AS HABITATS FOR PESTS

Species infesting stored products are food opportunists. Organisms such as insects, mites, rodents, birds, and microbes are opportunistic feeders and commonly infest stored products, leading to significant losses. Among these, insects cause the most substantial damage, both qualitatively and quantitatively, to stored grains and other commodities (Mariadoss and Alice 2018). The presence of insect pests results in increased expenses for grain handlers, as they incur costs for pest control measures. Grain-handling organizations also face additional expenses in controlling pests in bulk storage facilities. The number of grain storage insects has been on the rise in recent years (Bell 2014; Gottlieb et al. 2018). The economic damage caused by insects inhabiting grain storage facilities can be substantial. In developing countries, the damage can range from 30 to 50 percent, while developed nations experience grain weight losses ranging from 5 to 10 percent after prolonged storage. This threat is particularly significant in warmer climatic regions where favorable ecological conditions promote pest activity. The challenge of controlling these pests has led to an increased dependence on chemical pesticides, with efforts focused on raising the quantities and frequency of pesticide applications.

Stored product insects are known for their ability to survive for extended periods without food. They exhibit a wide tolerance range for environmental factors and can efficiently utilize available resources to sustain themselves. However, the optimal population density for many species occurs at a certain level, beyond which individual fitness decreases as the population grows or shrinks. Biosecurity concerns have become increasingly alarming in relation to stored product insects. More than 300 species of Coleopteran beetles, belonging to 40 families, are responsible for approximately 85 percent of stored pest damage (Zhang et al. 2015). These beetles are of quarantine significance due to their small size, broad host range, high reproductive capacity, and limited dispersal ability. Grain warehouses provide a stable environment and abundant food, which facilitates the establishment of alien species. For instance, *Trogoderma granarium* Everts, originating from Asia, was first reported in California, where it caused a loss of 220 million dollars, accounting for 10 percent of agricultural product income (Chu, Li, and Huang 2008; Qin et al. 2018). International grain movement serves as a pathway for the entry of quarantine pests between countries. Ensuring the safety of food grains and exporting high-quality produce to the international market necessitates effective control measures against stored product pests.

5.2 FUMIGATION AS GRAIN PROTECTANT

In an effective pest management program, a combination of prevention and control methods is integrated to achieve maximum benefits. Practices such as sanitation, temperature control (high and

DOI: 10.1201/9781003609834-5

low), moisture control, aeration, and the use of protectants are effective in preventing and minimizing infestations. These methods aim to manage pests in a way that the harm they cause or the risk they pose remains acceptable. However, in cases where the pressure from insect infestation is high, chemical treatments, including fumigants, are employed to disinfect the grain in both advanced and emerging countries. Fumigation is a versatile and long-standing practice used worldwide to reduce pest incidence during commodity storage, regardless of the pest category. It is a curative treatment that helps maintain an infestation-free storage environment and improves storage hygiene (Chadda et al. 2007). Fumigants, when used at suitable temperature and pressure, exist in a gaseous state at lethal concentrations for the target pests. Fumigation is considered a quick and effective tool for controlling stored grain insect pests and plays a key role in managing infestations in stored commodities. Both private and government-owned warehouses store grains meant for consumption, and fumigation is conducted at regular intervals to protect them from pest infestation and ensure the supply of insect-free products to consumers (Rajendran 2016; Sankarganesh et al. 2020). Additionally, the concept of zero tolerance for insect pests in food commodities has made fumigants an indispensable tool (Singh and Sharma 2015). For example, in India, where more than 200 million tons (MT) of food grains are handled each year, readily available fumigants are used to preserve the grains.

Fumigants possess unique characteristics and capabilities that make them suitable for situations where other forms of pest control are not practical or feasible, providing efficient and cost-effective control. They have the ability to penetrate areas that may not be accessible to sprays or dusts, including burrows and tightly packed commodities. Fumigants find application in various scenarios: (1) Precautionary hygiene during storage: Fumigation can be employed as a preventive measure to ensure storage hygiene and prevent infestations. (2) Providing wholesome food for consumers: Fumigation helps maintain the quality and safety of stored food products, ensuring they are free from pests. (3) Mandatory requirements in trade: Fumigation may be required by regulatory bodies or international standards to meet trade and quarantine regulations. (4) Quarantine measures: Fumigation is utilized as a quarantine measure to prevent the spread of pests across borders. Fumigants are broad-spectrum and volatile pesticides that act as respiratory poisons, anesthetics, narcotics, or enzyme poisons. The vapors of fumigants enter the insect's body through the body wall or respiratory system. Despite their chemical simplicity, these molecules have significant and potent effects on the target organisms. Insects can be affected at different life stages, and the entry of fumigant gas into larvae, pupae, and adults primarily occurs through the spiracles, which are located on the body's lateral surfaces. For insect eggs, gases diffuse through the egg's chorion or specialized respiratory channels. While some gases may also diffuse through the insect's integument, the extent of this route for fumigant entry is not well-established. The susceptibility of an insect to fumigation is influenced by its rate of respiration. Factors that increase the rate of respiration tend to make the insect more susceptible to poisoning by fumigants (Bond and Monro 1961).

5.3 ISSUES AND CONCERNS IN CURRENT FUMIGATION STRATEGIES

Successful fumigation operations depend on several factors, including proper monitoring and maintaining adequate fumigant concentrations. Due to the high toxicity of fumigants, fumigation operations require highly skilled, professionally trained, and authorized personnel who are experienced in conducting field operations. In fumigation, the temperature is the most significant environmental factor that affects the action of fumigants on insects. Within the typical fumigating temperature range of 10–35°C, the concentration of fumigant required to kill the target species decreases as the temperature rises. This decrease in concentration is primarily due to the increased rate of respiration in insects in response to higher temperatures (Sun 1946). Temperature also influences the fumigation performance through the physical sorption of fumigating materials. In some cases, higher temperatures require less fumigant gas to kill certain species, while with others, toxicity declines as the temperature decreases (e.g., with methyl bromide, toxicity decreases below its boiling point). Conversely, abnormally high or low temperatures during fumigation can result in sub-lethal concentrations of the fumigant gas. At low temperatures, the vaporization and diffusion of fumigants slow down, and the metabolism and activity of insects are also reduced. These factors can hinder the killing activity of the fumigant, particularly at temperatures below 10°C (50°F). Therefore, successful fumigation at lower temperatures may require higher dosages and longer exposure times. On the other hand, if the fumigation site is not perfectly sealed, excessive temperatures can cause fumigants to vaporize and dissipate too quickly, leading to inadequate maintenance of lethal concentrations for a sufficient duration to kill pests. Temperature exerts a far-reaching impact on all the factors that contribute to the successful outcome of fumigation.

All known fumigants possess a certain level of toxicity. Therefore, their application is limited to areas, spaces, items, or commodities that can be tightly sealed to prevent their escape. In fumigated buildings, occupants must evacuate, sometimes for a period of seven days or more, and the structures need to be tightly sealed. It is important to note that some fumigants can cause damage to items in the fumigated area. Extensive research has been conducted on the effects of fumigants, and certain chemicals have been found to have serious impacts on human health. As a result, some countries have prohibited or restricted the use of fumigants with high hazard potential (Ducom 2006). Some gases have been eliminated from use due to their undesirable characteristics, including chemical instability and destructive effects on materials. Excessively corrosive compounds can damage containers or spoil the structure and fittings of fumigation chambers or other treated spaces. Reactive chemicals can form irreversible compounds that remain as residues in products. In foodstuffs, such reactions can result in taint or the formation of undesirable residues, rendering them unfit for consumption. Physiologically active compounds like carbon tetrachloride can harm growing plants, fruits, or vegetables and negatively affect seed germination capacity (Bond and Monro 1961). They can interfere with plant physiology, leading to growth retardation, shortened storage life, delayed ripening, and stimulation of storage disorders. Fumigants can also produce foul or unpleasant odors in materials stored in nearby premises. Additionally, their chemical effects can spoil certain products, and their reaction with lubricants can cause interference with machinery. The selection of a fumigant for a specific pest control problem is influenced by various properties of the compound, the type of pests involved, and the nature of the commodity being treated (Heuser 1975).

In recent years, there has been increased attention on pesticide residues in food and the potential risks they pose to human health (Tsai 2010; WHO 2016). It is essential for an ideal fumigant not to alter the treated product in any way and should not leave any harmful residues that could pose risks to consumers or affect food processing. The Environmental Protection Agency (EPA) determines the acceptable levels of pesticide residues that can remain on agricultural products and processed foods. The EPA also establishes exposure limits, which define the concentration of fumigant above which it is not safe for workers to be exposed. Fumigants used for insect control on foodstuffs or other commodities can form residues. They can react with plant or animal constituents and leave behind significant residue. After treatment, unreacted fumigant may persist in some materials for a certain period. Generally, the residual issue decreases gradually over time, except for highly sorptive fumigants such as carbon tetrachloride, ethylene dibromide, and hydrogen cyanide. These fumigants can persist in some materials for weeks or even months after aeration (Amuh and Ghana Atomic Energy Commission 1975; Jagielski, Scudamore, and Heuser 1978; Mohapatra, Kar, and Giri 2015). Fumigants with higher boiling points tend to be more sorbed and leave residues for longer periods compared to volatile compounds. For example, acrylonitrile was found to stay in wheat for an extended period, while methyl bromide dissipated within a few hours (Dumas and Bond 1977). Some materials have a greater capacity to sorb and retain fumigants than others. Food items with high oil and fat content may retain more residue than cereals. For instance, methyl bromide is easily absorbed by lipid materials, so it is avoided in foods with high-fat content, such as butter, cheese, margarine, and meat (Rhodes et al. 1975). Milled fractions of treated wheat, particularly the bran, have been found to contain significant amounts of residual carbon tetrachloride (Jagielski, Scudamore, and Heuser 1978; Savi, Piacentini, and Scussel 2015).

Similarly, ethylene oxide has the potential to react with chlorides and bromides present in food, forming toxic compounds called chlorohydrins and bromohydrins (Scudamore and Heuser 1971). Methyl bromide, when used on wheat, can decompose and form non-toxic derivatives (Winteringham et al. 1955). Hydrogen cyanide can react with sugars in dried fruit, leading to the formation of levulose cyanohydrin (Page and Lubatti 1948). Additionally, certain formulations of fumigants, such as aluminum phosphide and calcium cyanide, can leave residues on food materials. After the production of phosphine gas, a small amount of non-decomposed aluminum phosphide and residue of aluminum hydroxide may remain. Calcium cyanide leaves behind a buildup of calcium hydroxide after the release of hydrogen cyanide (Bond and Monro 1961).

The presence of residual fumigants can have a significant impact on treated items, particularly when they are stored in tightly sealed containers or enclosed spaces like ship's holds. The amount of residue left behind can be influenced by factors such as dosage and timing of exposure during treatment. Higher temperatures generally lead to greater chemical reactions and increased residue, even if the dosage is reduced. The dissipation of fumigants is also influenced by atmospheric humidity and moisture. For example, higher humidity levels can result in greater desorption of ethylene dibromide from wheat compared to lower humidity levels (Dumas and Bond 1979). The moisture content of the commodity and the relative humidity of the environment both affect the retention of sorbed gases and the reactions of fumigants with components (Vardell 1975; Mohapatra, Kar, and Giri 2015). When commodities are re-fumigated due to the risk of re-infestation, the level of residue may be expected to increase. In such cases, precautions should be taken to ensure that residue levels do not exceed permitted levels (Banks, Desmarchelier, and Elek 1976). The conditions during fumigation and subsequent treatment play a role in determining the amount of residue remaining in fumigated materials (Lindgren, Sinclair, and Vincent 1968). Considering the various factors that contribute to residue accumulation before treatment can help minimize residue levels in some cases. However, there is a serious concern about potential long-term effects that may arise from fumigant residues. The presence of residuals in treated materials can pose risks to consumers who consume treated foods, as well as to workers in the workplace.

Methyl bromide (MBr) and phosphine (PH_3) are widely used fumigants for protecting stored grains in India and other countries. In addition to methyl bromide and phosphine, there are several other fumigants available for use in different countries, such as sulfuryl fluoride, propylene oxide, carbon disulfide, ethyl formate, carbonyl sulfide, hydrogen cyanide, methyl iodide, cyanogen, chloropicrin, and ozone treatment (Navarro 2006). However, the use of fumigants presents challenges with regard to environmental safety, efficacy, cost, and other factors that determine their value as fumigants.

5.3.1 Methyl Bromide

Methyl bromide (MBr) is an effective and versatile fumigant that has gained popularity for pest and disease control worldwide since its introduction in the 1930s, mainly due to its ability to penetrate large consignments of sacks or bags and its quick action (Singh and Sharma 2015). However, MBr is exceptionally harmful if inhaled and is odorless, so formulations containing chloropicrin are used as a warning agent on non-food products. MBr has been primarily used as a space and soil fumigant to control soil-borne pests, diseases, and weeds in the production of high-value crops like strawberries, tomatoes, peppers, cut flowers, and tobacco seedlings. It has also been widely used for post-harvest fumigation of durable commodities (such as grains, dried fruit, and other foodstuffs), quarantine fumigations of fresh fruit, and structural fumigation (Shaaya and Kostyukovsky 2009; Pizano 2014). It has found extensive application in factories, warehouses, ships, railroad cars, mills, and other industrial settings. MBr offers several advantages, including rapid treatment speed, low infrastructure requirements, recognition by quarantine authorities, broad registration for use, and quick dissipation after exposure. It is especially useful for treatments at low temperatures. After fumigating foodstuffs, most of the methyl bromide is desorbed and diffuses away rapidly, so gaseous methyl bromide does not typically present a residue problem. However, the chemical reaction between MBr and some constituents of materials may leave behind a small, variable amount of permanent residue (Heywood 1966; Tsai 2010).

MBr has a long and successful history as a quarantine fumigant and is often the only treatment approved by national quarantine authorities. There is a wealth of information and data on the response of pests to MBr (Bond and Monro 1961). It is non-corrosive and can be effectively applied to shipping containers or sacked/bulk commodities. Treatments with MBr may be applied before shipment to prevent specific quarantine pests, or upon receipt of a quarantine, the pest is detected or suspected. These treatments comply with international and national agreements and regulations, including the International Plant Protection Convention (IPPC) and various national quarantine regulations. In 1992, MBr was listed as an ozone-depleting substance (ODS) under the Montreal Protocol, an international treaty signed by 175 countries in 1987 to protect the ozone layer (UNEP 2018). The protocol aims to reduce the production of ODSs. Consequently, the use of MBr was classified as controlled use under the Montreal Protocol, with provisions for countries to apply for time-limited critical use exemptions (CUE) in cases where finding alternatives to MBr was particularly challenging. MBr is used for quarantine and pre-shipment (QPS) treatment to prevent the introduction and establishment of quarantine pests that could harm various productive sectors in many countries.

5.3.2 Phosphine

Globally, grain-handling organizations rely on phosphine gas (PH_3) or hydrogen phosphide as the most efficient method for eliminating insect pests from stored food grains. Phosphine is a low molecular weight and low boiling point compound that rapidly diffuses and deeply infiltrates materials, such as large bulks of grain or tightly packed materials. It is registered in India and used worldwide for disinfecting durable commodities, ranking as one of the most toxic fumigants for stored product insects (Sakka et al. 2023). This fumigant plays a crucial role in ensuring food security as it is the only general-purpose fumigant available and can be used on a variety of durable goods, including grains, dried fruits, silos, and ships. Countries such as Turkey, Indonesia, Iran, Jamaica, Vietnam, the Philippines, Thailand, and Cameroon have successfully replaced methyl bromide due to ozone depletion concerns. Phosphine has the potential to directly replace methyl bromide in many situations (Pizano 2014). It is a highly toxic fumigant used at low concentrations. Although its action against pests is generally slower than that of methyl bromide, it remains effective with a longer exposure period. Phosphine penetrates commodities well and can be rapidly removed through aeration after treatment. It is widely used to fumigate stacked materials by covering them with a tarpaulin. The phase-out of methyl bromide has significantly increased the use of phosphine not only in quantity but also in a variety of stored products beyond food grains, such as spices, cocoa beans, dried fruit, nuts, and even fresh fruits (Horn, Horn, and Sullivan 2005).

Phosphine has shown promise in effectively controlling many stored grain insect pests, making it a popular phytosanitary treatment during trade (Sakka et al. 2023). Its ease of use, low cost, and residue-free properties contribute to its popularity in various applications. Phosphine also helps provide non-infested material for the food industry with a relatively quick treatment time (Ducom, Roussel, and Stefanini 2004). These benefits have led to an increased reliance on this fumigant (Chaudhry 2000; Hasan and Reichmuth 2004; Wang, Collins, and Gao 2006). Commercial formulations are commonly available in the form of aluminum phosphide or, less commonly, magnesium phosphide formulated with ammonium carbamate or urea to reduce flammability risks. Similar to aluminum phosphide, magnesium phosphide reacts with water to produce hydrogen phosphide gas. Magnesium phosphide is more effective for warehouse and processing plant fumigations rather than grain storage. Recently, generators and cylindered phosphine formulations containing carbon dioxide (ECO2FUME®) or nitrogen (FRISIN®) have also become available, allowing the controlled release of phosphine gas and avoiding some problems associated with residue disposal in tablet formulations. Cylinder-based formulations immediately release the gas, resulting in rapidly increasing concentrations (Ducom 2006). In solid formulations, it is necessary to introduce the required quantity all at once, considering sorption and leaks. With cylinder-based formulations, the dosage can be adjusted periodically to maintain concentrations above the minimum required, reducing the total quantity delivered.

Phosphine gas has remarkable penetrating and distributing power; however, its insecticidal action is distinct from that of other fumigants due to the non-linear relationship between its dosage, concentration, and time of exposure. The toxicity of phosphine to insects declines as the temperature falls; to achieve the desired level of control, it is necessary to fumigate with several days of exposure. It is generally ineffective when the fumigation condition is <15°C and low relative humidity. High concentrations can have a narcotic effect on most species of stored product insects, potentially shielding them from the toxic effects (Winks 1973; Chaudhry 2000). Additionally, adequate moisture is required for the release of the actual toxicant, phosphine, in aluminum phosphide and magnesium phosphide products. At relative humidity below 25 per cent or grain moisture below 10 percent, the release of phosphine requires more time. Besides, phosphine is reactive to some metallic salts which are contained in sensitive items like photographic film and some inorganic pigments. It is known to erode copper and its alloys; electrical and electronic items need special protection from its exposure. When phosphine reacts with copper, particularly in humid conditions, the metal exhibits corrosive effects. Furthermore, many deaths have been reported in India, where phosphine tablets have been used as suicidal weapons (Garg, Chanana, and Gargi 2009). Phosphine is very toxic to all forms of animal life. Hence, exposure, even in small amounts, should be avoided.

5.3.3 Sulfuryl Fluoride

Sulfuryl fluoride (SF) is an inorganic, non-flammable, odorless, and colorless gas that has been used as a structural fumigant for termite control for over half a century. It is employed to fumigate various materials such as construction materials, transport vehicles, wood, flour mills, food industries, dried fruits, and cereal grains (Cox 1997; Bell, Savvidou, and Wontner-Smith 1998; Navarro 2006). SF has the potential to replace methyl bromide with a similar exposure time of 24 hours under normal conditions (Emekci 2010). It exhibits faster diffusion rates in treated commodities. While SF is highly toxic to post-embryonic stages of insects (Bond and Monro 1961), controlling the eggs of most insects at desired dosages, especially at lower temperatures, is challenging, thus limiting the use of sulfuryl fluoride. Combining sulfuryl fluoride with other fumigants such as hydrogen cyanide (HCN), CO_2, phosphine, or heat may be feasible to achieve satisfactory control at all stages (Zettler and Arthur 2000). However, its registration is not widespread in developing nations, it has the potential to act as a greenhouse gas (Tsai 2010), and it also leaves fluoride residues on treated materials (Rajendran and Sriranjini 2007), which restricts its use as a fumigant.

5.3.4 Hydrogen Cyanide

Hydrogen cyanide (HCN) is a colorless, flammable liquid and one of the earliest fumigants widely used under modern conditions. It is registered in India, New Zealand, and with certain restrictions in Germany (Navarro 2006). HCN can be utilized for the fumigation of various dry foodstuffs, grains, and seeds. It exhibits rapid paralyzing action, but sub-lethal concentrations may result in apparent death in insects. When using hydrogen cyanide (HCN), studies have shown that exposing certain species to sub-lethal concentrations before applying the full concentration renders the fumigation less effective compared to subjecting the insects to the full concentration from the beginning. A similar protective effect can also occur with the fumigant phosphine if insects are exposed to excessive concentrations during treatment (Bond and Monro 1961; Winks 1973). Additionally, insects that have been narcotized by sub-lethal concentrations of HCN are protected from lethal treatments with other fumigants, such as methyl bromide (Bond and Monro 1961) and phosphine

(Bond, Robinson, and Buckland 1969). It is generally not recommended for use on moist materials due to the acid formed when it combines with water, which has a burning effect. Furthermore, HCN lacks effective penetration due to high sorption at atmospheric pressure (Emekci 2010). The high dermal toxicity of HCN gas makes it hazardous for applicators.

5.3.5 Carbonyl Sulfide

Carbonyl sulfide (COS) is a colorless gas and a major sulfur compound naturally present in the atmosphere, with traces found in grains and seeds in the range of 0.05–0.1 mg/kg (Wright 2000). Laboratory trials suggest that COS is effective against a wide range of storage pests at concentrations ranging from 10 to 40 g/m^3, with an exposure time of one to five days at temperatures $\geq 5°C$ (Desmarchelier 1994). COS is used as a fumigant for durable commodities and structures, and it has been trademarked in Australia as COSMIC-TM since 1992. However, studies have pointed out some negative effects of COS, such as its impact on cereal germination, off-odors in walnuts and milled rice, and color changes in soybeans (Navarro 2006).

5.3.6 Carbon Disulfide

Carbon disulfide (CS_2) is an old fumigant used for on-farm storage in certain parts of Australia and, to a limited extent, in China. When fumigating grain, carbon disulfide is often combined with non-flammable ingredients. This fumigant has only a minor effect on germination, but residues of carbon disulfide persist in treated commodities for a longer period (Haritos, Ren, and Desmerchelier 1998). However, it carries risks of flammability, persistence in treated commodities, absence of residue limits set by Codex Alimentarius, and high human toxicity (Navarro 2006). It poses a risk due to its potential to burn or explode, and its use as a fumigant has been associated with numerous explosions.

5.3.7 Methyl Iodide

Methyl iodide (MI) was used as a pre-plant soil fumigant for the control of a wide range of organisms, including insects, nematodes, fungi, and weeds (Grech, Ohr, and Sims 1996; Ohr et al. 1998). MI is most toxic to eggs and least toxic to adults of *Sitophilus* spp., *Tribolium confusum*, and *Plodia interpunctella* (Did et al. 2004). MI fumigation carries the potential to increase human health risks and is considered a carcinogenic compound (Ashworth et al. 2011). Nevertheless, the US EPA has registered it as a soil fumigant since 2007 (Pelley 2009).

5.3.8 Chloropicrin

Chloropicrin (PS) is a heavy, colorless, and non-flammable chemical that is extremely toxic to insects, vertebrates, and many soil microbes, including fungi. It is also harmful to plants, even when used as a warning agent in other fumigants. The use of chloropicrin on foodstuffs is prohibited due to its adsorption potential. Currently, chloropicrin is primarily used as a warning agent during fumigations with sulfuryl fluoride or methyl bromide.

5.3.9 Ethane Dinitrile

Ethane dinitrile (EDN), also known as cyanogen (C_2N_2), is a colorless gas with an almond-like odor and is an environmentally safer broad-spectrum fumigant. It acts quickly against stored product insects and has good penetration capabilities through the grain mass (Ducom 2006). EDN has great potential for use in space fumigations and in flour/rice mills (Navarro 2006). However, the phytotoxic properties of EDN hinder the germination capacity of seeds.

5.3.10 Propylene Oxide

Propylene oxide (PPO) is a colorless, flammable liquid with a relatively low boiling point of 35°C. It is commonly used in the chemical industry (Weast 1986). PPO is considered a safe fumigant and has been registered and used in the USA since 1984 as a sterilant for commodities such as dry and shelled walnuts, spices, cocoa powder, and nutmeats. However, PPO is flammable in air concentrations ranging from 3 to 37 percent, so it must be used under low pressures or in CO_2-enriched atmospheres to prevent flammability. PPO with low pressure can be utilized for QPS treatments where there is technical and economic feasibility (Isikber et al. 2006).

5.3.11 Ethyl Formate

Ethyl formate (EF) is a highly flammable volatile solvent that boils at 55°C and vaporizes rapidly at normal temperatures (Emekci 2010). Laboratory tests against insect pests of food commodities and field trials on cereals, spices, pulses, dry fruits, and oilcakes have shown positive results for the use of EF as a fumigant. The effective dosage for commodities ranged from 300 to 400 g/m^3 with a 72-hour exposure period (Rajendran 2001). Ethyl formate has been suggested as a fumigant for grain storage due to its low residue concentration, which remains below the permissible limit of 250 ppm. However, inadequate dosage and lack of airtightness during fumigation can render it ineffective and may pose adverse effects on humans and the environment (Muthu et al. 1984; Mohapatra, Kar, and Giri 2015). One promising alternative to methyl bromide for fresh fruits is cylinderized liquid EF formulated with CO_2. However, adopting such commercial practices can be expensive and also involves workplace safety issues, such as handling heavy cylinders and limiting CO_2 emissions, especially for large-scale commercial fumigation.

5.3.12 Carbon Dioxide

Carbon dioxide (CO_2) is a colorless, odorless, and tasteless gas that is approximately 1.5 times heavier than air. It is poisonous at higher concentrations and is used for fumigating food products at around 60 percent concentration. The use of carbon dioxide is desirable because treated materials do not retain any toxic residues. Additionally, CO_2 does not affect the germination potential of treated grain and does not produce any unpleasant flavors or odors. However, fumigation with carbon dioxide requires long exposure periods to be effective. Low temperatures significantly reduce its effectiveness, and fumigation times of three to four weeks or more may not be practical.

5.3.13 Ethylene Dibromide

In 1925, the insecticidal properties of ethylene dibromide (EDB) were discovered. It gained prominence as an insecticidal fumigant due to its effectiveness in destroying fruit flies in fruit and as a fumigant for grain in tropical regions. EDB is also used as an ingredient in various spot and liquid grain fumigants worldwide. However, EDB has a high boiling point and is adsorbed by many materials, limiting its penetration capabilities. As a result, its utility is more limited compared to other fumigants.

5.3.14 Ozone

Ozone (O_3), a powerful oxidant and sterilant, has shown great potential for use as an insect control agent and as an inhibitor of mold spore development in stored grain at levels below 45 ppm (Rajendran 2001; Pimentel et al. 2009; Tiwari et al. 2010; McDonough et al. 2011). Ozone can be easily generated from atmospheric oxygen on-site, making it safer for the environment. Studies on ozone treatments against stored product insects conducted in different parts of the world have shown significant results (Savi, Piacentini, and Scussel 2015; Singh and Sharma 2015). However, a major drawback of ozone is its corrosive nature toward most metals (Mason et al. 1998). Effective ozone treatment of storage facilities requires a special ozone air delivery and return system.

Fumigation may not always be a practical solution, as its effectiveness can be influenced by various factors. Successful grain fumigation depends on factors such as the type and condition of the grain, including the size, shape, and permeability of the kernels, as well as the presence of dockage (chaff, dust, and other materials) in the grain. In cases where there are extensive infestations in the field and the grain has been internally consumed by pests before harvest, and storage, disinfection during storage becomes necessary. If the infestation is widespread throughout the grain mass, controlling it later can be challenging. To ensure effective fumigation, the fumigant must be properly applied to circulate and reach all the pests. For a fumigant to work effectively, it is crucial to maintain the right concentration of gas in the atmosphere surrounding the target pests. Depending on the commodities and pests being treated, fumigants are applied in various formulations and dosages. Several factors can affect the concentration of fumigant molecules, including the sorptive quality of the treated product (either through absorption into the commodity or adsorption on its surface), moisture, humidity, and temperature during treatment, diffusion speed of the fumigant through the commodity, interactions with synthetic compounds or articles in the treated area, amount of fumigant applied, susceptibility of the target pests, and the pressure of gas in the chamber. In structural fumigation, proper sealing of the structure is of utmost importance. If the structure is sealed tighter, the fumigation process will be more efficient.

5.4 BASIS OF PHOSPHINE RESISTANCE IN INSECT PESTS

Pest control measures after harvest play a crucial role in safeguarding the global food supply (Shen et al. 2023). Phosphine, a fumigant, is widely used for the disinfestation of stored grains and commodities and is unique in its effectiveness as a phytosanitary treatment against various insect pests worldwide. However, the development of phosphine resistance in target insect pests has emerged as a significant challenge in post-harvest pest control (Talukder 2005). The indiscriminate and prolonged use of phosphine has led to the development of resistance in insect populations, which poses a serious concern. The importance of phosphine in protecting stored grain products has increased following international agreements phasing out the use of the fumigant methyl bromide (Bell 2000; Zettler and Arthur 2000; Rajendran 2001). Improper fumigation practices, including the lack of ideal airtight conditions in storage units, contribute to frequent control failures and the need for more frequent applications (Lorini et al. 2007). These factors collectively exert selection pressure, promoting the development of phosphine resistance (Bengston et al. 1999; Collins et al. 2002).

Reports of stored product beetles exhibiting resistance to phosphine have been documented for over five decades (Hassan et al. 2019). Long-term use of a single fumigant for an extended period increases the risk of the appearance of resistant pest populations (Nayak, Daglish, and Byrne 2005; Pimentel et al. 2008). Extensive research conducted by various groups worldwide has thoroughly examined this phenomenon (Nayak et al. 2013; Agrafioti, Athanassiou, and Nayak 2019; Sakka and Athanassiou 2021; Sakka et al. 2023). It has been observed that certain insect stages, such as eggs and pupae, exhibit higher tolerance to phosphine compared to other stages (Nakakita and Winks 1981). Phosphine resistance has been identified in multiple insect species, posing a significant threat to its sustainable use (Ahmad et al. 2013).

The first global survey on phosphine resistance was conducted in 1970 by Champ and Dyte (1976). They utilized a discriminating dose bioassay on adult insects and documented the presence of phosphine resistance in several key species of stored grain pests across multiple countries. Phosphine resistance has been detected in major stored product insect species, including *Rhyzopertha dominica* (F.), *Tribolium* spp. Macleay, *Sitophilus* spp. Schoenherr, *Oryzaephilus surinamensis* Linnaeus, and *Cryptolestes* spp. Ganglbauer. In the earliest known resistance study, it was found that out of 849 strains of insects collected from 82 countries, 20 percent of the insects exhibited resistance to phosphine fumigant. The highest level of resistance (10–12 times) was observed in the lesser grain borer *R. dominica* (Champ and Dyte 1976). In 2012, highly resistant strains of *Tribolium castaneum* Herbst and *R. dominica* were identified from Oklahoma, USA, with resistance levels up to 200-fold and 1500-fold, respectively, compared to susceptible laboratory strains of the same species.

Notably, strong resistance to phosphine (337 times) in *Sitophilus oryzae* Linnaeus was first reported in China in 1997, followed by India (425 times) in 1988 (Rajendran 1999). Substantial increases in phosphine resistance among stored product pests have been reported worldwide, including in Malaysia (Rahim, Faridah, and Rasali 2004), Brazil (Lorini et al. 2007), Australia (Nayak et al. 2013), and India (Kaur et al. 2015). The exact distribution of phosphine resistance is unknown, but it is likely not uncommon in countries where the fumigant has been extensively used. Increased selection pressure in some countries is believed to contribute to the higher reported incidence of phosphine resistance (Friendship, Halliday, and Harris 1986). Resistance to phosphine has been particularly high in the lesser grain borer *R. dominica* (Huang et al. 2019; Aulicky, Stejskal, and Frydova 2019) and *Tribolium* spp. (Collins et al. 2002; Lorini et al. 2007). Indiscriminate use of phosphine fumigation and sub-lethal dose applications may contribute to an increased risk of control failures and the development of resistance (Konemann et al. 2017). The effectiveness of phosphine as a control method is significantly reduced by the development of resistance in insects, posing a threat to its continued use (Schlipalius, Ebert, and Collins 2014). The susceptibility of insects to phosphine can be influenced by the quality of the food they consume, the type and dose of fumigant used for control, and other factors. Surviving populations, if exposed to higher doses, may develop even higher levels of resistance than would occur with the recommended treatment doses. Understanding the mechanisms of phosphine resistance in these insects is crucial for their management and may help identify synergists or novel alternative compounds (Bell 2000; Chaudhry et al. 2004). Addressing phosphine resistance is essential for sustainable pest management and the long-term viability of the agricultural and food industries (Deeksha et al. 2023).

5.5 FACTORS IMPACTING SUSTAINABILITY AND ADOPTION OF ALTERNATIVES

Under the right conditions, grains can be stored for extended periods with minimal quality loss. But the prevailing climatic conditions promote insect multiplication and the inherent vulnerabilities of grain storage systems can lead to rapid infestation of fumigated commodities. Improved methods of handling, storing, and processing commodities can reduce the need for pesticides. Fumigants should be reserved for situations where infestations are unavoidable or inevitable. Among the available fumigants, phosphine has shown promise for effectively controlling various stored grain insect pests and is commonly used as a phytosanitary treatment during trade. In contrast, in relation to environmental hazards and the development of resistance, the use of conventional fumigants has become very challenging. Excessive or improper use of fumigants can have adverse effects, potentially leading to increased restrictions on their use. However, with regard to the kind of storage facilities in India, at present, phosphine is the primary fumigation technique adopted for disinfestations of godowns, even at the farm level, to save the stored grains from damage by insect pests in the most economical way. Despite existing drawbacks in the storage practices, the use of phosphine fumigant has facilitated the preservation of food grains while meeting consumer demands for quality. It is also possible that phosphine-resistant strains could be controlled if fumigation is conducted properly (Hassan et al. 2019).

5.6 CONCLUSION

The methyl bromide phase-out process has been instrumental in raising awareness about ozone depletion and the steps necessary for its protection with alternative chemicals. However, implementing fumigation techniques effectively requires a well-established system with proper infrastructure, equipment, and supply chains. A comprehensive approach is necessary, including the registration and commercial availability of effective fumigants at reasonable costs (Pizano 2014). Changes in label registration, specific procedures, fumigant application methods, and information about controlling pest populations are just a few of the variables that can affect how fumigant insecticides are used to protect stored products. The successful adoption of alternative fumigants may also depend on their feasibility within local conditions, which is crucial for ensuring their effectiveness. Providing training and information on new developments and improved techniques is an essential component of promoting their successful implementation. When assessing the long-term viability of alternative fumigants, cost-effectiveness, technical feasibility (providing adequate pest control), and commercial availability should be considered. As new methods are learned and adapted to fit current commercial and regulatory environments, challenges and costs are inevitably involved. If the alternative fast-acting fumigants, like sulfuryl fluoride, are approved for use in foodstuffs and readily available, the change may be minor. On the other hand, heating and other non-chemical methods may necessitate significant investment and procedure modifications. The challenge lies not in simply replacing chemical fumigants that are unfit to the environment but rather in developing long-term and successful pest management strategies with available options to ensure sustainability.

REFERENCES

Agrafioti, P., C. G. Athanassiou, and M. K. Nayak. 2019. "Detection of Phosphine Resistance in Major Stored-Product Insects in Greece and Evaluation of a Field Resistance Test Kit." *Journal of Stored Products Research* 82 (June): 40–47. doi:10.1016/j.jspr.2019.02.004.

Ahmad, A., M. Ahmad, A. Mubarik, N. Ahmed, M. Ali, M. A. Noorullah, M. Abbas, and S. Arif. 2013. "Monitoring of Resistance against Phosphine in Stored Grain Insect Pests in Sindh." 1501–1507.

Amuh, J. K. A., and Ghana Atomic Energy. Commission. 1975. "Persistence of Ethylene Dibromide Residues in Fumigated Maize." *Ghana Journal of Science* 15 (1).

Ashworth, D. J., L. Luo, R. Xuan, and S. R. Yates. 2011. "Irrigation, Organic Matter Addition, and Tarping As Methods of Reducing Emissions of Methyl Iodide from Agricultural Soil." *Environmental Science & Technology* 45 (4): 1384–1390. doi:10.1021/es1032822.

Aulicky, R., V. Stejskal, and B. Frydova. 2019. "Field Validation of Phosphine Efficacy on the First Recorded Resistant Strains of *Sitophilus granarius* and *Tribolium castaneum* from the Czech Republic." *Journal of Stored Products Research* 81 (March): 107–113. doi:10.1016/j.jspr.2019.02.003.

Banks, H. J., J. M. Desmarchelier, and J. A. Elek. 1976. "Determination of Bromide Ion Content of Cereals and Other Foodstuffs by Specific Ion Electrode." *Pesticide Science* 7 (6): 595–603. doi:10.1002/ps.2780070611.

Bell, C. H. 2000. "Fumigation in the 21st Century." *Crop Protection* 19 (8–10): 563–569. doi:10.1016/S0261-2194(00)00073-9.

Bell, C. H. 2014. "A Review of Insect Responses to Variations Encountered in the Managed Storage Environment." *Journal of Stored Products Research* 59 (October): 260–274. doi:10.1016/j.jspr.2014.06.004.

Bell, C. H., N. Savvidou, and T. J. Wontner-Smith. 1998. "The Toxicity of Sulfuryl Fluoride (Vikane) to Eggs of Insect Pests of Flour Mills." In *Proceedings of the Seventh International Working Conference on Stored Product Protection*, 345–350.

Bengston, M., P. J. Collins, G. J. Daglish, V. L. Hallman, R. M. Kopittke, and H. Pavic. 1999. "Inheritance of Phosphine Resistance in *Tribolium castaneum* (Coleoptera: Tenebrionidae)." *Journal of Economic Entomology* 92 (1): 17–20. doi:10.1093/jee/92.1.17.

Bond, E. J., and H. A. U. Monro. 1961. "The Toxicity of Various Fumigants to the Cadelle, Tenebroides Mauritanicus1." *Journal of Economic Entomology* 54 (3): 451–454. doi:10.1093/jee/54.3.451.

Bond, E. J., J. R. Robinson, and C. T. Buckland. 1969. "The Toxic Action of Phosphine Absorption and Symptoms of Poisoning in Insects." *Journal of Stored Products Research* 5 (4): 289–298. doi:10.1016/0022-474X(69)90002-2.

Chadda, I. C., K. Jayraj, and A. V. Dhuri. 2007. "New Method of Phosphine and Carbon Dioxide Application and Its Optimization." In *Proc. Int. Conf. Controlled Atmosphere and Fumigation in Stored Products*, no. 8–13 August 2004, 455–465.

Champ, B. R., and C. E. Dyte. 1976. *Report of the FAO Global Survey of Pesticide Susceptibility of Stored Grain Pests.* FAO Plant Protection Bulletin, 25, 49–67.

Chaudhry, M. Q. 2000. "Phosphine Resistance." *Pesticide Outlook* 11 (3): 88–91. doi:10.1039/b006348g.

Chaudhry, M. Q., H. A. Bell, N. Savvidou, and A. D. MacNicoll. 2004. "Effect of Low Temperatures on the Rate of Respiration and Uptake of Phosphine in Different Life Stages of the Cigarette Beetle, *Lasioderma serricorne* (F.)." *Journal of Stored Products Research* 40 (2): 125–134. doi:10.1016/S0022-474X(02)00093-0.

Chu, W. J., W. F. Li, and X. L. Huang. 2008. "Potential Distributions of *Trogoderma granarium* by Means of Semi-Quantitative Analysis." *Entomological Journal of East China* 17: 287–292.

Collins, P. J., G. J. Daglish, M. Bengston, T. M. Lambkin, and H. Pavic. 2002. "Genetics of Resistance to Phosphine in *Rhyzopertha dominica* (Coleoptera: Bostrichidae)." *Journal of Economic Entomology* 95 (4): 862–869. doi:10.1603/0022-0493-95.4.862.

Cox, C. 1997. "Sulfuryl Fluoride." *Journal of Pesticide Reform*," 17–20.

Deeksha, M. G., S. M. Nebapure, V. K. Kalia, D. Sagar, R. Bhattacharya, A. Dahuja, and S. Subramanian. 2023. "Comparison of Phenotypic and Genotypic Frequency of Phosphine Resistance in Select Field Populations of *Tribolium castaneum* from India." *Molecular Biology Reports* 50 (8): 6569–6578. doi:10.1007/s11033-023-08605-z.

Desmarchelier, J. M. 1994. "Carbonyl Sulphide as a Fumigant for Control of Insects and Mites," 78–82.

Did, M. G., O. Noboru, H. N. Id, and Y. S. Db. 2004. "Susceptibility of Four Stored Grain Insects to Methyl Iodide." *Research Bulletin of the Plant Protection Service Japan* 6 (40): 2–7.

Ducom, P. J. F. 2006. "The Return of the Fumigants." In *9th International Working Conference on Stored Product Protection*, 510–516.

Ducom, P., C. Roussel, and V. Stefanini. 2004. "Quick Stored Products Disinfestation before Processing One or Two Day Phosphine Fumigation," no. August 2004, 47–52.

Dumas, T., and E. J. Bond. 1977. "Penetration, Sorption, and Desorption of Fumigant in the Treatment of Food Materials with a Methyl Bromide-Acrylonitrile Mixture." *Journal of Agricultural and Food Chemistry* 25 (3): 677–680. doi:10.1021/jf60211a041.

Dumas, T., and E. J. Bond. 1979. "Relation of Temperature to Ethylene Dibromide Desorption from Fumigated Wheat." *Journal of Agricultural and Food Chemistry* 27 (6): 1206–1209. doi:10.1021/jf60226a049.

Emekci, M. 2010. "Quo Vadis the Fumigants?" In *10th International Working Conference on Stored Product Protection;* Estoril, Portugal; 2010.06.27-07.02. doi: 10.5073/jka.2010.425.167.

EPA. 2009. "Volume IV – Part 1 Office of Environmental Health Hazard Assessment" IV (August).

Friendship, C. A. R., D. Halliday, and A. H. Harris. 1986. "Factors Causing Development of Resistance to Phosphine by Insect Pests of Stored Produce." In *In GASGA Seminar on Fumigation Technology*, 141–149.

Garg, S., H. R. A. Chanana, and J. Gargi. 2009. "Fatal Period in Murder-Suicide Celphos (Aluminium Phoshide) Poisoning," 92–95.

Gottlieb, D., E. Qvinn, M. Nega, A. Rapaport, J. Doron, and M. Kostyukovsky. 2018. *Climate Change and Its Implications on Stored Food Grains.* Julius-Kühn-Archiv. doi:10.5073/jka.2018.463.022.

Grech, N. M., H. D. Ohr, and J. J. Sims. 1996. "Methyl Iodide as a Soil Fumigant." U.S. Patent 5,518,692, Issued May 21, 1996.

Haritos, V. S., Y. L. Ren, and J. M. Desmerchelier. 1998. "Regulatory Toxicology of Alternative Fumigants." In *Proceedings of the Seventh International Working Conference on Stored Product Protection*, Beijing, China, 356–363.

Hasan, M. M., and C. Reichmuth. 2004. "Relative Toxicity of Phosphine against the Bean Bruchid *Acanthoscelides obtectus* (Say) (Col., Bruchidae)." *Journal of Applied Entomology* 128 (5): 332–336. doi:10.1111/j.1439-0418.2004.00850.x.

Hassan, R. S., E. A. Mahmoud, T. M. Sileem, and W. A. A. Sayed. 2019. "Evaluation of Fast Neutron Irradiation as a New Control Method against the Indian Meal Moth, *Plodia interpunctella* (Hübner)." *Journal of Radiation Research and Applied Sciences* 12 (1): 38–44. doi:10.1080/16878507.2019.1594133.

Heuser, S. G. B. 1975. "Factors Influencing Dosage and Choice of Toxicant in Stored-Product Fumigation," 246–253.

Heywood, B. J. 1966. "Pesticide Residues in Total Diet Samples: Bromine Content." *Science* 152: 1408–1408.

Horn, F., P. Horn, and J. Sullivan. 2005. "Current Practice in Fresh Fruit Fumigation with Phosphine in Chile."

Huang, Y., F. Li, M. Liu, Y. Wang, F. Shen, and P. Tang. 2019. "Susceptibility of *Tribolium castaneum* to Phosphine in China and Functions of Cytochrome P450s in Phosphine Resistance." *Journal of Pest Science* 92 (3): 1239–1248. doi:10.1007/s10340-019-01088-7.

Isikber, A. A., S. Navarro, S. Finkelman, M. Rindner, and R. Dias. 2006. "Propylene Oxide as a Potential Quarantine Fumigant for Insect Disinfestation of Nuts" 27 (1999): 630–634.

Jagielski, J., K. A. Scudamore, and S. G. Heuser. 1978. "Residues of Carbon Tetrachloride and 1,2-dibromoethane in Cereals and Processed Foods after Liquid Fumigant Grain Treatment for Pest Control." *Pesticide Science* 9 (2): 117–126. doi:10.1002/ps.2780090205.

Kaur, R., M. Subbarayalu, R. Jagadeesan, G. J. Daglish, M. K. Nayak, H. R. Naik, S. Ramasamy, C. Subramanian, P. R. Ebert, and D. I. Schlipalius. 2015. "Phosphine Resistance in India Is Characterised by a Dihydrolipoamide Dehydrogenase Variant That Is Otherwise Unobserved in Eukaryotes." *Heredity* 115 (3): 188–194. doi:10.1038/hdy.2015.24.

Konemann, C. E., Z. Hubhachen, G. P. Opit, S. Gautam, and N. S. Bajracharya. 2017. "Phosphine Resistance in *Cryptolestes ferrugineus* (Coleoptera: Laemophloeidae) Collected From Grain Storage Facilities in Oklahoma, USA." *Journal of Economic Entomology* 110 (3): 1377–1383. doi:10.1093/jee/tox101.

Lindgren, D. L., W. B. Sinclair, and L. E. Vincent. 1968. "Residues in Raw and Processed Foods Resulting from Post-Harvest Insecticidal Treatments." In *Residue Reviews/Rückstands-Berichte*, 1–121. New York, NY: Springer US. doi:10.1007/978-1-4615-8431-5_1.

Lorini, I., P. J. Collins, G. J. Daglish, M. K. Nayak, and H. Pavic. 2007. "Detection and Characterisation of Strong Resistance to Phosphine in Brazilian *Rhyzopertha dominica* (F.) (Coleoptera: Bostrychidae)." *Pest Management Science* 63 (4): 358–364. doi:10.1002/ps.1344.

Mariadoss, A. and R.P.S.J Alice. 2018. *Standard operating procedures for stored grain pest management in food grain storage depots.* National Institute of Plant Health Management, Hyderabad, India, pp. 1–42.

Mason, Linda J., C. Allen Strait., Charles P. Woloshuk, and Dirk E. Maier. 1998. "Controlling Stored Grain Insects with Ozone Fumigation." In *In 7th International Working Conference on Stored-Product Protection*, 14–19.

McDonough, M. X., C. A. Campabadal, L. J. Mason, D. E. Maier, A. Denvir, and C. Woloshuk. 2011. "Ozone Application in a Modified Screw Conveyor to Treat Grain for Insect Pests, Fungal Contaminants, and Mycotoxins." *Journal of Stored Products Research* 47 (3): 249–254. doi:10.1016/j.jspr.2011.04.001.

Mohapatra, D., A. Kar, and S. K. Giri. 2015. "Insect Pest Management in Stored Pulses: An Overview." *Food and Bioprocess Technology* 8 (2): 239–265. doi:10.1007/s11947-014-1399-2.

Muthu, M., S. Rajendran, T. S. Krishnamurthy, K. S. Narasimhan, J. R. Rangaswamy, M. Jayaram, and S. K. Majumder. 1984. "Ethyl Formate as a Safe General Fumigant," 369–393. doi:10.1016/B978-0-444-42417-4.50037-2.

Nakakita, H., and R. G. Winks. 1981. "Phosphine Resistance in Immature Stages of a Laboratory Selected Strain of *Tribolium castaneum* (Herbst) (Coleoptera: Tenebrionidae)." *Journal of Stored Products Research* 17 (2): 43–52. doi:10.1016/0022-474X(81)90016-3.

Navarro, S. 2006. "New Global Challenges to the Use of Gaseous Treatments in Stored Products." In *9th International Working Conference on Stored Product Protection*, 495–509.

Nayak, M. K., G. J. Daglish, and V. S. Byrne. 2005. "Effectiveness of Spinosad as a Grain Protectant against Resistant Beetle and Psocid Pests of Stored Grain in Australia." *Journal of Stored Products Research* 41 (4): 455–467. doi:10.1016/j.jspr.2004.07.002.

Nayak, M. K., J. C. Holloway, R. N. Emery, H. Pavic, J. Bartlet, and P. J. Collins. 2013. "Strong Resistance to Phosphine in the Rusty Grain Beetle, *Cryptolestes ferrugineus* (Stephens) (Coleoptera: Laemophloeidae): Its Characterisation, a Rapid Assay for Diagnosis and Its Distribution in Australia." *Pest Management Science* 69 (1): 48–53. doi:10.1002/ps.3360.

Ohr, H. D., N. M. Grech, and J. J. Sims. 1998. "Methyl Iodide as a Fumigant." U.S. Patent 5,753,183, Issued May 19, 1998.

Page, A. B. P., and O. F. Lubatti. 1948. "Some Effects of Fumigants on Food." *Chemistry and Industry* 46: 723–727.

Pimentel, M. A. G., L. R. D'Antonino Faroni, M. D. Batista, and F. H. da Silva. 2008. "Resistance of Stored-Product Insects to Phosphine." *Pesquisa Agropecuária Brasileira* 43 (12): 1671–1676. doi:10.1590/S0100-204X2008001200005.

Pimentel, M. A. G., L. R. D. A. Faroni, R. N. C. Guedes, A. H. Sousa, and M. R. Tótola. 2009. "Phosphine Resistance in Brazilian Populations of *Sitophilus zeamais* Motschulsky (Coleoptera: Curculionidae)." *Journal of Stored Products Research* 45 (1): 71–74. doi:10.1016/j.jspr.2008.09.001.

Pizano, M. 2014. *Phasing-Out Methyl Bromide in Developing Countries – A Success Story and Its Challenges.* UNEP Division of Technology, Industry and Economics (UNEP DTIE), 1–54.

Qin, Y., L. Wang, V. Stejskal, and Z. Li. 2018. "Global Establishment Risk of Stored Products Beetles." *Julius-Kühn-Archiv* 463.

Rahim, M., M. E. Faridah, and M. Rasali. 2004. "Current Status of Phosphine Resistance in Stored Grain Insects in Malaysia." *Journal of Tropical Agriculture and Food Science* 32 (1): 101–107.

Rajendran, S. 1999. "Resistência à Fosfina Em Pragas de Insetos de Grãos Armazenados Na Índia." In *Proceedings of the 7th International Working Conference on Stored Product Protection*, 635–641.

Rajendran, S. 2001. "Alternatives to Methyl Bromide as Fumigants for Stored Food Commodities." *Pesticide Outlook* 12 (6): 249–253.

Rajendran, S. 2016. "Status of Fumigation in Stored Grains in India." *Indian Journal of Entomology* 78 (special): 28. doi:10.5958/0974-8172.2016.00022.5.

Rajendran, S., and V.-R. Sriranjini. 2007. "Use of Fumigation for Managing Grain Quality." *Stewart Postharvest Review* 3 (6): 1–8. doi:10.2212/spr.2007.6:9.

Rhodes, D. N., R. L. S. Patterson, D. J. Puckey, S. G. Heuser, H. E. Wainman, B. Chakrabarti, and E. N. W. Allan. 1975. "Exposure of Meat to Methyl Bromide during Refrigerated-Store Pest Control; Residues in Beef and Lamb and Effect on Quality." *Journal of the Science of Food and Agriculture* 26 (9): 1375–1380. doi:10.1002/jsfa.2740260917.

Sakka, M. K., and C. G. Athanassiou. 2021. "Population-Mediated Responses of Lasioderma Serricorne (Coleoptera: Anobiidae) to Different Diagnostic Protocols for Phosphine Efficacy." Edited by James Campbell. *Journal of Economic Entomology* 114 (2): 885–890. doi:10.1093/jee/toaa300.

Sakka, M. K., C. T. Nakas, D. Bochtis, and C. G. Athanassiou. 2023. "Quick Knockdown Results in High Mortality: Is This Theory Correct? A Case Study with Phosphine and the Red Flour Beetle." *Pest Management Science* 79 (10): 3740–3748. doi:10.1002/ps.7555.

Sankarganesh E., Girish AG, Alice R. P. Sujeetha, and Mariadoss A. 2020. "Influence of Relative Humidity on Phosphine Concentration during Aluminium Phosphide (ALP) Fumigation in Pigeon Pea." *International Journal of Chemical Studies* 8 (1): 2082–2084. doi:10.22271/chemi.2020.v8.i1ae.8572.

Savi, G. D., K. C. Piacentini, and V. M. Scussel. 2015. "Reduction in Residues of Deltamethrin and Fenitrothion on Stored Wheat Grains by Ozone Gas." *Journal of Stored Products Research* 61 (March): 65–69. doi:10.1016/j.jspr.2014.12.002.

Schlipalius, D. I., P. R. Ebert, and P. J. Collins. 2014. "An Emerging International Picture of Phosphine Resistance: Opportunities for Global Cooperation." In *Proceedings. 11th International Working Conference on Stored Product Protection*, 10–14.

Scudamore, K. A., and S. G. Heuser. 1971. "Ethylene Oxide and Its Persistent Reaction Products in Wheat Flour and Other Commodities: Residues from Fumigation or Sterilisation, and Effects of Processing." *Pesticide Science* 2 (2): 80–91. doi:10.1002/ps.2780020207.

Shaaya, E., and M. Kostyukovsky. 2009. "The Potential of Biofumigants as Alternatives to Methyl Bromide for the Control of Pest Infestation in Grain and Dry Food Products." In *Recent Advances in Plant Biotechnology*, 389–403. Boston, MA: Springer US. doi:10.1007/978-1-4419-0194-1_16.

Shen, X., M. Che, H. Xu, X. Zhuang, E. Chen, P. Tang, and K. Wang. 2023. "Insight into the Molecular Mechanism of Phosphine Toxicity Provided by Functional Analysis of Cytochrome B5 Fatty Acid Desaturase and Dihydrolipoamide Dehydrogenase in the Red Flour Beetle, *Tribolium castaneum*." *Pesticide Biochemistry and Physiology* 194 (August): 105482. doi:10.1016/j.pestbp.2023.105482.

Singh, G., and R. K. Sharma. 2015. "Alternatives to Phosphine Fumigation of Stored Grains: The Indian Perspective." *Himachal Journal of Agricultural Research* 41 (2): 104–113.

Sun, Y. P. 1946. "An Analysis of Some Important Factors Affecting the Results of Fumigation on Insects." In *St. Paul. Minnesota. Agric. Exp. Stat. Tech. Pou.*

Talukder, F. 2005. "Insects and Insecticide Resistance Problems in Postharvest Agriculture," 207–211.

Tiwari, B. K., C. S. Brennan, T. Curran, E. Gallagher, P. J. Cullen, and C. P. O. Donnell. 2010. "Application of Ozone in Grain Processing." *Journal of Cereal Science* 51 (3): 248–255. doi:10.1016/j.jcs.2010.01.007.

Tsai, W.-T. 2010. "Environmental and Health Risks of Sulfuryl Fluoride, a Fumigant Replacement for Methyl Bromide." *Journal of Environmental Science and Health, Part C* 28 (2): 125–145. doi:10.1080/10590501.2010.481806.

UNEP. 2018. *Handbook for the Montreal Protocol on Substances That Deplete the Ozone Layer. Twelfth Ed. United Nations Environment Programme.* Ozone Secretariat, UNEP/Earthprint, Nairobi, Kenya.

Vardell, H. H. 1975. "Methyl Bromide: Effect of Multiple Fumigation on Residues in Flour2." *Journal of Economic Entomology* 68 (1): 69–70. doi:10.1093/jee/68.1.69.

Wang, D., P. J. Collins, and X. Gao. 2006. "Optimising Indoor Phosphine Fumigation of Paddy Rice Bag-Stacks under Sheeting for Control of Resistant Insects." *Journal of Stored Products Research* 42 (2): 207–217. doi:10.1016/j.jspr.2005.02.001.

Weast, R. C. 1986. *CRC Handbook of Chemistry and Physics.* Taylor & Francis.

WHO. 2016. Summary Report from the May 2016 Joint FAO/WHO Meeting on Pesticide Residues (JMPR), 1–6.

Winks, R. G. 1973. *Characteristics of Response of Grain Pests to Phosphine, 1974 Canberra, Commonwealth Scientific and Industrial Research Organization, Division of Entomology, Annual Report – 74.S.*

Winteringham, F. P. W., A. Harrison, R. G. Bridges, and P. M. Bridges. 1955. "The Fate of Labelled Insecticide Residues in Food Products. II.—The Nature of Methyl Bromide Residues in Fumigated Wheat." *Journal of the Science of Food and Agriculture* 6 (5): 251–261. doi:10.1002/jsfa.2740060503.

Wright, E. J. 2000. "Carbonyl Sulfide: Progress in Research and Commercialization of a New Commodity Fumigant." In *Proc. Ann. Int. Res. Conf. Methyl Bromide Alternat. Emission Reduct*, 86–91.

Zettler, J. L., and F. H. Arthur. 2000. "Chemical Control of Stored Product Insects with Fumigants and Residual Treatments." *Crop Protection* 19 (8–10): 577–582. doi:10.1016/S0261-2194(00)00075-2.

Zhang, S. F., Y. Fan, X. H. Gao, and G. H. Zhan. 2015. *Beetles of Stored Products.* Beijing: Science Press, 372 pp. (In Chinese)

6 Conventional Non-Chemical Approaches for Stored Product Pest Management

C. Kathirvelu and N. Muthukrishnan
Stored Product Research Laboratory, Department of Entomology, Faculty of Agriculture, Annamalai University, Annamalainagar, Tamil Nadu, India

L. Gopianand
Insect Taxonomy Laboratory, Department of Entomology, Faculty of Agriculture, Annamalai University, Annamalainagar, Tamil Nadu, India

6.1 INTRODUCTION

Cereals and legumes, commonly called grains, form a crucial part of the diet for the majority of the global population (Duranti 2006; Rajashekar et al. 2016). Large quantities of food grains suffer damage following harvest primarily due to insufficient storage and processing facilities. Insufficient storage capacity, insect and pest infestations, and other factors contribute to 10–20% of grain losses in total production (Phillips and Throne 2010; Mishra and Panigrahi 2017). According to FAO, worldwide annual losses in stored produce are approximately 10% of all stored grain (Parfitt, Barthel, and Macnaughton 2010). Additionally, variations in climatic factors, such as floods, droughts, temperature fluctuations, and irregular rainfall patterns, can significantly impact agricultural production (Khatri-Chhetri et al. 2017). Stored-product insects can cause significant post-harvest losses, ranging from around 9% in developed countries and more than 20% in developing nations. There are no granaries on the planet that can be filled with grains without the presence of insects. Insects continue to cause problems in storage by inflicting qualitative and quantitative damage, whether the food grains are stored on the farm, at home, or in warehouses. In India, cereals experienced storage losses of approximately 0.75–1.21%, while pulses and oilseeds encountered losses within the range of 1.18–1.67% and 0.22–1.61%, respectively (Jha et al. 2015). The harvested produce inevitably contained eggs, larvae, or pupae due to field carryover infestation, which was unavoidable in India. Farmers have a long battle history against stored product insect pests, using accessible control methods. Furthermore, these insects contaminate food products with live insects, insect products (e.g., chemical excretions or silk), dead insects, and fragments of their bodies. As a result of pest control measures employed to combat these stored product insects, there is a potential risk of human exposure to dangerous chemicals, especially when using hazardous fumigants (Damalas and Eleftherohorinos 2011). Therefore, the non-chemical or insecticide-free approach helps to maintain the nutritional quality of grains, effectively reduces health risks and ensures food safety by preventing the presence of pesticide residues in the grains (Srivastava and Mishra 2021).

6.2 MECHANICAL CONTROL

Mechanical methods prove to be highly practical among various pest control approaches. Numerous mechanical devices have been designed and developed for both monitoring and mass trapping of

DOI: 10.1201/9781003609834-6

stored product insects, taking advantage of their behavioral patterns. Various devices and their applications will be discussed in this section.

6.2.1 Mechanical Devices

Entoleters are mechanical-rotary apparatuses employed for killing the stored product insect from the grains. The process involves depositing grain onto the disc's central point, which has a high-speed horizontal disc, and propelling it outward in a radial manner through the application of centrifugal force. This device extends flour shelf life by swiftly killing larvae with its swirling rotor pin. The high-velocity collision proves to be highly effective in killing both immature and adult insects present in the grain. However, a small portion of grains gets broken and leads to economic losses (Brabec et al. 2015). The grains like oats were first dehulled using an Entoleter Centrifugal Machine at 2,113 rpm with one pass before packaging to remove foreign materials including insect debris. The rpm of the Entoleter was adjusted to achieve an acceptable level of breakage up to 3% (Morales-Quiros et al. 2019). Enrolled flour can be packed and stored for two to three years.

The indicator device is a cone-shaped perforated cup (3 mm perforation) with a lid at the top, fixed at the bottom with a container and circular dish smeared with sticky material. Farmers place 200 g of pulses inside before storage. Emerging beetles enter the perforations, get trapped in the sticky substance, and can be easily located and removed. The device with 2 mm perforations is suitable for cereals, helping eliminate the initial beetle population and preserving stored pulses. Timely detection aids in preserving valuable pulses during storage (Mohan 2000).

Another device is named an automatic insect removal bin, which consists of four parts: outer container, inner perforated container, collection vessel, and lid. It utilizes the wandering behavior of stored product insects and their movement toward well-aerated regions (Mohan 2000). Grains are held in the specially designed inner perforated container, with aeration provided between the inner and outer containers. Insects wandering toward aeration slip through the perforations and fall into the collection vessel via a pitfall mechanism. Perforated rods (2 mm) are fixed in the inner container to collect emerging insects quickly. The container is suitable for storing rice, wheat, broken pulses, coriander, etc. It effectively removes common grain attackers like rice weevil, lesser grain borer, red flour beetle, and saw-toothed beetle. Within ten days, over 90% of insects can be removed from the grains. The containers are available in various capacities, from 2 to 500 kg. The device can also remove free-living larvae and emerging adults (Mohan 2000; Mohan and Rajesh 2016).

6.2.2 Traps

Several traps are utilized to manage the stored product insects (Table 6.1). Among those, probe traps are the most commonly used methods, but they are laborious and time-consuming. With this approach, grains (0.5–1 kg) are drawn from the stored bin using probes, and sieves are utilized to separate the insects from the grains. The probes remain in the grain storage bins for extended periods, requiring manual inspection by an inspector to visually examine them (Banga et al. 2018). TNAU (Coimbatore, India) has developed various types of traps (Mohan et al. 1994). These devices are useful in the timely detection and monitoring of insect infestation in stored food grains. These traps are developed based on the principle that insects have a natural tendency to wander toward the air. Probe traps consist of a main tube with an insect trapping tube and a detachable cone at the bottom. Additionally, equi-spaced perforations of 2 mm diameter are made in the main tube (Srivastava, 2017). Pitfall traps capture insects on the grain surface and are used for monitoring and mass trapping. The two-in-one probe trap demonstrates high efficiency by combining both probe and pitfall trap functionalities. This design is particularly well-suited for trapping pulse

TABLE 6.1
Various Traps for Trapping Stored Product Insect Pests

Different Traps	Target Pests
Bait traps	Inactive and hidden insects
Entoleter	Kills all the stages of insects and mites
Indicator device	Eliminating the initial population of general storage pests
Light trap	Lesser grain borer, red flour beetle, saw-toothed beetle, and Psocids
Pheromone traps	Almond moth, angoumois grain moth, rice moth, lesser grain borer, red flour beetle, saw-toothed grain beetle, and khapra beetle
Pitfall traps, standard model probe trap, TNAU model probe trap	Used for capturing insects active on the grain surface
Pulse beetle trap	Pulse beetle
TNAU automatic insect removal bin	Rice weevil, lesser grain borer, red flour beetle, and saw-toothed beetle

Sources: Loschiavo and Atkinson (1967), White et al. (1990), and Mohan and Rajesh (2016).

beetles, which tend to wander on the grain surface. The trap incorporates an indicator device, which includes a perforated cone-shaped cup with a lid at the top. This cup is fixed at the bottom along with a container and circular dish coated with a sticky material (Mohan, 2007; Mohan and Rajesh, 2016). Probe traps should be placed at the top 6 in. of the grain, where the insect activity is seen during the early period of storage. They can remove >80% of the insects within 10–20 days. Insects like *Sitophilus oryzae*, *Rhyzopertha dominica*, *Tribolium castaneum*, *Oryzaephilus surinamensis*, and *Sitotroga cerealella* can effectively be trapped and monitored by probe traps.

The pulse beetle trap is a probe trap model designed for pulse beetles, consisting of a 200 ml plastic container with 3.5 mm diameter perforations to match beetle size. Its round bottom enables trapped insects to move, mate, and lay eggs within the trap, preventing escape, and placing traps on the sides and center of storage bins effectively collects emerging beetles. The standard model trap has two parts, namely, a perforated lid (2 or 3 mm) and a cone-shaped bottom portion. The application of a special coating with sticky material on the inner side of the cone to hold trapped insects is necessary in this model and this procedure is tedious. TNAU model has a perforated lid, and cone-shaped bottom which tapers into a funnel-shaped trapping tube and no sticky material needs to be applied in this model (Mohan et al. 2007).

The UV light trap consists of a 4 W germicidal lamp emitting ultraviolet rays around 250 nm. The lamp is placed at the center of a funnel (310 mm diameter at the top, 35 mm diameter at the bottom) and attached to a transparent plastic container for collecting trapped insects. Hooks and a tripod stand facilitate hanging at desired points, preferably around warehouse corners at 1.5 m above ground level during the night. It attracts various stored product insects, including paddy pests and Psocids (Kirkpatrick, Yancey, and Marzke 1970). Recommended usage is two traps per 60 × 20 m (L × B) godown with 5 m height. The trap is effective for long-term grain storage, after fumigation, and for monitoring infested areas. Sticky traps are also utilized with pheromones for monitoring aerial populations of stored product insects. Furthermore, bait traps are also designed to monitor stored product coleopteran pests. The bait bag is a welded plastic mesh container, 200 mm square, with over 19 perforations/cm^2 and a 2 mm aperture. Add 100 g of brown rice or crushed groundnut, close with staples, and hang on grain stack sides. Perforated cylindrical rods with crushed groundnut attractant can be inserted between gunny bags to trap the insect in the inner layer.

Various models of pheromone traps have been designed for managing stored product pests. In lepidopteran moths, females produce chemical substances that attract males. In major beetle families, adult males release aggregation pheromones that attract both females and males. For example, Male *T. castaneum* emits the aggregation pheromone 4,8-dimethyldecanal (DMD) during active feeding at a moderately low population density. Suzuki (1981) first identified and synthesized this pheromone from males of both *T. castaneum* and *T. confusum*. The effectiveness of these pheromones in stored product pest management has been demonstrated by Doud and Phillips (2020); Cai et al. (2022).

6.3 PHYSICAL CONTROL

Physical control methods have emerged as a promising and effective alternative for managing storage pests. These methods involve manipulating the storage environment or using mechanical barriers to hinder pest activities, disrupt their life cycles, and prevent their access to stored commodities. Several physical measures are enlisted in Table 6.2.

6.3.1 Temperature

Temperature is vital for controlling storage pests, and its effect may vary depending on species, development stages, and moisture levels (Fields and Muir, 1996) (Figure 6.1). Most of the storage pests are unable to withstand extreme temperatures, whether it be cold or heat. Elevating the grain

TABLE 6.2
Effectiveness of Different Physical Treatments on Stored Insect Pests

Insect Species	Physical Treatment	Susceptible Stage	Reference
	(a) Temperature		
Callosobruchus maculatus	Microwave power level (28 seconds with 400 W)	Eggs mortality	Purohit et al. (2013)
Callosobruchus chinensis	2,450 MHz microwave radiation for 100 seconds at 700 W	All stages	Singh (2013)
Rhyzopertha dominica, *Tribolium castaneum*, and *Oryzaephilus surinamensis*	Fluidized bed heat treatment (60°C for 120 seconds)	All stages with 100% mortality	Boonsanga (2023)
	(b) Irradiation		
C. chinensis	Gamma radiation at 300 Gy	Female fecundity, hatchability, and adult fertility reduced	Chiluwal et al. (2019)
Sitophilus granarius, *T. castaneum*, and *R. dominica*	Combination of 100 Gy gamma-radiation exposure with DE @ 1 g/kg	All stages	Sileem, Mohamed, and Mehany (2019)
Plodia interpunctella	Fast neutron irradiation at the dose of 256 Gy	Preventing the development of all immature stages	Hassan et al. (2019)
	(c) Modified Atmospheric Packaging		
Sitophilus spp.	Modified atmosphere packaging storage (40% CO_2 treatment after 5 days)	All stages	Carli et al. (2010)
	(d) Inert Dust		
R. dominica and *Sitophilus oryzae*	Diatomaceous earth, silicid (1000 ppm) – 1 kg of DE formulation/ton of soft wheat for 7 days	98% mortality in lesser grain borer and 83% mortality in rice weevil	Agrafioti et al. (2023)

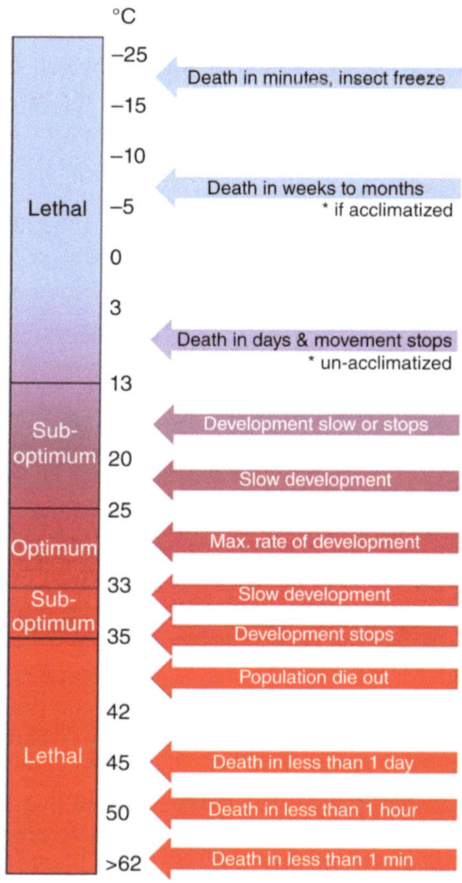

FIGURE 6.1 Influence of temperature on the development of stored-product insects.

temperature to 55–65°C for a duration of 10–12 hours has proven to be an effective method for eliminating all life stages of stored grain pests. Likewise, maintaining low temperatures provides long-term protection for stored seeds, keeping them free from insect infestations. Low temperatures hinder insect development and effectively eliminate a significant number of immature stages of stored grain insects (Guru et al. 2022). In recent studies, non-thermal heating techniques have demonstrated significant potential exists for the effective management of pests in stored grain, offering more quality grains as well as enhanced safety for nature (Paul et al. 2020). However, advancements in this field have introduced innovative techniques like dielectric heating using microwave (MW) heating and radio frequency (RF) (Das, Kumar, and Shah. 2013). These radiations are not only shown to be fatal to insects but also to cause sterility in the surviving individuals (Paul et al. 2020).

Sweet potato weevil (SPW) pupae in storage roots were successfully eliminated using vapor heat treatment (Shimabukuro et al. 1997). Additionally, sweet potato roots infected with SPW adults or immatures were subjected to controlled atmospheres with reduced oxygen and elevated carbon dioxide levels, along with a balanced amount of nitrogen, for a duration of up to 10 days at temperatures of 25 and 30°C. This treatment effectively eradicated the adult weevils (Delate and Brecht 1989). Specifically, when exposed to an atmosphere consisting of 8% oxygen and 40–60% carbon dioxide at 30°C, the adult weevils were eliminated within a short period of four to eight days (Delate, Brecht, and Coffelt 1990).

6.3.2 Irradiation (Ionizing Radiations)

Infrared (IR) heating involves IR heating elements through which energy is transferred from a higher temperature body to a lower temperature body without the need for physical contact, using electromagnetic radiation. IR radiation wavelengths typically range from 780 nm to 1 mm (Jun and Irudayaraj 2003). Grain disinfestation also involves the use of ionizing radiations, such as X-rays, gamma rays (from cobalt-60 and cesium-137), or electric beams (Hallman and Blackburn 2016). These radiations operate at the molecular level and can disrupt the DNA structure. Consequently, DNA damage results in insect sterilization, abnormalities, mutations, and mortality.

6.3.3 Modified Atmospheric Storage (Hermetic Storage)

Insects rely on oxygen (O_2) for their developmental metabolism, taking in O_2 and exhaling carbon dioxide (CO_2). By modifying the concentrations of these gases, in conjunction with other fumigants such as phosphine, storage pests can be effectively eliminated (Constantin et al., 2020). This method, used to manage insects in storage, is known as modified or controlled atmospheric storage. Exposing insects to low O_2 levels (<3%) and elevated CO_2 levels (>50%) for more than 24 hours causes significant distress, and prolonged exposure can lead to the mortality of remaining insect populations (Tutuncu and Emekci 2019). The Greek term "hermetic" is derived from the Greek god "Hermes". Hermetic means "airtight". Hermetic storage (HS) technology is a non-chemical-based system of storage. HS involves sealing the container completely, creating a modified atmosphere with low oxygen (or no oxygen) and high carbon dioxide composition (Obeng-Ofori 1995). This phenomenon is effectively utilized in the development of storage structures with this hermetic technology.

Purdue Improved Crop Storage (PICS) bag was a three-layer hermetic bag system that forms a barrier against oxygen influx and carbon dioxide escape. The use of PICS bags was studied by Williams, Baributsa, and Woloshuk (2014). The study revealed that keeping maize in PICS bags is a good way to reduce aflatoxin formation in storage. PICS also acts as a moisture barrier, limiting rewetting and drying, according to the researchers. These PICS bags are widely used in Latin America and Africa (De Bruin et al. 2012). Metal silos, which also utilize this hermetic principle for effectively controlling the rice weevil, *Sitophilus zeamais* and larger grain borer, *Prostephanus truncates* (De Groote et al. 2013).

6.3.4 Inert Dust

An alternative to conventional insecticides in controlling storage pests is inert dust (Losic and Korunic 2018). Its effectiveness is increased with a decrease in relative humidity. Based on their physical and chemical composition, categorized into zeolite (Eroglu, Emekci, and Athanassiou 2017), attapulgite (Lampiri et al. 2022), graphene (Moisidis et al. 2022), wood ash (Bohinc and Trdan 2017), and diatomaceous earth (DE) (Athanassiou et al. 2009). Among this, DE, which was derived from the fossils of unicellular algae (diatoms) was widely utilized in the direct application of the grains for the control of storage pests. DE, on contact with the insect's body, attaches to the waxy cuticle layer, leading to water loss, desiccation, and ultimately death (Shah and Khan 2014). DE offers several advantages, including being stable for extended periods, having low mammalian toxicity, and being easily separable from the treated grains (Losic and Korunic 2018). The disadvantage of using DE is that it affects the physical properties of grains like bulk density (Korunic et al. 1996).

6.3.5 Plant Products

For pulses, the recommended method is to mix coconut, groundnut, or mustard oil at a rate of 1 kg for every 100 kg of pulses. These oils act as natural repellents and create a protective coating around

TABLE 6.3
Natural Plant Products for Grain Protection against Target Pests

Crop	Plant Products	Target Pest	Mode of Action	Period
Ragi	Dried neem leaves and thumbai leaves are mixed with the grains and stored	Lesser grain borer, saw-toothed grain beetle, and flat grain borer.	Repellent activity	Up to 6 months
Pulses	20 ml of neem oil is mixed with 1 kg of pulses	Weevils, red flour beetle, long-headed flour beetle, fig moth	Repellent, feeding, and oviposition deterrent, growth inhibition.	Up to 1 year
Paddy	Pungam leaves placed in between stored jute bags	Angumouis grain moth and rice weevil	Repellent activity	Up to 6 months
Pulses and oilseeds	Neem leaves placed along with grains and stored	Pulse beetle, lesser grain borer, and other borers	Repellent activity	Up to 1 year
Pulses and cereals	Mixing of turmeric powder with grains and stored	General storage pests	Turmerones and arturmerone have the repellent activity	Up to 6–8 months
Pulses, ragi, paddy, maize	Garlic cloves are kept in layers in storage bins filled with seeds	General storage pests	Antifeedant, bactericidal, fungicidal, insecticidal, nematicidal, and repellent properties	Up to 1 year
Maize	Mixed with dried seeds and leaves of tulsi/basil	Weevil pests	Antifeedant, repellent, and larvicidal activity	Up to 1 year

Sources: Karthikeyan et al. (2009) and Prakash (2016).

the pulses, making them unattractive to pulse beetles and preventing infestations during storage. Similarly, for cereal grains, the use of neem seed kernel is recommended. Neem has insecticidal properties that can deter pests like grain weevils and moths (Prakash 2016). The suggested dosage is to mix 1 kg of neem seed kernel for every 100 kg of cereal grains before storing them. By employing these plant-based products, farmers and grain handlers can avoid the use of synthetic chemical pesticides while still effectively protecting their pulses and grains from pest damage for an extended period of up to six months (Karthikeyan et al. 2009). This approach is more environmentally friendly and poses fewer health risks, making it a preferable option for long-term grain storage. Some of the natural plant products that are currently used for grain protection against stored product pests are furnished in Table 6.3.

6.3.6 Activated Clay

The application of activated clay, specifically kaolin, as a means of safeguarding grains from insect pests constitutes a physical approach to pest control. In Egypt, rock phosphate was also used as a grain protectant. Additionally, some local farmers in West Africa utilized ashes, lime, and sand dust to protect grains against pest infestation. Comprised mainly of alumina and silicon dioxide, kaolin exhibits remarkable effectiveness against numerous storage pests while remaining non-toxic to higher animals. To harness its potential, kaolinite clay undergoes a series of activation steps involving acid and heat treatments. Upon activation, the finely divided clay particles act as both an abrasive agent and an absorbent. They efficiently abrade the protective cuticle of the insects and simultaneously absorb the lipoid particles present in their epicuticle. As a result, the pests undergo desiccation, ultimately leading to their mortality. This method offers a safe and environmentally friendly solution for protecting stored grains, ensuring their preservation without resorting to harmful chemicals.

6.4 CULTURAL CONTROL

Cultural control is a pest management approach that involves modifying the environment of the host to reduce the spread, growth, and proliferation of stored-product pests. Various cultural operations are employed, including the elimination of infested grains before storing new ones. Cultural control measures for storage product pests mainly concentrated on the carryover pests from the field, where the infestation begins in the field itself. Additionally, disinfestation and protective measures like white-washing and painting the walls, floors, and ceilings of vacant stores with insect-repellent paints, such as coal tar, are implemented. These techniques make the environment more difficult and unfavorable for storage insects' survival, dissemination, growth, and reproduction (Casida and Quistad 1998).

6.4.1 Containers/Bags

Traditionally, farmers have commonly adopted the practice of not cleaning empty food grain bags, intending to reuse them in the following season. However, this habit can lead to a significant issue as eggs and larvae of insects tend to remain concealed within the seams and mesh of these bags. During the off-season, these insects feed on grain fragments and other residues that may be sticking to the bags. Consequently, when grains are filled into these reused bags, the infestation reappears due to the presence of these hidden pests. Triple-layer bags, comprising a 20 μm + 80 μm Polyethylene bag + woven polypropylene bag, were found to incur less damage from rice weevils compared to double- and single-layer bags. This result was based on data collected from each storage bag after 90 days, with an inoculation ratio of 30 pairs of five-day-old adult *S. oryzae* per 1 kg of rice grains (Figure 6.2).

6.4.2 Deep Planting and Hilling Up

Deep planting and proper hilling of the potato tubers during sowing reduced the PTM population by keeping the tubers beyond the access of adult pests (Ali 1993; Sileshi and Teriessa 2001). Unless the

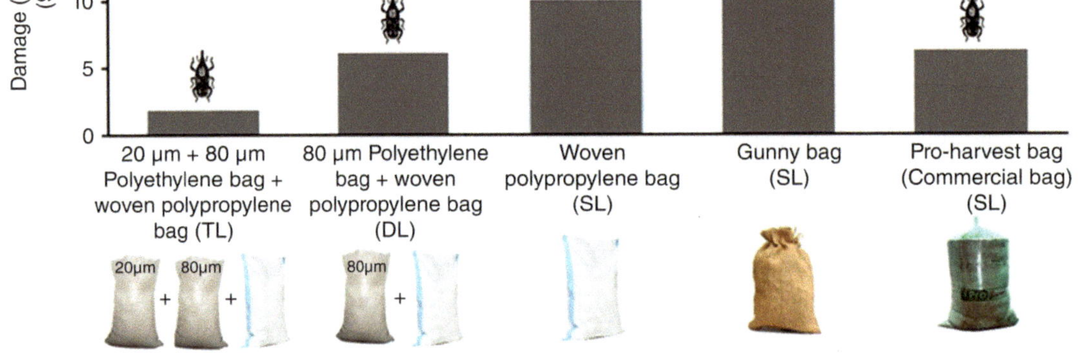

FIGURE 6.2 Efficacy of triple, double, and single layers of storage bags for the management of rice weevil at 90 DAT.

soil is extensively cracked, adult moths cannot easily reach tubers covered with more than 5 cm of soil. Potato tuber moth (PTM) infestation is inversely proportional to planting depth and the number of hilling operations. Deep planting into the hill or planting into the furrow and earthing up after six weeks provides protection from SPWs (MacFarlane 1987).

6.4.3 Earthing Up

PTM prefers green foliage for feeding and oviposition. Tuber infestation increases when the foliage begins to fall. Adults enter the soil through cracks to seek refuge from light and deposit eggs on tubers, whereas larvae burrow for sustenance. Therefore, the duration between drying and harvesting is of utmost importance. The extended presence of deceased vines and unharvested tubers in the field increases the probability of tuber infestation. Tubers that are uncovered or situated near the surface face a heightened risk of PTW damage. Applying a 1–2-in. layer of soil to the hills right after vine termination substantially decreases the incidence of tuber infestation (Tsedaley 2015). Das (1998) documented that sweet potato roots piled on the ground and covered with a 2 cm thick layer of local materials remained uninfected when covered with sand and rice husk or showed minimal infestation with sawdust and ash during a three-month storage period.

6.4.4 Irrigation

Maintaining regular irrigation until harvest reduced soil cracking and thus prevented the exposure of tubers to PTM female moths (Hanafi 1999). Maintaining soil moisture levels through overhead irrigation helps prevent soil cracking, while the application of 0.1 in. of water daily using a center pivot irrigation system, starting from the vine kill until harvest, reduces the occurrence of PTM (Tsedaley 2015). The extent of PTM infestation increased in direct correlation with the duration of irrigation intervals (Shelton and Wyman 1979). Furrow irrigation is a common practice in parts of Himachal Pradesh. With furrow irrigation, furrows and hills should be well-maintained to prevent washout and exposure of tubers.

6.4.5 Mixed Cropping

Sweet potato and other crops such as colocasia, rice, okra, ginger, maize, and yam are grown in mixed cropping systems. These systems exhibited a low occurrence of SPW, *Cylas formicarius* (Rajasekhara Rao 2005). In low-land rice fields in India, crop rotations like rice-sweet potato-cowpea, rice-rice-sweet potato, and rice-sweet potato-rice proved to be successful in mitigating SPW infestations (Pillai et al. 1996).

6.4.6 Mulching

The wheat straw acted as a mechanical barrier for PTM, and *Phthorimaea operculella* adults to reach tubers. Additionally, the use of rice straw or black plastic mulch was effective in reducing the SPW infestations in the root area. Covering tubers with about a 1 cm layer of wheat straw afforded significant control in stored potatoes. Dry leaves of different plants (*Lantana* and *Eupatorium*) used to cover tubers reduced the infestation considerably. They have dual functions of forming a physical barrier to the egg-laying moths and also acting as a repellent (Kroschel 1995).

6.4.7 Paired Row System

Planting sweet potato and yam/potato beans in paired rows at a ratio of 2:1 or in alternating rows proved to be an effective strategy for minimizing tuber infestations caused by the SPW. The practice of planting alternating rows of sweet potatoes and marigolds in a 1:1 ratio led to reduced tuber infestation.

6.4.8 Planting and Crop Rotation

Farmers should avoid using infested seed tubers and ensure potato fields are located at a considerable distance from previous plantings. It is crucial to eliminate any volunteer potato plants and other Solanaceae plants found in uncultivated areas, as they serve as breeding grounds for PTM populations in the following year (Hanafi 1999).

6.4.9 Re-Ridging

Performing re-ridging during the tuber formation stage effectively deters weevils from laying eggs and grubs from entering the tubers. Forming five re-ridgings at 10-day intervals between 50 and 90 days after planting substantially decreases damage caused by weevils, including PTM and SPW, to the tubers (Palaniswami and Mohandas 1994).

6.4.10 Resistant Varieties/Cultivars

Plant resistance is an efficient approach for controlling pests indirectly. The process of resistance involves several factors. In cereals, specific grain characteristics such as hard grains, the presence of glumes, a finely corneous endosperm, and a compact arrangement of pericarp layers are crucial for resistance. Additionally, small oval seeds with a rough and light testa, displaying a creamy white color, are less favored by pulse beetles for oviposition. It is observed that local or traditional cultivars generally experience fewer pest attacks compared to introduced or exotic genotypes. Furthermore, biochemical factors also contribute to the deterrence of oviposition by pests. Table 6.4 presents a compilation of significant resistant varieties/cultivars of various crops against stored product insect pests.

TABLE 6.4
Resistant Varieties/Cultivars of Various Crops against Stored Product Pests

Crop	Associated Pest	Resistant Variety/Cultivar	References
Pulses			
Chick Pea	*Callosobruchus maculatus, Callosobruchus analis*	Bittle-98, Punjab-91, Dasht, Parbat, CH-52/02, B-8/03, CM 3142–3/92, CM 3142–2/92, CM 88, CM72 and Pb91	Shafique and Ahmad (2005); Aslam, Shaheen, and Ayya (2006)
Pigeon pea	*Callosobruchus chinensis, C. maculatus*	*Cajanus cajanifolius* (ICPW030, 031), *C. platycarpus* (ICPW067, 068) and *C. scarabaeoides* (ICPW-094, 095)	Jadhav, Mallikarjuna, and Rao (2012)
Cowpea	*C. maculatus*	UPC 628, Pant Lobia-1	Mishra and Panigrahi (2017)
Black gram	*C. analis, C. chinensis, C. maculatus, Callosobruchus phaseoli and Zabrotes subfasciatus*	Var. *silvestris* (VM2011, VM 2164, VM3529)	Tomooka, Kashiwaba, and Vaughan (2000); Kashiwaba et al. (2003); Souframanien, Gupta, and Gopalakrishna (2010)
Mung bean	*C. chinensis, C. maculatus*	Var. *sublobata* (V1128, V2709, V2802)	Talekar and Lin (1992); Sugawara et al. (1996); Somta et al. (2007)
Cereals			
Rice	*Sitophilus oryzae*	Kalogira, BR 3, BR 11	Rashid et al. (2009)
Tuber crops			
Potato	*Phthorimaea operculella*	Kenya Baraka, King Edward, Russett Burbank, and Kufribadshah	Ojero and Mueke (1985); Tsedaley (2015)
Sweet potato	*Cylas formicarius*	Selopia and H-268, H85-168, CO-3, PI 508523, PI 54116, PI 564107, RS-92, RS-47, IGSP-14, S-1-60.	Palaniswami and Mohandas (1992)

6.4.11 Sanitation

Tuber inspection and removal of infested tubers at periodic intervals were important factors affecting PTM infestation in storage. Sorting of black-spotted tubers every three to four weeks from the harvested potatoes will control the PTM in storage. Removal of senescing vines has been found to affect the level of infestation as it exposes the tubers. When the vines deteriorate, PTM larvae migrate from the vines and infest the tubers underneath (Chandel, Chandla, and Garg 2008). Food grains are harvested and threshed in the fields before being transported to godowns using various means such as trunks, trollies, and bullock carts. However, these transportation methods serve as significant sources of cross-infestation. Insects that may be present in the joints or corners of these carriers can migrate to the stored food grain, leading to infestations. If storage godowns or containers are not properly whitewashed or plastered and thoroughly cleaned, some grains may remain in cracks and crevices. Insects present in the stores can continue to feed and multiply even during this period. When fresh stocks are stored in such infested conditions, they can also get infected. Insects from the old stocks can crawl or fly to the fresh stocks, leading to the infestation of the new batches. This phenomenon is referred to as "cross infestation."

6.4.12 Splitting of Pulses

Pulse beetle is categorized under the primary pests, which prefer to feed mainly on whole pulses. The pulse beetle does not prefer split pulses. Therefore, the splitting of pulses can be followed before storage to protect the pulses from pulse beetle damage (Devadas and Amuthan 2000).

6.4.13 Time of Harvesting

Numerous insects pose a threat to crops in the field, and there's also a risk of insects from nearby storage areas flying to the fields and laying eggs on harvested or nearly ripe material. However, timely and prompt harvesting can effectively prevent such infestations. Failure to do so may result in the infestation being transported directly to threshing yards and subsequently to storage facilities.

6.5 CONCLUSION

Food grains serve as the primary source of caloric requirements for both humans and animals globally, particularly in developing countries. The spoilage of grains during storage has led to widespread starvation, affecting millions of people worldwide, including those suffering from undernourishment. Thereby, this chapter discusses the various conventional non-chemical measures that are not only cost-effective but also environmentally friendly and significantly extend the shelf life of stored food grains by effectively reducing or preventing insect infestation. Additionally, non-chemical measures not only hinder resistance development in insects but also reduce the environmental impact by minimizing chemical pesticide use to preserve a healthier ecosystem.

REFERENCES

Agrafioti, P., M. Vrontaki, M. Rigopoulou, E. Lampiri, K. Grigoriadou, P. M. Ioannidis, C. I. Rumbos, and C. G. Athanassiou. 2023. "Insecticidal Effect of Diatomaceous Earth Formulations for the Control of a Wide Range of Stored-Product Beetle Species." *Insects* 14 (7): 656. doi:10.3390/insects14070656.

Ali, M. A. 1993. "Effects of Cultural Practices on Reducing Field Infestation of Potato Tuber Moth (*Phthorimaea operculella*) and Greening of Tubers in the Sudan." *The Journal of Agricultural Science* 121 (2): 187–192. doi:10.1017/S0021859600077042.

Aslam, M., F. A. Shaheen, and A. Ayya. 2006. "Management of *Callosobruchus chinensis* Linnaeus in Stored Chickpea through Interspecific and Intraspecific Predation by Ants." *World Journal of Agricultural Sciences* 2 (1): 85–89.

Athanassiou, C. G., F. H. Arthur, G. P. Opit, and J. E. Throne. 2009. "Insecticidal Effect of Diatomaceous Earth against Three Species of Stored-Product Psocids on Maize, Rice, and Wheat." *Journal of Economic Entomology* 102 (4): 1673–1680. doi:10.1603/029.102.0435.

Banga, K. S., N. Kotwaliwale, D. Mohapatra, and S. K. Giri. 2018. "Techniques for Insect Detection in Stored Food Grains: An Overview." *Food Control* 94 (December): 167–176. doi:10.1016/j.foodcont.2018.07.008.

Bohinc, T., and S. Trdan. 2017. "Comparison of Insecticidal Efficacy of Four Natural Substances against Granary Weevil (*Sitophilus granarius*) Adults: Does the Combined Use of the Substances Improve Their Efficacy?" *Spanish Journal of Agricultural Research* 15 (3): e1009.

Boonsanga, K.. 2023. *Fluidized Bed Heating Technique for Controlling of Post-Harvest Insect Pests in Paddy and Milled Rice.* Chiang Mai: Graduate School, Chiang Mai University.

Brabec, D. L., T. C. Pearson, E. B. Maghirang, and P. W. Flinn. 2015. "Detection of Fragments from Internal Insects in Wheat Samples Using a Laboratory Entoleter." *Cereal Chemistry* 92 (1): 8–13. doi:10.1094/CCHEM-08-13-0173-R.

Cai, L., S. Macfadyen, B. Hua, H. Zhang, W. Xu, and Y. Ren. 2022. "Identification of Biomarker Volatile Organic Compounds Released by Three Stored-Grain Insect Pests in Wheat." *Molecules* 27 (6): 1963. doi:10.3390/molecules27061963.

Carli, M. D., B. Bresolin, C. P. Z. Noreña, I. Lorini, and A. Brandelli. 2010. "Efficacy of Modified Atmosphere Packaging to Control *Sitophilus* spp. in Organic Maize Grain." *Brazilian Archives of Biology and Technology* 53 (6): 1469–1476. doi:10.1590/S1516-89132010000600024.

Casida, J. E., and G. B. Quistad. 1998. "Golden Age of Insecticide Research: Past, Present, or Future?" *Annual Review of Entomology* 43 (1): 1–16. doi:10.1146/annurev.ento.43.1.1.

Chandel, R. S., V. K. Chandla, and I. D. Garg. 2008. "Integrated Pest Management of Potato Tuber Moth in India." In *Integrated Pest Management for the Potato Tuber Moth, Phthorimaea operculella (Zeller)-a Potato Pest of Global Importance.* ed. J. Kroschel and L. A. Lacey. Wapato, WA, USA: USDA-ARS, 127–138.

Chiluwal, K., J. Kim, S. D. Bae, G. H. Roh, H. J. Park, and C. G. Park. 2019. "Effect of Gamma Irradiation on Fecundity, Sterility, and Female Sex Pheromone Production of *Callosobruchus chinensis* (Coleoptera: Bruchidae)." *Journal of Economic Entomology* 112 (1): 156–163. doi:10.1093/jee/toy317.

Damalas, C. A., and I. G. Eleftherohorinos. 2011. "Pesticide Exposure, Safety Issues, and Risk Assessment Indicators." *International Journal of Environmental Research and Public Health* 8 (5): 1402–1419. doi:10.3390/ijerph8051402.

Das, G. P. 1998. "The Control of the Sweetpotato Weevil in Storage." *Tropical Science* 38 (4): 196–197.

Das, I., G. Kumar, and N. G. Shah. 2013. "Microwave Heating as an Alternative Quarantine Method for Disinfestation of Stored Food Grains." *International Journal of Food Science* 2013: 1–13. doi:10.1155/2013/926468.

De Bruin, T., S. Navarro, P. Villers, and A. Wagh. 2012. "Worldwide Use of Hermetic Storage for the Preservation of Agricultural Products." Proc 9th. Int. Conf. on Controlled Atmosphere and Fumigation in Stored Products, no. October: 15–19, 2012.

De Groote, H., S. C. Kimenju, P. Likhayo, F. Kanampiu, T. Tefera, and J. Hellin. 2013. "Effectiveness of Hermetic Systems in Controlling Maize Storage Pests in Kenya." *Journal of Stored Products Research* 53: 27–36. doi:10.1016/j.jspr.2013.01.001.

Delate, K. M., and J. K. Brecht. 1989. "Quality of Tropical Sweet Potatoes Exposed to Controlled-Atmosphere Treatments for Postharvest Insect Control." *Journal of the American Society for Horticultural Science* 114 (6): 963–968. doi:10.21273/JASHS.114.6.963.

Delate, K. M., J. K. Brecht, and J. A. Coffelt. 1990. "Controlled Atmosphere Treatments for Control of Sweetpotato Weevil (Coleoptera: Curculionidae) in Stored Tropical Sweet Potatoes." *Journal of Economic Entomology* 83 (2): 461–465. doi:10.1093/jee/83.2.461.

Devadas, C. T., and G. Amuthan. 2000. Status Report on Storage of Pulses. *Agricultural Engineering Today.* 24.

Doud, C. W., and T. W. Phillips. 2020. "Responses of Red Flour Beetle Adults, *Tribolium castaneum* (Coleoptera: Tenebrionidae), and Other Stored Product Beetles to Different Pheromone Trap Designs." *Insects* 11 (11): 733. doi:10.3390/insects11110733.

Duranti, M.. 2006. "Grain Legume Proteins and Nutraceutical Properties." *Fitoterapia* 77 (2): 67–82. doi:10.1016/j.fitote.2005.11.008.

Eroglu, N., M. Emekci, and C. G. Athanassiou. 2017. "Applications of Natural Zeolites on Agriculture and Food Production." *Journal of the Science of Food and Agriculture* 97 (11): 3487–3499. doi:10.1002/jsfa.8312.

Guru, P. N., D. Mridula, A. S. Dukare, B. M. Ghodki, A. U. Paschapur, I. Samal, M. N. Raj, V. K. Padala, M. Rajashekhar, and A. R. N. S. Subbanna. 2022. "A Comprehensive Review on Advances in Storage Pest Management: Current Scenario and Future Prospects." *Frontiers in Sustainable Food Systems* 6 (November). doi:10.3389/fsufs.2022.993341.

Hallman, G., and C. Blackburn. 2016. "Phytosanitary Irradiation." *Foods* 5 (1): 8. doi:10.3390/foods5010008.

Hanafi, A. 1999. "Integrated Pest Management of Potato Tuber Moth in Field and Storage." *Potato Research* 42 (2): 373–380. doi:10.1007/BF02357863.

Hassan, R. S., E. A. Mahmoud, T. M. Sileem, and W. A. A. Sayed. 2019. "Evaluation of Fast Neutron Irradiation as a New Control Method against the Indian Meal Moth, *Plodia interpunctella* (Hübner)." *Journal of Radiation Research and Applied Sciences* 12 (1): 38–44. doi:10.1080/16878507.2019.1594133.

Jadhav, D. R., N. Mallikarjuna, and G. V. R. Rao. 2012. "*Callosbruchus maculatus* Resistance in Some Wild Relatives and Interspecific Derivatives of Pigeonpea." *Indian Journal of Plant Protection* 40 (1): 40–44.

Jha, S. N., R. K. Vishwakarma, T. Ahmad, A. Rai, and A. K. Dixit. 2015. *Report on Assessment of Quantitative Harvest and Post-Harvest Losses of Major Crops and Commodities in India.* ICAR-All India Coordinated Research Project on Post-Harvest Technology, ICAR-CIPHET, P.O.-PAU, Ludhiana-141004. p. 162.

Jun, S., and J. Irudayaraj. 2003. "A Dynamic Fungal Inactivation Approach Using Selective Infrared Heating." *Transactions of the ASAE* 46 (5). doi:10.13031/2013.15435.

Karthikeyan, C., D. Veeraragavathatham, D. Karpagam, and A. Firdouse. 2009. "Traditional Storage Practices." *Indian Journal of Traditional Knowledge* 8 (4): 564–568.

Kashiwaba, K., N. Tomooka, A. Kaga, O.-K. Han, and D. A. Vaughan. 2003. "Characterization of Resistance to Three Bruchid Species (*Callosobruchus* Spp., Coleoptera, Bruchidae) in Cultivated Rice Bean (*Vigna umbellata*)." *Journal of Economic Entomology* 96 (1): 207–213. doi:10.1093/jee/96.1.207.

Khatri-Chhetri, A., P. K. Aggarwal, P. K. Joshi, and S. Vyas. 2017. "Farmers' Prioritization of Climate-Smart Agriculture (CSA) Technologies." *Agricultural Systems* 151 (February): 184–191. doi:10.1016/j.agsy.2016.10.005.

Kirkpatrick, R. L., D. L. Yancey, and F. O. Marzke. 1970. "Effectiveness of Green and Ultraviolet Light in Attracting Stored-Product Insects to Traps." *Journal of Economic Entomology* 63 (6): 1853–1855. doi:10.1093/jee/63.6.1853.

Korunic, Z., P. G. Fields, M. I. P. Kovacs, J. S. Noll, O. M. Lukow, C. J. Demianyk, and K. J. Shibley. 1996. "The Effect of Diatomaceous Earth on Grain Quality." *Postharvest Biology and Technology* 9 (3): 373–387. doi:10.1016/S0925-5214(96)00038-5.

Kroschel, J.. 1995. Integrated Pest Management in Potato Production in the Republic of Yemen with Special Reference to the Integrated Biological Control of the Potato Tuber Moth *(Phthorimaea operculella* Zeller*)*. Tropical Agriculture, 8, Margraf Verlag, Weikersheim, Germany, pp. 227.

Lampiri, E., P. Agrafioti, I. Vagelas, and C. G. Athanassiou. 2022. "Insecticidal Effect of an Enhanced Attapulgite for the Control of Four Stored-Product Beetle Species." *Agronomy* 12 (7): 1495. doi:10.3390/agronomy12071495.

Loschiavo, S. R., and J. M. Atkinson. 1967. "A Trap for the Detection and Recovery of Insects in Stored Grain." *The Canadian Entomologist* 99 (11): 1160–1163. doi:10.4039/Ent991160-11.

Losic, D., and Z. Korunic. 2018. "Diatomaceous Earth, A Natural Insecticide for Stored Grain Protection: Recent Progress and Perspectives." 219–247. doi:10.1039/9781788010160-00219.

MacFarlane, D. R. 1987. "Physical Aspects of Vitrification in Aqueous Solutions." *Cryobiology* 24 (3): 181–195. doi:10.1016/0011-2240(87)90022-8.

Mishra, S. K., and J. Panigrahi. 2017. "Identification of Inter-Specific Hybrid between *Cajanus cajan* (L.) and *C. cajanifolius* (H.) Using Cyto-Morphological and DNA Markers." *Electronic Journal of Plant Breeding* 8 (1): 63. doi:10.5958/0975-928X.2017.00009.6.

Mohan, S. 2000. "Insect Removal Bin Could Help Farmers in Developing Countries." *Magazine of American Association of Agricultural and Biological Engineers* 7: 69–70.

Mohan, S., M. Gopalan, P. C. Sundarababu, and V. V. Sreenarayanan. 1994. "Practical Studies on the Use of Light Traps and Bait Traps in the Management of *Rhyzopertha dominica* (F.) in Rice Warehouses." *International Journal of Pest Management* 40 (2): 148–152. doi:10.1080/09670879409371873.

Mohan, S., and A. Rajesh. 2016. "Tools for Stored Product Insects Management and Technology Transfer." *Indian Journal of Entomology* 78 (special): 59. doi:10.5958/0974-8172.2016.00026.2.

Mohan, S., S. S. Sivakumar, S. R. Venkatesh, and G. S. V. Raghavan. 2007. "Penetration of Polyethylene Sheets Coated with Protein-Enriched Pea Flour Solution by Two Stored-Product Insects." *Journal of Stored Products Research* 43 (2): 202–204. doi:10.1016/j.jspr.2006.01.002.

Moisidis, I. C., M. K. Sakka, R. Karunagaran, D. Losic, and C. G. Athanassiou. 2022. "Insecticidal Effect of Graphene against Three Stored-Product Beetle Species on Wheat." *Journal of Stored Products Research* 98 (September): 101999. doi:10.1016/j.jspr.2022.101999.

Morales-Quiros, A., C. Campabadal, D. E. Maier, S. M. N. Lazzari, F. A. Lazzari, and T. W. Phillips. 2019. "Chilled Aeration to Control Pests and Maintain Grain Quality during Summer Storage of Wheat in the North Central Region of Kansas." *Applied Engineering in Agriculture* 35 (4): 657–688. doi:10.13031/aea.13252.

Obeng-Ofori, D. 1995. "Plant Oils as Grain Protectants against Infestations of *Cryptolestes pusillus* and *Rhyzopertha dominica* in Stored Grain." *Entomologia Experimentalis et Applicata* 77 (2): 133–139. doi:10.1111/j.1570-7458.1995.tb01993.x.

Ojero, M. F. O., and J. M. Mueke. 1985. "Resistance of Four Potato Varieties to the Potato Tuber Moth, *Phthorimaea operculella* (Zell.) in Storage." *Insect Science and Its Application* 6 (2): 205–207. doi:10.1017/S1742758400006640.

Palaniswami, M. S., and N. Mohandas. 1992. "Search for Resistance to *Cylas formicarius* F. in Sweet Potato." *Journal of Root Crops* 18: 41–52.

Palaniswami, M. S., and N. Mohandas. 1994. "Reridging as a Cultural Method for the Management of Sweet Potato Weevil *Cylas formicarius* F." *Journal of Root Crops* 20: 101–105.

Parfitt, J., M. Barthel, and S. Macnaughton. 2010. "Food Waste within Food Supply Chains: Quantification and Potential for Change to 2050." *Philosophical Transactions of the Royal Society B: Biological Sciences* 365 (1554): 3065–3081. doi:10.1098/rstb.2010.0126.

Paul, A., M. Radhakrishnan, S. Anandakumar, S. Shanmugasundaram, and C. Anandharamakrishnan. 2020. "Disinfestation Techniques for Major Cereals: A Status Report." *Comprehensive Reviews in Food Science and Food Safety* 19 (3): 1125–1155. doi:10.1111/1541-4337.12555.

Phillips, T. W., and J. E. Throne. 2010. "Biorational Approaches to Managing Stored-Product Insects." *Annual Review of Entomology* 55: 375–397.

Pillai, K. S., M. S. Palaniswami, P. Rajamma, C. S. Ravindran, and T. Premkumar. 1996. *An IPM Approach for Sweet Potato Weevil.* In *Tropical Tuber Crops: Problems, Prospects and Future Strategies.* ed. G.T. Kurup, M. S. Palaniswami, V. Potty, G. Padmaja, S. Kabeerathumma and S. V. Pillai. New Marine Lines, Churchgate, Mumbai: Oxford & IBH Publishing Company, 329–339.

Prakash, B. G. 2016. "Indigenous Practices for Eco-Friendly Storage of Food Grains and Seeds." *Advances in Plants & Agriculture Research* 3 (4). doi:10.15406/apar.2016.03.00101.

Purohit, P., D. S. Jayas, B. K. Yadav, V. Chelladurai, P. G. Fields, and N. D. G. White. 2013. "Microwaves to Control *Callosobruchus maculatus* in Stored Mung Bean (Vigna Radiata)." *Journal of Stored Products Research* 53 (April): 19–22. doi:10.1016/j.jspr.2013.01.002.

Rajasekhara Rao, K. 2005. "Systems Approach for Management of Insect Pest Problem in Tuber Crops by Farmers of Meghalaya." *CTCRI News 22.*

Rajashekar, Y., N. Tonsing, T. Shantibala, and J. R. Manjunath. 2016. "2, 3-Dimethylmaleic Anhydride (3, 4-Dimethyl-2, 5-Furandione): A Plant Derived Insecticidal Molecule from *Colocasia esculenta* Var. *esculenta* (L.) Schott." *Scientific Reports* 6 (1): 20546. doi:10.1038/srep20546.

Rashid, M. H., M. A. Haque, M. M. Rahman, and A. F. M. S. Ahsan. 2009. "Study on Resistance of Different Rice Varieties against Rice Weevil." *Int. J. Sustain. Crop Prod* 4 (1): 35–40.

Shafique, M., and M. Ahmad. 2005. "Chickpea Grains Resistance to Pulse Beetle, *Callosobruchus analis* (F.) (Coleoptera: Bruchidae)." *Pakistan Journal of Zoology* 37 (2): 123–126.

Shah, M. A., and A. A. Khan. 2014. "Use of Diatomaceous Earth for the Management of Stored-Product Pests." *International Journal of Pest Management* 60 (2): 100–113. doi:10.1080/09670874.2014.918674.

Shelton, A. M., and J. A. Wyman. 1979. "Time of Tuber Infestation and Relationships Between Pheromone Catches of Adult Moths, Foliar Larval Populations, and Tuber Damage by the Potato Tuber Worm." *Journal of Economic Entomology* 72 (4): 599–601. doi:10.1093/jee/72.4.599.

Shimabukuro, S., A. Ishikawa, M. Iwata, T. Sakaguchi, S. Makiguchi, and H. Katsumata. 1997. "Efficacy of Vapor Heat Treatment on Sweet Potato Infested with Sweet Potato Weevil, *Cylas formicarius (*Fabricius) (Coleoptera: Brentidae), West Indian Sweet Potato Weevil, *Euscepes postfasciatus* (Fairmaire) (Coleoptera: Curculionidae), and Sweet Potato." *Research Bulletin of the Plant Protection Service, Japan* 33: 35–41.

Sileem, T., S. Mohamed, and A. Mehany. 2019. "First Trial to Assess Gamma-Irradiation and Diatomaceous Earth Combination on Mortality of Some Stored Product Insects." *Arab Journal of Nuclear Sciences and Applications* 52 (3): 115–120. doi:10.21608/ajnsa.2019.6159.1134.

Sileshi, G., and J. Teriessa. 2001. "Tuber Damage by Potato Tuber Moth, *Phthorimae operculella* Zeller (Lepidoptera: Gelechiidae), in the Field in Eastern Ethiopia." *International Journal of Pest Management* 47 (2): 109–113. doi:10.1080/09670870151130552.

Singh, R. P. 2013. *Status Paper on Pulses*. Government of India ministry of agriculture, department of agriculture & cooperation, Bhopal, Madhya Pradesh, pp. 215.

Somta, P., C. Ammaranan, P. A.-C. Ooi, and P. Srinives. 2007. "Inheritance of Seed Resistance to Bruchids in Cultivated Mungbean (Vigna Radiata, L. Wilczek)." *Euphytica* 155 (1–2): 47–55. doi:10.1007/s10681-006-9299-9.

Souframanien, J., S. K. Gupta, and T. Gopalakrishna. 2010. "Identification of Quantitative Trait Loci for Bruchid (*Callosobruchus maculatus*) Resistance in Black Gram [Vigna Mungo (L.) Hepper]." *Euphytica* 176 (3): 349–356. doi:10.1007/s10681-010-0210-3.

Srivastava, S., and H. N. Mishra. 2021. "Ecofriendly Nonchemical/Nonthermal Methods for Disinfestation and Control of Pest/Fungal Infestation during Storage of Major Important Cereal Grains: A Review." *Food Frontiers* 2 (1): 93–105. doi:10.1002/fft2.69.

Sugawara, F., M. Ishimoto, N. Le-Van, H. Koshino, J. Uzawa, S. Yoshida, and K. Kitamura. 1996. "Insecticidal Peptide from Mungbean: A Resistant Factor against Infestation with Azuki Bean Weevil." *Journal of Agricultural and Food Chemistry* 44 (10): 3360–3364. doi:10.1021/jf960166c.

Suzuki, T. 1981. "Identification of the Aggregation Pheromone of Flour Beetles *Tribolium castaneum* and *T. confusum* (Coleoptera: Tenebrionidae)." *Agricultural and Biological Chemistry* 45 (6): 1357–1363. doi:10.1080/00021369.1981.10864714.

Talekar, N. S., and C. P. Lin. 1992. "Characterization of *Callosobruchus chinensis* (Coleoptera: Bruchidae) Resistance in Mungbean." *Journal of Economic Entomology* 85 (4): 1150–1153. doi:10.1093/jee/85.4.1150.

Tomooka, N., K. Kashiwaba, and D. A. Vaughan. 2000. "The Effectiveness of Evaluating Wild Species: Searching for Sources of Resistance to Bruchid Beetles in the Genus Vigna Subgenus Ceratotropis," 27–41. doi: /10.1023/A:1003906715119.

Tsedaley, B. 2015. "Integrated Management of Potato Tuber Moth (*Phthorimaea operculella*) (Zeller) in Field and Storage" 5 (3): 134–145.

Tutuncu, S., and M. Emekci. 2019. "Comparative Efficacy of Modified Atmospheres Enriched With Carbon Dioxide against *Cadra (= Ephestia) cautella*." *Journal of the Science of Food and Agriculture* 99 (13): 5962–5968. doi:10.1002/jsfa.9871.

White, N. D. G. R. T., P. G. F. Arbogast, S. R. R. C. Hillmann, B. Loschiavo, J. E. Subramanyam, J. E. Throne, and V. F. Wright. 1990. "The Development and Use of Pitfall and Probe Traps for Capturing Insects in Stored Grain." *Journal of the Kansas Entomological Society 63* (4): 506–525.

Williams, S. B., D. Baributsa, and C. Woloshuk. 2014. "Assessing Purdue Improved Crop Storage (PICS) Bags to Mitigate Fungal Growth and Aflatoxin Contamination." *Journal of Stored Products Research* 59: 190–196. doi:10.1016/j.jspr.2014.08.003.

7 Bio-Based Approaches

Irengbam Barun Mangang
Kelappaji College of Agricultural Engineering and Food Technology, Kerala Agricultural University, Tavanur, Kerala, India

M. Loganathan and R. Jagan Mohan
National Institute of Food Technology Entrepreneurship and Management, Thanjavur, Tamil Nadu, India

7.1 PROBLEMS

Regarding the study of eradication and management measures that can be applied in the food industry, it is necessary to understand the stored product insects. They are the major problem during the storage of food grains like rice, wheat, sorghum, barley, maize, pulses, beans and gram in the CWC (Central Warehousing Corporation), FCI (Food Corporation of India), CSC (Civil Supplies Corporation) and farm storage units since early times (Pourya et al. 2018).

Stored grain insects are classified into two types: internal and external feeders. The internal feeder infests the whole grains that have not been damaged, while the external feeder feeds on the grains, damaged grains and processed food products (Aulicky et al. 2015). Packaged food can be damaged by two types of stored product insects: penetrators and invaders (Stejskal et al. 2017; Arthur 2018). Penetrators are those insects that can make holes in the packaging material and infest the food commodity. Examples include *Sitophilus oryzae*, *Rhyzopertha dominica* and *Cadra cautella*. Invaders are those insects that infest the food commodity in the packaging material by invading through already existing holes, *viz. Tribolium castaneum*, *Tribolium confusum* and *Plodia interpunctella*. The majority of stored grain insects comprise beetles and moths. The growth of insects is largely influenced by the temperature and relative humidity of the product (Athanassiou et al. 2016). However, the growth of beetles is majorly influenced by the ambient temperature of the food commodity and moisture content rather than the humidity of the environment. The growth of moths is more dependent on the relative humidity of the commodity. The stored product insects originated from the tropical and sub-tropical regions of the world. However, these insects migrated to the temperate regions of the world during transportation and trading of stored food commodities between different countries. Insects have the least tendency to control their body temperatures. They tend to reproduce more rapidly at suitable temperatures. On the other hand, these insects get inactivated at lower temperatures (10–15°C). When the temperatures get chilled (0–4°C), they tend to die (Andreadis and Athanassiou 2017). They are unable to hibernate or remain inactivated except for *P. interpunctella* and *Trogoderma granarium*. The temperature range for the growth of insects differs for each species. The optimum temperature required for the highest survival of *T. castaneum* eggs to the adult ranges from 25 to 28°C.

7.1.1 Management Techniques

The use of chemical insecticides for eradication and prevention of insects from damaging the stored products is effective. However, studies on chemical insecticides were reported for their harmfulness to workers while handling. There are cases of resistance development to the chemicals by the stored product insects (Ramya, Srivastava, and Subramanian 2018; Malekpour et al. 2018). Many countries have started banning chemical insecticides like dichlorvos. There is a need for the development of

 DOI: 10.1201/9781003609834-7

new insecticides. The use of chemicals leaves residues on the grain. Consumption of these grains can have adverse effects on humans. The world has now diverted toward organic foods, and people have shifted their habits for a healthier lifestyle. During this era, the repeated use of chemical insecticides for the management of stored product insects seems reduntant.

Many researchers worked on physical methods like radio frequency and infrared disinfestation techniques. The development of hermetic storage bags is also a huge achievement. Every method has advantages and disadvantages. There are certain criteria that need to be met before using hermetic storage. The grains need to be properly dried at equilibrium moisture content, without which fungal growth occurs during the storage period. The changes that take place during its storage in hermetic bags cannot be monitored without breaking the packaged commodity.

Considering the health and environmental hazards given by chemical insecticides, bioinsecticides are a safer choice. It is the use of medicinal plants in making different plant-based insecticides. They were used traditionally in every house of farmers because of their effectiveness. Neem, chili and garlic bulbs are some of the commonly used medicinal plants for managing stored product insects by farmers. Research to screen different medicinal plants was also conducted to determine their efficacy against stored product insects. Regardless of the successful research, there is still a gap in the applicability of using medicinal plants in managing the stored product insects. The reason is the low duration of efficacy. To overcome such scenarios, research needs to be focused on the application of medicinal plants rather than the current trend of determining new active components in a medicinal plant. Medicinal plants with the potential to possess effective plant-based insecticides that are safe for humans is the key area that needs to be focused. In this book chapter, the scope of medicinal plants and their potential applicability in managing stored product insects were discussed.

7.2 BIO-BASED APPROACHES

Bioinsecticides for repelling the insects from the field crops are employed successfully on a large scale, and they are widely available in the market. Bioinsecticides are plant-based phytochemicals that act on insects without compromising the health of an individual being. Plants possess a range of phytochemicals that are used for pharmaceuticals, cosmetics and the agriculture industry. The phytochemicals may be carbohydrates, lipids, alkaloids, phenolics, terpenoids or a combination of these chemicals. They are Generally Regarded as Safe (GRAS). Most of these plants are aromatic and have the ability to be repellent to insects. They leave no residues. These plants have a long history of use as medicine, food and cosmetics. They are biodegradable. They possess a lesser amount of toxicity toward mammals and are safe for humans. They are natural substances of plant origin. Each phytochemical has variance in its structure, providing different functions and one such property is its toxicity to insects. They also possess either repellent, feeding deterrents, chemosterilants or growth-inhibiting effects based on different studies in laboratory studies. These properties were properly defined and described in Table 7.1. Traditionally, leaves, flowers, peel, bulbs, rhizome, roots or seeds of the plant or its extracts were applied near the grain either to the walls or to the grain bags during storage by farmers. However, these bioinsecticides are not employed yet in grain storage organizations for the management of stored product insects. The main reason is that continuous reapplication consumes more labor costs, making bioinsecticides inefficient in the long run. It is not easy to adopt this technology despite its effectiveness. Following this problem, it is of great importance to understand the mechanism that makes traditional ways effective for a short duration.

7.2.1 Mechanism

The mechanism is that the active component present in the plant possessed either repellent, toxicity, feeding deterrents, chemosterilants or growth-inhibiting effects based on different studies in the laboratory studies. Different plants have effective active components in different parts of the plant, either in the leaves, flowers, peel, bulb, rhizome, root or seeds, as showcased in Table 7.2.

TABLE 7.1
Classification of Bioinsecticides Based on Their Effective Properties

Sl. No.	Bio-Based Approaches	Definition
1	Repellents	Repellents are the volatile components undesired by the insects that act locally or at a distance, causing the insects to repel from the treated parts.
2	Antifeedants/Feeding deterrents	Antifeedants or 'feeding deterrents' are compounds that inhibit or disturb the feeding of insects by making the treated materials unpalatable.
3	Toxicity	Toxicity is of two types, *viz.* contact and fumigant toxicity. Contact toxicity means insects in direct contact with components of the plants, leading to mortality, while fumigant toxicity means insects in indirect contact with components of the plants (volatile), leading to mortality.
4	Reproduction inhibitors/ Chemosterilants	Reproduction inhibitors or 'chemosterilants' are the compounds that inhibit egg hatchability postembryonic development, leading to less progeny production.
5	Insect growth or development inhibitor	Insect growth inhibitors are compounds that inhibit the development of insects from one stage to another stage.

TABLE 7.2
Different Parts of the Plant Possess Insecticidal Properties

Scientific Name	Part of the Plant	Active Components	References
Tanacetum cinerariifolium	Dried flowers	Pyrethrum	Ramirez et al. (2013)
Azadirachta indica	Leaves	Azadirachtin	Biswas et al. (2002)
Chenopodium ambrosioides	Leaves	Carene, cymol, terpinene	Langsi et al. (2018), Dufera et al. (2019)
Sophora flavescens	Dry roots	Matrine, Quinolizidine	Bloomquist et al. (2018), Saleem et al. (2019)
Citrus sinensis	Peel	d-Limonene	El-Bakry et al. (2016)
Allium sativum	Bulb	Diallyl disulfide and diallyl sulfide	Meriga, Mopuri, and MuraliKrishna (2012)
Ocimum basilicum	Leaves	Estragole	Mishra, Tripathi, and Tripathi (2012)
Mentha spicata	Leaves	Carvone	Mackled et al. (2019); Pang et al. (2020)
Curcuma longa	Rhizome	Tumerone and curcumin	Abdelgaleil et al. (2019), Lithi et al. (2019)
Andrographis paniculate	Leaves	Andrographolide	Baliyarsingh, Mishra, and Rath (2021)
Veratrum nigrum	Roots	Celangulin and dihydroagarofuran	Vanichpakorn, Ding, and Cen (2010)
Cymbopogon citratus	Leaves	Citral	Hernandez-Lambrano et al. (2015), Feroz (2020)
Acorus calamus	Leaves	α and β-asarone	Abhijith, Mohan, and Nelson (2019), Dhivya et al. (2019), Dhivya (2019)
Annona squamosa and *Annona reticulata*	Leaves	Squamocin	Anita, Sujatha, and Prabhudas (2012), Malathi et al. (2019)

FIGURE 7.1 Screening studies conducted to identify the potential of plants in managing stored product insects. (Mangang et al. 2020.)

The active components can be identified through gas chromatography-mass spectroscopy (GC-MS) analysis. The active components are further individually extracted to cross-check their effectiveness against target insects. Many screening studies were conducted in various research to select the effective insecticide that can replace the use of chemical insecticides. This individual analysis is performed to precisely evaluate that the extracted active components from the effective plant are the main reason behind the effectiveness of the plant against the target insects. In a study, ten medicinal plants were screened by extracting the active components using three solvents (polar, mid-polar and non-polar), as shown in Figure 7.1. In the 21st century, plant products serve as the striving platform for the production of bioinsecticides, which may replace the need for chemicals or may act synergistically (Marrone 2019). However, these safe insecticides have not been used in stored product insect management since the duration of effectiveness is short in terms of a few days. This is due to limited control release technologies adopted and validation studies performed. Instead of screening novel plants and identifying new active components present in them, the era has come for utilizing the potentially proven effective plants and making them applicable using encapsulation.

7.3 REMEDIES

There are different encapsulation techniques available and discussed in detail by Anandharamakrishnan (2015). One such technique to increase the number of days that the bioinsecticides work is encapsulation using the emulsification technique. Emulsification is the dispersion of two immiscible liquids by stabilizing with the help of an emulsifier, thereby forming an emulsion. When the dispersed phase is oil and the dispersion medium is water, it is called oil in water emulsion. The concept is water; when mixed with protein, it forms a protein solution, and this protein solution binds to the hydrophilic end of the emulsifier. The oils that need to be encapsulated bind to the lipophilic end. Emulsifier also reduces the surface tension, thereby allowing the formation of new surfaces. Hereafter, when these immiscible liquids are centrifuged at a high speed, a larger number of emulsified oil droplets are formed, and the process is depicted in Figure 7.2. The solution is dried further using spray drying, freeze drying or any form of drying based on the desired outcome. Even the raw materials used can be modified or added. Hence, depending on the desired dependent values of the product, the independent variables and process can be optimized.

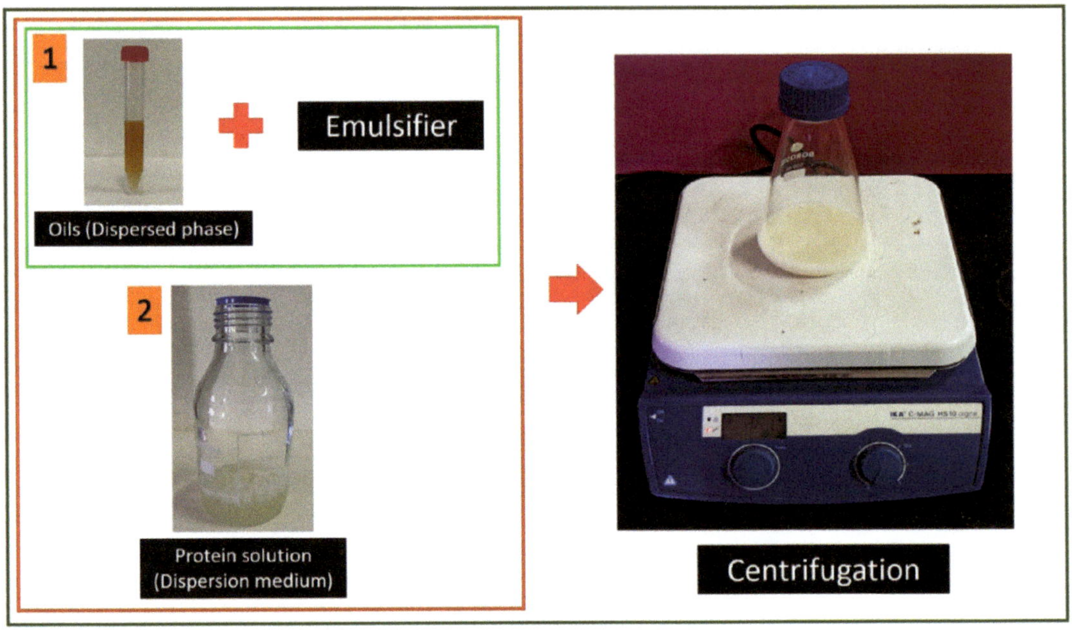

FIGURE 7.2 The basic process used in emulsification.

7.3.1 Nano-emulsions Studied for Control Release of Active Components

The use of bioinsecticides is a sustainable approach that might reduce the use of chemical insecticides. The problem that arises during the use of bioinsecticides is continuous reapplication to increase the duration of efficacy. This leads to an increase in labor. It is difficult to adopt this technology despite its effectiveness. Hence, researchers started working on producing nano-emulsions of plant-based insecticides for control release of the active components. Golden et al. (2018) encapsulated pulegone for control release and were characterized using spectroscopic and microscopic techniques. The nano-emulsions of pulegone were used for determining the insecticidal activity against rice weevil and red flour beetle. It was found that 90 percent mortality of the adults was observed using the formulated nano-emulsion for up to five weeks. This research showcased that the developed nano-emulsion can be used as a fumigant. Campolo et al. (2020) worked on increasing the stability of nano-emulsions formed using essential oils and surfactants. Tween 20, tween 80, span 20 and span 80 were the surfactants compared for the study. It was found that tween 80 gave the highest stability of up to 28 days. This research suggested that the selection of the right surfactant for nano-encapsulating the essential oil is the primary approach needed for the development of plant-based insecticides. Palermo et al. (2021) used commercially available essential oils of garlic, anise, fennel, artemisia, lavender, mint and rosemary to make nano-emulsions. The developed nano-emulsions were checked for repellency and acute toxicity effects against *T. confusum*. The developed nano-emulsions were effective in an area preference test conducted. However, the duration of efficacy remained a research gap. There is also a need for scale-up to meet the real scenario. Lima et al. (2021) studied plant-based nanoproducts as a sustainable alternative to the use of chemical insecticides. The study was conducted to observe the developed nano-emulsion of *Baccharis reticularia* essential oil for the management of red flour beetle. The droplet size of essential oil nano-emulsions was retained below 200 nm for up to 150 days. The repellent action of the developed nano-emulsions was effective; however, a study on the effective duration was unavailable. Nano-emulsions were made from fennel, mint and sweet orange and the formulation was tested against *R. dominica* (Giunti et al. 2021). The behavioral response of the developed nano-emulsion showed the repellent properties. However, the responsiveness of the adults was recovered after a 24-hour interval. The research suggested that motor and sensory fatigue when a new habituate is introduced could influence the study. Draz et al. (2022) conducted a study to check the insecticidal properties of nano-emulsions prepared from anise, thyme and coriander essential oils separately. The toxicity of the developed nano-emulsions was checked against *S. oryzae* and *T. castaneum.* It was found that nano-emulsion made from thyme essential oil was the most effective. Tanveer et al. (2022) worked on developing a new formulation that makes use of each essential oil (*Cymbopogon citratus*, *Ocimum basilicum*, *Curcuma longa* and *Trachyspermum ammi*) with diatomaceous earth separately. The developed emulsion was checked for its toxicity against *T. castaneum.* It was found that the use of emulsion increased the toxicity in a 72-hour study. Another study was conducted using citronella essential oil emulsified to increase efficacy against *Oryzaephilus surinamensis* (Gharsan et al. 2022). The study revealed that the developed formulation was twice as effective as that of pure oil. The research recommends that the formulation is an alternative to chemical insecticides. However, the toxicity study needs to be conducted for longer days to clarify the doubt of increased duration in terms of efficacy. These plants have potential applicability in the food industry. Some of the bio-approaches include the use of these potential plants in making packaging material, nettings and insect-repellent pellets.

7.3.2 Packaging Material

Anti-insect packaging is an important part that needs to be discussed while looking at packaging material. A packaging material needs to have this feature under the primary function which is protection. It is not the protection from the air or moisture but from the insects. Sometimes, already-infested cereals can seldom get detected if insects are present in the early stage (egg). Possessing a small amount of egg can lead to serious infestation, rendering the packaged food product unfit for consumption. Insects can penetrate the packaging material from the outside, too. However, a properly packaged food product is unavailable for invasion by insects if properly stored in a good packaging container like glass or tin. Even though these containers are available, the affordability of the packaged goods by the consumers was considered while manufacturing the goods. Hence, different polyethylene bags were used to store food commodities. Earlier studies, as shown in (Table 7.3),

TABLE 7.3
Recent Advances in Essential Oil Incorporated Chitosan Films for Active Packaging

Essential Oil	The Potential of Developed Film	Reference
Thyme	Antimicrobial	Lian et al. (2019)
Apricot kernel	Antioxidant and antimicrobial	Priyadarshi et al. (2018)
Basil	Antimicrobial	Amor et al. (2021)
Eugenol and ginger	Antioxidant	Bonilla et al. (2018)
Clove	Antioxidant	Lee, Kim, and Park (2018)
Mandarin lime	Antimicrobial	de Oliveira Filho et al. (2020)
Apricot kernel	Antimicrobial	Wang et al. (2020)
Fennel and peppermint	Antioxidant	Liu et al. (2020)
Nisin and perilla	Antioxidant and antimicrobial	Wang et al. (2021)
Cinnamon	Antimicrobial	Xu et al. (2021)
Ginger	Antioxidant	Al-Ali, Al-Hilifi, and Rashed (2021)
Rosemary	Antimicrobial	Walid et al. (2022)
Cinnamaldehyde and thymol	Antioxidant and antimicrobial	Bai et al. (2022)

mainly focused on the use of medicinal plants with the only aim of increasing the antimicrobial and antioxidant effects. Considering the damage that can be caused by insect infestation, there is a need to produce anti-insect packaging.

7.3.2.1 Some Approaches to Producing Anti-Insect Packaging Film

Hou, Fields, and Taylor (2004) compared packaging materials treated with neem, protein-enriched pea flour, defatted pea flour, pea protein extract and DEET for their repellent property, inhibiting penetration and invasion activity against *S. oryzae*, *T. castaneum*, *Cryptolestes ferrugineus* and *O. surinamensis*. It was found that DEET was better than the plants. However, in another study by Wong et al. (2005), carton boxes with zeolite and citronella prevented insect infestation up to 50 percent, which was pretty low. Similar repellent action ranging from 53 to 87 percent was observed during an area preference test against *T. castaneum* (Licciardello et al. 2013). The quantity and active components of the oil (citronella, rosemary, or oregano) played an important role during the analysis.

Chung et al. (2013) encapsulated the thyme oils by polymerization using pre-polymer treated with emulsifiers like tween 80 and pluronic F-127 as coating material and evaluated for the repellent behavior of packaging materials. It was found that the releasing profile of thyme oil depends on the emulsifier and the microencapsulating material. The food commodity, which was kept together with the encapsulated thyme oil, expressed the highest repellency of 90 percent for four weeks. Kim et al. (2013) worked on insect-repellent film. The encapsulation was achieved by incorporating cinnamon oil in gum arabic, whey protein isolates or polyvinyl alcohol and coating separately to form insect-repellent packaging material. The microencapsulation efficiency was determined by checking the repellent effect against *P. interpunctella*. The result showed that cinnamon oil encapsulated with whey protein isolate gave the highest repellent activity of 94.6 percent followed by gum arabic with 90.4 percent. There are other wall materials available for the encapsulation of different active components as shown in Figure 7.3.

Even though many plant-based insecticides were screened (Kamaraj, Rahuman, and Bagavan 2008; Arivoli and Tennyson 2013; Mangang et al. 2020; Subrahmaniyan, Mangang, and Manickam 2022), the use of these insecticides for the development of packaging materials was minimal. There are different researches with successful incorporation of essential oils in the developed film with the potential to be used for antimicrobial and antioxidant effects as given in Table 7.3. These essential oils were already proven effective in many screening studies performed against insects (Negahban, Moharramipour, and Sefidkon 2007; Kim and Lee 2014; Plata-Rueda et al. 2018; Barros et al. 2022). These developed packaging materials can be analyzed against different stored product insects for their effectiveness in protecting processed food products.

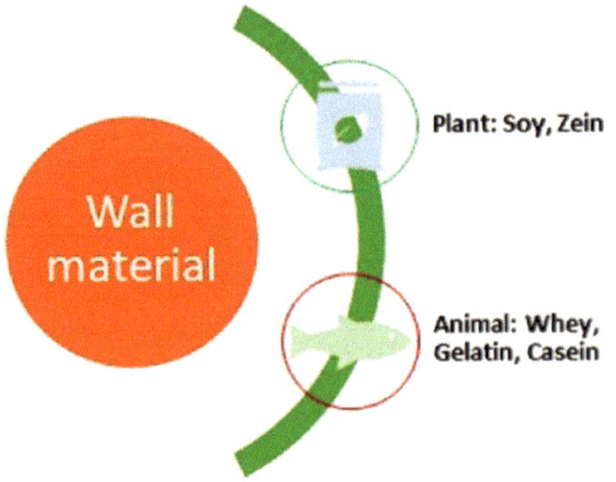

FIGURE 7.3 Proteinaceous wall materials available for emulsification. (Suyanto et al. 2019.)

7.3.3 Nettings

The use of nettings was basically for the prevention of malaria, a disease caused by mosquito bites. Long-lasting insecticide nettings (LLIN) coated with pyrethroid worked against *Anopheles* mosquitoes by providing physical and chemical barriers. The nettings not only protect humans from malaria, they also kill the mosquitoes exposed to the nettings. However, the LLIN induced resistance since the *Anopheles* mosquito evolves over time (Wondji et al. 2012). A second-generation LLIN was introduced with the addition of piperonyl butoxide (PBO) to the first-generation LLIN. PBO is an oil that reduces the defense system of mosquitoes, thereby making pyrethroid more effective. There is still a search for more effective chemicals to be used along with LLIN. These LLINs obtained commercial success. Hence, different research studies were conducted to protect grains from stored product insects using LLIN.

7.3.3.1 Some Approaches to Producing Anti-Insect Nettings

Msango and Longwe (2013) worked on developing an effective LLIN to cover the phosphine-fumigated area of a warehouse in Malawi for controlling tobacco beetle, Lasioderma *serricorne*. Traps were used to check the level of infestation after fumigation, determining the ability of LLIN to prevent tobacco beetle. The results showed the effectiveness of LLIN. Morrison Iii and Wilkins (2018) studied the effect of LLIN on the behavior of *R. dominica* and *T. castaneum*. The adults were exposed to LLIN for a period of time in a glass petri plate, and their flight activity was separately observed. The results described that the dispersal of the insects was reduced after its exposure. The mobility and dispersal of *T. castaneum* and *Trogoderma variabile* were also determined when exposed to LLIN (Wilkins et al. 2020). In this study, the larvae and adults were analyzed by video tracking their movement in a PVC pipe after exposure to LLIN for a few minutes. The results showed that the larval stage was more tolerant to LLIN than the adults. The results determined that the migration of insects can be reduced significantly. However, the insects were forcefully exposed to these settings for 1, 5 and 10 min, wherein the insects will move away from an unwanted substance in a few seconds in real case scenario.

Wilkins et al. (2021) conducted another research to prove that LLIN is effective for the management of insects. They reported that LLIN can be used for intercepting immigration of insects and can be used to cover gaps in the storage house (like windows and vents). Another possibility is to attract and kill using LLIN along with traps. The study determined (1) the ability of the trap to capture the insects, (2) whether LLIN prevented the stored product insects from migration and (3) the success rate of using LLIN alone, trap alone or combined. The study gave a clear representation that over 3000 insects were captured in the interception traps over a period of 48 hours as observed for over two years. The insects include 14 stored product insects. The research suggested that the effect of LLIN may influence the ability of insects to reach certain heights of the traps. LLIN also brought a reduction in progeny production by 98–100 percent. Since the use of LLIN in stored product insects has elevated in the past few years, Scheff et al. (2021) incorporated deltamethrin in LLIN. The effect of the developed LLIN was observed for its effectiveness against three stored product insects *viz. T. castaneum*, *R. dominica* and *S. oryzae*. In a wind tunnel assay, the repellent activity was tested, wherein the results showed ineffectiveness against each species. The movement assay was also determined by exposing for 5 min in deltamethrin-treated LLIN and found that *T. castaneum* was the most resistive insect traveling the most distance at high velocity, and they were found to be highly mobile. The mortality of insects was also tested using the developed LLIN after exposure for 60 min. After seven days of observation with and without food, 42 and 95 percent of adult *T. castaneum* were found dead, respectively. However, 52 percent of *T. castaneum* adults recovered after keeping along with food. The research suggested that the insect's duration of lethal exposure to LLIN was increased as the developed LLIN was devoid of repellent activity. Nettings have to be significantly improved as stored product insects are minute and are able to pass through it. The question is whether the insects die even if they happen to cross the nettings.

In a recent study, Mangang et al. (2022) formulated an emulsion of neem oil and thyme essential oil and incorporated it to form models of bioinsecticide nettings. These nettings were called insect-repellent nettings (IRNs). The development of insect-repellent nettings showed that nettings could prevent the movement of insects by up to 70 percent. It has repellent properties; however, *T. castaneum* adults were minute enough to pass through the holes of the nets. It can prevent insect movement from another facility by acting as a primary barrier. However, further studies are needed to incorporate better bioinsecticides that possess contact toxicity to the stored product insects.

7.3.4 Novel Bio-Based Approach

There is a wide applicability of naphthalene balls against microbes and insects (Gurmit 1987; Supriya, Patra, and Bisweswar 2010). They are used to keep the garments away from termites and cockroaches and also in washrooms. Studies were there for the management of stored product insects using these naphthalene balls (Pajaro-Castro, Caballero-Gallardo, and Olivero-Verbel 2017). However, cases of naphthalene ball poisoning causing methemoglobinemia (Kapoor et al. 2014; Deo et al. 2016) and other acute toxicity make it unsafe for use as a repellent in stored product pest management. In a recent study, insect-repellent pellets (IRPs) were developed using bioinsecticides (Mangang et al. 2022), as shown in Figure 7.4. It had the potential to repel insects by hindering insect movement toward the grains. The oils of lemongrass, eucalyptus and neem leaves were encapsulated for the controlled release of fumes, avoiding the need for reapplication after a few days. The active components (phenol, 2,4-di-tert-butyl-, citral, neral, geraniol, n-hexadecanoic acid) present in IRP during the initial stage were also found after a storage period of 35 days. The active components present in IRPs have antifungal, repellent and fumigant properties. It can be used in a grain storage system by keeping it in a sachet and storing it along with the grains. The IRPs can thus be termed potent botanical insecticides and are an alternative to synthetic insecticides.

7.4 CONCLUSION

Even though processed food products are packaged in glass or tin, the food can still be infested with insects if there are small numbers of minute eggs present during packaging. The eggs will gradually turn into adults and then produce offspring, which will eventually infest the whole food product inside the packaging before the packaged good is opened by the consumer. The question is, who would eat a processed food commodity if it is already consumed by the insect? The use of packaging films incorporated with plant-based insecticides is the potential method that has

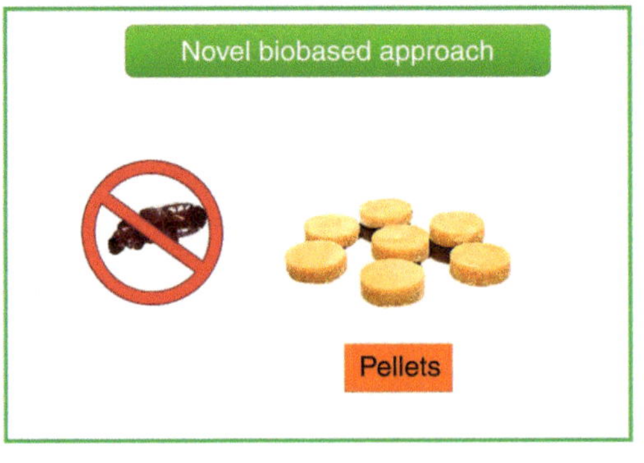

FIGURE 7.4 Insects repellent pellets which possessed repellent, fumigant and anti-fungal properties.

reproduction-inhibiting, growth-regulating effects without affecting the health of the consumers. The research work carried out on the use of chitosan is broad in terms of food packaging applications. It was either used as part of antimicrobial or antioxidant potential. Future research needs to be conducted using these packaging materials to solve the problems that occur with stored products. There are also scopes for producing insect-repellent pellets in the market if the study can be scaled up.

REFERENCES

Abdelgaleil, S. A. M., A. El-Bakry, A. A. M. Zoghroban, and S. M. I. Kassem. 2019. "Insecticidal and Antifungal Activities of Crude Extracts and Pure Compounds from Rhizomes of *Curcuma longa* L. (Zingiberaceae)." *Journal of Agricultural Science and Technology* 21 (4): 1049–1061.

Abhijith, N., S. Mohan, and S. J. Nelson. 2019. "Insecticidal Activity of Sweet Flag *Acorus calamus* Against *Lasioderma serricorne* (F.) and *Cryptolestes ferrugineus* (Stephens)." *Indian Journal of Entomology* 81 (1): 66–68.

Al-Ali, R. M., S. A. Al-Hilifi, and M. Rashed. 2021. "Fabrication, Characterization, and Anti-Free Radical Performance of Edible Packaging-Chitosan Film Synthesized from Shrimp Shell Incorporated with Ginger Essential Oil." *Journal of Food Measurement and Characterization* 15 (4): 2951–2962.

Amor, G., M. Sabbah, L. Caputo, M. Idbella, V. De Feo, R. Porta, T. Fechtali, and G. Mauriello. 2021. "Basil Essential Oil: Composition, Antimicrobial Properties, and Microencapsulation to Produce Active Chitosan Films for Food Packaging." *Foods* 10 (1): 121.

Anandharamakrishnan, C.. 2015. *Spray Drying Techniques for Food Ingredient Encapsulation*. John Wiley & Sons.

Andreadis, S. S., and C. G. Athanassiou. 2017. "A Review of Insect Cold Hardiness and Its Potential in Stored Product Insect Control." *Crop Protection* 91: 93–99.

Anita, S., P. Sujatha, and P. Prabhudas. 2012. "Efficacy of Pulverised Leaves of *Annona squamosa* (L.), *Moringa oleifera* (Lam.) and *Eucalyptus globulus* (Labill.) Against the Stored Grain Pest, *Tribolium castaneum* (Herbst.)." *Recent Research in Science and Technology* 4 (2).

Arivoli, S., and S. Tennyson. 2013. "Screening of Plant Extracts for Oviposition Activity Against *Spodoptera litura* (Fab).(Lepidoptera: Noctuidae)." *International Journal of Fauna and Biological Studies* 1 (1): 20–24.

Arthur, F. H.. 2018. "Structural Pest Management for Stored Product Insects." In *Recent Advances in Stored Product Protection*, 65–81. Springer.

Athanassiou, C. G., N. G. Kavallieratos, A. Chiriloaie, T. N. Vassilakos, V. Fatu, S. Drosu, M. Ciobanu, and R. Dudoiu. 2016. "Insecticidal Efficacy of Natural Diatomaceous Earth Deposits from Greece and Romania Against Four Stored Grain Beetles: the Effect of Temperature and Relative Humidity." *Bulletin of Insectology* 69 (1): 25–34.

Aulicky, R., V. Stejskal, Z. Kucerova, and P. Trematerra. 2015. "Trapping of Internal and External Feeding Stored Grain Beetle Pests with Two Types of Pitfall Traps: A Two-Year Field Study." *Plant Protection Science* 52 (1): 45–53.

Bai, M.-Y., Q. Zhou, J. Zhang, T. Li, J. Cheng, Q. Liu, W.-R. Xu, and Y.-C. Zhang. 2022. "Antioxidant and Antibacterial Properties of Essential Oils-Loaded β-Cyclodextrin-Epichlorohydrin Oligomer and Chitosan Composite Films." *Colloids and Surfaces B: Biointerfaces* 215: 112504.

Baliyarsingh, B., A. Mishra, and S. Rath. 2021. "Evaluation of Insecticidal and Repellency Activity of Leaf Extracts of *Andrographis paniculata* Against *Tribolium castaneum* (Red Flour Beetle)." *International Journal of Tropical Insect Science* 41 (1): 765–773.

Barros, F. A. P., M. Radünz, M. A. Scariot, T. M. Camargo, C. F. P. Nunes, R. R. de Souza, I. K. Gilson, H. C. S. Hackbart, L. L. Radünz, and J. V. Oliveira. 2022. "Efficacy of Encapsulated and non-Encapsulated Thyme Essential Oil (*Thymus vulgaris* L.) in the Control of *Sitophilus Zeamais* and Its Effects on the Quality of Corn Grains Throughout Storage." *Crop Protection* 153: 105885.

Biswas, K., I. Chattopadhyay, R. K. Banerjee, and U. Bandyopadhyay. 2002. "Biological Activities and Medicinal Properties of Neem (*Azadirachta indica*)." *Current Science* 82: 1336–1345.

Bloomquist, J. R., S. Jiang, J. Taylor-Wells, L. Yang, and Y.-X. Li. 2018. "Insecticidal Activity and Physiological Actions of Matrine, a Plant Natural Product." In *Advances in the Biorational Control of Medical and Veterinary Pests*, 175–186. ACS Publications.

Bonilla, J., T. Poloni, R. V. Lourenço, and P. J. A. Sobral. 2018. "Antioxidant Potential of Eugenol and Ginger Essential Oils with Gelatin/Chitosan Films." *Food Bioscience* 23: 107–114.

Campolo, O., G. Giunti, M. Laigle, T. Michel, and V. Palmeri. 2020. "Essential Oil-Based Nano-Emulsions: Effect of Different Surfactants, Sonication and Plant Species on Physicochemical Characteristics." *Industrial Crops and Products* 157: 112935.

Chung, S. K., J. Y. Seo, J. H. Lim, H. H. Park, M. J. Yea, and H. J. Park. 2013. "Microencapsulation of Essential Oil for Insect Repellent in Food Packaging System." *Journal of Food Science* 78 (5): E709–E714.

de Oliveira Filho, J. G., I. P. B. de Deus, A. C. Fernandes Valadares, C. C. Fernandes, E. B. B. Estevam, and M. B. Egea. 2020. "Chitosan Film with *Citrus limonia* Essential Oil: Physical and Morphological Properties and Antibacterial Activity." *Colloids and Interfaces* 4 (2): 18.

Deo, P., K. K. Sahu, D. P. Dhibar, and S. C. Varma. 2016. "Naphthalene Ball Poisoning: a Rare Cause of Acquired Methaemoglobinaemia." *Case Reports* 2016:bcr2016215102.

Dhivya, V. 2019. "Development of *Acorus calamus* L. Nano Emulsion and Their Insecticidal Activity Against Pulse Beetle (*Callosobruchus maculatus* F.)." *Inter International Journal of Agriculture Sciences* 7: 0975–3710.

Dhivya, V., S. J. Nelson, K. S. Subramanian, Y. S. J. T. Edward, K. Rajamani, V. P. Santhanakrishnan, and S. Sithanantham. 2019. "Formulation of Sweet Flag Oil (*Acorus calamus*) Nanoemulsion by Spontaneous Emulsification Method for the Management of *Sitophilus oryzae*." *International Journal of Chemical Studies* 7: 2072–2076.

Draz, K. A., R. M. Tabikha, M. I. Eldosouky, A. A. Darwish, and M. Abdelnasser. 2022. "Biotoxicity of Essential Oils and Their Nano-Emulsions Against the Coleopteran Stored Product Insect Pests *Sitophilus oryzae* L. and *Tribolium castaneum* Herbst." *International Journal of Pest Management* 70: 1–15.

Dufera, L. T., A. Y. Tadesse, W. S. Gobena, and C. G. Kuyu. 2019. "On Farm Evaluation of *Eucalyptus globulus* Labill Leaf and *Chenopodium ambrosioides* L. Whole Plant Powder Against Storage Insect Pests in Stored Maize at Sokoru District in Jimma Zone of Oromia Regional State, Ethiopia." *Psyche* 2019: 1–7.

El-Bakry, A. M., N. F. Abdel, Aziz, E. A. Sammour, and S. A. M. Abdelgaleil. 2016. "Insecticidal Activity of Natural Plant Essential Oils Against Some Stored Product Insects and Their Side Effects on Wheat Seed Germination." *Egyptian Journal of Biological Pest Control* 26 (1): 83.

Feroz, A. 2020. "Efficacy and Cytotoxic Potential of Deltamethrin, Essential Oils of *Cymbopogon citratus* and *Cinnamonum camphora* and Their Synergistic Combinations Against Stored Product Pest, *Trogoderma granarium* (Everts)." *Journal of Stored Products Research* 87: 101614.

Gharsan, F. N., W. M. Kamel, T. S. Alghamdi, A. A. Alghamdi, A. O. Althagafi, F. J. Aljassim, and S. N. Al-Ghamdi. 2022. "Toxicity of Citronella Essential Oil and Its Nanoemulsion Against the Sawtoothed Grain Beetle *Oryzaephilus surinamensis* (Coleoptera: Silvanidae)." *Industrial Crops and Products* 184: 115024.

Giunti, G., O. Campolo, F. Laudani, L. Zappalà, and V. Palmeri. 2021. "Bioactivity of Essential Oil-Based Nano-Biopesticides Toward *Rhyzopertha dominica* (Coleoptera: Bostrichidae)." *Industrial Crops and Products* 162: 113257.

Golden, G., E. Quinn, E. Shaaya, M. Kostyukovsky, and E. Poverenov. 2018. "Coarse and Nano Emulsions for Effective Delivery of the Natural Pest Control Agent Pulegone for Stored Grain Protection." *Pest Management Science* 74 (4): 820–827.

Gurmit, S. 1987. "Naphthalene Balls for the Protection of Coconuts and Oil Palms Against *Oryctes rhinoceros* (L.)." *Planter* 63 (736): 286–292.

Hernandez-Lambrano, R., N. Pajaro-Castro, K. Caballero-Gallardo, E. Stashenko, and J. Olivero-Verbel. 2015. "Essential Oils from Plants of the Genus Cymbopogon as Natural Insecticides to Control Stored Product Pests." *Journal of Stored Products Research* 62: 81–83.

Hou, X., P. Fields, and W. Taylor. 2004. "The Effect of Repellents on Penetration into Packaging by Stored-Product Insects." *Journal of Stored Products Research* 40 (1): 47–54.

Kamaraj, C., A. A. Rahuman, and A. Bagavan. 2008. "Screening for Antifeedant and Larvicidal Activity of Plant Extracts Against *Helicoverpa armigera* (Hübner), *Sylepta derogata* (F.) and *Anopheles stephensi* (Liston)." *Parasitology Research* 103 (6): 1361–1368.

Kapoor, R., P. Suresh, S. Barki, M. Mishra, and M. K. Garg. 2014. "Acute Intravascular Hemolysis and Methemoglobinemia Following Naphthalene Ball Poisoning." *Indian Journal of Hematology and Blood Transfusion* 30 (1): 317–319.

Kim, I. H., J. Han, J. H. Na, P. S. Chang, M. S. Chung, K. H. Park, and S. C. Min. 2013. "Insect-Resistant Food Packaging Film Development Using Cinnamon Oil and Microencapsulation Technologies." *Journal of Food Science* 78 (2): E229–E237.

Kim, S.-I., and D.-W. Lee. 2014. "Toxicity of Basil and Orange Essential Oils and Their Components Against Two Coleopteran Stored Products Insect Pests." *Journal of Asia-Pacific Entomology* 17 (1): 13–17.

Langsi, D. J., C. N. Fokunang, C. Suh, K. Eloh, P. Caboni, and E. N. Nukenine. 2018. "Insecticidal Activity of Essential Oils of *Chenopodium ambrosioides* and *Cupressus sempervirens* and Their Binary Combinations on *Sitophilus zeamais*." *GSC Biological and Pharmaceutical Sciences* 3 (2).

Lee, M. H., S. Y. Kim, and H. J. Park. 2018. "Effect of Halloysite Nanoclay on the Physical, Mechanical, and Antioxidant Properties of Chitosan Films Incorporated with Clove Essential Oil." *Food Hydrocolloids* 84: 58–67.

Lian, H., Y. Peng, J. Shi, and Q. Wang. 2019. "Effect of Emulsifier Hydrophilic-Lipophilic Balance (HLB) on the Release of Thyme Essential Oil from Chitosan Films." *Food Hydrocolloids* 97: 105213.

Licciardello, F., G. Muratore, P. Suma, A. Russo, and C. Nerín. 2013. "Effectiveness of a Novel Insect-Repellent Food Packaging Incorporating Essential Oils Against the Red Flour Beetle (*Tribolium castaneum*)." *Innovative Food Science & Emerging Technologies* 19: 173–180.

Lima, L. A., P. S. Ferreira-Sa, M. D. N. Garcia Jr, V. L. P. Pereira, J. C.T. Carvalho, L. Rocha, C. P. Fernandes, R. N. P. Souto, R. S. Araujo, and G. Botas. 2021. "Nano-Emulsions of the Essential Oil of *Baccharis reticularia* and Its Constituents as Eco-Friendly Repellents Against *Tribolium castaneum*." *Industrial Crops and Products* 162: 113282.

Lithi, U. J., M. Faridullah, V. C. Roy, K. C. Roy, and A. K. M. N. Alam. 2019. "Efficiency of Organic Pesticides, Turmeric (*Curcuma longa*) and Neem (*Azadirachta indica*) Against Dry Fish Beetle (*Dermestes* sp.) During Storage Condition." *Journal of the Bangladesh Agricultural University* 17 (1): 110–116.

Liu, T., J. Wang, F. Chi, Z. Tan, and L. Liu. 2020. "Development and Characterization of Novel Active Chitosan Films Containing Fennel and Peppermint Essential Oils." *Coatings* 10 (10): 936.

Mackled, M. I., M. EL-Hefny, M. Bin-Jumah, T. F. Wahba, and A. A. Allam. 2019. "Assessment of the Toxicity of Natural Oils from *Mentha piperita, Pinus roxburghii*, and *Rosa* spp. Against Three Stored Product Insects." *Processes* 7 (11): 861.

Malathi, S., G. Rameshkumar, R. L. Rengarajan, T. Rajagopal, S. Muniasamy, and P. Ponmanickam. 2019. "Phytofabrication of Silver Nanoparticles Using *Annona reticulata* and Assessment of Insecticidal and Bactericidal Activities." *Journal of Environmental Biology* 40 (4): 626–633.

Malekpour, R., M. A. Rafter, G. J. Daglish, and G. H. Walter. 2018. "The Movement Abilities and Resource Location Behaviour of *Tribolium castaneum*: Phosphine Resistance and Its Genetic Influences." *Journal of Pest Science* 91: 739–749.

Mangang, I. B., K. A. Appugol, R. Vidhya Lakshmi, V. Meenatchi, and L. Manickam. 2022. "Insect Repellent Nettings: An Application of Plant-Based Insecticides: Preventing the Dispersal of the Red Flour Beetle." *Journal of the Science of Food and Agriculture* 102 (14): 6696–6706.

Mangang, I. B., A. Tiwari, M. Rajamani, and L. Manickam. 2020. "Comparative Laboratory Efficacy of Novel Botanical Extracts Against *Tribolium castaneum*." *Journal of the Science of Food and Agriculture* 100 (4): 1541–1546.

Marrone, P. G. 2019. "Pesticidal Natural Products–Status and Future Potential." *Pest Management Science* 75 (9): 2325–2340.

Meriga, B., R. Mopuri, and T. MuraliKrishna. 2012. "Insecticidal, Antimicrobial and Antioxidant Activities of Bulb Extracts of *Allium sativum*." *Asian Pacific Journal of Tropical Medicine* 5 (5): 391–395.

Mishra, B. B., S. P. Tripathi, and C. P. M. Tripathi. 2012. "Repellent Effect of Leaves Essential Oils from *Eucalyptus globulus* (Mirtaceae) and *Ocimum basilicum* (Lamiaceae) Against Two Major Stored Grain Insect Pests of Coleopterons." *Nature and Science* 10 (2): 50–54.

Morrison III, W. R., and R. V. Wilkins. 2018. "The Use of Long-Lasting Insecticide Netting to Prevent Dispersal of Stored Product Insects." *Julius-Kühn-Archiv* 463: 172–176.

Msango, K. R., and L. L. Longwe. 2013. "Efficacy of Long-Lasting Insecticidal Net (LLIN) in Controlling Tobacco Beetle (*Lasioderma Serricorne*) in Malawi." *Tobacco Science* 50 (50): 31–33.

Negahban, M., S. Moharramipour, and F. Sefidkon. 2007. "Fumigant Toxicity of Essential Oil from *Artemisia sieberi* Besser Against Three Stored-Product Insects." *Journal of Stored Products Research* 43 (2): 123–128.

Pajaro-Castro, N., K. Caballero-Gallardo, and J. Olivero-Verbel. 2017. "Toxicity of Naphthalene and Benzene on *Tribolium castaneum* Herbst." *International Journal of Environmental Research and Public Health* 14 (6): 667.

Palermo, D., G. Giunti, F. Laudani, V. Palmeri, and O. Campolo. 2021. "Essential Oil-Based Nano-Biopesticides: Formulation and Bioactivity Against the Confused Flour Beetle *Tribolium confusum*." *Sustainability* 13 (17): 9746.

Pang, X., Y.-X. Feng, X.-J. Qi, Y. Wang, B. Almaz, C. Xi, and S.-S. Du. 2020. "Toxicity and Repellent Activity of Essential Oil from *Mentha piperita* Linn. Leaves and Its Major Monoterpenoids Against Three Stored Product Insects." *Environmental Science and Pollution Research* 27 (7): 7618–7627.

Plata-Rueda, A., J. M. Campos, G. da Silva Rolim, L. C. Martínez, M. H. Dos Santos, F. L. Fernandes, J. E. Serrão, and J. C. Zanuncio. 2018. "Terpenoid Constituents of Cinnamon and Clove Essential Oils Cause Toxic Effects and Behavior Repellency Response on Granary Weevil, *Sitophilus granarius*." *Ecotoxicology and Environmental Safety* 156: 263–270.

Pourya, M., A. Sadeghi, H. Ghobari, C. N. T. Taning, and G. Smagghe. 2018. "Bioactivity of *Pistacia atlantica* Desf. Subsp. Kurdica (Zohary) Rech. F. and *Pistacia khinjuk* Stocks Essential Oils Against *Callosobruchus maculatus* (F, 1775) (Coloeptera: Bruchidae) Under Laboratory Conditions." *Journal of Stored Products Research* 77: 96–105.

Priyadarshi, R., B. Kumar, F. Deeba, A. Kulshreshtha, and Y. S. Negi. 2018. "Chitosan Films Incorporated with Apricot (*Prunus armeniaca)* Kernel Essential Oil as Active Food Packaging Material." *Food Hydrocolloids* 85: 158–166.

Ramirez, A. M., N. Saillard, T. Yang, M. C. R. Franssen, H. J. Bouwmeester, and M. A. Jongsma. 2013. "Biosynthesis of Sesquiterpene Lactones in Pyrethrum (*Tanacetum cinerariifolium*)." *PLoS One* 8 (5): e65030.

Ramya, R. S., C. Srivastava, and S. Subramanian. 2018. "Monitoring of Phosphine Resistance in Indian Populations of *Tribolium castaneum* (Herbst) from Stored Wheat." *Indian Journal of Entomology* 80 (1): 19–23.

Saleem, M. S., T. S. Batool, M. F. Akbar, S. Raza, and S. Shahzad. 2019. "Efficiency of Botanical Pesticides Against Some Pests Infesting Hydroponic Cucumber, Cultivated Under Greenhouse Conditions." *Egyptian Journal of Biological Pest Control* 29 (1): 1–7.

Scheff, D. S., A. R. Gerken, W. R. Morrison, J. F. Campbell, F. H. Arthur, and K. Y. Zhu. 2021. "Assessing Repellency, Movement, and Mortality of Three Species of Stored Product Insects after Exposure to Deltamethrin-Incorporated Long-Lasting Polyethylene Netting." *Journal of Pest Science* 94 (3): 885–898.

Stejskal, V., M. Bostlova, M. Nesvorna, V. Volek, V. Dolezal, and J. Hubert. 2017. "Comparison of the Resistance of Mono-and Multilayer Packaging Films to Stored-Product Insects in a Laboratory Test." *Food Control* 73: 566–573.

Subrahmaniyan, M., I. B. Mangang, and L. Manickam. 2022. "Evaluation of Antioxidant Properties, Total Phenolic and Flavonoid Content of Six Medicinal Plants." *Uttar Pradesh Journal of Zoology* 43: 32–39.

Supriya, B., P. S. Patra, and M. Bisweswar. 2010. "Efficacy of Some Vegetable Oils and Naphthalene in Storage Condition Against *Callosobruchus chinensis* L. Infestation." *Green Farming* 1 (2): 174–176.

Suyanto, A., E. Noor, M. S. Rusli, and F. Fahma. 2019. "Nano-emulsion and Nano-encapsulation of Fruit Flavor." International Conference on Food Science & Technology, IOP Conf. Series: Earth and Environmental Science 292: 012025

Tanveer, M., S. Ejaz, S. M. Zaka, M. Batool, T. Zahra, Q. Muhammad Saghir, and Saeed. 2022. "Toxicology of Diatomaceous Earth, Phyto Oils and Their Admixed Emulsions Against Adults of *Tribolium castaneum* (Herbst)." *Toxicology Reports* 9: 1172–1179.

Vanichpakorn, P., W. Ding, and X.-X. Cen. 2010. "Insecticidal Activity of Five Chinese Medicinal Plants Against *Plutella xylostella* L. Larvae." *Journal of Asia-Pacific Entomology* 13 (3): 169–173.

Walid, Y., N. Malgorzata, R. Katarzyna, B. Piotr, O. L. Ewa, B. Izabela, A. W. Wissem, H. Majdi, J. Slim, and H. N. Karima. 2022. "Effect of Rosemary Essential Oil and Ethanol Extract on Physicochemical and Antibacterial Properties of Optimized Gelatin–Chitosan Film Using Mixture Design." *Journal of Food Processing and Preservation* 46 (1): e16059.

Wang, D., Y. Dong, X. Chen, Y. Liu, J. Wang, X. Wang, C. Wang, and H. Song. 2020. "Incorporation of Apricot (*Prunus armeniaca*) Kernel Essential Oil into Chitosan Films Displaying Antimicrobial Effect Against *Listeria Monocytogenes* and Improving Quality Indices of Spiced Beef." *International Journal of Biological Macromolecules* 162: 838–844.

Wang, H., L. Guo, L. Liu, B. Han, and X. Niu. 2021. "Composite Chitosan Films Prepared Using Nisin and *Perilla frutescense* Essential Oil and Their Use to Extend Strawberry Shelf Life." *Food Bioscience* 41: 101037.

Wilkins, R. V., J. F. Campbell, K. Y. Zhu, L. A. Starkus, T. McKay, and W. R. Morrison III. 2021. "Long-Lasting Insecticide-Incorporated Netting and Interception Traps at Pilot-Scale Warehouses and Commercial Facilities Prevents Infestation by Stored Product Beetles." *Frontiers in Sustainable Food Systems* 4: 561820.

Wilkins, R. V., K. Y. Zhu, J. F. Campbell, and W. R. Morrison III. 2020. "Mobility and Dispersal of Two Cosmopolitan Stored-Product Insects Are Adversely Affected by Long-Lasting Insecticide Netting in a Life Stage-Dependent Manner." *Journal of Economic Entomology* 113 (4): 1768–1779.

Wondji, C. S., M. Coleman, I. Kleinschmidt, T. Mzilahowa, H. Irving, M. Ndula, A. Rehman, J. Morgan, K. G. Barnes, and J. Hemingway. 2012. "Impact of Pyrethroid Resistance on Operational Malaria Control in Malawi." *Proceedings of the National Academy of Sciences* 109 (47): 19063–19070.

Wong, K. K. Y., F. A. Signal, S. H. Campion, and R. L. Motion. 2005. "Citronella as an Insect Repellent in Food Packaging." *Journal of Agricultural and Food Chemistry* 53 (11): 4633–4636.

Xu, Y., K. Hou, C. Gao, X. Feng, W. Cheng, D. Wu, L. Meng, Y. Yang, X. Shen, and Y. Zhang. 2021. "Characterization of Chitosan Film with Cinnamon Essential Oil Emulsion Co-Stabilized by Ethyl-Nα-Lauroyl-l-Arginate Hydrochloride and Hydroxypropyl-β-Cyclodextrin." *International Journal of Biological Macromolecules* 188: 24–31.

8 Use of Inert Materials for the Management of Stored Product Insects

Monisha Chandran and M. Loganathan
National Institute of Food Technology Entrepreneurship and Management, Thanjavur (NIFTEM-T), Thanjavur, Tamil Nadu, India

8.1 INTRODUCTION

The stored products play a crucial role in agriculture to ensure food safety and security for future generations. Numerous biotic and abiotic factors might cause the food grains and other agricultural produce to deteriorate. These include the increased humidity, temperature, rodents, insects, mites, and microorganisms (Ganie 2012). The insect infestation in stored commodities causes significant quantitative and qualitative damage to agricultural produce, primarily grains and legumes. As a whole, pest damage was accountable for almost 20–30 per cent of tropical zones and 5–10 per cent of temperate zones of the world's agricultural production (Phillips and Throne 2010; Mohammed et al. 2013). Insect infestations may lead small-scale farmers to lose up to 80 per cent of their stock after six to eight months of storage (Nguemtchouin et al. 2013). Meeting the escalating demands of a growing global population necessitates reducing or controlling insect pest populations during food storage. Currently, the primary approach to managing pests relies heavily on synthetic insecticides and fumigants. This reliance has led to the emergence of resistance to commonly employed grain protectants and fumigants (Golob et al. 1982). However, their use poses a negative impact on environmental and ecological risks, such as disrupting natural predator-prey relationships, fostering insect resistance and resurgence, polluting soil, contaminating water and air, and leaving residual effects that can lead to various health issues in animals and humans (Akbar et al. 2021).

Consequently, there is an increasing demand for an environmentally sustainable, safe, and economically viable alternative to chemical pesticides in safeguarding stored grains and seeds from these insect pests (Ibrahim and Salem 2019). Inert materials, which are non-toxic materials, are an essential part of integrated pest management strategies. Therefore, it's necessary to develop safe, affordable alternatives for controlling insects. During the history of mankind, numerous techniques have been developed to combat insect infestations. These include airtight storage and the utilization of materials derived from plants, animals, and minerals, some of which were discovered in archaeological investigations (Hakbijl 2002). Earlier applications of chemically inert dust, such as the traditionally utilized ashes, sand, fine charcoal, powdered minerals, diatoms, clays, lime, and soil, are the remedies particularly for protecting stored commodities (Ibrahim and Salem 2019). As an insecticide, ashes and other inert dust can be used effectively against stored pests. The insecticidal action for insect mortality is caused by desiccation, which is triggered by dust particles destroying the protective epicuticular lipid layer. Dusts are used by numerous cultures to protect their stored goods (Subramanyam and Roesli 2000; Park, Briscoe, and Alexander 2001; Hakbijl 2002). Reduced grain bulk density and flowability, dustiness after application, low insecticidal efficacy, and decreased effectiveness at higher moisture contents are the main limitations (Kljajić et al. 2011; Korunić et al. 2016).

By combining or integrating with other techniques, inert dusts like kaolin (KA), zeolite (ZE), and diatomaceous earth (DE) have been used in integrated pest management strategies for stored

 DOI: 10.1201/9781003609834-8

grain insects (Atta et al. 2023; Gad, Al-ayat, and Abdelgaleil 2023). This chapter summarizes that the use of inert dust to store products is an eco-friendly, affordable, and efficient technique since these resources are readily available locally, have low costs, and are easily adaptable, all of which encourage farmers to adopt this promising technology.

8.1.1 Inert Materials for Disinfestation of Stored Products

For decades synthetic pesticides have been employed to manage insects, including those affecting long-lasting food grains during storage. However, persistent use of these pesticides can lead to significant issues such as insect resistance development, environmental pollution, and risks associated with handling toxic substances. Therefore, there is a necessity to explore alternative, cost-effective, and safe insect control methods (Golob et al. 1982). Given the adverse impacts of using pesticides, it is anticipated that employing natural resources such as inert dust will provide an effective alternative method of controlling pests in stored goods. The effects of inert dust such as clay, rice husk ash (RHA), volcanic ash, massive bamboo leaves ash, wood ash, corncob ash, coconut shells ash, and cow dung ash on adult mortality as well as the suppression of *Rhyzopertha dominica* population growth on rice during storage was reported (Shah and Khan 2014; Astuti et al. 2020).

8.2 INERT DUST

Inert dust possesses insecticidal properties, which are chemically nonreactive or non-toxic and kill insects through physical means instead of chemical approaches. The inert dusts are generally classified into four groups such as earth, DE, silica aerogels, and non-silica dusts (Banks 1976; Cofie-Agblor et al. 1996; Ibrahim and Salem 2019). In addition, the inert materials are broadly (Subramanyam and Roesli 2000) categorized into four distinct classes. Clays, sand, wood ash, and paddy husk ash are in the first group. The second category consists of mineral dusts that were not previously employed in commercial silos and stores, such as dolomite, magnesite, copper oxychloride, katelsous limestone (calcium carbonate), and salt (sodium chloride). Inert dust containing silicon dioxide, often known as synthetic silica, belongs to the third group. The final category consists of inert dusts like zeolite and DE that naturally contain silica.

8.3 TRADITIONAL INSECTICIDES

In numerous regions worldwide, locally sourced materials are extensively utilized to safeguard stored grains from insect damage. Small-scale farmers commonly mix kitchen fire ashes with selected cereal and pulse grains intended for seed storage (Golob et al. 1982). Similarly, tobacco growers utilize dust derived from ground, lower-quality leaves for producing ashes. Nevertheless, there is a scarcity of quantitative studies substantiating the efficacy of these materials or comparing them to conventional insecticides. The utilization of these agricultural wastes would be a safer, more affordable, and environmentally beneficial option to toxic or destructive synthetic insecticides (Olorunmota et al. 2017) when used as contact insecticides or seed protectants in the stored grains.

8.3.1 Clay

Earlier, indigenous peoples in North America and Africa utilized clays as an insecticide. In addition to clays, sands and earth have been used in stored seeds as a top layer (Golob et al. 1982). The clay which is commonly found in soils and sediments is an essential component of many formulations of drugs, pesticides, and cosmetics. As an inert dust, clay alone can be utilized in crop protection to assist control of pest infestation (Kéita et al. 2001). The clay minerals with unique structures of layers had a vital utilization as porous adsorbents for organic compounds or as catalysts for longer periods of time. Numerous researchers have focused on utilizing different polymer matrices that interact with the clay surface in the interlayer area to achieve a high degree of exfoliation of the clay particles (Bergaya and Lagaly 2006). The insecticidal activity on clay was due to decomposition through catalytic action. Reducing the acidity of the clays by treating them with urea or hexamethylene tetramine, or volatile organic bases, slows down the pace of decomposition. One molecule of the suitable base for each sodium ion is usually sufficient to neutralize sodium clays (Fowkes et al. 1960).

After being treated with heat and acid, kaolinic clay destroyed all storage insects (*R. dominica*, *Sitophilus oryzae* L., *Callosobruchus chinensis* L., and *Tribolium castaneum* H.), leading to 100 per cent mortality in less than 24 hours at a minimal dosage of 10 mg, insects are rendered dry by the activated clay's adsorption of lipid particles from their epicuticular wax (Swamiappan 1975). In recent years, researchers have focused on environmentally safe pest management and have been made to employ modified clays as new natural insecticides and montmorillonite and kaolinite with particle size less than 50 µm as effective materials for the insecticidal activity on maize weevil (*Sitophilus zeamais*) (Nguemtchouin et al. 2010). Similarly, the aid of a modified clay with a different formulation of essential oil (*Ocimum gratissimum*) was found to have enhanced absorption capacity, and it also contributes to environmentally safe pest management (Nguemtchouin et al. 2013).

8.3.2 Wood Ash

Wood ash has been exploited as an insecticide since prehistoric times pesticide (Boiteau et al. 2012). At times, the interstitial space in grain bulks is filled with substantial amounts of chemically inert or less active elements, such as wood ashes, sand, other minerals, powders, or plant products (Jean et al. 2015). The findings of Golob et al. (1982) depict the efficacy of incorporating locally available ash powders into maize at concentrations of 1, 5, 15, and 30 per cent (*w/w*) and found that wood ash inhibited infestation with 30 per cent efficient as pirimiphos-methyl. The application of it is less practicable due to the enormous amounts of wood ash required to preserve stored grains against insect infestations adequately. Similarly, Jean et al. (2015) conducted research on DE and wood ash powder (*Hymenocardia acida*) against *S. zeamais*. The results revealed that DE exhibits 100 per cent mortality and wood ash with 87.11 per cent mortality at dose of 40 g/kg for seven days of exposure time.

The combined impact of tamboti wood ash and burnt goat-dropping ash against cowpea weevil (*Callosobruchus maculatus* Fabr.) revealed a negative effect on progeny and effective mortality with 80 per cent in wood ash and 70 per cent goat-dropping ash in stored cowpea seeds. Thus, these are capable of alternative insecticides instead of synthetic insecticides, which are costly and harmful to the environment. However, further research on these ash dosages and the correct proportion for enhanced control of pest management (Tiroeselea and Ranthoakgalea 2019).

Boiteau et al. (2012) depicted that wood ash causes insecticidal effects in the Colorado potato beetle. Larvae and beetles treated with wood ash repeatedly for up to ten days exhibited insect mortality; however, following four consecutive topical administrations, the mortality rate dropped to 22 per cent.

8.3.3 Paddy Husk and Rice Husk Ash (RHA)

RHA is a significant source of Si and K with a high specific surface area and is one of the byproducts of milling rice. It provides long-term benefits and is helpful for stabilizing soil. RHA is a commonly available, less expensive form of silicon that may be accessible (Jeer et al. 2018). It was found that the mortality of *R. dominica* with five traditional inert ashes, such as RHA, corncob ash, groundnut ash, coconut shell ash, and volcanic ash, in stored rice. The bamboo leaves resulted in 100 per cent mortality with a dose rate of 8 g/kg after three (DAI) days after infestation. Similarly rice husk had complete mortality with 4 and 8 g/kg concentration after 7 (DAI) and gave a positive approach for alternative insecticide (Astuti et al. 2020). Another research has been conducted with rice husk, sand, saw dust, deltamethrin, and carbaryl dust to eradicate the potato tuber moth (*Phthorimaea operculella*). The findings revealed 100 per cent mortality and no infestation for two-month period using RHA, sand, deltamethrin, and carbaryl-treated tubers. The treatments with RHA, sand, and sawdust showed a positive approach even after five months of storage period with lesser infestations of 2.16, 1.74, 6.59 per cent, respectively (Das and Rahman 2010). According to Hossain et al. (1994), sawdust, sand, and ash were also beneficial for reducing PTM damage, and rice bran shielded the tubers against PTM infestations for up to three months during storage. It has been suggested that silicon (Si) decreases the damage caused by rice pests. Olorunmota et al. (2017) found an effective insecticide control of *C. maculatus* using RHAs and melon shell powders. Following RHA, which achieved 100 per cent pest mortality at 1.0–2.0 g/20 g in cowpea, as per the standard, were melon shell ash (MSA) and rice husk powder (RHP), which produced 93.33 and 70 per cent mortality at 2.0 g/20 g, respectively. RHA exhibited the highest percentage of inhibition of oviposition while MSA and RHP followed closely behind.

8.3.4 Diatomaceous Earth

The industrial preparation of DE is done by quarrying, drying, and pulverizing with specific mean aggregate particle size and moisture content in order to get suitable powder for commercial use. A fine, talc-like dust with a mean particle size distribution ranging from 0.5 μm to more than 100 μm (mostly between 10 and 50 μm) is the final output of the particles through this procedure. The DE is typically made up of amorphous silicon dioxide with 80 and 95 per cent purity (Quarles 1992; Subramanyam and Roesli 2000). Owing to the possible existence of toxic residues in food and the resistance of storage insects, the use of direct contact insecticides and fumigants is a vital part of the chemical control of storage insects. Therefore, additional efforts are focused on introducing alternative protective tactics, of which inert dust, which is previously referred to as sorptive dust (Ebeling 1971), resembles for effectiveness and safety. DE is the most renowned inert dust; its products are utilized in various applications across the world (Subramanyam and Roesli 2000; Galović et al. 2017).

Considering the concerns of consumers to pesticide residues in processed foods and the evolution of resistance among stored grain insects to the organophosphate grain protectants like malathion, pirimiphos-methyl, and chlorpyrifos-methyl, as well as fumigant phosphine, thus the novel DEs formulations as potential alternatives to synthetic pesticides for managing stored grain insects (Subramanyam et al. 1994; Islam 2017). As DEs are derived from fossilized diatoms, they operate through a physical mode of mechanism, exhibit minimal toxicity to mammals, leave behind no harmful chemical residues, and present lower risks to both the environment and human health compared to synthetic pesticides. Additionally, they demonstrate stability and persistence across a range of temperatures and are effective against insect species that are resistant to chemical treatments (Collins and Cook 2006).

The DE is an inert dust composed of fossilized residue of diatoms, which are unicellular and microscopic and have a fine particle made up of silica. In its native form, DE can be identified as soft, chalky rock deposits (Korunić et al. 2016). The DE is one of the most potent and safest naturally occurring insecticides. The mode of action is physical and chemically unreactive and has an insecticidal effect by desiccation. The DE toxicity for oral toxicity in rat LD_{50} of 3160 mg/kg is generally regarded as safe (Ebeling 1971; Subramanyam et al. 1994). The DE is currently approved in a number of countries (Australia, Canada, China, Croatia, Germany, the USA, and other Asian countries) to be utilized as a structural treatment or as a grain protectant (Korunić et al. 2016).

The stored-product insects are susceptible to the effects of DE due to their xeric ecology. The DE could exert an abrasive effect, even if its specific mode of action is uncertain. Further, desiccation could result from the mineral particles adhering to and being absorbed by the insect epicuticle's surface wax; the insects will lose so much water that they will die from desiccation (Korunic and

Fields 2020; Vargas-abasolo et al. 2024). The primary drawbacks include reduced grain bulk density and flowability, dustiness during application, limited insecticidal efficiency, and decreased performance at elevated moisture contents (Korunić et al. 2016).

Another study has been conducted for the combination effect of three insecticide formulations using dill essential oil (Dill), bait, silica gel and DE, disodium octaborate tetrahydrate (DOT), and pyrthrins (PY) were combined with Dill-DE, Py-Dill-DE, and DOT-PY-DILL-DE for a synergistic effect on the insecticidal effect on *S. oryzae, Sitophilus granarius, R. dominica, and T. castaneum* (Table 8.1). The DE and hydrophilic silica could synergize with Py-Dill-DE, increasing the amount of pyrethrin that reaches the insects (Korunic and Fields 2020). This combination may also work synergistically because once the pyrethrin is bound on the sodium channel, the nerves get hyperexcitation, which leads to paralyzed and the insect is unable to feed, depriving it of food's water content and the water needed for digestion, which exacerbates the desiccation caused by DE.

8.3.5 Silica Aerogels

Synthetic silicas and silica gels are composed using 99.5 per cent silicon dioxide. These synthetic silicas include silica gels and tricalcium trisilicophosphates. Silicophosphates are effective against stored-product insects at a rate of 10–15 g/kg of grain and have a bulk density ranging from 140 to 160 g/l (Quarles 1992). To produce silica aerogel, alcohol is added to precipitated silicas instead of water, and the alcohol is then evaporated in a vacuum. Sand and chlorine gas react to make gaseous sodium tetroxide, which is oxidized to silicon dioxide at high temperatures, resulting in fumed or pyrogenic silica. In general, the bulk density of fumigated silica is smaller than that of precipitated silica (Subramanyam and Roesli 2000).

Aerogels are low-density, extremely porous materials made of inorganic oxides, including tungsten oxide, silica, alumina, zirconia, stannic oxide, or a combination of these oxides. It has been reported that silica aerogel can kill wood termites more effectively than the widely employed insecticide DDT. The rapid loss of bodily fluids from the organic tissue causes death. From an environmental perspective, aerogels' physical in nature, as opposed to chemical insecticidal activity would be a safe and desirable feature (Gesser and Goswami 1989).

The three main parameters to take into consideration for silica gels with superior effectiveness are the volume of silica, the type of adhesive, and the exposure period (Loschiavo 1988). In a similar vein, when silica gel was treated with *Oryzaephilus mercator*, total death had been found in 48 hours. The findings indicated that the insect's stored epicuticular wax layer would selectively sorption of moisture from the insect. To increase the effectiveness of DE in getting rid of *R. dominica* and *Tribolium confusum*, silica gel was utilized. For ten days of exposure, the combined effect of 20 per cent silica gel and DE 200 ppm showed the highest mortality rate of 97 and 82 per cent in *T. confusum* and *R. dominica*, respectively (Delgarm, Ziaee, and McLaughlin 2020).

The major limitation in utilizing silica gels was the dust formulations, which adhere to kernels and are regarded as contaminants by the trade to control insects in stored grain. In a similar vein, it is undesirable to utilize silica aerogels in dwellings since the fine particles drift through the air and land on countertops, floors, and furniture. More significantly, breathing in these floating particles with very low size, 5 µm over an extended period of time, may result in respiratory ailments (Loschiavo 1988). The remarkable advantage is less toxicity, and they do not decompose and also the insects do not become resistant to silica gels. Thus, the silica gels can be combined with other inert dusts to overcome the limitation caused by their adhesive nature (Vrba, Arai, and Nosal 1983).

8.3.6 Zeolites with Insecticidal Property

Similar to DE, zeolites are microporous chemically composed of crystalline aluminosilicate minerals that are found in volcanogenic sedimentary rocks (Eroglu 2015; Ibrahim and Salem 2019). These zeolites constitute pure silica, and the Si/Al ratio is an essential characteristic of zeolite.

TABLE 8.1
The Effect of Inert Material on Stored Insect Grains

Inert material	Formulations	Insect	Dosage	Mortality	Inference	Reference
Silico-sec®	Pure DE contains 90% SiO_2	*Sitophilus granarius, Lepidoglyphus destructor* and *larvae of Ephestia kuehniella.*	10 g/m²	The mortality was 80–95, 100, and 93–100% for Silico-sec®	Silico-sec® dust was the most potent towards all the pest species, causing higher or equal mortalities at a lower dose.	Collins and Cook (2006)
Diasecticide™	Pure DE contains 83.68% SiO_2		20 g/m²	The mortality was observed 11, 49, and 68–93% for Diasecticide™		
Diatomaceous earth (Celatom MN 23)	Different combinations were done using DE, silica gel, pyrethrin, dill essential oil, disodium octaborate tetrahydrate, and Yeast.	*Sitophilus oryzae, S. granarius, Rhyzopertha dominica* and *Tribolium castaneum*	Py-Dill-DE (150 ppm) DOT-Py-Dill-DE (200 ppm)	After three days mortality was 100% and after seven days the mortality was 96–100%	The combination of formulation leads a synergistic effective in insect mortality. The pyrethrin leads death of insect by nerve excitation and paralysis.	Korunic and Fields (2020)
			DE-Dill (300 ppm)	DE-Dill less effective and survival of insect after seven days		
Botanicals and inert dusts	*Acorus calamus, Vitex negundo, Adhatoda vasica, Calotropis gigantea*, wheat straw, kaolinite clay, and DE were assessed at different doses	*S. oryzae* L.	1, 5, and 10 g per 100 g of botanicals and inert dust in stored wheat	The average mean adult mortality for botanicals are 100%, 95%, 98%, 96%, and 100% for inert dusts at two to six month of storage period	*A. calamus* was the most effective treatment in limiting seed damage and weight loss, followed by *A. vasica* and *Calotropis gigantea*, and on other hand *V. negundo* was the least effective treatment.	Akbar et al. (2021)
Inert dust	Rock dust Charcoal powder Diatomaceous earth	*T. castaneum*	Each treatment was blended with canola seeds (12 ± 1% moisture) in a plastic bag at a 2% dosage	The mortality was attained in 14th day of exposure was 25%, 60%, 100%.	The charcoal dust and DE showed a promising results to control stored grain pest	Cremonez et al. (2020)
Silica powders	Silica powder 1–94% SiO_2 with 26.4 µm particle diameter. Silica powder 2–97% SiO_2 with 31.4 µm particle diameter.	*T. castaneum*	The silica powders are formulated with 12 different concentrations (0.5–5.0 g/m²)	The 100% mortality in silica 1 at concentration 1.5–5 g/m² (36 hours) and silica 2 range from 0.75 to 5 g/m² (48 hours)	After exposure of silica powder to insects' bodies, it became brittle, primarily as a result of desiccation and a subsequent decrease in water content.	Manivannan, Subramanyam, and Siliveru (2024)

(Continued)

TABLE 8.1 *(Continued)*
The Effect of Inert Material on Stored Insect Grains

Inert material	Formulations	Insect	Dosage	Mortality	Inference	Reference
Nano zeolite	The nano zeolite was synthesized by ball milling machine with average particle size of 40–45 nm	Tribolium *confusum*	The doses at 250, 500, 600, 700, and 800 ppm of nano zeolite powder is mixed in wheat grains	The mortality of 96.66%, 92%, 80.23%, 66.66%, 37% at 800, 700, 600, 500, 250 ppm, respectively, was recorded after two weeks of exposure	Nano zeolite affixed to the entire insect body with scratches and cracks on the cuticle that result to the dehydration. As the dose and exposure intervals of *T. confusum* and *C. maculatus* adults were increased, the mortality of the seeds of wheat and cowpea treated with nano zeolite increased significantly.	Ibrahim and Salem (2019)
		Callosobruchus *maculatus*	The doses at 100, 200, 300, 400, and 500 ppm of nano zeolite powder is mixed in seeds of cowpea	The mortality of 100%, 88.33%, 70.66%, 55%, 44% was recorded at 500, 400, 300, 200, and 100 ppm, respectively, after two weeks of exposure		
Natural zeolites	The natural zeolite (NZ), modified natural zeolite (MNZ), fine granulated natural zeolite (FGNZ)	*S. oryzae*, *R. dominica* and *T. castaneum.*	The NZ, MNZ, FGNZ were applied at the rate of 0.25–0.75 g/kg of wheat	The efficacy was found to be 74, and 100% and inhibition rate was observed to be 95%, 89%, and 90% in wheat after 21 days of exposure	The factors affecting the efficacy of zeolite was temperature, humidity, substrate and air quality, length of exposure, SiO_2, and size-shape of the particles contained in the product.	Kljajić et al. (2011)
Natural Zeolite (Minazel plus, Minazel), and DE	SiO_2 (63–68%), Al_2O_3 (11–14%), CaO (2.5–4.5%), and Fe_2O_3 (0.8–2.5%), and MgO, Na_2O, K_2O (<1.5%). Ammonium ion is an added element in Minazel Plus that ensures improved adhesiveness to the substrate. Particle size (>85%) are 100 µm in size. DE particle size was 5 µm with SiO_2 (83%), silica aerogel (10%), and Al_2O_3 (3%).	*S. oryzae*, *R. dominica* and *T. castaneum.*	The Minazel plus, Minazel dust were formulated at 0.25, 0.50, 0.75 g/kg and DE for 0.15, 0.20, and 0.30 g/kg, respectively, in wheat at 13% moisture content.	The Minazel at 0.75 g/kg against *S. oryzae* resulted 86% mortality and DE with 0.30 g/kgresulted 83% mortality against *T. castaneum*. The mortality rate were found to be Minazel plusm < Minazel particularly in *S. oryzae* with rate of 0.50 g/kg of zeolite treatment.	Minazel plus contains ammonium which act as an ion-exchange process in zeolite surface binding regions. Thus the modified zeolite have decreased sorption since less free active surface which results in less effectiveness.	Kljajić et al. (2010)

(Continued)

TABLE 8.1 *(Continued)*
The Effect of Inert Material on Stored Insect Grains

Inert material	Formulations	Insect	Dosage	Mortality	Inference	Reference
Kaolinitic clay	It constitutes SiO_2 – 46%, and Al_2O_3 – 38%.	*R. dominica, S. oryzae, Callosobruchus chinensis* and *T. castaneum*	Dose – 10 mg	The kaolinitic clay found to have 100% mortality within 24 hours of time period.	According to the bioassay the activated kaolin resulted in 100% mortality of all four species of insects within 24 hours, even at the lowest level of 10 mg.	Swamiappan (1975)
Wood ash and botanical	The neem extract was prepared using aqueous extraction (100% w/v) and locally purchased wood ash from residents of Ekona	Banana borer weevil (*Cosmopolites sordidus*)	The concentration of neem extract was 25%, 50%, 75%, 100% (w/v) The wood ash concentration was 20% (w/v)	The neem extract with 50%, 75%, 100% concentration leads 68–79% mortality and similarly wood ash causes 68% of mortality at 14 days of exposure in corns	Wood ash and neem were particularly successful in lowering weevil oviposition due to less freedom of motion and mating due to these treatments' repulsive qualities.	Fontem and Edison (2022)
Diatomaceous earth and some conventional inert dusts	The inert materials was kaolin powder, paddy husk ash, coal ash, alluvial soil, china clay, carbaryl powder against pulse beetle	Pulse beetle (*C. chinensis*)	The bioassays were performed with varied concentration for each component based on the nature of the inert material	The mortality was recorded for 24 and 48 hours. The insecticides toxicity sequence was as follows: carbaryl > diatomaceous earth > paddy husk ash, china clay > coal ash > kaolin powder	This research suggested that the predominant way inert dusts functions is by abrasion. For adults of this species, higher dosage rates or longer periods of exposure are required to achieve 100% mortality.	Taylor, Shah, and Khan (2014)

Researchers have identified more than 50 mineral groups that are naturally occurring zeolites; the most well-known ones are reportedly clinoptilolite, erionite, chabazite, heulandite, mordenite, stilbite, and phillipsite (Ramesh and Reddy 2011). The pure zeolites' ability to eradicate insects relies on a number of factors, including their molecular structure (cations, ion-exchange capacity, bulk density, crystalline, or amorphous structure), SiO_2 percentage, dimensions, size and presence of particles, porosity, pH levels, absorption capacity (dehydration-rehydration), and locality (Eroglu 2015).

Zeolites have the potential to be used in the protection of stored products and are known to be as toxic as DE. Zeolites is one of the approved compounds by the Codex Alimentarius Commission in 1999 to manage plant pests and diseases in food commodities (Eroglu 2015). Nonetheless, an IARC investigation revealed no effects on human carcinogenicity. Research surprisingly found that the ion-exchange process in zeolite surface binding regions effectively inhibits the insects (Table 8.1). The sole process relevant for an efficacious modification is the competitive adsorption of ammonium ions and lipid molecules onto the changed zeolite surface, as it is improbable that the enormous insect epicuticular lipid molecules will enter the zeolite's interior channels. As a result of this, there is less free active surface available for the sorption of epicuticular lipid molecules on the modified zeolite as part of its active surface sites have already been taken up by ammonium ions. This could lead to a reduction in Minazel Plus' efficacy (Kljajić et al. 2011).

8.3.7 Kaolin

Grain that has been stored can be protected from pests through the application of minerals like DE, fine sand, lime, bentonite, and specific varieties of kaolin clay. When the minerals are incorporated with the grain that has been threshed, they will cover the gaps between the grain kernels and stop insects from moving around and scattering inside the grain that has been stored (Ali, Ahmed, and Abdel-aziz 2017). A study was conducted on with kaolin, DE and other inert dusts against a pulse beetle (*C. Chinensis*). The results revealed that it has the least toxicity when compared with DE and the LD_{50} for kaolin powder (23,436.40 ppm) and DE (591.98 ppm) after 24 hours of treatments against pulse beetle (Shah and Khan 2014). To eradicate groundnut weevil using non-toxic insecticides few combined materials have been opted for a synergistic effect, i.e., the kaolin powder was flavoured using a few essential oils. The results show that kaolin powder blended with *Lantana camara* L. (Verbenaceae) leads to 97.22 per cent mortality after 36 hours with 50 µl/g DOE and *Annona senegalensis* Pers. leads to 100 per cent mortality after 48 hours with the same concentration of 50 µl/g against ground weevil (Gueye, Thiaw, and Sembene 2012). Another study was conducted on the synergistic effect of cypermethrin, dichlorvos, ash, and kaolin (Karimzadeh, Javanshir, and Hejazi 2020). The combination of LC_{25} cypermethrin and LD_{25} ash led to 80 per cent mortality, whereas the synergistic blend of LC_{25} cypermethrin and LD_{25} kaolin resulted in 78.33 per cent mortality. These were considerably higher than the total of their individual mortality rates, indicating that cypermethrin synergistically exacerbated the effects of ash and kaolin. An interesting study was done using entomopathogenic fungi (EPF) with kaolin as an active carrier agent (Saeed, Laing, and Miller 2022). The strain from *Beauveria bassiana* (Balsamo) Vuillemin is the EPF that has been tested as the most widely utilized to control stored-grain insects. The results were found to be *B. bassiana* Strain MS-8 with a kaolin dose of 2 g/kg was accountable for the highest degree of larval death (90.0%) and the lowest number of adults that emerged (1.6 insect/100 g of rice grain).

8.4 INSECTICIDAL ACTION MECHANISM

Several hypotheses have been given out to clarify why oxygen deprivation, or anoxia, results in higher mortality rates; nevertheless, desiccation tends to provide the best reason for the results that were observed under the conditions and methods employed to eradicate museum pests (Selwitz and Maekawa 1999; Valentin, Alguerő, and De Hijas 2016). In general, there are three types of testimony. Firstly, the respiratory system of insects is well characterized by their physiology, which causes them to desiccate more quickly when exposed to low concentrations of carbon dioxide or oxygen. Secondly, as conditions shift in favour of dehydration, death rates typically increase. Thus, anoxia is generally a more effective death process when the temperature is raised, or the humidity is lowered. The third significant point of evidence is the positive correlation between mortality rates and weight loss, which in anoxic conditions can only happen via water loss (Selwitz and Maekawa 1999).

The dusts are much more effective at low relative humidity and seem to be an efficient long-term grain protectant in dry tropical or subtropical regions. As relative humidity increases, inert dusts become less effective as the impact is triggered by desiccation. They continue to function effectively under these conditions for extended periods of time (Banks 1976). Once the application of dust on the body of insects, it eventually becomes dehydrated and collapses. The inert dust is the powder that directly scratches (DE) or absorbs (silica aerogel) the waxy coating layer of the insect, which actually prevents an excessive amount of moisture from escaping into the dry grain and intergranular air (Figure 8.1) (El-aziz 2015).

The catalytic action of the clay diluent causes a breakdown in wettable powders and dust-containing insecticides. Sometimes, the heat produced can cause an ignition to occur on its own. As the acidity of the clay surface increases, it also affects the rate at which these insecticides react when mixed with various clay diluents (Fowkes et al. 1960). Similarly, the insecticidal impact on wood ash, as inert dust, is due to its physical characteristics as it induces insect mortality by causing the cuticle to dry out, thereby suffocating the insect (Jean et al. 2015). Because the silica action absorbs the lipid layer on cuticles, a high level of silica (SiO_2) in inert dust material may kill insects. Researchers found several inert materials with substantial concentrations of silica content, such as volcanic ash (45.70%), rice husks (79%), huge bamboo leaves (58.30%), corncobs (47.78%), and coconut shells (45.05%) (Astuti et al. 2020) which can effectively act as insecticides. Wood ash acts like diatomaceous silica-based dust, depleting the insect's cuticle, absorbing the protective wax layer that typically hinders body water loss, and finally causing the insect to die by desiccation. This clarification would explain the long exposure time needed for wood ash to cause Colorado potato beetle mortality (Boiteau et al. 2012).

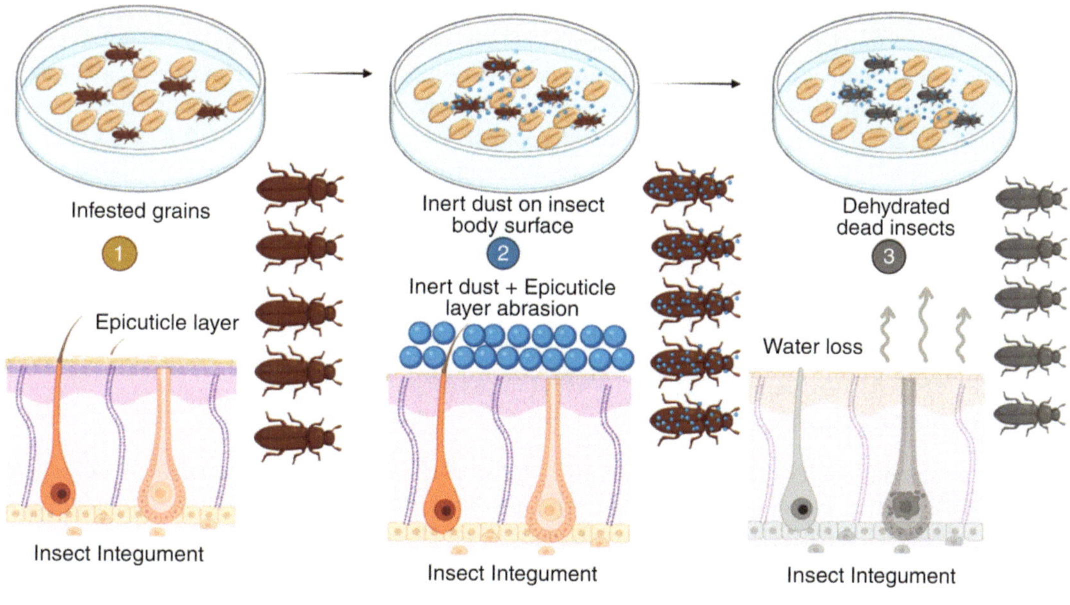

FIGURE 8.1 Inert material mode of action on insect surface.

8.4.1 Factors Affecting Inert Materials

The effectiveness of inert dust on storage insects was influenced by an array of factors, including particle size, temperature, relative humidity, grain moisture content, time of exposure, and treatment of application.

8.4.2 Particle Size

The key variables are intrinsic hardness and particle size. The particles that are greater than 15 µm in carborundum have no effect, most likely due to their inability to stick to insects. The maximum efficiency of the substance is attained when the particle size reduces from 10 µm to approximately 2 µm (Alexander, Kitchener, and Briscoe 1944). Even though the size of the particles plays a significant role in efficacy, dust with smaller particles, like Aerosil R974, is regarded as respirable dust and could pose a health risk to users. Nevertheless, this issue does not affect Dryacide, insect, or other dusts with bigger particle sizes, which are similar to those of dust formulations of traditional insecticides. However, when there is too much application of dust, it is safe to use proper face masks to avoid breathing in dust. The USA Environmental Protection Agency has approved insect registration for use on stored grain and in grain-handling facilities because it considers these materials to be safe (Golob, Avenue, and Maritime 1997).

8.4.3 Temperature, Relative Humidity, and Grain Moisture Content

A key element influencing the effectiveness of inert dust is temperature and relative humidity. Low relative humidity makes insects more susceptible to the effects of insecticidal silicas, and low moisture content grain enables improved pest control (Mutambuki 2013; Floros et al. 2018) depicted the impact of temperature and relative humidity on lethal effects of natural zeolite on adult bean weevils (*Acanthoscelides obtectus*).

When grain moisture rises beyond 14 per cent, the insecticidal action of amorphous silica incorporated into the grain rapidly diminishes. The susceptibility to silica dust was expected when they crawl among kernels (Subramanyam and Roesli 2000). When the insecticidal effects depend on temperature for effective mortalities of stored pests, few studies depict that fatalities were extremely low at 15°C, 20°C, and 25°C, ranging from 7.5 to 22.5 per cent after 24–48 hours of exposure at 0.5 g/kg beans. After 144 hours, the rate of deaths increased steadily to 70 per cent. After 240 hours of exposure in all temperatures, they attained mortality (Floros et al. 2018). The insecticidal effect of three different wood ash, such as common beech, Norway spruce, and black locust, were mixed with wheat grains. The treatments were conducted with varying temperature (15°C, 20°C, 25°C) and relative humidity (55% and 75%) with two concentrations (10 and 20 g/m^2). All three wood ashes exhibited over 95 per cent mortality for beetles after seven days at 25°C and RH-55 per cent. However, even after seven days of exposure 75 per cent RH revealed low mortality in all the three treated ashes (Bohinc, Horvat, and Andri 2018). The vulnerability to insecticides elevates with an increase in temperature and decrease in relative humidity.

8.5 RECENT ADVANCEMENTS IN INERT MATERIALS USING BINARY COMBINATIONS

Inert materials that are less harmful or non-toxic to people, animals, and the environment include kaolin, DE, and other inert dust. They have been effectively used as grain protectors against several kinds of insects which attack stored grains (Arnaud et al. 2005; Korunić et al. 2016; Galović et al. 2017; Ziaee, Ebadollahi, and Wakil 2019). Nevertheless, these substances have certain drawbacks, such as altering the treated grains' bulk density and physical characteristics when used in massive amounts (Golob, Avenue, and Maritime 1997; Korunić et al. 2016). In comparison to using these compounds independently, researchers discovered that combined treatments exploiting binary mixes were more effective in controlling stored grain insects (Gad, Al-ayat, and Abdelgaleil 2023).

According to Gad et al. (2023), integrating DE and kaolin (inert dust) with a binary combination of two insecticides, such as abamectin (AB) and deltamethrin (DL), leads to high efficacy and insecticidal activity than single treatments against *Trogoderma granarium*. When *T. granarium* was treated with a combination of KA (1000 mg/kg) and AB (1.0 mg/kg), 100 per cent larval mortality was attained. However, the treatments, including binary combinations of DL with DE or KA, had the highest larval mortality of *T. granarium* with 70 per cent after 21 days of treatment.

The insecticidal activity of wheat grains against *T. granarium* using a combination of three organophosphorus (OP) insecticides combined with inert dust such as DE, zeolite, and kaolin (Atta et al. 2023) revealed that when OPs and inert dust were combined in binary mixtures, the mortality rate of khapra beetle larvae was higher compared to when they were used individually. The complete mortality of the larvae was observed with the combination of CH and IDs at a concentration of 5 ppm. Additionally, mixtures of either OPs with inert dusts resulted in significant mortality of over 82 per cent in *T. granarium* larvae.

A positive approach was found towards the integration of inert dust, such as zeolite and kaolin ash, with carbon dioxide (Gad et al. 2022). The adult mortality of *C. maculatus and C. chinensis* insect species exposed to cowpeas treated with individual and binary treatments enhanced with an increase in dosage and the duration of exposure. The results depicted 100 per cent mortality against both the insects by integrating treatments zeolite (1000 mg/kg) and 20% carbon dioxide, and similarly, kaolin (100 mg/kg) with 20% carbon dioxide resulted in complete mortality against *C. chinensis* after seven days of exposure.

The khapra beetle is difficult to control due to its high tolerance activity in stored grains. There is an optimistic response to using binary treatments to eradicate and manage the stored pests. In a study, Gad, Al-ayat, and Abdelgaleil (2023) found that integrating chitin synthesis inhibitors and inert dust (DE, ZE, KA) leads to significant mortality of *T. granarium* in stored wheat than individual insecticides. The mixture of insecticides with 5.0 mg/kg chitin inhibitors with 1000 mg/kg DE resulted in complete larval mortality within 14 days.

8.6 MERITS AND LIMITATIONS FOR OPTING INERT MATERIALS

Inert dust exhibits non-toxic or inert with two mammalian toxicity with DE rat oral LD_{50}, >5000 mg/kg body weight. As long as the dusts are dry, they offer protection and are stable on the grain. They do not leave harmful residues, in contrast to grain protectants made of organophosphate. Insects might not be able to adapt to inert dust (Ebeling 1971). The majority of the dust particles can be removed from grain treated with inert dust by cleaning it beforehand before milling. According to El-aziz et al. (2015), inert dusts are ovicidal, and treating abandoned warehouses, mills, and storage facilities on the exterior may be a beneficial use for them.

The primary limitation was a decline in flowability (Subramanyam and Roesli 2000). The primary issues with using inert dusts are that they can be ineffective in certain situations, are dusty to apply, and reduce the bulk density and flowability of grain. As a result of inert dust adhering to the kernel surface and raising intergranular friction, grain does not flow as freely, increasing angles of repose and lowering bulk density (Golob, Avenue, and Maritime 1997). There will be amplification in angle of repose, rate of feed loading troubles, dusty environment and commodities, improper cleaning of mills, and leftover dust in grains are the major limitations for inert materials.

8.7 FUTURE PROSPECTS

The insect infestation in stored grains may lead to a high risk in quality and toxicity in grains and seed, reduction in germination and decreased nutritional value of grains. To protect stored products from insect attacks, a number of research papers suggested using synthetic alternatives or binary blends of chemical pesticides, which are less toxic and inert materials. These mixes have a number of benefits, including a reduction in insecticide rates and maintenance costs, in addition to an increase in insecticide effectiveness and residual effect (Atta et al. 2023). The binary combinations, such as the use of insecticides and inert dust blends might provide a remedy to these issues. Additionally, these combinations might increase the insecticides' ability to kill stored grain insects at modest treatment levels effectively. Furthermore, integrating insecticides with innocuous dust can lower application rates, prevent food contamination, prevent the evolution of resistance, and lessen environmental damage. The overall cost of pest management may be reduced by using this strategy as well. Furthermore, a few techniques have emerged in pest management strategies, such as nanoparticles as insecticides, hydrogels, electrostatic dust, edible coating of grains, and encapsulated baits were highly effective in controlling pests in stored commodities. However, the practical limitation of these techniques is more expensive, and this can be effectively overcome by the usage of inert materials that are economically feasible, environmentally friendly, and non-toxic to humans.

8.8 SUMMARY AND CONCLUSION

It can be concluded that insecticides should be safety and non-toxic for all living hood and the environment. The recent binary combination can support the mitigations and can be effectively utilized in integrated pest management. It is undeniable that incorporating IDIs into the IPM strategy can result in significant improvements in the management of stored grain and seed. The benefits include safety for mammals, overcoming chemical resistance in numerous insect species, reduced treatment costs, and great availability. They also significantly reduce chemical pollution of the commodity and the environment. Thus, inert materials have a potential role in integrated pest management programmes in grain storage.

REFERENCES

Akbar, R., M. Sharma, D. Sharma, S. Rafiq, and R. T. Nisa. 2021. "Management of Rice Weevil, *Sitophilus oryzae* L. (Coleopteran ; Curculinoidae) by Botanicals and Inert Dusts in Stored Wheat." *Biological Forum – An International Journal*, 13 (4): 367–372.

Alexander, P., J. A. Kitchener, and H. V. A. Briscoe. 1944. "Inert Dust Insecticides Part I. Mechanism of Action." *Annals of Applied Biology* 31 (2): 143–149. doi: 10.1111/j.1744-7348.1944.tb06225.x

Alexander, P., J. A. Kitchener, and H. V. A. Briscoe. 2001. "Inert Dust Insecticides Part 111. The Effect of Dusts on Stored Products Pests Other than Calandra Granariu." *Annals of Applied Biology* 31 (2): 156–159. doi: 10.1111/j.1744-7348.1944.tb06227.x

Ali, E. A., S. S. Ahmed, and S. Y. Abdel-aziz. 2017. "Efficacy of Some Natural Products Mixed with Wheat Flour on the Sur-Vival and Development of the Red Flour Beetle, Tribolium Castaneum (Coleoptera : Tenebrionidae)." *Egyptian Scientific Journal of Pesticides* 3 (2), 11–22. doi: 10.13140/RG.2.2.24615.96166.

Arnaud, L., H. Tran Thi Lan, Y. Brostaux, and E. Haubruge. 2005. "Efficacy of Diatomaceous Earth Formulations Admixed with Grain against Populations of Tribolium Castaneum." *Journal of Stored Products Research* 41 (2): 121–130. doi:10.1016/j.jspr.2003.09.004.

Astuti, L. P., R. Maula, A. Rizali, and M. B. Mario. 2020. "Effect of Five Types Inert Dust to *Rhyzopertha dominica* (Fabricius) (Coleoptera : Bostrichidae) in Stored Rice Seeds." *The Journal of Experimental Life Science* 9 (3): 164–169. doi:10.21776/ub.jels.2019.009.03.04.

Atta, A. A. M., H. A. Gad, A. A. Al-ayat, and S. A. M. Abdelgaleil. 2023. "Evaluation of Three Organophosphorus Insecticides Alone and in Combination with Inert Dusts for the Control of *Trogoderma Granarium* Everts (Coleoptera: Dermestidae)" *Journal of Stored Product Research* 102: 102104. doi:10.1016/j.jspr.2023.102104

Banks, H. J. 1976. "Physical Control of Insects-Recent Developments." *Australian Journal of Entomology* 15 (1), 89–100. doi: 10.1111/j.1440-6055.1976.tb02079.x

Bergaya, F., and G. Lagaly. 2006. "Chapter 1 General Introduction: Clays, Clay Minerals, and Clay Science." In *Developments in Clay Science*. doi:10.1016/S1572-4352(05)01001-9.

Bohinc, T., A. Horvat, and G. Andri. 2018. "Comparison of Three Different Wood Ashes and Diatomaceous Earth in Controlling the Maize Weevil under Laboratory Conditions." *Journal of Stored Products Research* 79: 1–8. doi:10.1016/j.jspr.2018.06.007.

Boiteau, G., R. P. Singh, P. C. Mccarthy, and P. D. Mackinley. 2012. "Wood Ash Potential for Colorado Potato Beetle Control," *American Journal of Potato Research* 89: 129–135. doi:10.1007/s12230-012-9234-7.

Cofie-Agblor, R., W. E. Muir, R. N. Sinha, and P. G. Fields. 1996. "Heat Production by Adult *Cryptolestes ferrugineus* (Stephens) of Different Ages and Densities." *Postharvest Biology and Technology* 7 (4): 371–380. doi:10.1016/0925-5214(95)00050-X.

Collins, D. A., and D. A. Cook. 2006. "Laboratory Evaluation of Diatomaceous Earths, When Applied as Dry Dust and Slurries to Wooden Surfaces, against Stored-Product Insect and Mite Pests." *Journal of Stored Products Research* 42 (2): 197–206. doi:10.1016/j.jspr.2005.01.005.

Cremonez, P. S. G., L. A. De A Moraes, G. S. De Aquino, R. R. Machado, R. Hayashida, V. Sousa, and P. M. O. J. Neves. 2020. "Bioassay Establishment and Control of Tribolium Castaneum (Herbst, 1797) (Coleoptera: Tenebrionidae) with Inert Dusts in Stored Canola," *Entomological Communications* 10–13. doi:10.37486/2675-1305.ec02001.

Das, G. P., and M. M. Rahman. 2010. "Effect of Some Inert Materials and Insecticides against the Potato Tuber Moth, *Phthorimaea operculella* (Zeller), in *Storage operculella* (Zeller), in Storage." *International Journal of Pest Management* 43 (3) no. October 2014: 23–25. doi:10.1080/096708797228753.

Delgarm, N., M. Ziaee, and A. McLaughlin. 2020. "Enhanced-Efficacy Iranian Diatomaceous Earth for Controlling Two Stored-Product Insect Pests." *Journal of Economic Entomology* 113 (1): 504–510. doi:10.1093/jee/toz261.

Ebeling, W. 1971. "Sorptive Dusts for Pest Control." *Annual Review of Entomology.* doi:10.1146/annurev.en.16.010171.001011.

El-aziz, S. E. A. 2015. "Control Strategies of Stored Product Pests," *Journal of Entomology*, 8 (2): 101–122. doi:10.3923/je.2011.101.122.

Eroglu, N. (2015). A review: Insecticidal potential of Zeolite (Clinoptilolite), toxicity ratings and general properties of Turkish Zeolites. In *Proceedings of the 11th International Working Conference on Stored Product Protection, Chiang Mai, Thailand* (pp. 24–28).

Floros, G. D., A. I. Kokkari, N. A. Kouloussis, N. A. Kantiranis, P. Damos, A. A. Filippidis, and D. S. Koveos. 2018. "Stored-Product Evaluation of the Natural Zeolite Lethal Effects on Adults of the Bean Weevil under Different Temperatures and Relative Humidity Regimes." *Journal of Economic Entomology* 111(1): 482–490. doi:10.1093/jee/tox305.

Fontem, L. A. and A. Edison. 2022. "Farmers 'Knowledge and Perceptions of Banana Borer Weevil and Insecticidal Farmers 'Knowledge and Perceptions of Banana Borer Weevil and Insecticidal Potential of Neem and Wood Ash for Its Management in Fako Division, Cameroon," *American Journal of Life Sciences* 6 (2): 29–38. doi:10.11648/j.ajls.20180602.11.

Fowkes, F. M., H. A. Benesi, L. B. Ryland, W. M. Sawyer, K. D. Detling, E. S. Loeffler, F. B. Folckemer, M. R. Johnson, and Y. P. Sun. 1960. "Clay-Catalyzed Decomposition of Insecticides." *Journal of Agricultural and Food Chemistry* 8 (3), 203–210.doi:10.1021/jf60109a011.

Gad, H. A., S. E. El-deeb, F. S. Al-anany, and S. A. M. Abdelgaleil. 2022. "Effectiveness of Two Inert Dusts in Conjunction with Carbon Dioxide for the Control of *Callosobruchus maculatus* and *C. chinensis* in Stored Cowpea Seeds." *Journal of Stored Products Research* 95 (September 2021). doi:10.1016/j.jspr.2021.101910.

Gad, H. A., A. A. Al-ayat, and S. A. M. Abdelgaleil. 2023. "Management of Khapra Beetle, *Trogoderma granarium* Everts, Using Binary Combinations of Chitin Synthesis Inhibitors and Inert Dusts." *Journal of Stored Products Research* 104: 102194. doi:10.1016/j.jspr.2023.102194.

Gad, H. A., A. A. Al-ayat, M. M. M. Hassuba, and S. A. M. Abdelgaleil. 2023. "Effectiveness of Binary Combinations of Abamectin and Deltamethrin with Two Inert Dusts for the Management of *Trogoderma granarium* Everts on Wheat Grains." *Journal of Stored Products Research* 100: 102071. doi:10.1016/j.jspr.2022.102071.

Galović, I., J. Halamić, A. Grizelj, V. Rozman, A. Liška, Z. Korunić, P. Lucić, and R. Baličević. 2017. "Croatian Diatomites and Their Possible Application as a Natural Insecticide." *Geologia Croatica* 70 (1): 27–39. doi:10.4154/gc.2017.04.

Ganie, S. A. 2012. "Pest Control and Disinfection in Rural Godowns" *Ecologically Based Integrated Pest Management.* New India Publishing Agency, New Delhi, 821–839.

Gesser, H. D., and P. C. Goswami. 1989. "Aerogels and Related Porous Materials." *Chemical Reviews* 89 (4): 765–788. doi:10.1021/cr00094a003.

Golob, P., C. Avenue, and C. Maritime. 1997. "Current Status and Future Perspectives for Inert Dusts for Control of Stored Product Insects." *Journal of Stored Products Research*, 33 (1): 69–79. doi: 10.1016/0022-474X(82)90004-2

Golob, P., J. Mwambula, V. Mhango, and F. Ngulube. 1982. "The Use of Locally Available Materials as of Maize Grain against Insect Infestation during Storage in Malawi." *Journal of Stored Products Research* 18 (2): 67–74.

Gueye, S., C. Thiaw, and M. Sembene. 2012. "Short Communication Insecticidal Effect of Kaolin Powder Flavoured with Essential Oils of *Lantana camara* L. (Verbenaceae) and *Annona senegalensis* Pers. (Annonaceae) on *Caryedon serratus* Olivier (Coleoptera-Bruchidae), a Groundnut Stock Pest." *International Journal of Biological and Chemical Sciences* 6 (4): 1792–1797. doi: 10.4314/ijbcs.v6i4.33

Hakbijl, T. 2002. "The Traditional, Historical and Prehistoric Use of Ashes as an Insecticide, with an Experimental Study on the Insecticidal Efficacy of Washed Ash," *Environmental Archaeology* 7 (1): 13–22. doi: 10.1179/env.2002.7.1.13

Hossain, S. M. Z., G. P. Das, and M. Z. Alam. 1994. "Use of various indigenous materials and insecticides in controlling potato tuber moth in storage". *Bulletin of the Institute of Tropical Agriculture* 17, 79–84.

Ibrahim, S. S., and N. Y. Salem. 2019. "Insecticidal Efficacy of Nano Zeolite against *Tribolium confusum* (Col., Tenebrionidae) and *Callosobruchus maculatus* (Col., Bruchidae)." *Bulletin of the National Research Centre* 43: 1–8.

Islam, W. 2017. "Eco-Friendly Approaches for the Management of Red Flour Beetle: Tribolium Castaneum (Herbst)," *Science Letters* 5 (2): 105–114.

Jean, W. G., N. E. Nchiwan, N. Dieudonné, S. Christopher, and C. Adler. 2015. "Efficacy of Diatomaceous Earth and Wood Ash for the Control of *Sitophilus zeamais* in Stored Maize." *Journal of Entomology and Zoology Studies* 3 (5): 390–397.

Jeer, M., K. Suman, T. U. Maheswari, S. R. Voleti, and A. P. Padmakumari. 2018. "Field Crops Research Rice Husk Ash and Imidazole Application Enhances Silicon Availability to Rice Plants and Reduces Yellow Stem Borer Damage." *Field Crops Research* 224: 60–66. doi:10.1016/j.fcr.2018.05.002.

Karimzadeh, R., M. Javanshir, and M. J. Hejazi. 2020. "Individual and Combined Effects of Insecticides, Inert Dusts and High Temperatures on *Callosobruchus maculatus* (Coleoptera : Chrysomelidae)." *Journal of Stored Products Research* 89: 101693. doi:10.1016/j.jspr.2020.101693.

Kéita, S. M., C. Vincent, J. P. Schmit, J. T. Arnason, and A. Bélanger. 2001. "Efficacy of Essential Oil of *Ocimum basilicum* L. and *O. gratissimum* L. Applied as an Insecticidal Fumigant and Powder to Control *Callosobruchus maculatus* (Fab.) [Coleoptera: Bruchidae]." *Journal of Stored Products Research* 37 (4): 339–349. doi:10.1016/S0022-474X(00)00034-5.

Kljajić, P., G. Andrić, M. Adamović, M. Bodroža-Solarov, M. Marković, and I. Perić. 2010. "Laboratory Assessment of Insecticidal Effectiveness of Natural Zeolite and Diatomaceous Earth Formulations against Three Stored-Product Beetle Pests." *Journal of Stored Products Research* 46 (1): 1–6. doi:10.1016/j.jspr.2009.07.001.

Kljajić, P., G. Andrić, M. Adamović, and M. P. Golić. 2011. "Possibilities of Application of Natural Zeolites in Stored Wheat Grain Protection against Pest Insects" *Journal on processing and Energy in Agriculture* 15 (1): 12–16.

Korunic, Z., and P. G. Fields. 2020. "Evaluation of Three New Insecticide Formulations Based on Inert Dusts and Botanicals against Four Stored-Grain Beetles." *Journal of Stored Products Research* 88: 101633. doi:10.1016/j.jspr.2020.101633.

Korunić, Z., V. Rozman, A. Liška, and P. Lucić. 2016. "A Review of Natural Insecticides Based on Diatomaceous Earths." *Poljoprivreda* 22 (1): 10–18. doi:10.18047/poljo.22.1.2.

Loschiavo, S. R. 1988. "Safe Method of Using Silica Aerogel to Control Stored-Product Beetles in Dwellings." *Journal of Economic Entomology* 81 (4): 1231–1236. doi:10.1093/jee/81.4.1231.

Manivannan, S., B. Subramanyam, and K. Siliveru. 2024. "Efficacy of Two Amorphous Silica Powders Applied to Soft Red Winter Wheat against the Lesser Grain Borer, *Rhyzopertha dominica* (F.) (Coleoptera : Bostrichidae)." *Journal of Stored Products Research* 106: 102264. doi:10.1016/j.jspr.2024.102264.

Mohammed, H. H. 2013. "Repellency of Ethanolic Extract of Some Indigenous Plants against *Tribolium confusum* (Duval) (Coleoptera: Tenebrionidae)." *IOSR Journal of Agriculture and Veterinary Science* 2 (6): 27–31.

Mutambuki, K. 2013. "The Influence of Grain Moisture Content on the Efficacy of Silica Dust on *Prostephnus truncatus* (Horn) (Coleoptera: Bostrichidae) and *Sitophilus zeamais* (Motsch) (Coleoptera: Curculionidae)." *Journal of Stored Products and Postharvest Research* 4 (2): 23–29. doi:10.5897/jsppr12.005.

Nguemtchouin, M. G. M., M. B. Ngassoum, P. Chalier, R. Kamga, L. S. T. Ngamo, and M. Cretin. 2013. "*Ocimum gratissimum* Essential Oil and Modi Fi Ed Montmorillonite Clay, a Means of Controlling Insect Pests in Stored Products." *Journal of Stored Products Research* 52: 57–62. doi:10.1016/j.jspr.2012.09.006.

Nguemtchouin, M. M. G., M. B. Ngassoum, L. S. T. Ngamo, X. Gaudu, and M. Cretin. 2010. "Insecticidal Formulation Based on *Xylopia aethiopica* Essential Oil and Kaolinite Clay for Maize Protection." *Crop Protection* 29 (9): 985–991. doi:10.1016/j.cropro.2010.06.007.

Olorunmota, R. T., T. I. Ofuya, J. E. Idoko, and B. A. Ogundeji. 2017. "Effect of Rice Husk and Melon Shell Wastes as Possible Grain Protectants in Cowpea Storage." *Journal of Advances in Biology & Biotechnology* 16 (2): 1–9. doi:10.9734/JABB/2017/37114.

Phillips, T. W., and J. E. Throne. 2010. "Biorational Approaches to Managing Stored-Product Insects." *Annual Review of Entomology* 55: 375–397.

Quarles, W. 1992. "Diatomaceous Earth for Pest Control." *IPM Practitioner* 14 (5/6): 1–11.

Ramesh, K., and D. D. Reddy. 2011. "Zeolites and Their Potential Uses in Agriculture." *Advances in Agronomy* 113: 219–241.. doi:10.1016/B978-0-12-386473-4.00004-X.

Saeed, M. B., M. D. Laing, and R. M. Miller. 2022. "Short- and Long-Term Evaluation of a Fixed Dose of *Beauveria bassiana* Strain MS-8 Diluted in Various Doses of Kaolin as a Powder Formulation Applied to Rice Grains to Control Almond Moth, *Ephestia cautella* Walker (Lepidoptera : Pyralidae)." *Microorganisms* 10 (10): 1971. doi: 10.3390/microorganisms10101971

Selwitz, C., and Shin Maekawa. 1999. *Inert Gases in the Control of Museum Insect Pests.* Getty Publications.

Shah, M. A., and A. A. Khan. 2014. "Use of Diatomaceous Earth for the Management of Stored-Product Pests." *International Journal of Pest Management* 60 (2): 100–113. doi:10.1080/09670874.2014.918674.

Subramanyam, B., and R. Roesli. 2000. "Inert Dusts." In *Alternatives to Pesticides in Stored-Product IPM*, 321–380. doi:10.1007/978-1-4615-4353-4_12.

Subramanyam, B., C. L. Swanson, N. Madamanchi, and S. Norwood. 1994. "Effectiveness of Insecto, a New Diatomaceous Earth Formulation, in Suppressing Several Stored-Grain Insect Species." In *Proceedings of the 6th International Working Conference on Stored-Product Protection. 2*, 650–659.

Swamiappan, B. M. 1975. "Effect of Activated Kaolinitic Clay on Some Storage Insects." *Zeitschrift für Angewandte Entomologie* 80 (1976): 385–389.doi: 10.1111/j.1439-0418.1976.tb03341.x

Tiroeselea, B., and G. Ranthoakgalea. 2019. "Tamboti Wood Ash and Burnt Goat Dropping Ash, Safe Alternatives to Control Cowpea Weevils, *Callosobruchus maculatus* (Fabr.) (Coleoptera : Bruchidae) during Storage for Subsistence Farming." *Environmental Monitoring and Assessment* 191: 1–9. doi: 10.1007/s10661-019-7632-8

Valentin, N., Alguerö, M., & Martin de Hijas, C. (1992). Evaluation of disinfection techniques for the conservation of polychrome sculpture in Iberian museums. *Studies in Conservation* 37 (1): 165–167. doi:10.1179/sic.1992.37.s1.035.

Vargas-abasolo, R., D. Rivera-duarte, V. M. Almaraz-valle, and M. Mejia. 2024. "Acta Tropica Thinking Green: Insecticidal Effect of Biorational Solutions against *Triatoma pallidipennis* Stal (Hemiptera: Triatominae)." *Acta Tropica* 252 (October 2023). doi:10.1016/j.actatropica.2024.107152.

Vrba, C. H., H. P. Arai, and M. Nosal. 1983. "The Effect of Silica Aerogel on the Mortality of *Tribolium confusum* (Duval) as a Function of Exposure Time and Food Deprivation." *Canadian Journal of Zoology* 61 (7): 1481–1486. doi: 10.1139/z83-199.

Ziaee, M., A. Ebadollahi, and W. Wakil. 2019. "Integrating Inert Dusts with Other Technologies in Stored Products Protection." *Toxin Reviews* 40 (4): 404–419. doi:10.1080/15569543.2019.1633673.

9 Grain Aeration

V. Chelladurai and D. Praveen Kumar
Department of Agricultural Engineering, Bannari Amman Institute of Technology, Sathyamangalam, Erode, Tamil Nadu, India

T. Senthilkumar
University of Prince Edward Island, Charlottetown, PEI, Canada

9.1 INTRODUCTION

Global grain production has been on the rise each year, emphasizing the importance of investing in post-harvest systems. Fluctuations in product prices and options for raw material holders are influenced by various factors such as public policies, government interventions, and trade barriers (Ziegler, Paraginski, and Ferreira 2021). Adequate storage is essential to facilitate commercialization after harvest while preserving grain quality. However, the increasing production levels have led to a shortage in static storage capacity, resulting in higher costs and losses along the marketing chain.

According to the Food and Agriculture Organization (FAO) of the United Nations, nations should ideally have a static storage capacity higher than their production capacity to guarantee food security (FAO 2022). Both quantitative and qualitative losses are frequently the outcome of insufficient storage capacity. Moreover, the improper structural design of a large number of current storage units makes proper storage difficult (Coradi et al. 2020). Establishing core guidelines and procedures is essential to guaranteeing appropriate grain storage all along the way (Aldaco et al. 2020). While advancements in agricultural technologies have boosted food grain production, other aspects of the agro-industrial chain, from harvesting to retailing, have not seen significant progress (Coradi et al. 2020). The recent COVID-19 pandemic has further exacerbated food supply chain disruptions, leading to increased food losses and waste at both pre-harvest and post-harvest stages, thereby threatening national food security and hindering progress toward sustainable development goals (SDGs) such as "zero hunger (SDG 2)" and "Responsible consumption and production (SDG 12)" (Parvin, Gawanmeh, and Venkatraman 2018; Tushar et al. 2023). Globally, approximately one-third of agricultural production is wasted annually due to inadequate post-harvest management, with poor grain storage management contributing to 20–50% of post-harvest losses in emerging economies (Singh and Fielke 2017; Hellegers 2022).

Storage is crucial for ensuring a consistent supply of grains to consumers. Proper post-harvest storage is essential to maintain the quality and nutritional value of grains over time (Mobolade et al. 2019). However, insufficient storage capacity and maintenance can lead to spoilage, resulting in the development of harmful substances, nutrient loss, resource wastage, and decreased market value (Ziegler, Paraginski, and Ferreira 2021). Silos or warehouses can be used to store grains, whether they are imported or harvested. Grain storage can be affected by a variety of climatic and biological conditions, including pests like fungi, insects, mites, and rodents, as well as variations in temperature, humidity, CO_2 levels, and moisture content (Mutungi et al. 2019; Anukiruthika, Jian, and Jayas 2021). These elements play a major role in the reduction in grain quantity and quality.

In order to preserve and maintain the quality of grains, drying and aerating them after harvest are essential processes. In addition to ensuring nutritional value and preventing spoiling, proper drying and aeration also increase the market value of the product. In today's modern agriculture, effective grain drying and aeration procedures are essential due to the growing demand for food security and sustainability. Grain drying techniques, such as air and sun drying, have been used for generations.

DOI: 10.1201/9781003609834-9

TABLE 9.1
Differences between Grain Drying and Grain Aeration

Grain Drying	Grain Aeration
The primary objective of grain drying is to reduce moisture content to safe storage levels, thereby preventing spoilage and maintaining grain quality	In contrast, grain aeration focuses on controlling temperature and moisture levels within the storage facility to prevent moisture migration, reduce temperature differentials, and minimize the risk of spoilage
Grain drying involves the application of heat to evaporate moisture from grains, either through convective, microwave, or infrared methods	Grain aeration, on the other hand, relies on the circulation of ambient air through the grain mass, either through natural convection or forced means using mechanical fans
Grain drying is typically performed immediately after harvest when grain moisture levels are high	Grain aeration is an ongoing process that continues throughout the storage period to maintain optimal storage conditions
Grain drying requires specialized equipment such as dryers, heaters, and fans capable of generating and controlling heat and airflow	Grain aeration equipment includes fans or blowers, ducting systems, and control systems for regulating airflow rates and distribution within the storage facility

These techniques, however, are frequently cumbersome, ineffective, and weather dependent. More control over temperature, ventilation, and humidity is possible with mechanical systems used in modern grain drying procedures, which leads to faster and more consistent drying times. Hot air is used in convective drying to remove moisture from grains. Because of its scalability and efficiency, this approach is commonly employed. The creation of automated control systems and high-capacity dryers with increased energy efficiency are examples of technological advancements in convective drying. In order to reduce drying time and energy consumption, research has also concentrated on optimizing drying characteristics such as air temperature, velocity, and relative humidity.

Maintaining consistent cool, dry temperatures is crucial for effective grain preservation, and this can be accomplished by aeration (Thorpe 2022). One common environmentally friendly method for preserving grains during storage is aeration. Grain aeration is a passive cooling technique that regulates temperature and moisture content by forcing ambient air through grains that have been stored. Grain shelf life is prolonged when proper aeration inhibits the formation of mold, insect infestation, and moisture movement. The goals of recent developments in grain aeration technology are to increase monitoring and control capabilities, lower energy usage, and improve efficiency. Grain aeration and drying are, therefore, two crucial steps in the post-harvest treatment of grains, each with a specific function in the management of grain storage. Although their goals are the same to maintain grain quality and reduce moisture content, they use distinct methods for different purposes. Table 9.1 summarizes the main distinctions between grain aeration and drying.

9.2 GRAIN AERATION

Aeration refers to the process of circulating ambient air through a mass of grain to alter its microclimate, creating conditions unfavorable for pests' development when temperature and moisture levels are suitable. It is widely employed to preserve stored grain by modifying its microclimate, discouraging harmful organisms' growth, and maintaining grain quality. Viewing the grain mass as an ecosystem where grain, microflora, and insects interact, significant storage losses often stem from favorable moisture conditions for microflora and destructive insect infestation. These losses result from the dynamic interactions among ecosystem components influenced by grain and ambient conditions. Aeration's role is to "condition" stored grain by improving existing conditions by circulating suitable air through the grain mass.

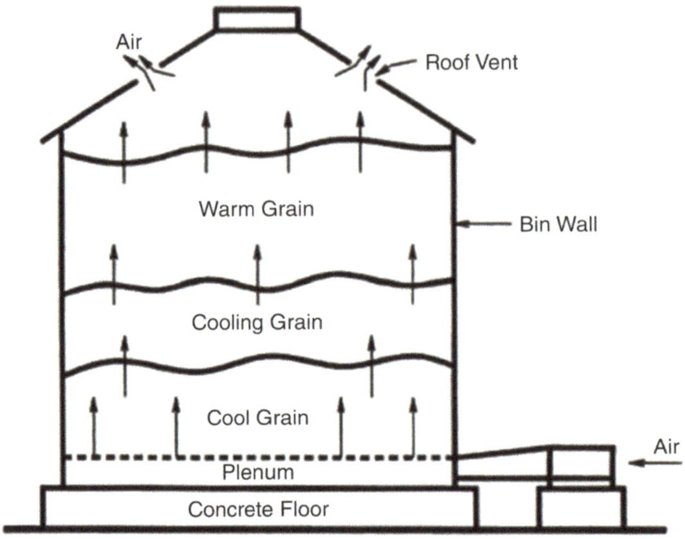

FIGURE 9.1 Representation of basic aeration system design. (© 2002 From Aeration Systems Design by Jayas and Muir. Reproduced by permission of Taylor and Francis Group, LLC, a division of Informa plc., Figure 5.1, pp. 198.)

In a typical aeration system setup, the fundamental components consist of a storage bin equipped with either a fully or partially perforated false floor, creating an air plenum, or with perforated ducts installed in-floor or on-floor; a fan linked to the plenum or duct network to push air through the grain; and one or more roof vents serving as outlets or inlets for the air as shown in Figure 9.1. Aeration aims to adjust the grain's temperature to align closely with the ambient or chilled air circulating through it. However, when the fan initiates, the entire grain mass doesn't cool simultaneously. Instead, a temperature gradient, typically a cooling front, begins forming where the air enters the grain. Consequently, the grain cools gradually in layers, resulting in distinct zones of cooled, cooling, and warm grain within the bulk. Within the cooling zone, grain temperatures vary gradually between the cooled grain temperature and the initial warm grain temperature. In the middle of this zone, the grain temperature typically falls midway between the warm and cool grain temperatures. Thus, the aeration system must remain operational until the cooling front has traversed the entire grain mass, ensuring uniform cooling throughout (Jayas, White, and Muir 1994; Navarro and Noyes 2001).

9.3 OBJECTIVES OF AERATION

Aeration aims to enhance and sustain the condition of bulk grain in storage by circulating air with desired properties until a new microclimate is established to prevent grain deterioration (Jayas and Muir 2002). Although aeration primarily aims to maintain rather than improve grain quality attributes, it can improve storage conditions in various ways based on the air properties and grain condition. Therefore, understanding aeration's effects on grain is essential before implementation (Figure 9.2).

9.3.1 Primary Objectives of Aeration

The primary objectives of aeration are as follows:

1. Cooling the grain to ambient temperature to slow or halt deterioration.
2. Achieving uniform grain temperature throughout the storage bin to prevent moisture migration caused by temperature variations.

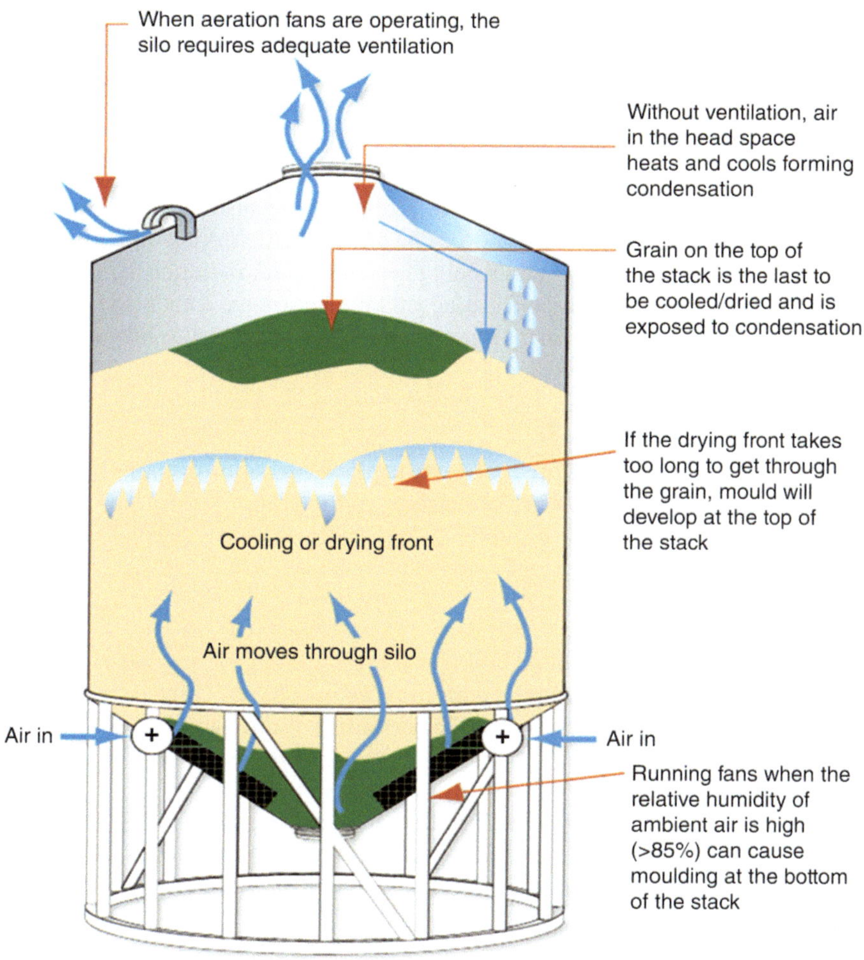

FIGURE 9.2 Schematic representation of air movement inside grain silo. (Source: Grains Research and Development Corporation, Australia Grain Storage Fact Sheet 2014.)

9.3.2 Secondary Objectives of Aeration

Additional objectives of aeration extend beyond merely adjusting grain temperature. Aeration systems can effectively mitigate odors stemming from the growth of fungi, insects, and mites, which are critical as grain quality can suffer due to undesirable smells like mustiness. Moreover, aeration allows for the incorporation of fumigants to control insects if the storage bin can be properly sealed. Aeration systems with intergranular air and fumigant circulation mechanisms provide even fumigant distribution throughout the grain bulk. The recirculation rate is typically 10% of the average airflow rate that is utilized for aeration. Aeration helps to release fumigant gases after fumigation. Furthermore, in areas with harsh winters, aeration alone can substantially minimize or completely eradicate insect infestation by cooling the grain throughout the winter months.

9.4 GRAIN AERATION TECHNIQUES

9.4.1 Natural Aeration

An economical and sustainable way to regulate the moisture content and temperature of stored grain bulk is through natural aeration. In contrast to forced aeration, which makes use of blowers or mechanical fans, natural aeration makes use of air's inherent buoyancy to generate airflow within the grain mass. The tools, equipment, and operating theories of natural aeration systems utilized in grain storage facilities are discussed below.

9.4.1.1 Principles of Natural Aeration

The idea of thermal buoyancy, which states that temperature variations between the bulk of the grain and the surrounding air produce a pressure differential that causes airflow, is the foundation of natural aeration (de Carvalho Lopes et al. 2006). The air inside the storage facility rises during the day as a result of the surface of the grain bulk being heated by solar radiation. A natural airflow pattern is produced as the warmer air rises and is dragged into the storage area by the colder outside air. The process is reversed at night when airflows in the other direction due to the grain mass cooling.

9.4.1.2 Tools and Equipment

9.4.1.2.1 Roof Vents

Grain storage facilities frequently feature roof vents to allow for natural ventilation. In order to create a chimney effect that encourages circulation through the grain mass, these vents are carefully positioned on the roof to let hot, stale air escape. Adjustable apertures on roof vents allow you to regulate the ventilation rate according to the surrounding air quality.

9.4.1.2.2 Eave Openings

Another passive ventilation element found in grain storage buildings is eave apertures. These holes, which let in fresh air, are found close to the building's sides or eaves. Insect screens or louvers can be added to eave openings to keep bugs out while permitting ventilation.

9.4.1.2.3 Ducting Systems

Grain storage facilities can benefit from the installation of ducting systems to improve natural ventilation. These systems are made up of channels or ducts that bring outside air into the storage area. Dampers or valves can be installed in ducting systems to control the distribution and rate of airflow.

9.4.1.2.4 Temperature Sensors

In order to keep an eye on the environmental conditions in grain storage facilities, temperature sensors are necessary. To measure temperature gradients and trends, these sensors are usually positioned at different points throughout the storage structure. Temperature sensor data can be utilized to determine any problems like hot patches or moisture buildup as well as to evaluate how well natural aeration is working (Liu and Li 2021).

9.4.1.3 Working of Natural Aeration System

Wind speed, humidity, ambient temperature, and storage facility design are some of the variables that affect the successful working of natural aeration systems. Natural aeration may efficiently remove moisture and heat from the grain bulk in hot, dry conditions, assisting in the preservation of quality and preventing spoiling. Natural aeration, however, could not be as efficient in humid climates or during times of low airflow; therefore, extra steps like extra chilling or drying might be required. Passive air exchange between the grain bulk and the surrounding environment is the foundation of natural aeration systems' operation. Warmer, lighter air within the storage facility is replaced by colder, denser air when temperature differentials cause airflow. This constant air exchange helps control humidity and temperature, resulting in ideal circumstances for grain storage.

9.4.2 Forced Aeration

One common technique for regulating the temperature and moisture content in grain storage facilities is forced aeration. Forcible aeration uses mechanical fans or blowers to actively circulate air through the grain mass, in contrast to natural aeration, which depends on passive airflow caused by temperature differentials (Binelo et al. 2019). The tools, equipment, and working principles of forced aeration systems used in grain storage facilities are given below.

9.4.2.1 Principles of Forced Aeration

Forced convection, in which heat and moisture are removed from grain mass by forcing air through it mechanically, is the basis for forced aeration systems. A positive pressure differential is created in the storage facility by carefully placing fans or blowers, which force air into the grain mass through ducts or channels. The air flows through the mass of the grain, absorbing heat and moisture to aid in drying and cooling. The airflow cycle is subsequently finished when the exhaust air is released from the building.

9.4.2.2 Tools and Equipment

9.4.2.2.1 Centrifugal Fans

In forced aeration systems for grain storage, centrifugal fans are frequently employed (Figure 9.3). These fans use centrifugal force to force air out of their housing, which encloses a rotating impeller. Air enters the wheel axially at first, and the centrifugal force produced by the revolving wheel's blades propels the air outward from the center. Centrifugal fans are preferred for their high airflow rates and pressure capabilities, making them suitable for large storage facilities with multiple

FIGURE 9.3 Centrifugal fans were installed in the grain bins for forced aeration.

aeration ducts. These fans typically feature three primary types of blade configurations: forward curved, straight bladed, and backward curved (Jayas and Muir 2002).

a. Forward-curved fans
   Forward-curved centrifugal fans generate greater air volume compared to radial (straight) bladed or backward-curved fans of equivalent diameter and power. However, each fan wheel is tailored for specific low-pressure air handling applications like air conditioning systems, furnaces, and specialized uses. Due to their limited capability in handling low static pressure ranges and potential overloading issues, forward-curved fans are not suitable for use in aeration systems.
b. Straight-bladed fans
   The straight-bladed centrifugal fan, also known as a radial-bladed or paddlewheel fan, is suitable for tall upright or vertical structures with low capacity where high pressures are necessary to produce relatively small airflows. In this fan type, the blades align along the radius of the wheel. While radial-blade fans can generate significant pressure compared to forward-curved fans, they are less efficient and generally unsuitable for aeration purposes. They are more suited for handling materials or operating in environments with dirty air due to their self-cleaning blade design (Jayas and Muir 2002). Notably, straight-bladed fans require the greatest power when delivering maximum air volume. To prevent overloading, a restriction should be applied on the discharge or intake side of the fan, such as a shutter, to reduce airflow rates when operating in aeration systems with static pressures lower than their design specifications.
c. Backward-curved fan
   The backward-curved (inclined) centrifugal fan is typically the preferred choice for large, relatively deep aeration systems where airflow variations may be substantial. Its impeller blades are curved plates, airfoils, or flat plates that tilt backward. Usually, the point of maximum power absorption for this fan type falls within its normal operating range, making it generally non-overloading. Among backward-inclined centrifugal fans, the airfoil-blade design tends to be the most efficient albeit more costly. The thin steel curved-blade centrifugal fan is less efficient and cheaper than the airfoil version but more efficient and expensive than the flat-blade one. The backward-inclined flat-blade centrifugal fan is the least efficient but most cost-effective option. However, the efficiency differences among these three designs are relatively minor. All three designs perform effectively in aeration systems when appropriately specified for the required pressure and airflow volumes. If the airflow increases due to operation against lower static pressure, the power absorbed decreases. This scenario resembles operating the fan in a system with a partially filled bin or aerating grain with lower airflow resistance than initially planned (Jayas and Muir 2002).

9.4.2.2.2 Axial Fans

Axial fans are another type of fan used in forced aeration systems. These fans feature a propeller-style blade that rotates around an axis, generating airflow parallel to the fan shaft (Figure 9.4).

Axial fans are often used in smaller storage facilities or for portable aeration systems due to their compact size and ease of installation. While axial fans typically have lower pressure capabilities than centrifugal fans, they are more energy-efficient and cost-effective for certain applications. Axial-flow fans direct airflow parallel to the centerline of their tubular housing. The main types used in grain aeration are vane-axial and tube-axial fans, both enclosed within a tubular casing containing the fan motor and impeller wheel. These fans are categorized as low-pressure and are suitable for horizontal storage and shallow-depth grain bins. Despite their ability to deliver the same volume as centrifugal fans, axial fans produce intense noise due to higher fan blade tip speeds. Vane-axial fans, more efficient yet costlier than tube-axial ones, feature straightening vanes bolted

FIGURE 9.4 Axial fans are installed in the grain bins for forced aeration.

or welded beside the motor, which rectifies the swirling airflow pattern. This design enhances airflow delivery for the same power compared to tube-axial fans, as the straightening vanes recover energy from turbulence. The majority of axial aeration fans adopt the vane-axial design.

9.4.2.2.3 Ducting Systems

In order to distribute airflow throughout the grain storage facility, ducting systems are crucial parts of forced aeration systems. Ducts are usually positioned above or below the grain mass and can be composed of metal or flexible materials like PVC or polyethylene. Figure 9.5 depicts the ducts that are situated beneath the grain bulk. To control airflow rates and distribution, ducting systems may use dampers, valves, or louvers.

9.4.2.2.4 Control Systems

Forced aeration system operation is monitored and controlled by control systems. These systems could have humidity and temperature sensors in addition to controllers that change the direction and speed of the fans. Automated warning systems, remote monitoring capabilities, and interfaces with other grain management technologies are some of the features of advanced control systems.

9.4.2.3 Working of Forced Aeration System

Grain moisture content, duct design, fan capacity, and ambient conditions are some of the variables that affect the efficiency of forced aeration systems. In the process of operating, fans force

FIGURE 9.5 The V-shaped duct system is placed below the grain mass with a mesh for easy airflow.

air through the grain bulk by creating positive pressure inside the storage facility. Moving through the grain mass, the air takes in heat and moisture, which helps to dry and cool the grain mass. It is imperative to exercise caution in regulating the airflow rate and aeration duration to avoid excessive or uneven drying of the grain mass.

9.4.3 Smart Aeration System

Grain storage technology has advanced significantly with smart aeration systems, which combine automation, data analytics, and sensor technologies to maximize aeration efficiency. These systems improve operational efficiency and lower post-harvest losses by enabling real-time monitoring and control of grain temperature and moisture within grain storage facilities. At grain storage facilities, smart aeration systems use a network of sensors to track the temperature, humidity, moisture content, and general weather. Real-time data is gathered by these sensors, and data analytics algorithms are used to evaluate the status of the grain and forecast possible problems like spoiling or insect infestation (Priyadarsani, Pradhan, and Jena 2023). Automated control systems modify aeration settings, including fan speed, airflow direction, and duration, based on this data to maximize grain storage conditions.

9.4.3.1 Tools and Equipment

Giving real-time information on the environmental conditions inside grain storage facilities, sensor networks are the basis of smart aeration systems. These networks could have plenum sensors, temperature, humidity, CO_2, grain moisture, and plenum sensors in addition to a weather station to gather data on ambient weather. In order to record spatial differences in temperature and moisture levels, sensors are usually placed throughout the storage structure. Sensor networks that are wireless are frequently utilized to save installation costs and enable remote monitoring. In order to find patterns, trends, and anomalies suggestive of problems with grain storage, data analytics platforms process and examine sensor data. To create predictive models for grain condition and spoiling risk, these platforms might make use of machine learning techniques. Proactive management of grain storage facilities is made possible by advanced analytics capabilities, which give operators the ability to take preventative measures to reduce risks and improve storage conditions. Aeration parameters must be changed by control systems in accordance with predetermined operational objectives and data analytics insights. These systems could be distributed control systems (DCS), cloud-based control platforms, or programmable logic controllers (PLCs). In order to minimize energy usage and maintain target temperature and moisture levels, control algorithms optimize fan operation, airflow distribution, and aeration schedules (Ahmad et al. 2020).

9.4.3.2 Working of Smart Aeration System

Smart aeration systems operate on the principle of closed-loop feedback control, where sensor data is continuously analyzed to adjust aeration parameters in real time (Figure 9.6) (Li et al. 2020; Panigrahi, Senthilkumar, and Singh 2024). The working principle of smart aeration systems can be summarized in the following steps:

1. *Data acquisition*: Sensor networks collect (1) grain temperature, humidity, and moisture content data from sensor cables from various locations and depths within the grain storage facility, (2) plenum pressure, temperature, and humidity data from the fan plenum sensor, and (3) ambient weather data from a weather station.
2. *Data analysis*: Data analytics platforms process sensor data to assess grain condition, predict spoilage risk, and identify potential issues such as hot spots or moisture accumulation. Machine learning algorithms may be used to develop predictive models based on historical data and environmental factors.
3. *Decision-making*: Based on the data analysis results, ambient weather conditions, and plenum sensor data, control systems determine the appropriate aeration parameters to maintain optimal storage conditions. This may involve operating the fans, adjusting fan speed, airflow direction, or aeration schedules to mitigate risks and optimize grain quality.
4. *Control implementation*: Control systems implement the selected aeration parameters by adjusting fan operation and airflow distribution within the grain storage facility. Automated control algorithms ensure that aeration processes are optimized in real time, minimizing manual intervention, and human error.
5. *Monitoring and feedback*: Smart aeration systems continuously monitor environmental conditions and aeration performance to validate control decisions and identify deviations from desired outcomes. Operators may receive real-time alerts or notifications regarding potential issues, allowing for timely intervention if necessary.

FIGURE 9.6 Smart aeration system. Grain bin installed with 1. Temperature sensor cable. 2. Moisture sensor cable. 3. Plenum sensor. 4. Weather station. 5. Smart fan control systems.

9.5 CATEGORIES OF AERATION SYSTEM

9.5.1 Based on Direction of Airflow

Aeration can be categorized into up-flow (pressure) and down-flow (suction) aeration based on the direction of airflow.

9.5.1.1 Pressure Airflow

Pressure airflow involves fans that compress ambient air, leading to a rise in air temperature due to the generated heat (Figure 9.7). In temperate climates where cooling conditions are prevalent, pressure airflow systems are commonly utilized. Typically, these systems heat ambient air by approximately 2–5°C (4–9°F), making them suitable for such regions (Jayas and Muir 2002).

9.5.1.2 Suction Airflow

Suction aeration, commonly employed in cooler regions, serves to prevent potential condensation of moisture in the exhaust air beneath the roof of the bin (Figure 9.8). In subtropical climates, downward aeration is advisable, especially when the grain is warm at the onset of the cool season. Suction airflow facilitates rapid "early cooling" of the top 1–3 m (3–10 ft) of grain, where insect populations tend to be most concentrated. Swift cooling of the surface and upper grain layers can effectively manage insects, potentially reducing or eliminating the need for fumigation and residual pesticides.

9.5.2 Based on Types of Storage

Differences exist in the aeration methods employed for short upright structures (bins), tall vertical structures (silos), and horizontal structures (flat storages). Additionally, airflow rate requirements vary among these structures based on their orientation (upright, vertical, or horizontal).

9.5.2.1 Aeration of Upright and Vertical Structures

Upright structures, such as bolted steel bins or welded steel tanks, have overall heights less than their diameters. Vertical structures, like concrete or steel silos, have overall heights greater than their diameters. In both upright and vertical structures, aeration can commence as soon as the grain covers the aeration ducts to a height, allowing adequate air distribution. In some cases where

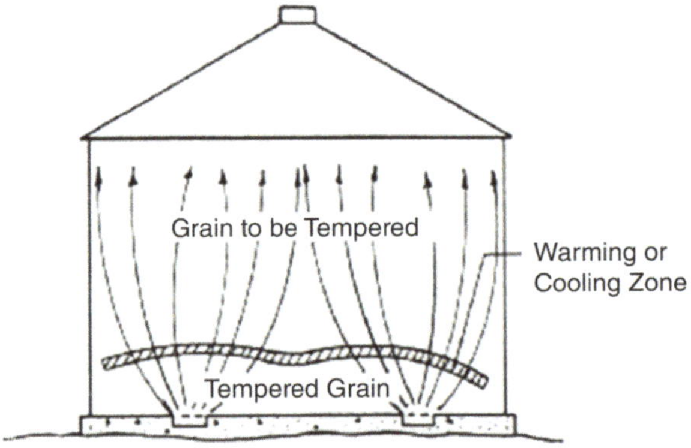

FIGURE 9.7 Up-flow/positive airflow system. (Source: McKenzie and Fossen 2002.)

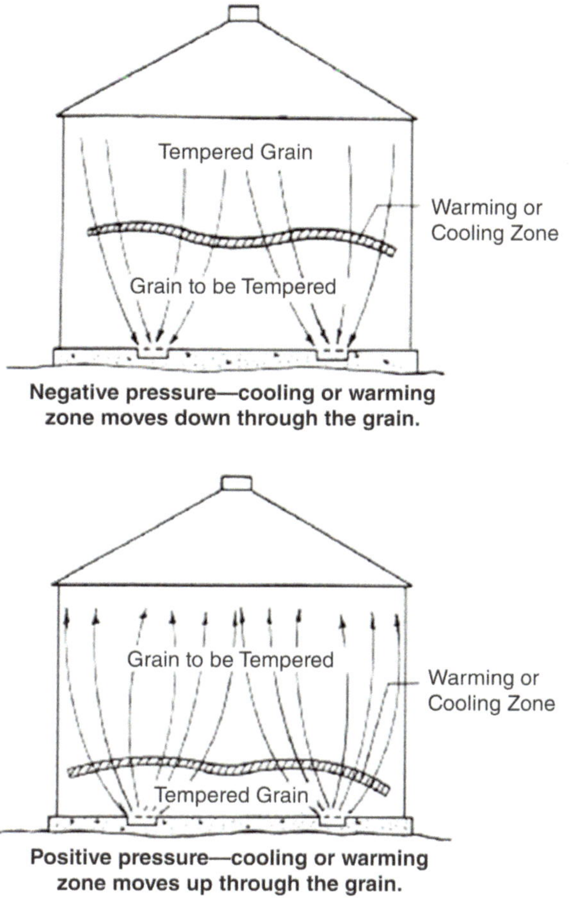

FIGURE 9.8 Down-flow/suction airflow system. (Source: McKenzie and Fossen 2002.)

maximum airflow rate is required, maintaining a lower bulk depth than the full bin can reduce airflow resistance and increase air volume. This increased airflow at shallower grain depths is particularly crucial when grain moisture exceeds safe storage levels. For storing wet grain in structures designed for the recommended dry grain aeration airflow rates, it is advisable to keep grain depths shallow (around ¼–½ of bin depth) to maintain airflow at higher-than-normal rates.

9.5.2.2 Aeration of Horizontal Structures

In horizontal structures, the arrangement of ducts in relation to the loading schedule is a crucial aspect of design. In large bins or warehouses where loading may span several weeks, the process requires careful consideration. If longitudinal ducting is present, grain can be loaded in shallow layers along the building's length to cover all duct surface areas. Once the aeration duct is covered with 1.5–2 m of grain, aeration can commence immediately. Some long warehouses with longitudinal ducts opt for two separate ducts for each half of the warehouse, with fans connected at each end. This allows the aeration system to be activated once half of the warehouse is filled. Alternatively, a duct pattern utilizing shorter ducts laid across the width of the warehouse facilitates filling to full depth starting from one end. Loading can then progress gradually to the entire length of the structure. This method is less labor-intensive than a full-length duct system. If each duct has its own fan, any part of the warehouse where the ducts are adequately covered can be aerated as needed.

9.6 ADVANTAGES OF GRAIN AERATION

1. *Moisture control*: Controlling the moisture content of stored grains is one of the main benefits of grain aeration. Aeration helps remove excess moisture from grain by circulating ambient air through the grain mass, which inhibits the growth of mold and lowers the danger of spoiling. Sustaining grain quality and extending storage life depends on effective moisture management.
2. *Temperature regulation*: Grain aeration lowers the danger of grain spoiling by limiting excessive heat buildup in warm weather and regulating the temperature within storage facilities. Aeration can reduce grain temperature and improve storage conditions by forcing colder outside air through the grain bulk. Controlling the temperature is essential to keeping insects out of the grain and preserving its quality.
3. *Pest management*: By establishing conditions that are detrimental to their growth and reproduction, aeration aids in the management of insect pests. Reduced moisture and cooler temperatures deter insects, lowering the risk of infestation and grain damage. To guarantee grain quality and safety, proper aeration techniques can support other pest management techniques, including cleaning and fumigation.
4. *Energy efficiency*: Aeration is more energy-efficient than other grain management methods like drying. For example, natural aeration uses very little energy because it depends on natural airflow and temperature variations to keep storage conditions stable. With appropriate equipment selection and control procedures, even forced aeration systems that employ mechanical fans or blowers can be configured for maximum energy efficiency.
5. *Preservation of grain quality*: In general, grain aeration reduces the danger of spoiling, mold growth, and insect infestation by establishing ideal storage conditions. Grain that is properly aerated maintains its marketability, germination potential, and nutritional content, guaranteeing that it satisfies consumer preferences and quality standards. In order to maximize grain quality and value, post-harvest management measures must include grain aeration.

9.7 LIMITATIONS OF GRAIN AERATION

1. *Weather dependency*: Grain aeration's reliance on weather is one of its main drawbacks, especially for natural aeration systems. The efficiency of aeration can be impacted by variables such as the surrounding air temperature, humidity, wind direction, and speed. Natural aeration may not be as efficient at times of high humidity or low airflow, requiring the use of additional drying or cooling techniques.
2. *Inefficient cooling*: Aeration is a useful tool for controlling the temperature in storage facilities, although it cannot be as quick to cool down as other techniques like mechanical refrigeration or drying. In large storage facilities or in hot weather, aeration relies on artificial or natural airflow to transfer heat from the grain mass to the surrounding environment, which may not be adequate to achieve quick temperature reductions.
3. *Limited moisture removal*: Grain aeration may not be appropriate for eliminating moisture from wet or recently harvested grains, but it is primarily useful for regulating the moisture content of grains that are stored. Under ideal circumstances, aeration can aid in the redistribution of moisture within the grain mass and encourage drying; nonetheless, it might not be enough to lower the moisture content to safe storage levels. In these circumstances, further drying techniques could be needed before aeration can preserve grain quality in an efficient manner.
4. *Potential energy costs*: Even though forced aeration systems are less energy-intensive than other grain management approaches, they can still be costly, especially for big storage facilities or long storage times. Fuel or electricity is needed to run mechanical fans or

blowers, and the amount of energy used varies based on the fan size, airflow rates, and control techniques. To maximize the benefits of aeration while minimizing energy expenditures, careful planning and optimization are necessary.

5. *Operational complexity*: It can be difficult to implement and maintain aeration systems; meticulous design, equipment selection, and operational monitoring are necessary. For efficient temperature and moisture control, as well as uniform airflow distribution, proper ducting design, fan positioning, and control strategies are crucial. Furthermore, specialist knowledge and expertise are needed to monitor environmental conditions within storage facilities and modify aeration parameters in response to changing conditions.

9.8 CONCLUSION

To sum up, one of the most important post-harvest management procedures for grains that affect their quality, storability, and market value is grain aeration. Grain aeration provides many benefits, such as energy efficiency, insect control, temperature regulation, moisture control, and grain quality preservation, to help preserve grain quality and reduce spoiling in storage facilities. It does, however, have certain drawbacks, including reliance on the weather, ineffective cooling, restricted ability to remove moisture, possible energy expenses, and operating complexity. Grain growers may efficiently adopt and improve aeration systems to increase grain quality and storage life while minimizing operational obstacles and costs by being aware of these benefits and limitations. New developments in aeration technology have resulted in increased effectiveness, reduced energy use, and preservation of quality. To solve issues including energy consumption, system scalability, and climate variability, more study is necessary. In the face of changing global problems, the grain business can improve sustainability, food security, and economic viability by combining cutting-edge technologies, data-driven strategies, and best practices.

REFERENCES

Ahmad, F., M. S. Younis, R. U. Zahid, and L. A. Shahid. 2020. "Machine Learning Based Grain Moisture Estimation for Real-Time Monitoring of High-Temperature Paddy Drying Silo." In *2020 IEEE 23rd International Multitopic Conference (INMIC)*, 1–6. IEEE.

Aldaco, R., D. Hoehn, J. Laso, M. Margallo, J. Ruiz-Salmón, J. Cristobal, R. Kahhat, P. Villanueva-Rey, A. Bala, and L. Batlle-Bayer. 2020. "Food Waste Management during the COVID-19 Outbreak: A Holistic Climate, Economic and Nutritional Approach." *Science of the Total Environment* 742. Elsevier: 140524.

Anukiruthika, T., F. Jian, and D. S. Jayas. 2021. "Movement and Behavioral Response of Stored Product Insects under Stored Grain Environments-A Review." *Journal of Stored Products Research* 90. Elsevier: 101752.

Binelo, M. O., V. Faoro, O. A. Kathatourian, and B. Ziganshin. 2019. "Airflow Simulation and Inlet Pressure Profile Optimization of a Grain Storage Bin Aeration System." *Computers and Electronics in Agriculture* 164. Elsevier: 104923.

Coradi, P. C., M. B. de Oliveira, L. de Oliveira Carneiro, G. A. C. de Souza, M. C. Elias, A. Brackmann, and P. E. Teodoro. 2020. "Technological and Sustainable Strategies for Reducing Losses and Maintaining the Quality of Soybean Grains in Real Production Scale Storage Units." *Journal of Stored Products Research* 87. Elsevier: 101624.

de Carvalho Lopes, D., J. H. Martins, E. de Castro Melo, and P. M. de Barros Monteiro. 2006. "Aeration Simulation of Stored Grain under Variable Air Ambient Conditions." *Postharvest Biology and Technology* 42 (1). Elsevier: 115–120.

FAO. 2022. "Food and Agriculture Organization of the United Nations." doi:10.1201/b11064-23.

Hellegers, P. 2022. "Food Security Vulnerability Due to Trade Dependencies on Russia and Ukraine." *Food Security* 14 (6). Springer: 1503–1510.

Jayas, D. S., and W. E. Muir. 2002. "Aeration Systems Design." In *The Mechanics and Physics of Modern Grain Aeration Management*, 195–249. CRC Press Inc.

Jayas, D. S., N. D. G. White, and W. E. Muir. 1994. *Stored-Grain Ecosystems*, Vol. 39. CRC Press.

Li, X., Z. Han, Q. Lin, Z. Wu, L. Chen, and Q. Zhang. 2020. "Smart Cooling-Aeration Guided by Aeration Window Model for Paddy Stored in Concrete Silos in a Depot of Guangzhou, China." *Computers and Electronics in Agriculture* 173. Elsevier: 105452.

Liu, J., and P. Li. 2021. "Control and Real-Time Data Acquisition of an Experimental Platform for Stored Grain Aeration Study." *Sensors* 21 (16). MDPI: 5403.

(Source: McKenzie, B.A., and L.V. Fossen 2002. "Managing Dry Grain in Storage." In *Agricultural Engineers Digest (AED-20)*, Perdue State University. https://www.extension.purdue.edu/extmedia/AED/AED-20.html

Mobolade, A. J., N. Bunindro, D. Sahoo, and Y. Rajashekar. 2019. "Traditional Methods of Food Grains Preservation and Storage in Nigeria and India." *Annals of Agricultural Sciences* 64 (2). Elsevier: 196–205.

Mutungi, C., F. Muthoni, M. Bekunda, A. Gaspar, E. Kabula, and A. Abass. 2019. "Physical Quality of Maize Grain Harvested and Stored by Smallholder Farmers in the Northern Highlands of Tanzania: Effects of Harvesting and Pre-Storage Handling Practices in Two Marginally Contrasting Agro-Locations." *Journal of Stored Products Research* 84. Elsevier: 101517.

Navarro, S., and R. T. Noyes. 2001. *The Mechanics and Physics of Modern Grain Aeration Management.* CRC press.

Panigrahi, S.S., T. Senthilkumar, and C.B. Singh. 2024. "Automated Aeration of Canola in Southern Alberta (Canada). In Jayas D.S., Singh C.B., Jian F. (eds), *Proceedings of the 12th International Conference on Controlled Atmosphere and Fumigation in Stored Products (CAF2024)*, 54–60, CAF Permanent Committee Secretariat, Winnipeg, Canada. https://caf.phtcenter.com/wp-content/uploads/2024/11/CAF2024-Proceedings.pdf

Parvin, S., A. Gawanmeh, and S. Venkatraman. 2018. "Optimised Sensor Based Smart System for Efficient Monitoring of Grain Storage." In *2018 IEEE International Conference on Communications Workshops (ICC Workshops)*, 1–6. IEEE.

Priyadarsani, S., A. K. Pradhan, and P. C. Jena. 2023. "Smart Food Grain Storage System Using Internet of Things (IoT): A Review." *ORYZA-An International Journal of Rice* 60 (03): 371–387. https://epubs.icar.org.in/index.php/OIJR/article/view/143147.

Senthilkumar, T., S.S. Panigrahi, and C.B. Singh. 2024. "Continuous Carbon Dioxide Monitoring in a Commercial Bin Filled with Canola." In Jayas D.S., Singh C.B., Jian F. (eds), *Proceedings of the 12th International Conference on Controlled Atmosphere and Fumigation in Stored Products (CAF2024)*, 220–226, CAF Permanent Committee Secretariat, Winnipeg, Canada. https://caf.phtcenter.com/wp-content/uploads/2024/11/CAF2024-Proceedings.pdf.

Singh, C. B., and J. M. Fielke. 2017. "Recent Developments in Stored Grain Sensors, Monitoring and Management Technology." *IEEE Instrumentation & Measurement Magazine* 20 (3). IEEE: 32–55.

Thorpe, G. R. 2022. "On the Rate of Cooling of Aerated Food Grains." *Biosystems Engineering* 222. IAgrE: 106–116. doi:10.1016/j.biosystemseng.2022.08.001.

Tushar, S. R., M. F. B. Alam, S. M. Zaman, J. A. Garza-Reyes, A. B. M. M. Bari, and C. L. Karmaker. 2023. "Analysis of the Factors Influencing the Stability of Stored Grains: Implications for Agricultural Sustainability and Food Security." *Sustainable Operations and Computers* 4 (December 2022). Elsevier B.V.: 40–52. doi:10.1016/j.susoc.2023.04.003.

Ziegler, V., R. T. Paraginski, and C. D. Ferreira. 2021. "Grain Storage Systems and Effects of Moisture, Temperature and Time on Grain Quality - A Review." *Journal of Stored Products Research* 91. Elsevier Ltd: 101770. doi:10.1016/j.jspr.2021.101770.

10 Methods of Insect and Mite Management Using Extreme Temperatures

Fuji Jian
Department of Biosystems Engineering, University of Manitoba, Winnipeg, Manitoba, Canada

10.1 INTRODUCTION

Compared to stored grain mold, insects and mites have a narrower range of temperatures for their development and multiplication. Thus, controlling temperatures to minimize or eliminate insects and mites plays an important role in integrated pest management. Lower temperatures can not only slow the rate of insect development and multiplication but mold multiplication and chemical reaction can also be restrained, which results in a longer, safe storage time. Using low or elevated temperatures to control insects is not new, and it is a common practice in the grain industry. During the First World War, wheat was heated to 58–60°C for at least 3 minutes to disinfest insects. In the early 1900s, several food-processing companies in the United States used heat to control pests. Processed food and commodities such as tea, spicy herbs, processed flour, and rice should always be kept at low temperatures to maintain food quality and prevent insects from infesting and multiplication. It is a normal practice to aerate (cool) harvested grain and keep grain at 5–15°C. This technology has been renewed in recent years due to people being more interested in organic food and less chemical residue and pollution.

Extreme temperature for control insects and mites is not only used for processed and unprocessed foods but also for fresh vegetables and fruits. Cold storage is used as a quarantine treatment or insect management tool for any products. For example, various fruits are stored between 0.5°C and 1.7°C for three weeks to kill all fruit fly eggs and larvae (Sharp and Gould 1994). Citrus is heated to disinfest Mexican fruit flies and Mediterranean fruit flies as a quarantine tool (Hallman et al. 1994; Hallman 1994; Sharp and Gould 1994). There are extensive review papers and books on this topic (Sharp and Gould 1994; Mason et al. 1998). Therefore, this chapter focuses on the control of insects and mites in processed and unprocessed products (stored products insects and mites) at low and elevated temperatures.

Using low or elevated temperatures to control stored product insects has been extensively studied and reviewed. Strang (1992) reviewed the lethal temperatures for the control of pest insects in museums. Fields (1992) reviewed the control of stored-product insects and mites with extreme temperatures. Beckett (2011) summarized the conditions that reduce the population growth of a given species at or above 99% of mortality. Stejskal et al. (2019) summarized the minimal thermal requirements for the development and activity of stored product pests, including Acari, Coleoptera, Lepidoptera, Psocoptera, Diptera, and Blattodea. Andreadis and Athanassiou (2017) reviewed insect cold hardiness and its potential application. Jian and Jayas (2024) reviewed aeration and chilled aeration for insect, mite, and microflora management and grain quality preservation. Jian (2024) reviewed lethal and mobile variations of stored product insects under low temperatures. These reviews focused on the practical application of using temperatures to control insects and mites. The method of achieving these temperatures has not been summarized. This chapter will summarize the methods currently used by the grain industry to achieve desired temperatures for insect and mite control.

DOI: 10.1201/9781003609834-10

10.2 INSECT AND MITE BODY TEMPERATURE

Insects and mites cannot regulate their body temperature except for some species before and during flight (Heinrich 1974). Under high ambient temperatures, even though insects might and can increase water evaporation through spiracles, change functions of flight muscles and body color by cuticular pigment, and change their position according to heat stress, these responses have minimum effects on their body temperatures. Insect flying requires a higher temperature than other activities, but their body temperature will be lower than 40°C during flying (Heinrich 1974), and their body temperature often becomes practically equal to their ambient temperature within minutes after flying. There are reports that honeybees and bumblebees can slightly regulate their body temperature by shivering (Heinrich 1974). There is no report showing the stored product insects and mites can regulate their body temperatures. Insect bodies are small, they have a larger ratio of surface area to their volume than larger vertebrates, and they do not sweat at high temperatures. Therefore, body temperatures of stored products, insects, and mites are the same as their ambient temperatures at most, if not all, time. When an insect is exposed to a different temperature, its body temperature will be changed to the ambient temperature in less than a few seconds or minutes.

10.3 TEMPERATURE USED FOR INSECT AND MITE CONTROL

Insects and mites require a certain temperature for their development and multiplication. The optimal temperature for the development and multiplication of most insects is 25–33°C. Lower or higher optimal temperatures will result in an increase in mortality. At certain high suboptimal temperatures, insects might have a higher development and multiplication, but the mortality is high, so the finite rate of increase of their population is low. The high suboptimal temperature is 33–42°C. At a high lethal temperature, insects will die in minutes, hours, or days. The high lethal temperature is higher than 42°C for almost all insects and mites. When the temperature is lower, insects and mites have a low rate of development and multiplication and high mortality. Insects will die in days or months at lower than their suboptimal temperature, which is 8–25°C, and it is species-dependent. When the temperature is lower than 8°C, most insects will stop biological activity, such as moving, eating, and breeding, and gradually die. The lethal time is temperature-dependent. Insects and mites will die in one minute at their supercooling point (SCP) temperature. The SCP temperature of most insects is −10°C to −25°C and depends on the species and stages of the insect. Compared to the high lethal temperature, more time is usually needed for the insect to die at the low lethal temperature because it is difficult to cool all grain to SCP, and variations of the lethal time among insect species and stages are larger at lower temperatures (especially in the low suboptimal temperature zone) than that at the high temperatures (Figure 10.1). Therefore, elevated temperature is usually used to eliminate insects and mites in days, while low temperature (lower than 8°C) is usually used to control insects and mites in weeks or months.

10.4 INSECT AND MITE CONTROL BY ELEVATED TEMPERATURE

Regardless of the methods, the purpose of the heat treatment is the same: the treated materials or facilities are heated up to a target temperature of 52°C as soon as possible and kept at 52°C for 1 h (56°C for 0.5 h, or 47°C for 4 h) at any places. Elevated heat treatments must be precise because of the narrow margin between efficacy and commodity tolerance to the temperature. Reviews of heat treatments can be found in Sharp and Gould (1994), Mangan and Hallman (1998), Elhadi (2006), Tang et al. (2007), Hansen, Johnson, and Winter (2011). Khamis et al. (2011) and Subramanyam, Mahroof, and Brijwani (2011) reviewed heat treatment of grain-processing facilities for insect management. Hansen, Johnson, and Winter (2011) reviewed the history and use of steam to control insects. Loganathan et al. (2011) found that the pupae and adults of *Callosobruchus maculatus* (F.) were equally heat tolerant. The lethal time to reduce survival by 50% (LT50) at 42°C for eggs,

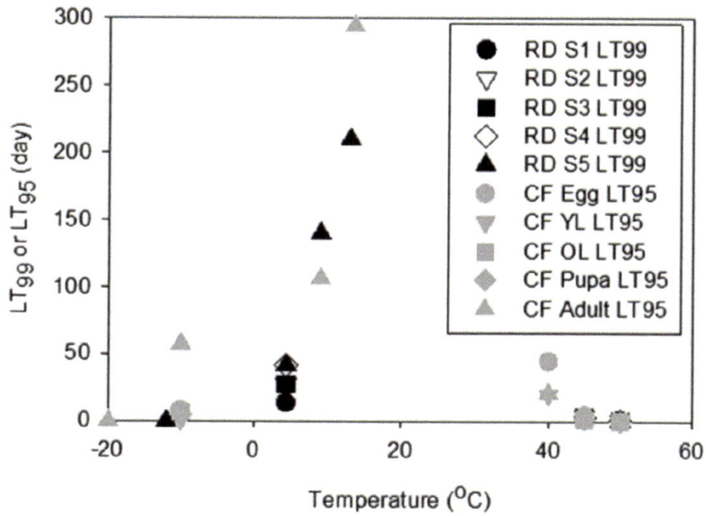

FIGURE 10.1 Lethal time (LT99 or LT95) of *Rhyzopertha dominica* and *Cryptolestes ferrugineus* (Jian 2024). In the legend, RD, *Rhyzopertha dominica*; CF, *Cryptolestes ferrugineus*; YL, young larvae; OL, old larvae, and S1 to S5 are the mixture of insect stages. S1 includes eggs and the first instars, S2 includes most of the first and second instars, S3 includes most of the second and third instars, S4 includes most of the third and fourth instars, and S5 includes the pupa, young adults, and old larvae. The number after LT is the mortality (%). Data source for *Rhyzopertha dominica*: at 50°C, 45°C, 4.4°C and 9°C and 13°C (Hagstrum, Dowdy, and Lippert 1994). Data source for *Cryptolestes ferrugineus*: at >40–10°C and 9°C and 13.5°C (Evans 1983).

larvae, pupae, and adults were 18, 57, 78, and 71 h, respectively. Pupa was the most cold-tolerant stage. The LT50 for pupae were 4907, 4262, 336, 36, and 13 min at the grain temperature of 42°C, 45°C, 50°C, 55°C, and 60°C, respectively. The LT50 of pupae at 0°C, −5°C, −10°C, and −15°C were 274, 122, 7, and 2 h, respectively. The following sections outline the most used methods of controlling insects by elevated temperature in the grain industry.

10.4.1 Methods of Heating Grain

10.4.1.1 Heating Grain Using Conventional Method

Heating grain using conventional methods includes using a dryer or fan plus a heater to directly heat grain. The grain, after heating, should be cooled so the quality of the grain can be maintained. Heating entire grain in a bin is usually not recommended because it takes a long time to heat all the grain bulk, and some places might not reach the target temperature while other places are too high and damage the grain (Qaisrani and Banks 2000). However, wet grain with insect infestation could be treated, so that drying and disinfestation can be achieved at the same time. This disinfestation is achieved by heating grain using a high airflow rate (HGHAF).

Heating stored grain for controlling insects was used in France as early as 1792 (Fields and White 2002). From the 1910s to the 1950s, this method was used by the grain industry in France, the United States, and Japan. Only Australia has used this method in the last 30 years. Several systems in Australia have been studied and used to quickly heat grain with hot air, such as the fluid bed, spouted bed, and pneumatic conveyor. All these systems are designed to pass the hot air through grain and quickly raise grain temperature to kill insects. To maintain grain quality, temperature, moisture monitoring, and air delivery are controlled. In a fluidized bed, the grain passes at a predetermined flow rate, and the air is at a velocity sufficient to suspend the grain mass above it, so a rapid heat transfer and good mixing exist. To further reduce the energy requirement, the fluid bed system can be modified to a spouted bed system, where air enters at a relatively high velocity through a nozzle in the bottom of a conical floor, causing the grain to "spout" above it. In a pneumatic conveyor, high-pressure air passes through the conveying pipe. A spouted bed can also use features of pneumatic conveying whereby the grain mass surrounds a central draft tube located above the nozzle.

The fundamental principle of HGHAF is to rapidly heat grain to above 60°C followed by standing and cooling (Dermott and Evans 1978; Claflin et al. 1986; Evans 1987). All stages of *Rhyzopertha dominica* (F.), *Sitophilus oryzae* (L.), and *Sitotroga cerealella* Olivier are disinfested when exposed to air of 80°C for 4 min (Dermott and Evans 1978). Raising grain temperature from 57°C to 64°C decreased $LT_{99.9}$ for *R. dominica* from 16.1 to 3.25 min. Holding grain at elevated temperatures for a limited time after heating (heat soak) can significantly increase insect mortality within the temperature range of 55–58°C (Evans 1981). The rate of heating is also related to insect mortality (Evans 1987). Therefore, Evans (1987) advocated that high rates of heating should be considered. Evans and Dermott (1981) recommended a process of heating the grain to 60°C rapidly and followed by cooling to about 25°C. To avoid over-drying the treated grain, water is usually added to increase the relative humidity of hot air to around 14% (Qaisrani and Banks 2000). The HGHAF is not widely applied in the grain industry due to the following issues: dust handling and management, heat recovery, cooling, post-treatment protection, high initial cost, and high energy consumption (Table 10.1).

Hansen, Johnson, and Winter (2011) summarized the study of the high temperature-controlled atmosphere for insect control. This method is similar to the method mentioned above because the result of both methods is to raise grain temperature to a lethal level and keep gain at the lethal temperature for a while (hours or minutes). Studies proved that insect mortality was higher in a controlled atmosphere at higher temperatures. However, this method is not widely accepted by the grain industry due to the same reasons mentioned above.

TABLE 10.1
Advantages and Disadvantages of Different Heat Treatment Methods

Method	Disadvantages	Advantages
Heating grain using a conventional method	Highest energy consumption, dust handling and management, grain quality might be reduced if the temperature is not well controlled, sophisticated equipment needed.	Grain can be dried at the same time.
Radio frequency and microwave treatments	Decrease baking and milling quality of some overheated materials due to the uneven heating (moisture content dependent). High initial cost.	Fast and selective heating and continuous processing of large amounts of grain. Competitive to fumigation.
Infrared heating	High initial cost, poor penetration, uneven heating and insect disinfestation. Mixing is usually required.	Simple equipment (installation and operation). High drying rate, fast disinfestation, high energy efficiency.
Steam treatment	High initial cost, increased moisture content of the treated materials, and reduced germination if treatment time is not carefully controlled.	Fast and selective heating and continuous processing of large amounts of grain. Internal and external insects, including eggs, can be killed.
Solar treatment	Unreliable (depending on the weather), various insect mortality, and labor-intensive.	Simple, inexpensive, and has potential in rural areas of developing countries.
Facility and building treatment	High energy consumption, need monitor coldest and hottest places, operation of mill and processing facility will be stopped during treatment (needs few days).	Medium cost. All insects could be killed if the desired temperature at any place is reached. Simple and well-developed technology.

10.4.1.2 Radio Frequency (RF) and Microwave (MW) Treatments

Radio frequency (RF) and microwave (MW) treatments have been widely used to disinfest insects in fresh fruits, fresh vegetables, dry fruits, dry vegetables, dairy products, rice bran, cereal grains, and oilseeds (Yadav et al. 2014). Under the MW or RF field, the energy is absorbed at a molecular level and manifests as changes in the vibrational energy of the molecules or heat by polarizing the guidance of polar diodes such as water or forced movement of ions. Thielens et al. (2018) exposed different insect species (Australian stingless Bee, Western honeybee, Desert locust, and Dor beetle) to frequencies of 2–120 GHz. All insects absorb the RF power and increase in absorbed RF power with the increase of frequency at and above 6 GHz. A shift of 10% of the incident power density to frequencies above 6 GHz would lead to an increase in absorbed power between 3% and 370%. Therefore, any RF and MW frequency can be used to disinfest insects if the RF and MW power is enough. Considering the supplied power level and practice, the MW frequency used is from 300 MHz to 300 GHz, which corresponds to a wavelength range of 1 m down to 1 mm, while RF is between 1 and 300 MHz. The US Federal Communications Commission (FCC) allows the following industrial applications: 13.56, 27.12, and 40.68 MHz for RF and 915, 2450, 5800, and 24,125 MHz for MW. The most used frequency for insect disinfestation is 27.12 MHz for RF and 2450 MHz for MW.

The longer wavelengths of RF compared with MV can penetrate further into the treated materials which results in more even heating. Both RF and MW have been widely used for the disinfestation of food materials in the grain industry, and many studies have shown that RF heating is preferred over MW heating for controlling insects due to RF heating's higher penetration depth and stronger selective heating effects (Nelson 1996; Wang et al. 2003). There are many review papers on this topic (Nelson 1996, Macana and Baik 2018; Wang et al. 2011).

The main disadvantage of RF and MW treatment is the nonuniform heating (Table 10.1). Several factors influencing the uniformity of electromagnetic heating include cavity effects and workload or product interaction, which are influenced by the thickness, geometry, and porosity of the materials. Nonuniformity increases the chance of insect pests surviving because mobile insects can run and avoid the RF and MW treatment and escape to colder spots. The treated grain usually does not have the same moisture content, and this results in the failure of insect treatment. Many studies have tried to solve this issue, such as RF or MW plus hot air and saline water and rotating and moving the treated materials. Improving the uniformity of the electromagnetic field in a cavity can be achieved by increasing the number of MW power ports, using various MW sources with different allowed frequencies, installing a mode stirrer, applying pulse (intermittent) MW heating, arranging for a mobile MW radiator, or designing the desired size and shape of the MW cavity. RF heating uniformity is improved by placing the samples in the middle of two plate electrodes and using a similar dielectric material around the samples (Jiao, Tang, and Wang 2014). MW heating in combination with conventional heating would result in more uniform heating of foods and destruction of fungi and bacteria (Fung and Cunningham 1980), and insects (Gunasekaran and Yang 2007). Gunasekaran and Yang (2007) concluded that pulsed (intermittent) MW heating resulted in more uniform temperature distribution in the samples than the continuous MW heating. Studies found that these combination treatments were more economical and effective for insect control. The control of stored-product insects in rye, corn, and wheat using a combination of MW and partial vacuum was studied by Tilton and Vardell (1982). They found that a population of insects (*S. cerealella* and *R. dominica*) were completely or partially (more than 95%) killed. Tilton and Brower (1987) treated the wheat grains with MW radiation before and after gamma radiation to control the insect population and suggested that the dose of gamma radiation could be reduced without a reduction in the actual mortality of the insects if a supplemental treatment is used. Valizadega, Pourmirza, and Safaraliza (2009) reported that a combination of MW radiation and cold storage (4°C up to 72 h) was highly compatible and synergistic for the control of *Oryzaephilus surinamensis* (L.). The efficiency of MW and IR radiation to control *S. oryzae* in soft winter wheat was studied by Kirkpatrick, Brower, and Tilton (1972), and they concluded that both MW and IR treatments can control insects but the temperature required to give full mortality was higher for MW energy. Therefore, it was suggested that the IR heating is preferable to MW heating. Gamma, infra-red, and MW radiation combinations have also been studied to control *R. dominica* in wheat by Kirkpatrick, Brower, and Tilton (1973), and the maximum reduction in the emergence was observed for gamma plus IR radiation (99%) followed by gamma plus MW (96%).

10.4.1.3 Material Surface Heating

There are several methods to heat up materials at their surface. The temperature inside the treated materials also increases after its surface temperature is raised. The methods include IR, steam, and solar heating.

IR heating: IR heating uses IR radiation energy in the IR wavelength to heat up the material, and insects are killed by the elevated temperature. The IR radiation does not heat up the medium (such as the air surrounding the treated materials); the temperature of the materials is not limited by the wet bulb temperature of the surrounding air; thus, the surface of the materials can be quickly heated to a high target temperature. Therefore, it can also be used for surface drying. Using IR radiation in the range of 0.3–430 THz for grain drying has been recognized before insect disinfestation has been applied (Kirkpatrick, Brower, and Tilton 1972). The most used IR is the medium and far IR sources, with wavelengths of 2–100 μm. In this range, different materials (including insects) have different maximum absorption wavelengths. The maximum absorption by medium rough rice occurs at 2.9 μm. Khamis et al. (2010) used a flameless catalytic IR emitter, which generated IR radiation in 3–7 μm wavelengths. Pei et al. (2018) found the adult *Sitophilus zeamais* Motschulsky and *Tribolium castaneum* (Herbst) had higher absorption in the wavelength of 5.6–7.6 μm than that of rough rice because insect consists of higher proteins and lipids than rough rice. Both rough rice and insects have a similar absorption in the wavelength of 2.8–3.3 μm, and this is related to the vibration of O–H, which is related to water. The heated temperatures of the adults are higher than that of rough rice because insects have a higher water content than rough rice.

Different methods can be used to produce IR radiation, and the configuration of IR radiation (e.g., distance and power) influences grain temperature, hence the insect mortality and grain quality after treatment. Gas-fired IR heaters produced by propane can be used to disinfest insects. However, this gas-fired IR emitter has an open flame and produces higher than 900°C temperatures. Therefore, its application is limited. Flameless catalytic IR radiation is a new technology and is now widely used in the natural gas industry to heat pipes and in the automobile industry to dry paints. Flameless IR radiation is emitted when propane or natural gas is combusted in the presence of a platinum catalyst, resulting in temperatures of about 400°C at the emitter surface. Pan et al. (2008) used a catalytic IR emitter to disinfest rough rice. The rice samples were heated for 60 s to 61.2°C, and the adults and eggs of lesser grain borers (*R. dominica*) and Angoumois grain moths (*S. cerealella*) were completely killed. Other emitters mostly use electricity and lamps, even though electricity energy is expensive. By using a laboratory benchtop flameless catalytic IR emitter, Khamis et al. (2011) reported higher grain temperatures were attained when using 113.5 versus 227.0 g of wheat and at 8.0 cm from the emitter versus 12.7 cm and during a 60 s exposure versus a 45 s exposure. The mean grain temperatures attained ranged from 108.4°C to 111.8°C. A 15-bu lot of wheat (13.5% moisture) is exposed to IR radiation, and the wheat has a maximum temperature of 48.6°C, 93% and 99% of the lesser grain borers and rice weevils, respectively, were killed in 24 h after treatment (Kirkpatrick 1975). Cogburn, Brower, and Tilton (1971) found a sequence of IR radiation and gamma radiation did not influence the effectiveness of the disinfestation. An average reduction in emergence of 52.2% and 55.3% was observed following gamma and IR radiation, respectively. When gamma treatment was followed by IR radiation or reversed sequence, the reduction in emergence was about 93.

Steam treatment: Steam is widely used for soil treatment (Hansen, Johnson, and Winter 2011) and not commercially used for grain because the grain could increase moisture content after treatment. However, empty buildings and equipment can be treated by steam. Chittenden (1896) recommended using steam on the flour mill machinery to control the Mediterranean flour moth, *Anagast kuehniella* (Zeller) (Lepidoptera: Pyralidae), and reported using 52–60°C for a few hours to kill insects. Divagar, Jian, and Cenkowski (2019) studied the effect of 105°C steam or hot air on adult mortality of three species of stored-product insect pests inside and outside of wheat kernels. The mortality of insects inside kernels was higher at lower moisture contents of wheat treated with hot air, whereas moisture content did not significantly affect the mortality of insects treated with steam. All adults of the three species outside kernels had 100% mortality in hot air for less than 75 s. In steam, the time to reach 100% mortality of insects outside or inside kernels was 3 s at any moisture content. The treatment to reach 100% mortality of insects inside kernels caused a 20% drop in germination in steam and an 81% drop in hot air. They recommend using steam instead of hot air. However, there is no further study and commercialization.

Solar heating: Even though solar heating has been widely used by farmers (Martin 1980), little has been studied using solar energy to control insects. The main reason might be due to its unreliable (depending on the weather) and various insect mortality, although this approach is simple, low cost, and has the potential for long-term storage in rural areas of developing countries (Table 10.1). Several studies have been conducted on using simple plastic solar heaters (Kitch et al. 1992), black plastic sheets (Chinwada and Giga 1996), solar heaters (Ntoukam et al. 1997), wooden surfaces (Arogba, Ugwu, and Abu 1998) to disinfest insects. Fawki et al. (2022) used clear and black polyethylene plastic bags and woven bags to treat wheat infested by *S. oryzae*. Different amounts of wheat (16, 21, and 25 kg) with 12.2% moisture content were tested under two field conditions: (1) wheat bags were kept at the field with no stacking or mixing; (2) wheat bags were mixed at the end of the day and stacked together during the night. The highest mortality achieved was 67.6% ± 30% in clear plastic bags and the grain in the bags was mixed and the bags were stacked together during the night.

10.4.2 Lethal Effect of Elevated Temperature when Different Methods Are Used

Regardless of the methods used to elevate temperature, these methods have no lethal effect on insects if the temperature is low. Therefore, the lethal effect is directly related to the temperature of the treated materials. All the methods used for insect control mainly kill insects by raising the medium (air or the treated materials) temperature. Lethal temperature and time for different insect species and stages have been extensively studied and reviewed (Stengaard Hansen et al. 2012; Fields 1992; Mangan and Hallman 1998; Tang et al. 2007). The temperature of the treated materials is influenced by the methods used for heating the material. Therefore, different treatment times and configurations are used for different heating methods.

10.4.2.1 RF and MW Heating

Exposure time and power level of RF and MW to achieve 100% mortality of insect pests depend on the treated materials (e.g., moisture content and chemical compounds), the configuration of the materials (e.g., thickness and shape), and the design and configuration of the MW or RF equipment (e.g., frequency and material handling) because these parameters influence the dielectric properties and heating. Macana and Baik (2018) and Hou, Johnson, and Wang (2016) reviewed the dielectric properties of food materials and insect pests. The dielectric constant increases with an increase in moisture content at any frequency. Insect bodies have 60–75% moisture content (Kröncke and Benning, 2022), and treated grain is usually less than 25% moisture content. Under RF or MV fields, insects can be heated up to their lethal temperature due to the difference in the moisture content between the insect and the treated materials, while the treated materials have minimum temperature change if the grain has a uniform moisture content. The temperature inside modeled insects is 1.4–1.7-fold higher than walnuts at 27 MHz (Wang et al. 2003).

Many studies prove that any stages and any species of insects in any food materials can be killed under RF and MW treatment (Yadav et al. 2014), even though some stages and species are more susceptible to RF or MW treatment (Phillips et al. 2003). Fully mortality of insects (Zhou et al. 2015) is achieved when rice is exposed to RF radiation with the intensity of 13.72 kW/kg for 4.3 min with corresponding rice temperatures ranging from 25°C to 50°C. Exposure times required to achieve 100% mortality of *Collosobruchus chinensis* (L.) at five different power levels (700, 560, 420, 280, and 140 W) in chickpeas are 100, 160, 200, 240, and 300 s, respectively; in pigeon peas, are 90, 140, 180, 240, and 280 s, respectively; and in green grams, are 70, 120, 160, 240, and 300 s, respectively (Singh, Singh, and Kotwaliwale 2012). Therefore, the required power level and treatment time for each equipment and material should be optimized for industrial application. The RF treatment protocol and optimum treatment parameters have been developed for food materials (Hou, Johnson, and Wang 2016).

10.4.2.2 Material Surface Heating

Even though the mortality of the insects is directly related to the treatment temperature of the materials (Khamis et al. 2011), the mortality of insects is influenced by the treatment methods because different methods might influence the temperature. For example, the right wavelength should be selected for the IR treatment, steam temperature and treatment time should be correctly selected for each material, and optimization of the sun heating procedure should be conducted. To achieve higher IR absorption for *S. zeamais* and *T. castaneum*, Pei et al. (2018) recommended wavelengths of 5.6–7.6 μm for disinfestation. Rough rice is heated to 60°C for 110 s by IR heating at the intensity of 2780 W/m^2, and all *S. zeamais* and *T. castaneum* are killed after 4 h (Pei et al. 2018). Tilton and Schroeder (1963) reported insects (*S. oryzae, R. dominica*, and *S. cerealella)* in or among kernels of rice were killed when grain temperature was higher than 65°C. Kirkpatrick, Brower, and Tilton (1972) reported IR treatment gave greater insect control (rice weevil) than a MW, and full mortality of *S. oryzae* can be reached after 2 min of IR treatment, and the eggs had less heat tolerance than adults (Duangkhamchan et al. 2017). Under a 250 W IR lamp, all insects can be killed in 412 s for 50 g of wheat (Keszthelyi, Lukács, and Pál-Fám 2021). Divagar, Jian, and Cenkowski (2019) recommended using 105°C steam and treating wheat kernels in 3 s to disinfest insects without significantly reducing wheat germination. To kill insects by using solar heating, the grain temperature must be higher than 42°C, and treatment time depends on the grain temperature (Fawki et al. 2022).

10.4.3 Other Benefits of Heat Treatment

Except to control insects, heat treatment might have the following direct benefits: materials drying, adult-perishing and progeny-suppressive, morphological changes of insects, and control of mold and bacteria. These heat treatment methods are also widely used for material drying due to the drying effect. Pan et al. (2008) found that rice was heated to 60°C for 60 s and removed 1.7% and 1.8% points of moisture content for the low (20.6%) and high (25.0%) moisture content rice, respectively, and all insects (adults and eggs of *R. dominica* and *S. cerealella*) were killed. Keszthelyi, Lukács, and Pál-Fám (2021) confirmed the adult-perishing and progeny-suppressive effects of IR irradiation. Rai et al. (1971) observed abnormal morphological changes in the adults of the yellow mealworm beetle from larval exposed to RF fields. Fungal spores and bacteria under RF and MW radiation can be destroyed because fungal and bacteria cells are sensitive to the heat produced during MW or RF treatment. An increase in MW heating resulted in the elimination of most fungi in 30 s (More, Magan, and Stenning 1992). Bacteria such as *Escherichia coli*, *Streptococcus faecalis*, *Staphylococcus aureus*, and *Salmonella* are destroyed by MV radiation (Fujikawa, Ushioda, and Kudo 1992; Heddleson, Doores, and Anantheswaran 1994; Atmaca et al. 1996).

10.4.4 Effect on Material Quality

The material quality is mainly influenced by the temperature of the treated material during and after the surface heating. Therefore, treatment temperature is the main factor that should be considered during heat treatment. To achieve a target temperature and uniform distribution of temperature, optimized procedures have been studied for different methods.

10.4.4.1 Conventional Heating Grain

Damage from poorly controlled grain dyer is a well-known issue in the grain industry. Therefore, treatment temperature and grain temperature after treatment are critical. The critical temperature damaging to the germination capacity of wheat is the function of the moisture contents of the grain, grain temperature at the end of the heat treatment, and the duration of heat treatment. The system studied can raise grain temperature to 65°C and air inlet temperature to 100°C. Evans (1981) reported that such treatment did not alter moisture contents or influence the baking quality of the wheat. Studies found the baking quality of dry wheat, the malting quality of paddy rice, and the oil quality of canola are all not significantly altered by a well-controlled heat treatment process at heat dosage that kills all stages of grain pest (Sutherland and Evans 1991). A large- and small-scale fluidized bed system with a throughput of 150 and 10 t/h, respectively, are operated successfully in Australia. The temperatures used in the grain industry vary from 51°C to 57°C for 12–24 h. It requires immediate cooling to avoid serious damage to stored grain.

10.4.4.2 Radio Frequency (RF) and Microwave (MW) Treatments

Macana and Baik (2018) reviewed more than 14 research publications and concluded that grain quality, including baking, milling, and storage quality, will not be significantly influenced if an optimized procedure is followed except for some minor changes, which include color, peroxide value, free fatty acid, germination, moisture content, flour protein, mixing development time, and mean peak height. After a uniform heating technique is used, it is possible to avoid these quality changes in the materials. However, if grain has a higher and uneven distribution of moisture content, damage to baking, milling, and storage (especially germination) quality might occur (Nelson 1996).

10.4.4.3 Material Surface Heating

Grain quality after surface heating is significantly influenced by the heating methods used. Wheat germination could be significantly reduced by more than 3 s steam treatment (Divagar, Jian, and Cenkowski 2019). Solar heating will usually not influence grain quality because it is difficult to achieve higher than 50°C by solar heating (Fawki et al. 2022) if no solar energy collector (heater) is used. Pan et al. (2008) concluded that rough rice heated by IR heating to 60°C followed by tempering and slow cooling can achieve simultaneous drying and disinfestation with high rice milling quality. However, Duangkhamchan et al. (2017) found that insect mortality could be achieved 100% after 2-min IR treatment (50–60°C), and the quality of heated rice could be significantly affected. Therefore, minimal changes in grain quality could be obtained using optimized temperature and exposure time. Duangkhamchan et al. (2017) found that the optimum temperature and time are 53.6°C and 1.2 min for the disinfestation of milled rice.

10.4.4.4 Facility and Building Heat Treatment

Heat treatment in food storage and processing facilities is now a common method of insect control in the world. Regardless of the shape, size, and configuration of the facility, the procedure is the same: raise the temperature at any place to 50–60°C and maintain the temperature for a while, e.g., 2–4 h after the temperature reaches 50°C for mill building, 6 h for an empty facility, and 24 h for an entire food-processing facility. A long heat treatment period is necessary for the heat to penetrate wall voids and equipment to kill insects harboring in them. Due to uneven heat distribution (especially in a large facility), fans should be used to even the temperature, so higher than 60°C at the top of the facility will not occur. At the bottom of the facility (especially since the facility has a concrete floor), the temperature could be lower than 42°C if the fan is not used (Figure 10.2) (Jian et al. 2012). Grain bulk inside a heat-treated building could not be heated to lethal temperature during the heat treatment period (24–36 h) (Jian et al. 2012), and grain residue with less than 0.01 m thick will not

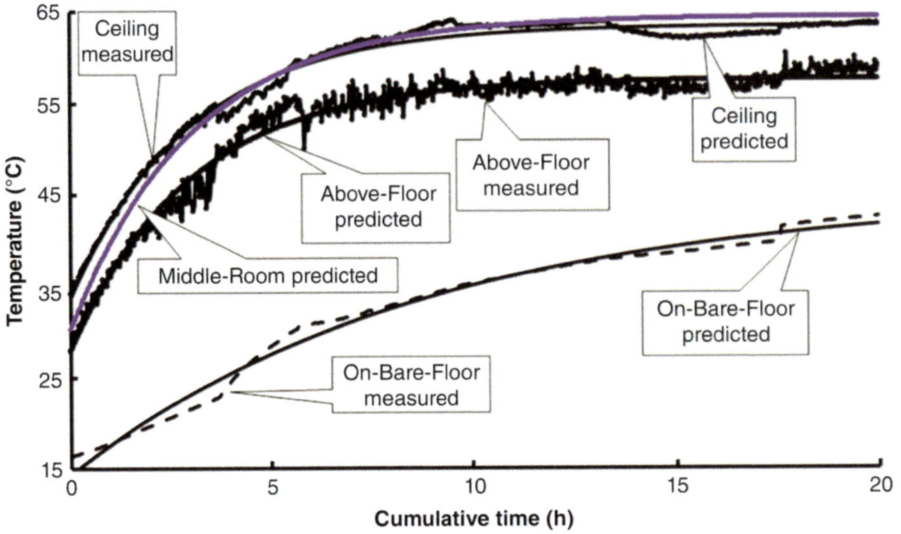

FIGURE 10.2 Measured and predicted temperatures at the South-West location (coldest) of the experimental area inside the building undergoing heat treatment. (Reprinted with permission from Fuji Jian and Jayas [2012].)

influence insect mortality (Jian et al. 2012). Inside heat treatment facilities, there are lots of heat-sensitive devices and equipment, and the building code requires the temperature not to be higher than 60°C. Therefore, temperature monitoring is important during heat treatment. Hammond (2014) outlined heat treatment protocols for different types of faculties. The protocols include temperature monitoring, heat input and sensor positioning, troubleshooting, risks, heating method, preparation and pre-treatment, different types of heat treatment, and finish of the treatment.

The effect of insect mortality is directly related to the temperature during and after heat treatment. The mortality and method of building treatment have been extensively studied. Beckett, Fields, and Subramanyam (2007) reviewed and compared different thermal procedures to treat stored product structures and equipment to control pests.

The commercialized procedure takes advantage of its medium cost and medium operation cost and tries to avoid its disadvantages (Table 10.1). Electric heaters, forced air gas heaters or steam heaters can be used to conduct a heat treatment. The most used heat source is produced by propane and natural gas because (1) the produced hot air can be directly forced into the treated facility; (2) with the forced air gas heaters, the building is placed under positive pressure during heat treatment (the entire air within the building is exchanged four to six times per hour); and (3) the forced air allows heat to reach gaps in the building and equipment much better than electric or steam heaters.

10.5 METHODS OF INSECT AND MITE CONTROL USING LOW-TEMPERATURE

10.5.1 Methods Used

The objective of any methods used is to cool products (grain) and facilities (including equipment and buildings) by using cold air. The air can be supplied by ventilation or directly expose the materials to cold air. These methods include (1) aeration (or chilled aeration), (2) facility treatment by using cold air, (3) exposing products to cold air such as turning, and (4) treatment of product at supercooling point (SCP) temperatures of the target insects. Turning, aeration, and chilled aeration (chilling) are the most used methods for controlling insects in stored grain bins. Forcing cold air into the building (facility) is the most used method for facility treatment. These methods are well-developed technology and have medium and low cost.

Stored grain in bins is usually not cooled to lower than 0°C because: (1) in the world, only a few countries located in the north of the temperate climates can apply this mothed; (2) it needs more than a few days to cool grain because the high thermal capacity of grain (so high energy consumption); (3) it might need to warm up the cold grain to avoid water condensation when the cold grain is exposed to warmer air during grain unloading; (4) condensation might also occur when lower than 0°C humid air is used to cool warm grain, and this condensed water might block the airway during grain cooling; and (5) aeration is needed if the grain is gradually warmed up by ambient air (to avoid the moisture migration due to convection currents).

Although typically not recommended, lower than 0°C air can be used to cool a small amount of bulk grain (less than a ton) under some conditions. Treatment of bagged products and grain at lower than 0°C is easily achieved, and it is a normal practice in labs and families for insect and mite disinfestation. For example, cool grain samples to −15°C and keep them at this temperature for 2 weeks or to −30°C and keep this temperature for a few minutes could kill all insects.

10.5.2 Lethal Effect and Operational Procedure

The lethal effect is directly influenced by the temperature insects exposed regardless of the methods used. The only difference among these methods is the time to reach the target temperature because different methods and treated materials have different cooling rates. There are extensive studies on the lethal effect of low temperature on insects. These studies have been extensively reviewed (Baust and Rojas 1985; Banks and Fields 1995; Strang 1992; Mason et al. 1998; Beckett 2011; Collins 2012; Fields and Abdelghany 2014; Pankaj, Shi, and Keener 2018). Aeration has been reviewed (Jian and Jayas 2024), and the operation strategy of aeration has been recommended and reviewed (Maier 1994). Therefore, this chapter will focus on some potential industrial applications.

10.5.3 Potential Industrial Applications

10.5.3.1 Small parcel/bag Treatment

Small bagged products (grain) can be treated by using low temperatures because (1) cold air can be easily distributed among small bags/parcels; (2) moisture migration usually is not an issue for bag/parcel storage; (3) condensation can be avoided because cold air will be distributed on the surface of bag/parcel at the beginning of cold treatment; and (4) pile of parcels/bags can be re-arranged. For example, cold winter air could be used to cool processed products before shipping because (1) most processed products are packed in bags or containers; (2) packed products can be piled in a shed or open if snow and rainwater are prevented; and (3) low temperature can extend the shelf life of the processed products. Packaged products often require extremely low temperatures or extremely long durations to reach temperatures required to kill insects because heat diffusion and transfer through the finished product are usually slow. Rearrange (turning) the pile could speed up the distribution of cold air. Spices, teas, medical herbs, and dried fruits can be treated by low-temperature air prior to processing.

10.5.3.2 Structure Treatment

At a place with lower than –15°C weather temperatures, structure treatment by forcing ambient air can be conducted. The recommended low temperature for the extermination of insect pests in the museum is –20°C for more than 48 h (Strang 1992). In Canada and the northern United States, air temperature in the middle of winter is sufficiently cold to control insects in food processing facilities. To maintain a longer enough cold temperature inside the treated structure, more than three continuous days at lower than –17°C are required (Fields and Abdelghany 2014). This three-day requirement is only for the structure treatment, and products (such as flour and bag stacks) must be removed from the facility because they need more time to be cooled. For the same reason, the equipment in the facility must be opened during the low-temperature treatment. Fans at different locations of the structure should be operated to ensure that sufficiently low temperatures are achieved at any location. Sensible equipment should be removed, and water must be drained to prevent water from freezing and cracking of pipe and equipment. Not like low-temperature treatment of stored grain, the temperature in the treated structure will return to room temperature one or a few days after cold treatment.

If lower than −17°C is not practicable, setting room temperature at lower than 18°C can also effectively reduce insect growth and multiplication because food processing equipment can produce a considerable amount of heat, causing ideal conditions for insect development and population growth if the produced heat is not removed efficiently and timely.

10.5.3.3 Treatment of Product at SCP Temperature

Killing all insects rapidly (e.g., less than one day) by using lower than SCP temperature is usually not recommended and only used for a small amount of product (such as less than a few hundred kg grains), surface treatment, or certain high-value products because (1) there is a wide variation of the SCP temperature due to different strain of insects, different acclimation histories and physical statuses of insects. For cold, hardy species such as rusty grain beetle and considering the above-mentioned variation, the applied temperature should be lower than –20°C to achieve a 100% mortality, which results in a high capital investment (grain chillers). (2) Even though insects and mites at their SCP temperature can be killed in few seconds, it is required one or few days to conduct this treatment because the insulation effect of the treated materials. (3) Treatments may be limited to certain times of year (such as winter). (4) Equipment or products may be damaged if the temperature is too low. (5) Facility operation will be stopped for insect treatment.

Ice-nucleating active bacteria have been used to increase the SCP temperatures of insect pests (Fields 1990; Lee et al. 1992), so insects can be killed in less time (e.g., less than a few hours) or higher than SCP temperatures. The SCPs of *Cryptolestes ferrugineus* (Stephens) adults increase from −17°C to −8°C when they are exposed to wheat that has been treated with commercial ice-nucleating active bacteria (Fields 1990). Ice-nucleating active bacteria have been reared commercially and used in snowmakers to increase the production of artificial snow. Also, the mortality of insects held at –10°C for 1 day increased from 8% to 81% with the application of the nucleators. Similar results are reported on the other four stored product insects (Lee et al. 1992). However, there is no study under field conditions.

For certain high-value products, flash freezing with liquid nitrogen to obtain –18°C is recommended. This method has been used for many years to disinfest herbs before processing (Adler 2010). Cold chambers at –20°C are used, and products are spread out in trays for treatment of about 24 h in Germany by using liquid air or nitrogen that can provide chamber temperatures of −80°C for a more rapid disinfestation in the core of a big bag (Adler 2010). Deep freezing (–22°C) facilities are used for long-term seed collection storage in agricultural genetic resource banks in the Czech Republic. Low temperature is also used to store mass-reared insects, which can be released at the right time later (Leopold 1999).

Surface treatment by using the Cryonite system has been evaluated to kill insects hiding in cracks and crevices (Jones et al. 2021). The Cryonite system turns liquid carbon dioxide to snow at an extremely low temperature (–79°C). Therefore, the temperature of insects hiding in cracks and crevices could be lower than their SCP temperatures after the snow is sprayed. The result of this method varies significantly depending on the materials and nature of the voids. A 5 s application (spraying CO_2 snow) using the jet nozzle gave 100% mortality within the top 10 cm of 5 mm wide cracks with no flour in a crack. The 30, 45, and 60 s applications achieved mortality within the top 10 cm of 79%, 97%, and 99%, respectively. Insects cannot be killed in cracks deeper than 10 cm or cracks filled with flour.

10.6 CONTROL MOLD AND INSECT SPREADING DURING HEAT AND COLD TREATMENT

10.6.1 Mold Spreading

Insect movement could spread microorganism distribution because many insects possess hairs and indentations on their exoskeletons that can act as mechanical vectors of pathogens. Maize weevils can carry numerous fungi such as *Aspergillus niger* (van Tieghem), *Aspergillus glaucus* (L.), *Aspergillus candidus* Link, *Penicillium islandicum* (Sopp), *Penicillium citrinum* (Thom), *Paecilomyces* spp., *Acremonium* spp., *Epicoccum* spp., *Fusarium semitectum* (Desm) Sacc., and yeasts (Dix 1984; Smalley 1989). Pre-harvest insect infestation can lead to the multiplication of fungi and an increase of aflatoxin in maize (Sauer, Storey, and Walker 1984).

Some insects disseminate mycotoxigenic species, and others are known to use spoilage molds as a food source, while some species avoid certain fungal species. Hairy fungus beetles feed fungi and aflatoxin and transmit *Salmonella enterica* (Minor & Popoff) over long distances (Tsai, Mason, and Woloshuk 2007). The mycotoxin, F-2, produced by the fungus *Fusarium roseum* var. *graminearum* Snyder and Hansen, is ingested and retained by confused flour beetle and lesser mealworm (Eugenio et al. 1970). There are reports on the association among fungi, mycotoxin, and insects, e.g., confused flour beetle with *Fusarium graminearum* (Schwein) Petch and *Fusarium tricinctum* (Corda) Sacc. (Wright, Casas, and Harein 1973) and hairy fungus beetles with *A. flavus* and *Penicillium purpurogenum* (Stoll.) (Tsai, Mason, and Woloshuk 2007). Ochratoxin A, citrinin, rubratoxin B, and patulin have little or no effect on the growth of confused flour beetles and slightly affect the larval growth of black flour beetles (Dunkel 1988). Lots of stored product insects feed on mycotoxin with no toxic effect on insects (Tsai, Mason, and Woloshuk 2007). Therefore, control of insect movement by aeration can also minimize fungi development and mycotoxin production because fungi can develop and multiply at lower than the minimum movement temperature of stored grain insects.

10.6.2 Insect Spreading

If the temperature is gradually decreased, insects gradually lose their mobility. Therefore, insects will be trapped at their locations during cold treatment. Not all insects and mites move to warmer and wet places and avoid cold or hot temperatures (Jian 2019; Anukiruthika, Jian, and Jayas 2021). Both flying and crawling can escalate the migration at warm locations. When the temperature is higher than 40–42°C, most insects will move away from the hot location because the high temperature has a lethal effect, and insects could detect this lethal temperature. If they can move, their response to hot temperatures might be similar for different species because they would run out of the hot place. Rusty grain beetles prefer 35–37.5°C and move away when the temperature is higher than 37.5°C. There are no insects found in the grain higher than 42°C in the grain columns, which also has cold grain (less than 42°C) at other locations (Jian et al. 2002). The adults of *Tribolium confusum* Duval had a similar preference for the grain surface or inside the grain when the temperature is ≤37.5°C (Lü, Jian, and Fields 2021). When at 40°C, *T. confusum* adults prefer the surface of the grain instead of staying inside the grain, and patch-searching behavior is disrupted. A half-hour after introduction, more than 70% of adults moved into the cold grain and would not move out of the grain at 45°C air temperature. Their refuge seeking increases with the increase of the air temperature. After the adults move into the grain pile, rising temperatures will not drive adults out of grain piles during the heat treatment (Lü, Jian, and Fields 2021) because the pile surface temperature is higher than the core temperature of the grain pile. At higher than 40°C, insects would run away from these high-temperature locations and take refuge if they can move before the heat knocks them down. Adults of *T. castaneum* aggregate in zones of lower temperatures during heat treatment, and the number of captured *T. castaneum* beetles in traps also spikes during the heat treatment (Semeao et al. 2013). Therefore, insects fleeing during heat treatment should be considered before heat treatment.

10.7 EFFECT OF MOISTURE CONTENT (RELATIVE HUMIDITY)

Even though moisture contents (relative humidity, RH) have some influence on insect mortality during heat and cold treatment, the main factor resulting in insect death is the temperature because free water at lower than 0°C is iced (solid) water, and at higher lethal temperatures relative humidity is usually low except during steam treatment. Studies also confirmed that relative humidity did not affect the SCP, and cold-tolerance of eggs is not affected by relative humidity (Jacob and Fleming 1986). Jian et al. (2015) found that the two measured grain moisture contents (11.1% and 16.0%) did not affect the measured chill-coma temperatures for all tested four insect *species (C. ferrugineus, T. castaneum, S. oryzae,* and *S. zeamais*). Evans (1983) also found that chill-coma temperatures had no relationship with the relative humidity.

However, at an insect's low suboptimal temperatures, humidity plays an important role in insect development; insects can survive in a wide range of temperatures at their optimum RH, and vice versa. For example, the development threshold of *Cryptoleste pusilloides* is 90% and 70% RH at 11°C and 13°C, respectively (calculated using data published by Lefkovitch [1964]). At 50% RH, the species could only complete development at 27.5–30°C. At 90% RH, the thresholds for *Cryptolestes capensis* (Waltl) and *Cryptolestes turcicus* (Grouvelle) are 11°C and 15°C, respectively (calculated using data published by Lefkovitch (1962a). While *C. turcicus* has a 27.5°C threshold at 50% (Lefkovitch, 1962b), and *C. capensis* has a 30°C threshold at 10% RH (Lefkovitch 1962a). At 20°C or 40°C and higher than 70% RH, pupae of red flour beetles may be formed but cannot at <70% RH (Howe 1956). At 45% RH, the survival rate of *C. ferrugineus* increased 3.5-fold after acclimation to 9°C, while it increased 7.9-fold at 70% RH. Evans (1983) hold young adults of *C. ferrugineus, O. surinamensis, R. dominica, Sitophilus granarius* (L.), *S. oryzae,* and *T. castaneum* at 30°C or 32°C; and gradually cool to 13.5°C and held at that temperature; cooled gradually to 9°C and held at that temperature; or transferred directly to 9°C (temperature decrease rate was 3.5°C per week). The survival time of the young larvae was shorter in 45% RH than in 70% RH (Evans 1983). Therefore, the effect of relative humidity (grain moisture content) at insect's low suboptimal temperatures should be considered for controlling insect populations.

10.8 CONCLUDING REMARKS

There are clearly circumstances, where the ideal conditions for insect or mite control are difficult or impossible to achieve. Climate change and an increase in global temperature would increase the grain temperature at each geographical location. Extreme temperatures become more often, and climate change makes the climate temperature unstable. This change makes the application of insect and mite control at low and high temperatures more challenging. It is needed to study insect multiplication and development under this rapidly changing condition. Also, it is interesting to know the mortality of insects treated at elevated temperatures (e.g., higher than 45°C) and rapidly exposed to low temperatures (e.g., lower than 0°C).

Most methods discussed in this chapter have been widely used by the grain industry as effective methods to control insects. These methods, such as heat treatment and chilled aeration, usually have a high energy consumption. Energy becomes more expensive, along with climate change and government taxation. Recent interest in temperature disinfestation is to develop rapid, reliable, and economical disinfestation strategies and methods. Therefore, more affordable methods should be studied.

REFERENCES

Adler, C. (2010). "Physical control of stored product insects." *International European Symposium on Stored Product Protection "Stress on Chemical Products"*, 25–26 May, 2009, Berlin, Germany. https://ojs.openagrar.de/index.php/JKA/article/view/641

Andreadis, S. S., and C. G. Athanassiou. 2017. "A Review of Insect Cold Hardiness and Its Potential in Stored Product Insect Control." *Crop Protection* 91 (January): 93–99. doi:10.1016/j.cropro.2016.08.013.

Anukiruthika, T., F. Jian, and D. S. Jayas. 2021. "Movement and Behavioral Response of Stored Product Insects Under Stored Grain Environments - A Review." *Journal of Stored Products Research* 90 (January): 101752. doi:10.1016/j.jspr.2020.101752.

Arogba, S., F. Ugwu, and J. Abu. 1998. "The Effects of Sun-Drying Surfaces and Packaging Materials on the Storability of Cowpea (*Vigna unguiculata*) Seed." *Plant Foods for Human Nutrition* 53: 113–120. doi: https://doi.org/10.1023/A:1008059817139.

Atmaca, S., Z. Akdag, S. Dasdag, and S. Celik. 1996. "Effect of Microwaves on Survival of Some Bacterial Strains." *Acta Microbiologica et Immunologica Hungarica* 43: 371–378.

Banks, J., and P. Fields. 1995. *Physical Methods for Insect Control in Stored-Grain Ecosystem*. New York: Marcel Dekker.

Baust, J. G., and R. R. Rojas. 1985. "Review—Insect Cold Hardiness: Facts and Fancy." *Journal of Insect Physiology* 31 (10): 755–759. doi:10.1016/0022-1910(85)90067-8.

Beckett, S. J. 2011. "Insect and Mite Control by Manipulating Temperature and Moisture before and During Chemical-Free Storage." *Journal of Stored Products Research* 47 (4): 284–292. doi:10.1016/j.jspr.2011.08.002.

Beckett, S. J., P. G. Fields, and B. Subramanyam. 2007. "Disinfestation of Stored Products and Associated Structures Using Heat." In *Heat Treatments for Postharvest Pest Control: Theory and Practice*, 182–237. UK: CABI. doi:10.1079/9781845932527.0182.

Chinwada, P., and D. P. Giga. 1996. "Sunning as a Technique for Disinfesting Stored Beans." *Postharvest Biology and Technology* 9 (3): 335–342. doi:10.1016/S0925-5214(96)00010-5.

Claflin, J. K., D. E. Evans, A. G. Fane, and R. J. Hill. 1986. "The Thermal Disinfestation of Wheat in a Spouted Bed." *Journal of Stored Products Research* 22 (3): 153–161. doi:10.1016/0022-474X(86)90009-3.

Cogburn, R. R., J. H. Brower, and E. W. Tilton. 1971. "Combination of Gamma and Infrared Radiation for Control of the Angoumois Grain Moth in Wheat23." *Journal of Economic Entomology* 64 (4): 923–925. doi:10.1093/jee/64.4.923.

Collins, D. A. 2012. "A Review on the Factors Affecting Mite Growth in Stored Grain Commodities." *Experimental and Applied Acarology* 56 (3): 191–208. doi:10.1007/s10493-012-9512-6.

Dermott, T., and D. E. Evans. 1978. "An Evaluation of Fluidized-Bed Heating as a Means of Disinfesting Wheat." *Journal of Stored Products Research* 14 (1): 1–12. doi:10.1016/0022-474X(78)90012-7.

Divagar, D., F. Jian, and S. Cenkowski. 2019. "Control of Three Species of Stored-Product Insects in Wheat Treated With Steam and Hot Air." *Journal of Economic Entomology* 112 (4): 1964–1974. doi:10.1093/jee/toz080.

Dix, D. E.. 1984. *Interactive Bionomics of the Maize Weevil, Sitophilus Zeamais Motschulsky, and Aspergillus Flavus Link*. University of Georgia.

Duangkhamchan, W., A. Phomphai, R. Wanna, L. Wiset, J. Laohavanich, F. Ronsse, and J. G. Pieters. 2017. "Infrared Heating as a Disinfestation Method Against *Sitophilus oryzae* and Its Effect on Textural and Cooking Properties of Milled Rice." *Food and Bioprocess Technology* 10 (2): 284–295. doi:10.1007/s11947-016-1813-z.

Dunkel, F. V. 1988. "The Relationship of Insects to the Deterioration of Stored Grain by Fungi." *International Journal of Food Microbiology* 7 (3): 227–244. doi:10.1016/0168-1605(88)90042-6.

Elhadi, M. Y.. 2006. "Effects of Insect Quarantine Treatments on the Quality of Horticultural Crops." *Stewart Postharvest Review* 2 (1): 1–18. doi:10.2212/spr.2006.1.6.

Eugenio, C., E. De Las Casas, P. K. Harein, and C. J. Mirocha. 1970. "Detection of the Mycotoxin F-2 in the Confused Flour Beetle and the Lesser Mealworm123." *Journal of Economic Entomology* 63 (2): 412–415. doi:10.1093/jee/63.2.412.

Evans, D. E. 1981. "The Influence of Some Biological and Physical Factors on the Heat Tolerance Relationships for *Rhyzopertha dominica* (F.) and *Sitophilus oryzae* (L.) (Coleoptera: Bostrychidae and Curculionidae)." *Journal of Stored Products Research* 17 (2): 65–72. doi:10.1016/0022-474X(81)90018-7.

Evans, D. E. 1983. "The Influence of Relative Humidity and Thermal Acclimation on the Survival of Adult Grain Beetles in Cooled Grain." *Journal of Stored Products Research* 19 (4): 173–180. doi:10.1016/0022-474X(83)90005-X.

Evans, D. E. 1987. "The Influence of Rate of Heating on the Mortality of *Rhyzopertha dominica* (F.) (Coleoptera: Bostrychidae)." *Journal of Stored Products Research* 23 (1): 73–77. doi:10.1016/0022-474X(87)90038-5.

Evans, D. E., and T. Dermott. 1981. "Dosage-Mortality Relationships for *Rhyzopertha dominica* (F.) (Coleoptera: Bostrychidae) Exposed to Heat in a Fluidized Bed." *Journal of Stored Products Research* 17 (2): 53–64. doi:10.1016/0022-474X(81)90017-5.

Fawki, S., P. G. Fields, F. Jian, and A. Yousery. 2022. "Control of *Sitophilus oryzae* (Coleoptera: Curculionidae) in Bags of Wheat Using Solar Radiation." *Journal of Stored Products Research* 96 (March): 101941. doi:10.1016/j.jspr.2022.101941.

Fields, P. G. 1990. *The Cold-Hardiness of Cryptolestes Ferrugineus and the Use of Ice Nucleation-Active Bacteria as a Cold-Synergist.*

Fields, P. G. 1992. "The Control of Stored-Product Insects and Mites With Extreme Temperatures." *Journal of Stored Products Research* 28 (2): 89–118. doi:10.1016/0022-474X(92)90018-L.

Fields, P. G., and A. Abdelghany. 2014. "Low Temperature to Control Insects in Grain Bins and Flour Mills." In *Proceedings of the 11th International Working Conference on Stored-Product Protection*, 331–378. Chiang Mai, Thailand.

Fields, P. G., and N. D. G. White. 2002. "Alternatives to Methyl Bromide Treatments for Stored-Product and Quarantine Insects." *Annual Review of Entomology* 47 (1): 331–359. doi:10.1146/annurev.ento.47.091201.145217.

Fujikawa, H., H. Ushioda, and Y. Kudo. 1992. "Kinetics of *Escherichia coli* Destruction by Microwave Irradiation." *Apply and Environmental Microbiology* 58, 920–924.

Fung, D. Y. C., and F. E. Cunningham. 1980. "Effect of Microwaves on Microorganisms in Foods." *Journal of Food Protection* 43 (8): 641–650. doi:10.4315/0362-028X-43.8.641.

Gunasekaran, S., and H.-W. Yang. 2007. "Optimization of Pulsed Microwave Heating." *Journal of Food Engineering* 78 (4): 1457–1462. doi:10.1016/j.jfoodeng.2006.01.018.

Hagstrum, D. W., A. K. Dowdy, and G. E. Lippert. 1994. "Early Detection of Insects in Stored Wheat Using Sticky Traps in Bin Headspace and Prediction of Infestation Level." *Environmental Entomology* 23 (5): 1241–1244. doi:10.1093/ee/23.5.1241.

Hallman, G. J. 1994. "Mortality of Third-Instar Caribbean Fruit Fly (Diptera: Tephritidae) Reared at Three Temperatures and Exposed to Hot Water Immersion or Cold Storage." *Journal of Economic Entomology* 87 (2): 405–408. doi:10.1093/jee/87.2.405.

Hallman, G. J., M. O. Nisperos-Carriedo, E. A. Baldwin, and C. A. Campbell. 1994. "Mortality of Caribbean Fruit Fly (Diptera: Tephritidae) Immatures in Coated Fruits." *Journal of Economic Entomology* 87 (3): 752–757. doi:10.1093/jee/87.3.752.

Hammond, D.. 2014. *Heat Treatment for Insect Control: Developments and Applications.* Elsevier.

Hansen, J. D., J. A. Johnson, and D. A. Winter. 2011. "History and Use of Heat in Pest Control: A Review." *International Journal of Pest Management* 57 (4): 267–289. doi:10.1080/09670874.2011.590241.

Heddleson, R. A., S. Doores, and R. C. Anantheswaran. 1994. "Parameters Affecting Destruction of *Salmonella* spp. by Microwave Heating." *Journal of Food Science* 59, 447–451.

Heinrich, B.. 1974. *Thermoregulation in Endothermic Insects: Body Temperature Is Closely Attuned to Activity and Energy Supplies. Science185.*

Hou, L., J. A. Johnson, and S. Wang. 2016. "Radio Frequency Heating for Postharvest Control of Pests in Agricultural Products: A Review." *Postharvest Biology and Technology* 113 (March): 106–118. doi:10.1016/j.postharvbio.2015.11.011.

Jacob, T. A., and D. A. Fleming. 1986. The Effect of Temporary Exposure to Low Temperature on the Viability of Eggs of *Oryzaephilus surinamensis* (L.)(Col., Silvanidae).

Jian, F. 2019. "Influences of Stored Product Insect Movements on Integrated Pest Management Decisions." *Insects* 10 (4): 100. doi:10.3390/insects10040100.

Jian, F., V. Chelladurai, D. S. Jayas, and N. D. G. White. 2015. "Three-Dimensional Transient Heat, Mass, and Momentum Transfer Model to Predict Conditions of Canola Stored Inside Silo Bags Under Canadian Prairie Conditions: Part II. Model of Canola Bulk Temperature and Moisture Content." *Transactions of the ASABE*, August, 1135–1144. doi:10.13031/trans.58.11053.

Jian, F., P. G. Fields, D. S. Jayas, N. D. G. White, and M. Loganathan. 2012. "Measured and Predicted Temperatures in a Grain Processing Building Under Heat Treatment - 1. Temperature Profiles During Heat Treatment." *Canadian Biosystems Engineering/Le Genie Des Biosystems Au Canada* 54 (March 2016). doi:10.7451/cbe.2012.54.3.1.

Jian, F., and D. S. Jayas. 2012. "The Ecosystem Approach to Grain Storage." *Agricultural Research* 1 (2): 148–156. doi:10.1007/s40003-012-0017-7.

Jian, F., and D. S. Jayas. 2024. "Aeration and Chilled Aeration for Insect, Mite, and Microflora Management and Grain quality Preservation". In *Control and Management of Pests in Stored Products*, edited by D. S. Jayas, CRC Press. 10.1201/9781003309888-10.

Jian, F. 2024. "Lethal and mobile variation of stored product insects and mites under low temperatures." *Journal of Stored Products Research*, 105;102240. https://doi.org/10.1016/j.jspr.2023.102240

Jian, F., D. S. Jayas, N. D. G. White, and W. E. Muir. 2002. "Temperature and Geotaxis Preference by *Cryptolestes ferrugineus (*Coleoptera : Laemophloeidae) Adults in Response to 5°C/m Temperature Gradients at Optimum and Hot Temperatures in Stored Wheat and Their Mortality at High Temperature." *Environmental Entomology* 31 (5): 816–826. http://www.bioone.org/doi/full/10.1603/0046-225X-31.5.816

Jiao, Y., J. Tang, and S. Wang. 2014. "A New Strategy to Improve Heating Uniformity of Low Moisture Foods in Radio Frequency Treatment for Pathogen Control." *Journal of Food Engineering* 141 (November): 128–138. doi:10.1016/j.jfoodeng.2014.05.022.

Jones, K., W. Morse, N. White, and P. Fields. 2021. "Effectiveness of Cryonite System in Treatment of Stored Product Insects," 277–283.

Keszthelyi, S., H. Lukács, and F. Pál-Fám. 2021. "Effects of Different Infra-Red Irradiations on the Survival of Granary Weevil Sitophilus Granarius: Bioefficacy and Sustainability." *Insects* 12 (2): 102. doi:10.3390/insects12020102.

Khamis, M., B. Subramanyam, H. Dogan, P. W. Flinn, and J. A. Gwirtz. 2010. "Effectiveness of Flameless Catalytic Infrared Radiation against Life Stages of Three Stored- Product Insect Species in Stored Wheat." In *10th International Working Conference on Stored Product Protection*, 695–700. doi:10.5073/jka.2010.425.248.

Khamis, M., B. Subramanyam, H. Dogan, P. W. Flinn, and J. A. Gwirtz. 2011. "Effects of Flameless Catalytic Infrared Radiation on *Sitophilus oryzae* (L.) Life Stages." *Journal of Stored Products Research* 47 (3): 173–178. doi:10.1016/j.jspr.2010.11.002.

Kirkpatrick, R. L. 1975. "Infrared Radiation for Control of Lesser Grain Borers and Rice Weevils in Bulk Wheat (Coleoptera: Bostrichidae & Curculionidae)." *Journal of the Kansas Entomological Society* 48 (1): 100–104.

Kirkpatrick, R. L., J. H. Brower, and E. W. Tilton. 1972. "A Comparison of Microwave and Infrared Radiation to Control Rice Weevils (Coleoptera: Curculionidae) in Wheat." *Journal of Kansas Entomological Society* 45 (4): 434–438.

Kirkpatrick, R. L., J. H. Brower, and E. W. Tilton. 1973. "Gamma, Infra-Red and Microwave Radiation Combinations for Control of *Rhyzopertha dominica* in Wheat." *Journal of Stored Products Research* 9 (1): 19–23. doi:10.1016/0022-474X(73)90034-9.

Kitch, L. W., G. Ntoukam, R. E. Shade, J. L. Wolfson, and L. L. Murdock. 1992. "A Solar Heater for Disinfesting Stored Cowpeas on Subsistence Farms." *Journal of Stored Products Research* 28 (4): 261–267. doi:10.1016/0022-474X(92)90007-D.

Kröncke, N., and R. Benning. 2022. "Determination of Moisture and Protein Content in Living Mealworm Larvae (*Tenebrio molitor* l.) using Near-Infrared Reflectance Spectroscopy (NIRS)." *Insects* 13 (560). doi: 10.3390/insects13060560.

Lee, R. E., J. M. Strong-Gunderson, M. R. Lee, and E. C. Davidson. 1992. "Ice-Nucleating Active Bacteria Decrease the Cold-Hardiness of Stored Grain Insects." *Journal of Economic Entomology* 85 (2): 371–374. doi:10.1093/jee/85.2.371.

Lefkovitch, L. P. 1962a. "The Biology of *Cryptolestes capensis* (Waltl.) (Coleoptera: Cucujidae)." *Bulletin of Entomological Research* 53: 529–535.

Lefkovitch, L. P. 1962b. "The Biology of *Cryptolestes turcicus* (Grouvelle) (Coleoptera: Cucujidae), a Pest of Stored and Processed Cereals." *Proceeding of the Zoological Society of London* 138 (1): 23–35.

Lefkovitch, L. P. (1964). "The Biology of *Cryptolestes pusilloides* (Steel & Howe) (Coleoptera, Cucujidae), a Pest of Stored Cereals in the Southern Hemisphere." *Bulletin of Entomological Research* 54, 649–656.

Leopold, R. A. 1999. "Cold Storage of Insects for Integrated Pest Management." In *In Temperature Sensitivity in Insects and Application in Integrated Pest Management*, edited by G. J. Hallman and D. L. Denlinger., 235–267. CRC Press.

Loganathan, M., D. S. Jayas, P. G. Fields, and N. D. G. White. 2011. "Low and High Temperatures for the Control of Cowpea Beetle, *Callosobruchus maculatus* (F.) (Coleoptera: Bruchidae) in Chickpeas." *Journal of Stored Products Research* 47 (3): 244–248. doi:10.1016/j.jspr.2011.03.005.

Lü, J., F. Jian, and P. Fields. 2021. "Preference of Tribolium Confusum (Coleoptera: Tenebrionidae) Adults to Grain Patches Under Constant and Rising Temperatures." *Journal of Stored Products Research* 90 (January): 101756. doi:10.1016/j.jspr.2020.101756.

Macana, R. J., and O. D. Baik. 2018. "Disinfestation of Insect Pests in Stored Agricultural Materials Using Microwave and Radio Frequency Heating: A Review." *Food Reviews International* 34 (5): 483–510. doi:10.1080/87559129.2017.1359840.

Maier, D. E. 1994. "Chilled Aeration and Storage of U.S. Crops—A Review." In *Proceedings of the 6th International Working Conference on Stored-Product Protection.*

Mangan, R. L., and G. H. Hallman. 1998. *Temperature Sensitivity in Insects and Application in Integrated Pest Management.* UK: Westview Press.

Martin, W. 1980. "Solar Grain Drying." *Agricultural Research* 28 (9): 7–9.

Mason, L. J., C. A. Strait, C. P. Woloshuk, and D. E. Maier. 1998. "Controlling Stored Grain Insects with Ozone Fumigation." In *In 7th International Working Conference on Stored-Product Protection*, 14–19.

More, H. G., N. Magan, and B. C. Stenning. 1992. "Effect of Microwave Heating on Quality and Mycoflora of Sorghum Grain." *Journal of Stored Products Research* 28 (4): 251–256. doi:10.1016/0022-474X(92)90005-B

Nelson, S. O. 1996. "Review and Assessment of Radio-Frequency and Microwave Energy for Stored-Grain Insect Control." *Transactions of the ASAE* 39 (4): 1475–1484. doi:10.13031/2013.27641.

Ntoukam, G., L. W. Kitch, R. E. Shade, and L. L. Murdock. 1997. "A Novel Method for Conserving Cowpea Germplasm and Breeding Stocks Using Solar Disinfestation." *Journal of Stored Products Research* 33 (2): 175–179. doi:10.1016/S0022-474X(96)00042-2.

Pankaj, S. K., H. Shi, and K. M. Keener. 2018. "A Review of Novel Physical and Chemical Decontamination Technologies for Aflatoxin in Food." *Trends in Food Science & Technology* 71 (January): 73–83. doi:10.1016/j.tifs.2017.11.007.

Pan, Z., R. Khir, L. D. Godfrey, R. Lewis, J. F. Thompson, and A. Salim. 2008. "Feasibility of Simultaneous Rough Rice Drying and Disinfestations by Infrared Radiation Heating and Rice Milling Quality." *Journal of Food Engineering* 84 (3): 469–479. doi:10.1016/j.jfoodeng.2007.06.005.

Pei, Y., T. Tao, G. Yang, Y. Wang, W. Yan, and C. Ding. 2018. "Lethal Effects and Mechanism of Infrared Radiation on *Sitophilus zeamais* and *Tribolium castaneum* in Rough Rice." *Food Control* 88 (June): 149–158. doi:10.1016/j.foodcont.2018.01.012.

Phillips, T. W., S. Halverson, T. Bigelow, G. Mbata, W. Halverson, M. Payton, S. Forester, and P. Ryas-Duarte. 2003. "Microwave Treatment of Flowing Grain for Disinfestation of Stored Product Insects." In *Advances in Stored Product Protection. Proceedings of the 8th International Working Conference on Stored Product Protection, York, UK, 22-26 July 2002*, 626–628. Wallingford: CABI. doi:10.1079/9780851996912.0626.

Qaisrani, K., and J. Banks. 2000. "The Prospects for Heat Disinfestation of Grain." In *Australian Postharvest Technical Conference.*

Rai, P. S., H. J. Ball, S. O. Nelson, and L. E. Stetson. 1971. "Morphological Changes in Adult *Tenebrio molitor* (Coleoptera: Tenebrionidae) Resulting from Radio Frequency or Heat Treatment of Larvae or Pupae." *Annals of the Entomological Society of America* 64, 1116–1121.

Sauer, D. B., C. L. Storey, and D. E. Walker. 1984. "Fungal Populations in U.S. Farm-Stored Grain and Their Relationship to Moisture, Storage Time, Regions and Insect Infestation." *Phytopathology* 74: 1050–1053.

Semeao, A. A., J. F. Campbell, J. M. S. Hutchinson, R. J. Whitworth, and P. E. Sloderbeck. 2013. "Spatio-Temporal Distribution of Stored-Product Insects Around Food Processing and Storage Facilities." *Agriculture, Ecosystems & Environment* 165 (January): 151–162. doi:10.1016/j.agee.2012.11.013.

Sharp, J. L., and W. P. Gould. 1994. "Control of Caribbean Fruit Fly (Diptera: Tephritidae) in Grapefruit by Forced Hot Air and Hydrocooling." *Journal of Economic Entomology* 87 (1): 131–133. doi:10.1093/jee/87.1.131.

Singh, R., K. K. Singh, and N. Kotwaliwale. 2012. "Study on Disinfestation of Pulses Using Microwave Technique." *Journal of Food Science and Technology* 49 (4): 505–509. doi:10.1007/s13197-011-0296-1.

Smalley, E. B. 1989. "Identification of Mycotoxin Producing Fungi and Conditions Leading to Aflatoxin Contamination of Stored Food Grains Identification of Mycotoxin Producing Fungi and Conditions Leading to Aflatoxin Contamination of Stored Food Grains." In *In Mycotoxin Prevention and Control in Food Grains*, edited by R. L. Semple, A. S. Frio, P.A. Hicks, and J. V. Lozare.

Stejskal, V., T. Vendl, Z. Li, and R. Aulicky. 2019. "Minimal Thermal Requirements for Development and Activity of Stored Product and Food Industry Pests (Acari, Coleoptera, Lepidoptera, Psocoptera, Diptera and Blattodea): A Review." *Insects* 10 (5): 149. doi:10.3390/insects10050149.

Stengaard Hansen, L., M. Åkerlund, T. Grøntoft, M. Ryhl-Svendsen, A. L. Schmidt, J.-E. Bergh, and K.-M. Vagn Jensen. 2012. "Future Pest Status of an Insect Pest in Museums, Attagenus Smirnovi: Distribution and Food Consumption in Relation to Climate Change." *Journal of Cultural Heritage* 13 (1): 22–27. doi:10.1016/j.culher.2011.05.005.

Strang, T. J. K. 1992. "A Review of Published Temperatures for the Control of Pest Insects in Museums." *Collection Forum* 8: 41–67.

Subramanyam, B., R. Mahroof, and M. Brijwani. 2011. "Stewart Postharvest Review," no. December. doi:10.2212/spr.2011.3.1.

Sutherland, J. W., and D. E. Evans. 1991. "A Review of High-Temperature Disinfestation of Grain." In *Grain Industry Pest Control Conference*, 1–10.

Tang, J., E. Mitcham, S. Wang, and S. Lurie, eds. 2007. *Heat Treatments for Postharvest Pest Control: Theory and Practice.* UK: CABI. doi:10.1079/9781845932527.0000.

Thielens, A., D. Bell, D. B. Mortimore, M. K. Greco, L. Martens, and W. Joseph. 2018. "Exposure of Insects to Radio-Frequency Electromagnetic Fields from 2 to 120 GHz." *Scientific Reports* 8 (1): 3924. doi:10.1038/s41598-018-22271-3.

Tilton, E. W., and H. W. Schroeder. 1963. "Some Effects of Infrared Irradiation on the Mortality of Immature Insects in Kernels of Rough Rice." *Journal of Economic Entomology* 56 (6): 727–730. doi:10.1093/jee/56.6.727.

Tilton, E. W., and H. H. Vardell. 1982. "Combination of Microwaves and Partial Vacuum for Control of Four Stored-Product Insects in Stored Grain." *Journal of the Georgia Entomological Society* 17: 106–112.

Tilton, E. W., and J. H. Brower. 1987. "Ionizing Radiation for Insect Control in Grain and Grain Products." *Cereal Foods World* 32 (4): 330–333.

Tsai, W.-T., L. J. Mason, and C. P. Woloshuk. 2007. "Effect of Three Stored-Grain Fungi on the Development of Typhaea Stercorea." *Journal of Stored Products Research* 43 (2): 129–133. doi:10.1016/j.jspr.2006.01.001.

Valizadega, O., A. A. Pourmirza, and M. H. Safaraliza. 2009. "Combination of Microwaves Radiation and Cold Storage for Control of *Oryzaephilus surinamensis* (L.) (Col. Silvanidae)." *Journal of Biological Sciences* 9 (3): 231–236. doi:10.3923/jbs.2009.231.236.

Wang, S., J. Tang, R. P. Cavalieri, and D. C. Davis. 2003. "Differential Heating of Insects in Dried Nuts and Fruits Associated with Radio Frequency and Microwave Treatments." *Transactions of the ASAE* 46 (4). doi:10.13031/2013.13941.

Wang, Y., Y. Li, S. Wang, L. Zhang, M. Gao, and J. Tang. 2011. "Review of Dielectric Drying of Foods and Agricultural Products." *International Journal of Agricultural and Biological Engineering* 4: 1–19.

Wright, V., D. L. E. Casas, and P. K. Harein. 1973. "The Development and Reproduction of Confused Flour Beetles Exposed to the Mycotoxin, Trichothecene." *Proceeding of North Central Branch* 28: 175.

Yadav, D. N., T. Anand, M. Sharma, and R. K. Gupta. 2014. "Microwave Technology for Disinfestation of Cereals and Pulses: An Overview." *Journal of Food Science and Technology* 51 (12): 3568–3576. doi:10.1007/s13197-012-0912-8.

Zhou, L., B. Ling, A. Zheng, B. Zhang, and S. Wang. 2015. "Developing Radio Frequency Technology for Postharvest Insect Control in Milled Rice." *Journal of Stored Products Research* 62 (May): 22–31. doi:10.1016/j.jspr.2015.03.006.

11 Dielectric Heating and Advanced Thermal Technologies

Indumathi Chokkanathan, Ashok Suraj, Banavaram Obanna Gari Madhu, Anantika Sharma, and Saravanan Shanmugasundaram
National Institute of Food Technology Entrepreneurship and Management, Thanjavur, Tamil Nadu, India

11.1 INTRODUCTION

In recent decades, the agricultural sector has witnessed a steady increase in food production attributed to the adoption of modern tools, advanced technologies, and updated agricultural practices. However, there remains a question regarding the adequacy of preservation technologies for these food products. Notably, in developed countries, consumer preferences lean toward quality food free from insecticidal residues, molds, and contamination. In less developed regions, inadequate handling and storage practices in warm, humid climates accelerate the hastened decay of food items stored. Alternative control techniques are becoming more popular in place of contact pesticides and fumigants due to growing concerns about the harmful effects of pesticide residues. Remarkably, of the 14 fumigants enumerated by Bond., Alexander, and Monro (1984), only phosphine and methyl bromide continue to be extensively employed in contemporary times, mostly in underdeveloped nations. While methyl bromide, effective but ozone-depleting, was discontinued in developed countries by 2005, phosphine is still used in developing nations due to its ease of application. Nonetheless, during the last ten years, insect resistance to phosphine has grown. While insect infestation and microbial growth can be avoided, extended storage of food commodities may still result in qualitative losses, encompassing changes in visual appearance, nutritional decline, insect presence, mold contamination, and mycotoxin existence, some of which are challenging in visual detection.

In the evolving landscape of thermal processing technologies, recent research underscores the vital role of dielectric heating and other advanced thermal technologies, as shown in Figure 11.1, effectively disinfecting food commodities. These advancements not only facilitate the disinfestation of food materials but also ensure the preservation of nutritional and sensory properties with minimal alterations. Moreover, these technologies present a more effective approach, considering both energy efficiency and cost-effectiveness. Furthermore, dielectric heating and other advanced thermal technologies are utilized in the disinfestation of diverse food products, serving as valuable aids in processing during drying, thawing, baking, cooking, tempering, and roasting.

11.2 DIELECTRIC HEATING

Dielectric heating is a kind of electromagnetic energy that, as a unique thermal technique, heats food using its intrinsic qualities. Unlike typical thermal processing, dielectric heating creates heat within the product, allowing for faster volumetric heating and a shorter time to inactivate microorganisms as well as all life stages of the insects.

DOI: 10.1201/9781003609834-11

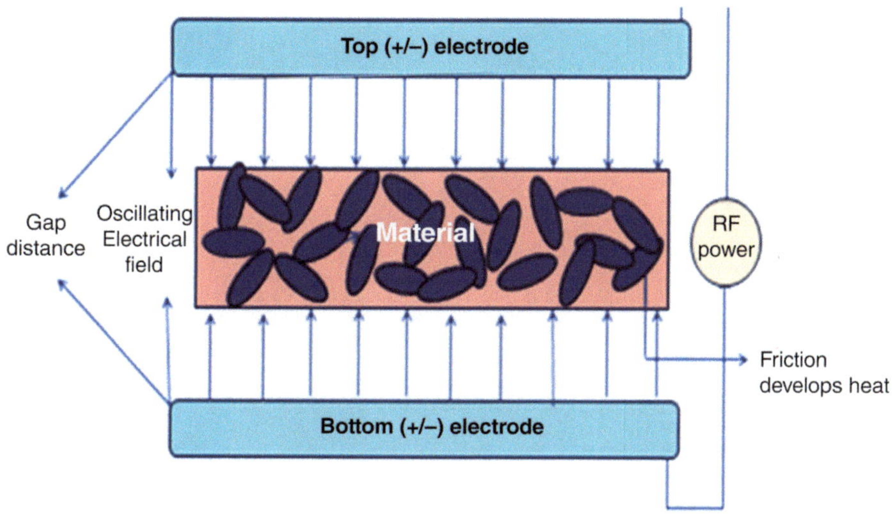

FIGURE 11.1 Thermal technologies for disinfecting stored products.

11.2.1 Principle of Dielectric Heating in Disinfestation

The heating of electrically insulating materials brought on by dielectric loss is known as dielectric heating. Materials having polar molecules will undergo continuous rotation by aligning with the oscillating field, such as an electromagnetic wave or a rapidly oscillating electric field. Dipole rotation, also referred to as dipolar polarization, is this. The molecules move in the opposite direction as the field changes. Electrical forces are used by rotating molecules to push, pull, and interact with other molecules, providing heat energy to nearby molecules and atoms in the material. During radio frequency (RF) treatments, heat is generated within the product due to molecular friction resulting from oscillating dipole molecules, space charge displacement, and migrating ions caused by the applied alternating electric field, as shown in Figure 11.2.

Insects are more vulnerable to enhanced dielectric heating than food grains because they have a greater moisture content and electrical conductivity (Paul et al. 2020). Thus, the disinfestation is attained in stored food products with optimized parameters of RF heating.

11.2.2 Radiofrequency Heating

RF is identified as a non-ionized waveform featuring a wavelength of up to 11 m and a frequency falling in the range of 1–300 MHz, as indicated by R. S. Hassan et al. (2019). These electromagnetic waveforms, referred to as RF waves, possess the capability to penetrate dielectric materials, developing volumetric heating through the dipole rotation process (Hou, Johnson, and Wang 2016; Hassan et al. 2019). The notable attributes of RF heating, including greater penetrative intensity, enhanced thermal effectiveness, minimized influence on product quality due to decreased thermal effects, reduced treatment time, user-friendly application, absence of ecological contaminants, and energy efficiency, position RF heating as a cutting-edge and advanced technology (Indumathi et al. 2021).

In the realm of RF heating, water molecules within insects undergo continuous rotation, creating friction with cell organelles and contributing to cell disruption, ultimately leading to the demise of the insects. Consequently, all stages of insects can be effectively eliminated without adversely affecting the product's quality, given the implementation of optimizing the current and exposure

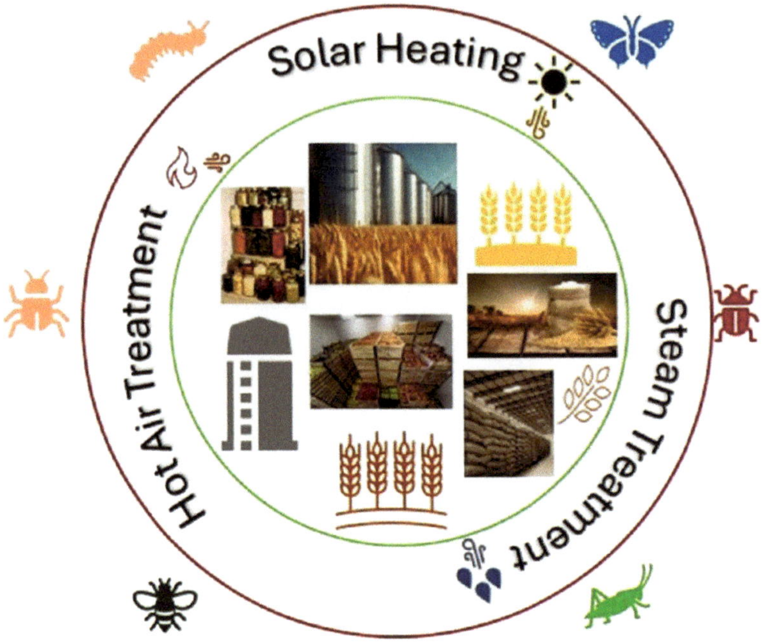

FIGURE 11.2 The working concept of RF heating.

time of RF electrodes to reach the lethal temperature for insects has been studied (Rashkovan et al. 2003). Several research efforts have concentrated on exploring RF heating for the purpose of suppressing insect proliferation in agricultural produce (Hou, Liu, and Wang 2019).

Yu, Shrestha, and Baik (2015) conducted an analysis revealing that insects exhibited a weaker electric field strength in comparison to canola seeds, with the range spanning from 0.46 to 0.73 times that of canola seeds. Notably, surpassing the power dissipation observed within canola seeds, the energy dissipation in insects exhibited a variance ranging approximately from 1.7 to 17.3 times. Furthermore, insects exhibited a dielectric loss factor that exceeded that of canola seeds, ranging from approximately 6.8 to 43.9 times higher. The authors emphasized how bulk densities, specific heats, and power dissipation of insects significantly influenced the temperature increase rate. When considering these aspects for both canola seeds and insects, it was determined that the temperature increment rate in insect bodies exceeded that of canola seeds by 1.4–10.1 times. In summary, by taking these variables into account for both canola seeds and insects, it was established that the calculated rate of temperature increase in insect bodies was 1.4–10.1 times greater than the observed increase in canola seeds.

For disinfecting semolina, characterized by limited storage capacity, an optimal and effective approach involves an 8-min exposure time with a 0.8 A RF current load. Achieving a complete eradication of *Tribolium castaneum* with a 100% disinfestation rate has been established using this approach while ensuring that color, moisture content, and gluten content remain unaltered. Under RF heating at 64.40°C ± 1.40°C, the mortality rate for all life stages of *T. castaneum* was 100%, with a susceptibility order observed as egg > pupa > larva > adult. Minimal changes in gluten were observed, as evidenced by scanning electron microscopy (SEM) micrographs revealing subtle structural distinctions when compared to untreated semolina (Indumathi et al. 2023).

In the case of the pulse beetle, *Callosobruchus maculatus*, a prevalent pest affecting stored legumes, exposure to both RF and MW irradiation at different electrode heights and power levels resulted in lethal times for achieving 99.99% mortality (LT99.99). The egg stage exhibited a time range of 3.62–8.23 min, while larvae experienced durations ranging from 4.70 to 7.27 min. Pupae displayed a variation of 7.10–8.96 min, and adults demonstrated times ranging from 5.40 to 7.73 min. Notably, samples of black gram treated with RF displayed no significant reduction in lysine, aspartic acid, and proline content. This underscores the efficacy of RF heating in disinfecting legumes without compromising their essential properties, thus addressing a limitation associated with MW heating (Tiwari et al. 2021).

In the investigation into the impact of RF treatment on peanuts and peanut oil quality, it was found that subjecting them to RF treatment at the least electrode height and conveyor speed of 180 mm and 5 m/h, respectively, led to a notable temperature increase of up to 89.96°C. This temperature rise had a significant effect on the quality of both peanuts and peanut oil. Specifically, it resulted in a considerable reduction in moisture content from 7.45% to 2.11%, as well as a decrease in protein content from 22.17% to 20.94%, while simultaneously increasing the oil extraction yield from 51.34% to 70.31%. Remarkably, there was no significant alteration observed in color values. During a storage period of up to two months at room temperature (35.30°C ± 2°C) in closed plastic containers, the quality of the oil experienced a decline. Nevertheless, all parameters remained within the acceptable range specified by CAC (Codex Alimentarius Commission). Moreover, the RF treatment exhibited complete mortality of *Caryedon serratus* and demonstrated a notable reduction in aflatoxin B1 and B2 contents by 92.64% and 90.51%, respectively. These findings underscore the potential of RF treatment as a feasible technique for implementation in peanut processing industries (Appugol et al. 2022).

The infested peanuts underwent RF exposure at selected electrode heights of 230 and 250 mm, paired with conveyor speeds ranging from 2.5 to 12.5 m/h for each height. Lethal times for 50 (LT_{50}) and 99% (LT_{99}) mortality were determined for both egg and adult stages of *Caryedon serratus*. Results revealed that the 230 mm electrode height yielded the shortest LT_{50} and LT_{99} values: 5.51 and 14.22 min for the egg stage and 3.01 and 11.60 min for the adult stage, respectively.

In the scenario of prolonged wheat flour storage without infestation, achieving complete mortality across every phase of *T. castaneum* was possible by subjecting them to RF heating with 0.5 kV and 0.9 A electrode current and a 15 min exposure time. The observed susceptibility order of *T. castaneum* to RF heating was adults < larvae < pupae < eggs. In-depth examinations utilizing both differential scanning calorimetry (DSC) and Fourier transform infrared (FTIR) spectroscopy demonstrated no adverse effects on wheat flour treated with RF. SEM analysis revealed the structural characteristics. Importantly, sensory evaluations of RF-treated wheat flour-made chapatti exhibited no significant impairment ($p < .05$) compared to chapatti produced from untreated wheat flour. Consequently, RF heating emerges as a viable alternative to pesticide-based disinfestation methods (Indumathi et al. 2021).

Food grains, while serving as dielectric materials, play a dual role as electric capacitors and resistors in RF processes. This enables the seamless storage and transformation of electrical energy into thermal energy when exposed to an electromagnetic field, as highlighted by Ling, Cheng, and Wang (2020). This ensures stable coupling power, fixed frequency maintenance, and control over power and feedback. Huang and colleagues demonstrated that 6 min of exposure to RF heating resulted in varying temperature increases (6.2°C, 6.6°C, and 5.9°C) for Indian meal moth larvae at different layers of soybeans when enclosed in plastic containers (Huang, Chen, and Wang 2015).

For industrial-scale disinfestation using RF processes, a greater rise in temperature is crucial (Ling, Cheng, and Wang 2020). In their 2017 study, Jiao and colleagues discovered that subjecting rough, brown, and milled rice to mild RF heating lasting 5 min efficiently managed immature rice weevils, including both eggs and adults (Jiao et al. 2017). Shrestha and Baik (2019) conducted a computer simulation of RF heating on wheat kernels and rusty grain beetles, revealing a temperature difference of 14°C between the adult rusty grain beetle and the wheat kernel. Non-uniform heating led to beetle migration to non-heated zones within the sample, reducing the selective heating impact of RF on pests (Ling, Cheng, and Wang 2020).

Turmeric powder underwent RF disinfestation to assess its efficacy. Various combinations of electrode height and conveyor speed were tested. Results indicated that an electrode height of

180 mm had the lowest LD 50 and $LD_{99.99}$ of 1.96 and 5.34 min, respectively. Complete mortality was achieved at an electrode height of 225 mm with a conveyor speed of 25 m/h, though with a higher $LD_{99.99}$. Interestingly, there was no significant reduction in curcumin content between untreated (3.10%) and treated (3.08%) turmeric powder. Essential components like ar-turmerone, turmerone, and curlone were present in both control and RF disinfested samples. RF disinfestation of turmeric powder is recommended for effective pest control while preserving nutrients and avoiding chemical insecticides (Mangang et al. 2022).

Numerous research endeavors have delved into the application of RF heating to combat the surge of pests in agricultural yields. Extensive investigations conducted in laboratories have reported a 100% success rate in disinfestation, particularly in in-shell walnuts plagued by *Amyelois transitella* (navel orange worm), *Cydia pomonella* (codling moth), and *Rhyzopertha dominica* (lesser grain borers). Likewise, rough rice harboring Indian meal moth (*Plodia interpunctella*) and cowpea weevil (*C. maculatus*) experienced total mortality after exposure to RF heating at temperatures ranging from 55°C to 58°C for 5–10 min. Remarkably, this treatment did not adversely affect legume germination or quality.

Effective strategies have also been devised to address *Sitophilus oryzae* infestation in brown, rough, and milled rice, showcasing the capability of RF energy to swiftly elevate temperatures in bulk materials like rice and wheat flour. Experimental studies on stored rapeseeds demonstrated a 100% mortality rate *for T. castaneum* at 80°C for adults and 60°C for larval phases across various moisture content levels, without compromising seed quality when employing suitable RF heating parameters.

The red flour beetle, *T. castaneum*, a globally significant post-harvest pest, found effective control in wheat flour through optimized RF protocols based on the correlation between temperature and disinfestation effects, as outlined by Indumathi, Devi, and Shanmugasundaram. (2021).

Treating maize grains affected by maize weevil within the temperature range of 50–60°C (with 14% moisture on a wet basis) exhibited no changes in the fundamental protein structure during RF treatment. However, it revealed advancements in emulsification and oil-holding characteristics, as reported by Hassan et al. in 2019. Likewise, subjecting rice moths to RF treatment at 57°C resulted in thorough disinfestation without impacting the fat, protein, or moisture levels in milled rice. This process led to heightened sensory qualities and enhanced cooking performance.

Notably, temperatures exceeding 100°C in RF treatment may be applied for developing the enhanced adsorbents and emulsions that involve refining their development process. Simultaneously, employing lower temperatures becomes a strategic approach to safeguard functional properties and counteract infestations. Thus, RF heating plays the crucial role of extending the shelf life of stored foods by disinfesting all life stages of insects without compromising food product quality under optimized RF parameters.

11.2.3 Microwave Heating

Exhibiting protective effects against insects due to their short-wavelength electromagnetic radiation, microwaves offer various advantages when employed for disinfestation. These benefits include a decrease in processing time, improved preservation of quality, lower energy consumption, and the absence of harmful residual products, as highlighted by Vadivambal, Jayas, and White (2007). Exploring the effects of microwaves on the life stages of *T. castaneum* in barley, the researchers investigated the vulnerability of eggs. The study found that higher power and longer exposure times notably influenced germination characteristics, shedding light on the impact of microwaves on the mortality of *T. castaneum* life stages in barley. Numerous studies have documented the successful application of microwave disinfestation across a range of agricultural commodities, including figs, raisins, cowpeas, green gram, chickpeas, fava beans, apricots, and dates. However, the use of microwave treatment can lead to impairment of germination in seeds and a decline in the quality of the grain (Dalmoro et al. 2018; Taheri, Brodie, and Gupta 2020). This detrimental impact is linked to non-uniform heating, resulting from variations in cold and hot spot temperatures during microwave treatment (Taheri, Brodie, and Gupta 2020). Microwaves selectively warm areas with dielectric fluids, generating hot spots. Ahmady, Mousa, and Zaitoun (2016) conducted bioassay treatments on the confused flour beetle, *T. confusum*, utilizing microwave technology. Their research revealed a complete elimination of the beetles when exposed to a household microwave system at 400 W for 25 s, achieving a 100% reduction in population.

Crucially, optimizing radiation uniformity and hot spot temperature is necessary to mitigate the adverse effects of microwave treatment. Strategies aimed at mitigating these effects involve employing and utilizing hot air for particle agitation (Kudra 1989), mode stirrer's application (Ye et al. 2017), and employing precision in managing the temperature of the product's surface and microwave output results (Koné et al. 2013). The moisture content of the product is a crucial factor in microwave disinfestation. Greater moisture content in the external grain layer, as opposed to the internal layer, results in heightened heating and subsequent disinfestation (Yadav et al. 2008; Vasilev, Vasilev, and Samarin 2019). In a particular investigation, researchers examined lentil seeds exposed to microwaves at different power levels and a fluidizing air temperature reaching 60°C, assessing factors such as antioxidant enzyme activity, macronutrient composition, and cooking quality (Taheri, Brodie, and Gupta 2020). Feng and Tang (1998) proposed using a microwave-fluidized bed for stirring grains, aiming to uphold consistent heat distribution. Microwave-fluidized bed drying presents clear advantages over traditional fluidized bed dryers, ensuring more efficient preservation of grain quality. Through the regulation of fan and conveyor speed, as well as operational temperature, uniform heat distribution is attained at every stage of the process (Manickavasagan et al. 2006), potentially preventing further infestation.

Microwave treatment has proven effective in various applications related to grain and pulse processing. Divekar et al. (2017) found that exposure of wet grains, including wheat, barley, and canola, to microwave power for 14–56 s within the range of 400–600 W resulted in modifications to the surface temperature, influenced by moisture content and grain size. This, in turn, led to slight alterations in the nutritional profile of the grains. Additionally, the study demonstrated that microwave treatment of corn grains at 300 W, with 50–60°C product temperature, had significant effects on the fatty acid composition, increasing oleic acid content while reducing linoleic acid content. The findings suggest that microwave heating less than 60°C shows potential for quality control, as indicated by Hassan, Pawelzik, and von Hoersten (Hassan, Pawelzik, and von Hoersten 2021), who highlighted its potential for disinfection and disinfestation at an industrial level. Overall, microwave technology emerges as a versatile tool with implications for both quality enhancement and industrial-level applications in the agricultural and food processing sectors.

11.3 INFRARED RADIATION

The electromagnetic spectrum, spanning from microwaves to the visible region, includes infrared wavelengths ranging from 0.5 to 100 μm. Renowned for its energy efficiency, environmental friendliness, and reduced water usage compared to traditional heating methods, infrared technology boasts advantages like swift heating, uniform heat distribution, high energy efficiency, greater heat transfer rates, and elevated quality of the product. These attributes have led to the widespread adoption of infrared technology in various food processing operations, including boiling, drying, polyphenol, peeling, and recovery of antioxidants; freeze-drying; heating; roasting; microbiological inhibition; and sterilization of juice, grains, and bread; and cooking (Aboud et al. 2019).

In the early 1960s and 1970s, the use of infrared irradiation for grain disinfestation gained prominence. Its remarkable working principle capabilities have established it as a preferred disinfection method in grain industries over conventional chemical approaches (Ramaswamy, Krishnamurthy, and Jun 2012). For microbial cell inactivation, infrared treatment employs three modes: thermal inactivation, induction heating, and DNA damage. Studies conducted by Ha and Kang (2013) and Meenu, Guha, and Mishra (2018) have shown that infrared heating induces structural damage to microbial cells, causing cytoplasm and cell membrane disruption. This disruption leads to the leakage of cellular contents and eventual rupture, resulting in fatal outcomes. Achieving a 2 log10 CFU/g[1] reduction of *Aspergillus niger* spores within 5 min (Jun and Irudayaraj 2003), effectively deactivated. Infrared treatment-induced thermal inhibition affects bacterial cell proteins, ribosomes, cell covers, RNA, and DNA. Particularly, proteins, cell walls, and RNA prove more susceptible to infrared heating than conductive heating (Sawai et al. 1995). Due to its limited penetration depth, infrared proves more effective for sterilizing food surfaces (Aboud et al. 2019). Infrared radiation elevates the surface temperature of grains, even without the use of conductors. Hamanaka et al. (2007) demonstrated that a 60-s exposure to infrared heating effectively reduced microbial content in wheat grains.

Erdoğdu et al. (2010) utilized a modified infrared system with a catalytic infrared emitter to target and eliminate adult insects, eggs, and larvae in rice. The catalytic infrared emitter's effectiveness extends to saw-toothed grain beetles and merchant grain beetles, requiring only a brief exposure of 60 s to control insects developing internally or externally in stored grains (Ramaswamy, Krishnamurthy, and Jun 2012). Effectively eliminating *Sitotroga cerealella* (angoumois grain moths and eggs) as well as adults of *R. dominica* (grain borers), Pan et al. (2006) discovered that heating a single layer of rough rice at 61°C for 1 min achieved this result. Simultaneously, the process reduced moisture content to 1.4–1.7% w.b., ensuring the preservation of grain quality. The success of insect elimination in this process is attributed to the insects' high moisture content and selective heating during the treatment (Erdoğdu et al. 2010).

11.4 ADVANCED THERMAL TECHNOLOGIES FOR ENHANCED DISINFESTATION OF FOOD PRODUCTS

11.4.1 Hot Air Treatment for Disinfestation of Stored Products

The utilization of heat treatment techniques proves to be straightforward, cost-effective, and environmentally friendly in pest control for stored grains. This method eliminates the need for harmful residues, though it may impact the quality of the treated commodity. The elevated temperatures employed during disinfestation have the potential to alter the structural integrity and other properties of the treated commodities. However, transitioning these heat methodologies to an industrial scale can be intricate, time-consuming, and expensive. Scientific evidence indicates that maintaining a temperature of 50°C for 24 h effectively in postharvest agricultural commodities eradicates all life stages of major stored pests (Shadia 2011). Historical reports, such as Winterbottom's account in 1922, highlight the successful disinfestation of over half a million tons of grain between 1915 and 1919. The current surge in interest surrounding heat disinfestation necessitates the development of a rapid, reliable, and cost-effective strategy for implementation.

Extreme temperatures lead to the demise of insects by altering and potentially coagulating proteins and essential enzymes, in addition to inflicting damage to cell walls. Opit et al. (2011) utilized elevated temperature treatments for decontaminating grain silos and containers, proposing that effective pre-cleaning of storage facilities enhances the process. Heat treatments, commonly involving hot water, are widely employed for purifying grain storage structures, removing particles, husks, and insect fragments. In bulk storage installations, fluidized-bed heating between 56°C and 72°C during grain translocation in a moving stream enables swift disinfestation. Although heating flour mills to 50°C is a recognized disinfection practice, its widespread adoption remains limited (Donahaye 2000). The tolerance of arthropod pest species to heat or cold is influenced by factors, such as life stage, species, and environmental conditions. *Stegobium paniceum* pupae, for example, exhibit higher sensitivity to elevated temperatures, while adults display greater cold tolerance. Abdelghany et al. (2010) observed that the presence of heat shock proteins in *T. castaneum* larval bodies contributes to their increased heat tolerance.

The application of direct heating or cooling to grains is a widely adopted method for controlling stored product insects. Insects at all developmental stages are highly susceptible to extreme temperatures, with lethality occurring rapidly at temperatures of −20°C or 55°C, typically within minutes. The effectiveness of lethal temperatures varies based on factors, such as insect species, developmental stage, acclimation, and humidity levels (Fields 1992). Research involving the exposure of various stages of *Callosobruchus chinensis* pupae to elevated temperatures revealed induced sterility when subjected to 45°C for 72 h. Utilizing temperatures just below 60°C, such as heating pulses to 55°C for a brief period, can eliminate insects at all life stages without compromising quality, particularly given the sensitivity of pulse proteins to higher temperatures (Saxena, Dixit, and Harshan 1992). Additionally, heating grain silos/bins before storing grains can offer insect control within storage structures and facilities. Maintaining silos/bins at a temperature of at least 50°C for 2–4 h proves effective in controlling certain storage insects, making it a viable alternative to fumigation (Tilley, Casada, and Arthur 2007). For *C. maculatus*, a lethal time of 18, 57, 78, and 71 h was observed for eggs, larvae, pupae, and adults, respectively, at a temperature of 42°C (Loganathan et al. 2011). While high temperatures are effective, care must be taken to avoid exposing grains to excessively high temperatures, as this could lead to protein and nutrient loss in food legumes during drying or other processes.

The significance of low-temperature insect control techniques is growing in the realm of grain preservation. At lower temperatures, insect cells experience internal fluid freezing, leading to cellular damage and eventual insect demise (Fields 1992). Despite the infrequent use of low temperatures in stored grain disinfection due to the associated high costs, Arthur et al. (2017) suggest the development of distinct protocols tailored to each stored product insect species when employing −18°C low-temperature treatments. Exposure to low temperatures prompts insects to develop cold hardiness, posing challenges for effective insect control methods (Andreadis and Athanassiou 2017). The intricacies of designing extreme temperature treatments, influenced by various biotic and abiotic conditions, make their application challenging in grains and grain products (Abdelghany et al. 2010; Athanassiou, Arthur, and Hartzer 2018). Given the increasing preference for organic foods and the safety of these treatments for applicators without regulatory constraints, low-temperature approaches gain relevance. Athanassiou, Arthur, and Hartzer (2018) found that treating *P. interpunctella and Liposcelis bostrychophila* at −10°C for seven days efficiently eliminated the cold-tolerant *Liposcelis* eggs. This approach effectively preserved grain quality under subzero temperatures. Continuous low-temperature treatments offer opportunities for further research on insect control by exploring variations in temperature, exposure duration, and their impact on major stored product pests.

Utilizing ambient or refrigerated aeration for grain cooling holds unique advantages in insect and mite control, as lower temperatures significantly influence population characteristics (Aspaly et al. 2007; Beckett 2011). While thermal disinfection methods are effective, careful monitoring is essential to avoid temperature and time combinations that may surpass specified limits, potentially compromising pulse quality.

11.4.2 Solar Heating

Before the implementation of chemical control methods in stored grain, heat treatment stood out as one of the limited options for pest disinfestation. Particularly in challenging climates, the practice of sun drying was commonly employed to eradicate pests. In some Asian countries, solar heating is sporadically utilized to administer initial heat treatments to grain, which is then maintained by placing the heated grain in insulated storage facilities. Sun drying, solar heating/drying, or simply "Sunning" refers to harnessing thermal heat energy from the Sun to treat the product. Given that all phases of insect life—egg, larva, pupae, and adults—are sensitive to heat, managing the storage environment's temperature has been a longstanding practice, often achieved through exposure to the Sun's radiation. Additionally, a solar-powered heating system, as demonstrated by Baoua et al. (2012), can function both as a disinfection and drying unit.

The effectiveness of solar heating in eliminating insects, particularly bruchids, is attributed to their sensitivity to elevated temperatures, with temperatures above 45°C proving lethal. Following sun drying, proper grain handling involves cleaning, cooling, and storing the grains in well-sealed storage structures. Different insect species exhibit varying degrees of temperature sensitivity. Solar heating is commonly achieved by spreading grains on surfaces like mats, threshing yards, black polyethylene sheets, or tarpaulins exposed to the sun. Elevating grain temperatures up to 52°C has been accomplished using concrete floors, black polyethylene sheets, or tarpaulins in direct sunlight, effectively eradicating all bruchid life stages. Research has shown that just 3 hours of sunlight exposure reduces bruchid oviposition and increases adult insect mortality. Sun-drying methods, including exposure periods, have been proven effective in controlling insect infestations in various pulses and beans. For instance, sun-drying cowpeas on wooden surfaces outperformed concrete and corrugated iron steel surfaces, reducing fungal and insect loads. Solar heating of pigeon pea seeds on polyethylene sheets can reach temperatures lethal to storage insects, such as bruchids, making the seeds insect-free for extended storage periods. Although sunning is a cost-effective and widely used renewable energy source, the practice requires careful conditioning and monitoring due to unpredictable weather and temperature fluctuations, ensuring grains maintain appropriate storage moisture content.

11.4.3 Steam Treatment

Zucchini (*Cucurbita pepo* L.) underwent a disinfestation process using high humidity hot air treatment (HT), reaching a fruit core temperature of 45°C for 30 min (Neves, Silva, and Vieira 2019). Microorganisms and certain molds are prone to developing on harvested food grains during storage and distribution. Superheated steam, surpassing water's boiling point in temperature, has emerged as an advanced technique to increase the mortality rate of microorganisms and insects in stored grains. This is crucial as many microorganisms, including lactic acid and coliform bacteria, can induce acid fermentation in stored grains, and spore-forming bacteria may survive conventional heating or cold storage (Sinha and Wallace 1966). When steam condenses on food, it releases substantial heat, reversing the moisture process. Steam processing, with higher enthalpy than saturated steam or hot air-drying processes, proves more efficient. This method offers advantages over conventional drying, conducted in the absence of oxygen to eliminate oxidative reactions, avoid hazardous reactions, and condense valuable volatile components. Higher moisture content in stored grains can lead to infestations and also fungal contamination, with *Penicillium*, *Fusarium*, and *Aspergillus* causing qualitative and quantitative changes.

Fungal contamination may result in hot spots, elevating temperatures and stimulating germination (Sashidhar, Ramakrishna, and Bhat 1992). Mycotoxins produced from fungal growth can be toxic, causing various health issues. Superheated steam has shown effectiveness in reducing mycotoxins. Introduced in 2015 by FBI Co. Ltd. in Tokyo, Japan, Ultra-Superheated Steam Technology (USHD) employs high-frequency induction heating at 300–500°C, generating ionized gaseous particles, free radicals, hydrogen peroxide, superoxide anion radical, and singlet oxygen. Positive and significant effects on decontamination and disinfestation were observed upon application of USHD on maize kernels, chickpeas, red wheat, and peanuts. Furthermore, treatment with USHD at temperatures exceeding 500°C for 15 s enhances grain storage for up to two years.

11.4.4 Synergistic Approaches with Combined Thermal Technologies

To safeguard the quality of stored food grains, it is imperative to eliminate infestations by insect pests. Various thermal methods, including hot air, solar, microwave heating, RF, and infrared heating, can be employed for disinfestation. Combining these diverse thermal practices can yield a synergistic effect, ensuring the efficacy of the disinfestation process while upholding food safety standards. Integrating thermal technologies for disinfestation in stored food products proves to be an effective strategy for maintaining food safety and preventing infestations without resorting to chemical treatments.

In a study conducted by Kirkpatrick, Brower, and Tilton (1973), different treatments involving the combination of gamma radiation, infrared radiation, and microwave radiation were applied to soft winter wheat infested with eggs, larvae, pupae, or adults of *R. dominica*. The results demonstrated that the combination of gamma and infrared radiation was the most effective treatment, requiring lower gamma dosages and temperatures for insect control. Examining infestations by *R. dominica* (lesser grain borer) and *S. oryzae* (rice weevil) in wheat, research revealed that complete pest control was achieved through a combination of microwave and vacuum drying.

The utilization of microwave and RF treatment holds great promise due to its non-reactive, swift, and secure operational characteristics. This method provides the benefit of efficiently managing all developmental stages of pests within shorter storage periods, leaving no harmful residues in the end product (Yadav et al. 2014). A novel approach involving simultaneous hot air-assisted RF drying and disinfestation was developed and applied to assess drying characteristics and disinfection efficacy for walnuts. While the RF drying process aims to preserve product quality by using minimal temperature, potentially leading to incomplete insect control, the demonstrated feasibility of separately conducting RF drying and disinfection suggests significant potential for an effective combined treatment for in-shell walnuts through systematic studies. This approach allows for the optimization of both drying and disinfection processes to achieve desired outcomes without compromising product quality (Mao and Wang 2021). Gamage et al. (2015) investigated the combined hot air treatment and microwave for fruit fly disinfestation in apples. The findings suggest that microwave heat treatment is effective for insect disinfestation in apples without adversely affecting the flesh, peel firmness, and total soluble solids of the treated fruit.

Hybrid drying systems, characterized by diverse drying results, negligible environmental footprint, and diminished power usage, have emerged as an excellent technique in this regard. Developed hybrid solar dryers, when combined with other heating sources, enhance efficiency, subsequently lowering drying costs and energy consumption (Inyang, Oboh, and Etuk 2017; Babar et al. 2020). Tamuli and Choudhury (2020) designed a solar-powered dryer with three operational modes: solar, electric, and hybrid. Among these, the hybrid mode demonstrated superior performance in moisture removal. Developed by Li et al. (2011), a solar-assisted heat pump in-store dryer was employed to decrease corn moisture from 16.6% to 14.5% (w.b.) over 240 h. The dryer consumed only 1.24 kWh of power, demonstrating a more energy-efficient approach compared to the official standard of 2.0 kWh for achieving a 1% moisture reduction. In a study by Mahdizadeh and Zomorodian (2009), a mixed-mode passive solar dryer was developed to assess dried rice quality in comparison to sun drying. The solar-dried rice exhibited a whiter appearance, while the flavor remained consistent with sun-dried rice. Infrared drying, known for its high efficiency, was combined with other methods. Kirkpatrick, Brower, and Tilton (1973) investigated gamma, infrared, and microwave radiations' impact on *R. dominica* species in wheat, achieving maximum mortality with gamma and infrared (99%) studied microwave and vacuum combinations for insect control in stored rye, corn, and wheat, achieving complete control with increased treatment dosage. Microwave and RF treatment emerged as a non-reactive, swift, and safe option for pest control, affecting all developmental stages and leaving no toxic residue (Yadav et al. 2014). Microwave radiation, an alternative to chemical fumigation, generates heat within the food, reducing heat loss (Çelen and Karataser 2019). Taheri, Brodie, and Gupta (2020) explored microwave-fluidized bed drying to reduce Botrytis grey mold (BGM) in red lentil seeds, observing a 30% reduction in seed infections with specific combinations of microwave power and fluidized bed temperature.

11.5 FUTURE TRENDS IN ADVANCED THERMAL TECHNOLOGY FOR THE DISINFESTATION OF STORED PRODUCTS

Controlling the insects in stored products is a crucial component of food safety, security, and preservation. The limitations of conventional procedures have led to the continuous advancement of more sophisticated approaches. Chemical pesticides, which use chemicals like fungicides, insecticides, and rodenticides to kill or repel pests, have long been the primary form of pest management in products that are kept in storage (Powell and Cuthbertson 2013). But because of their extensive usage, now there are worries about the potential health risks from chemical residues, the emergence of pest resistance, unintentional damage to creatures that are not targeted, and a strengthening of regulations as a result of the effects on the environment and human health (Rani et al. 2021; Ali et al. 2021). Physical techniques, which include cleaning, drying, freezing, and hermetic storage, provide an alternative. However, because of their labor-intensiveness, limited efficacy against specific pests or life stages, and need for maintaining ideal storage conditions, physical methods are difficult to implement and effective in a variety of storage environments. Despite these drawbacks, conventional techniques have long served as the foundation of pest control for goods that are kept in storage, but their flaws have shown that the modern food supply system needs more advanced and environmentally friendly methods of insect management. It was recommended that thermal technology be used as a workable solution to control insects related to grain storage (*Stewart Postharvest Review* 2011).

Continuous research and development have led to new thermal technologies, which provide creative answers to many of the drawbacks of conventional techniques. Microwave heating, RF heating, infrared heating, induction heating, and dielectric heating are some of these cutting-edge thermal technologies. Achieving effective disinfestation in grain storage necessitates careful regulation of maximum kernel temperatures and kernel residence times to avoid thermal damage to the kernels (Abd El-Aziz 2011). Each thermal technology offers distinct benefits such as rapid and even heating, focused treatment, lower energy usage, accurate temperature management, and compatibility with various stored goods. Despite the successful application of modern thermal technologies in laboratories to ensure food safety and provide advantages, their industrial implementation has been constrained by high capital costs and uneven heating of solid foods (Jangam 2011).

The physicochemical damage incurred by stored foods during thermal treatment, coupled with non-uniform heating and post-processing quality loss, has prompted investigations into the combination of different food safety methods. This approach aims to mitigate disadvantages and leverage the advantages of individual methods (Gamage et al. 2015). Furthermore, advancements in control and monitoring systems enable real-time data analytics, automation, and integration with other technologies, enhancing efficiency, effectiveness, and scalability in industrial applications. Despite challenges, these innovations represent a significant step toward sustainable pest management for stored items. They offer potential solutions to longstanding issues, paving the way for a future food supply chain that is safer, more resilient, and more efficient.

11.6 CONCLUSION

In conclusion, advancements in food processing technologies, particularly dielectric heating and other advanced thermal methods, hold significant promise for tackling key issues related to food spoilage and disinfestation. These technologies offer notable benefits, such as reduced processing times, improved energy efficiency, and the preservation of nutritional and sensory qualities. As consumer preferences shift toward minimally processed and premium foods, the role of these innovative methods becomes increasingly important.

Nevertheless, while dielectric heating and other advanced thermal technologies present effective solutions, there remain critical research gaps. Future studies should focus on optimizing these technologies for different food types, examining their long-term effects on food quality, and developing strategies to overcome existing limitations. Additionally, exploring the integration of these advanced methods with current food processing systems is essential for improving overall efficiency and effectiveness. Addressing these research needs and fostering collaboration among researchers, industry professionals, and technology developers will be crucial for advancing these technologies and meeting the evolving demands of the food industry.

REFERENCES

Abdelghany, A. Y., S. S. Awadalla, N. F. Abdel-Baky, H. A. El-Syrafi, and P. G. Fields. 2010. "Effect of High and Low Temperatures on the Drugstore Beetle (Coleoptera: Anobiidae)." *Journal of Economic Entomology* 103 (5): 1909–1914. doi:10.1603/EC10054.

Aboud, S. A., A. B. Altemimi, A. R. S. Al-HiIphy, L. Yi-Chen, and F. Cacciola. 2019. "A Comprehensive Review on Infrared Heating Applications in Food Processing." *Molecules* 24 (22): 4125. doi:10.3390/molecules24224125.

Ali, S., M. I. Ullah, A. Sajjad, Q. Shakeel, and Hussain A. 2021. "Environmental and Health Effects of Pesticide Residues." In Inamuddin, Mohd Imran Ahamed, Eric Lichtfouse (Eds.), *A HussainSustainable Agriculture Reviews 48: Pesticide Occurrence, Analysis and, 2021*, 311–336. Springer. doi:10.1007/978-3-030-54719-6_8.

Andreadis, S. S., and C. G. Athanassiou. 2017. "A Review of Insect Cold Hardiness and Its Potential in Stored Product Insect Control." *Crop Protection* 91 (January): 93–99. doi:10.1016/j.cropro.2016.08.013.

Appugol, K. A., I. B. Mangang, S. Shanmugasundaram, and L. Manickam. 2022. "Radiofrequency Heating: A Novel Thermal Treatment on the Quality of Peanut during Disinfestation of *Caryedon serratus* and Its Potential in Reducing Aflatoxin." *Journal of Food Processing and Preservation* 46 (11). doi:10.1111/jfpp.17029.

Arthur, F. H., K. L. Hartzer, J. E. Throne, and P. W. Flinn. 2017. "Freezing for Control of Stored-Product Psocids." *Journal of Stored Products Research* 72 (May): 166–172. doi:10.1016/j.jspr.2016.12.005.

Aspaly, G., V. Stejskal, S. Pekár, and J. Hubert. 2007. "Temperature-Dependent Population Growth of Three Species of Stored Product Mites (Acari: Acaridida)." *Experimental and Applied Acarology* 42 (1): 37–46. doi:10.1007/s10493-007-9074-1.

Athanassiou, C. G., F. H. Arthur, and K. L. Hartzer. 2018. "Efficacy of Low Temperatures for the Control of All Life Stages of *Plodia interpunctella* and *Liposcelis bostrychophila*." *Journal of Pest Science* 91 (4): 1363–1369. doi:10.1007/s10340-018-0982-0.

Babar, O. A., A. Tarafdar, S. Malakar, V. K. Arora, and P. K. Nema. 2020. "Design and Performance Evaluation of a Passive Flat Plate Collector Solar Dryer for Agricultural Products." *Journal of Food Process Engineering* 43 (10). doi:10.1111/jfpe.13484.

Baoua, I. B., L. Amadou, V. Margam, and L. L. Murdock. 2012. "Comparative Evaluation of Six Storage Methods for Postharvest Preservation of Cowpea Grain." *Journal of Stored Products Research* 49 (April): 171–175. doi:10.1016/j.jspr.2012.01.003.

Beckett, S. J. 2011. "Insect and Mite Control by Manipulating Temperature and Moisture before and during Chemical-Free Storage." *Journal of Stored Products Research* 47 (4): 284–292. doi:10.1016/j.jspr.2011.08.002.

Bond, E. J., H. Alexander, and U. Monro. 1984. "FAO Plant Production and Protection Paper." In *Manual of Fumigation for Insect Control*, Vol. 54, 432. Rome, Italy: FAO.

Çelen, S., and M. A. Karataser. 2019. "Investigation of the Performance of a Hybrid Dryer Designed for the Food Industry." *Foods* 8 (2): 81. doi:10.3390/foods8020081.

Dalmoro, A., C. Naddeo, S. Caputo, G. Lamberti, L. Guadagno, M. D'Amore, and A. A. Barba. 2018. "On the Relevance of Thermophysical Characterization in the Microwave Treatment of Legumes." *Food & Function* 9 (3): 1816–1828. doi:10.1039/C7FO01488K.

Divekar, M. T., C. Karunakaran, R. Lahlali, S. Kumar, V. Chelladurai, X. Liu, F. Borondics, S. Shanmugasundaram, and D. S. Jayas. 2017. "Effect of Microwave Treatment on the Cooking and Macronutrient Qualities of Pulses." *International Journal of Food Properties* 20 (2): 409–422. doi:10.1080/10942912.2016.1163578.

Donahaye, E. J. 2000. "Current Status of Non-Residual Control Methods against Stored Product Pests." *Crop Protection* 19 (8–10): 571–576. doi:10.1016/S0261-2194(00)00074-0.

Erdoğdu, B., I. H. Ekiz., G. G. F. Erdoğdu, Atungulu., and Z. Pan. 2010. Industrial Applications of Infrared Radiation Heating and Economic Benefits in Food and Agricultural Processing." In G. G. Atungulu and Z. Pan (Eds.), *Infrared Heating for Food and Agricultural Processing*. CRC Press/Taylor & Francis Group.

Fields, P. G. 1992. "The Control of Stored-Product Insects and Mites with Extreme Temperatures." *Journal of Stored Products Research* 28 (2): 89–118. doi:10.1016/0022-474X(92)90018-L.

Gamage, T. V., P. Sanguansri, P. Swiergon, M. Eelkema, P. Wyatt, P. Leach, D. L. J. Alexander, and K. Knoerzer. 2015. "Continuous Combined Microwave and Hot Air Treatment of Apples for Fruit Fly (*Bactrocera tryoni* and *B. Jarvisi*) Disinfestation." *Innovative Food Science & Emerging Technologies* 29 (May): 261–270. doi:10.1016/j.ifset.2015.02.009.

Ha, J-W., and D-H. Kang. 2013. "Simultaneous Near-Infrared Radiant Heating and UV Radiation for Inactivating *Escherichia coli* O157:H7 and *Salmonella enterica* Serovar Typhimurium in Powdered Red Pepper (*Capsicum annuum* L.)." *Applied and Environmental Microbiology* 79 (21): 6568–6575. doi:10.1128/AEM.02249-13.

Hamanaka, D., T. Uchino, A. Inoue, K. Kawasaki, and Y. Hori. 2007. *Development of the Rotating Type Grain Sterilizer Using Infrared Radiation Heating.*

Hassan, R. S., E. A. Mahmoud, T. M. Sileem, and W. A. A. Sayed. 2019. "Evaluation of Fast Neutron Irradiation as a New Control Method against the Indian Meal Moth, *Plodia interpunctella* (Hübner)." *Journal of Radiation Research and Applied Sciences* 12 (1): 38–44. doi:10.1080/16878507.2019.1594133.

Hassan, A. B., E. Pawelzik, and D. von Hoersten. 2021. "Effect of Microwave Heating on the Physiochemical Characteristics, Colour and Pasting Properties of Corn (*Zea mays* L.) Grain." *LWT* 138 (March): 110703. doi:10.1016/j.lwt.2020.110703.

Hou, L., J. A. Johnson, and S. Wang. 2016. "Radio Frequency Heating for Postharvest Control of Pests in Agricultural Products: A Review." *Postharvest Biology and Technology* 113 (March): 106–118. doi:10.1016/j.postharvbio.2015.11.011.

Huang, Z., L. Chen, and S. Wang. 2015. "Computer Simulation of Radio Frequency Selective Heating of Insects in Soybeans." *International Journal of Heat and Mass Transfer* 90 (November): 406–417. doi:10.1016/j.ijheatmasstransfer.2015.06.071.

Indumathi, C., D. Manoj, K. Devi, and S. Shanmugasundaram. 2021. "Influence of Different Moisture Content on Heating Rate of Rice during Radio Frequency Heating." ~ *1702 ~ The Pharma Innovation Journal* 10 (8): 1702–1705.

Indumathi, C., D. Manoj, M. Loganathan, and S. Shanmugasundaram. 2023. "Radio Frequency Disinfestation of *Tribolium castaneum* (Herbst) in Semolina: An Emerging Thermal Technique." *Journal of Food Process Engineering* 46 (12). doi:10.1111/jfpe.14443.

Inyang, U., I. Oboh, and B. R. Etuk. 2021. "Modelling of Thin Layer Drying of Cocoyam Slices of Varying Thickness at Different Temperatures." *Nigerian Research Journal of Engineering and Environmental Sciences* 6 (1).

Jiao, S., W. Sun, T. Yang, Y. Zou, X. Zhu, and Y. Zhao. 2017. "Investigation of the Feasibility of Radio Frequency Energy for Controlling Insects in Milled Rice." *Food and Bioprocess Technology* 10 (4): 781–788. doi:10.1007/s11947-017-1865-8.

Jun, S., and J. Irudayaraj. 2003. "A Dynamic Fungal Inactivation Approach Using Selective Infrared Heating." *Transactions of the ASAE* 46 (5). doi:10.13031/2013.15435.

Kirkpatrick, R. L., J. H. Brower, and E. W. Tilton. 1973. "Gamma, Infra-Red and Microwave Radiation Combinations for Control of *Rhyzopertha dominica* in Wheat." *Journal of Stored Products Research* 9 (1): 19–23. doi:10.1016/0022-474X(73)90034-9.

Koné, K. Y., C. Druon, E. Z. Gnimpieba, M. Delmotte, A. Duquenoy, and J-C. Laguerre. 2013. "Power Density Control in Microwave Assisted Air Drying to Improve Quality of Food." *Journal of Food Engineering* 119 (4): 750–757. doi:10.1016/j.jfoodeng.2013.06.044.

Kudra, T. 1989. "Dielectric Drying of Particulate Materials in a Fluidized State." *Drying Technology* 7 (1): 17–34. doi:10.1080/07373938908916572.

Li, Y., H. F. Li, Y. J. Dai, S. F. Gao, L. Wei, Z. L. Li, I. G. Odinez, and R. Z. Wang. 2011. "Experimental Investigation on a Solar Assisted Heat Pump In-Store Drying System." *Applied Thermal Engineering* 31 (10): 1718–1724. doi:10.1016/j.applthermaleng.2011.02.014.

Ling, B., T. Cheng, and S. Wang. 2020. "Recent Developments in Applications of Radio Frequency Heating for Improving Safety and Quality of Food Grains and Their Products: A Review." *Critical Reviews in Food Science and Nutrition* 60 (15): 2622–2642. doi:10.1080/10408398.2019.1651690.

Loganathan, M., D. S. Jayas, P. G. Fields, and N. D. G. White. 2011. "Low and High Temperatures for the Control of Cowpea Beetle, *Callosobruchus maculatus* (F.) (Coleoptera: Bruchidae) in Chickpeas." *Journal of Stored Products Research* 47 (3): 244–248. doi:10.1016/j.jspr.2011.03.005.

Mahdizadeh, Z. A. F., and A. Zomorodian. 2009. "A Study of the Effect of Solar Drying System on Rice Quality." *Journal of Agricultural Science and Technology* 11: 527–534.

Mangang, I. B., R. Minj, L. Manickam, J. Rangarajan, and S. Shanmugasundaram. 2022. "Turmeric—a Super Food—Prevention of Post-Harvest Losses Using Radio Frequency Disinfestation." *Archives of Phytopathology and Plant Protection* 55 (12): 1450–1463. doi:10.1080/03235408.2022.2104583.

Manickavasagan, A., D. S. Jayas, N. D. G. White, and F. Jian. 2006. "Thermal Imaging of a Stored Grain Silo to Detect a Hot Spot." *Applied Engineering in Agriculture* 22 (6): 891–897. doi:10.13031/2013.22243.

Mao, Y., and S. Wang. 2021. "Simultaneous Hot-Air Assisted Radio Frequency Drying and Disinfestation for in-Shell Walnuts Using a Two-Stage Strategy." *LWT* 151 (November): 112134. doi:10.1016/j.lwt.2021.112134.

Meenu, M., P. Guha, and S. Mishra. 2018. "Impact of Infrared Treatment on Quality and Fungal Decontamination of Mung Bean (*Vigna radiata* L.) Inoculated with *Aspergillus* spp." *Journal of the Science of Food and Agriculture* 98 (7): 2770–2776. doi:10.1002/jsfa.8773.

Neves, F. I. G., C. L. M. Silva, and M. C. Vieira. 2019. "Combined Pre-Treatments Effects on Zucchini (*Cucurbita pepo* L.) Squash Microbial Load Reduction." *International Journal of Food Microbiology* 305 (September): 108257. doi:10.1016/j.ijfoodmicro.2019.108257.

Opit, G. P., F. H. Arthur, E. L. Bonjour, C. L. Jones, and T. W. Phillips. 2011. "Efficacy of Heat Treatment for Disinfestation of Concrete Grain Silos." *Journal of Economic Entomology* 104 (4): 1415–1422. doi:10.1603/EC11104.

Pan, Z., R. Khir, R. Lewis, L. Godfrey, A. Salim, and J. F. Thompson. 2006. "Simultaneous Rough Rice Drying and Disinfestation Using Infrared Radiation." In *2006 Portland, Oregon, July 9–12, 2006.* St. Joseph, MI: American Society of Agricultural and Biological Engineers. doi:10.13031/2013.21455.

Paul, A., M. Radhakrishnan, S. Anandakumar, S. Shanmugasundaram, and C. Anandharamakrishnan. 2020. "Disinfestation Techniques for Major Cereals: A Status Report." *Comprehensive Reviews in Food Science and Food Safety* 19 (3): 1125–1155. doi:10.1111/1541-4337.12555.

Powell, M. E., and A. G. S. Cuthbertson. 2013. "Pest Control." *Biologist* 60 (2): 19–21.

Ramaswamy, R., K. Krishnamurthy, and S. Jun. 2012. "Microbial Decontamination of Food by Infrared (IR) Heating." In *Microbial Decontamination in the Food Industry*, 450–471. Elsevier. doi:10.1533/9780857095756.2.450.

Rani, L., K. Thapa, N. Kanojia, N. Sharma, S. Singh, A. S. Grewal, A. L. Srivastav, and J. Kaushal. 2021. "An Extensive Review on the Consequences of Chemical Pesticides on Human Health and Environment." *Journal of Cleaner Production* 283 (February): 124657. doi:10.1016/j.jclepro.2020.124657.

Rashkovan, V. M., N. A. Khizhnyak, A. V. Basteev, L. A. Bazyma, L. N. de Rivera, and I. A. Ponomaryova. 2003. "Interaction of Electromagnetic Waves with Granular Agriculture Product and Insects." *Journal of Microwave Power and Electromagnetic Energy* 38 (4): 225–235. doi:10.1080/08327823.2003.11688501.

Sashidhar, R. B., Y. Ramakrishna, and R. V. Bhat. 1992. "Moulds and Mycotoxins in Sorghum Stored in Traditional Containers in India." *Journal of Stored Products Research* 28 (4): 257–260. doi:10.1016/0022-474X(92)90006-C.

Sawai, J., H. Igarashi, A. Hashimoto, T. Kokugan, and M. Shimizu. 1995. "Evaluation of Growth Inhibitory Effect of Ceramics Powder Slurry on Bacteria by Conductance Method." *Journal of Chemical Engineering of Japan* 28 (3): 288–293. doi:10.1252/jcej.28.288.

Saxena, R. C., O. P. Dixit, and V. Harshan. 1992. "Insecticidal Action of Lantana Camara against *Callosobruchus chinensis* (Coleoptera: Bruchidae)." *Journal of Stored Products Research* 28 (4): 279–281. doi:10.1016/0022-474X(92)90009-F.

Shrestha, B., and O.-D. Baik. 2019. "Multi-Physics Computer Simulation of Radio Frequency Heating to Control Pest Insects in Stored-Wheat." *Engineering in Agriculture, Environment and Food* 12 (1): 71–80. doi:10.1016/j.eaef.2018.09.005.

Sinha, R. N., and H. A. H. Wallace. 1966. "Association of Granary Mites and Seed-Borne Fungi in Stored Grain and in Outdoor and Indoor Habitats." *Annals of the Entomological Society of America* 59 (6): 1170–1181. doi:10.1093/aesa/59.6.1170.

Stewart Postharvest Review. 2011. "Modern Perspectives on Stored-Product Insect Pest Management." *Stewart Postharvest Review* 7 (3): 1–3. doi:10.2212/spr.2011.3.1.

Taheri, S., G. Brodie, and D. Gupta. 2020. "Fluidisation of Lentil Seeds during Microwave Drying and Disinfection Could Prevent Detrimental Impacts on Their Chemical and Biochemical Characteristics." *LWT* 129 (July): 109534. doi:10.1016/j.lwt.2020.109534.

Tamuli, B. R., and P. K. Choudhury. 2020. "Comparative Performance Assessment of a Solar Hybrid Dryer with Traditional Drying Techniques." In *Advances in Energy Research, Vol. 2: Selected Papers from ICAER 2017*, 351–360. doi:10.1007/978-981-15-2662-6_32.

Tilley, D. R., M. E. Casada, and F. H. Arthur. 2007. "Heat Treatment for Disinfestation of Empty Grain Storage Bins." *Journal of Stored Products Research* 43 (3): 221–228. doi:10.1016/j.jspr.2006.04.005.

Tiwari, A., S. Shanmugasundaram, R. Jaganmohan, and L. Manickam. 2021. "Dielectric Heating-assisted Disinfestation of Black Gram and Its Effect on Protein Profile: A Comparative Study on Radio Frequency and Microwave Heating." *Legume Science* 3 (2). doi:10.1002/leg3.83.

Vadivambal, R., D. S. Jayas, and N. D. G. White. 2007. "Wheat Disinfestation Using Microwave Energy." *Journal of Stored Products Research* 43 (4): 508–514. doi:10.1016/j.jspr.2007.01.007.

Vasilev, A. A., A. N. Vasilev, and G. Samarin. 2019. "Substantiation of Automated Control Modes for Grain Disinfection." In 2019 International Russian Automation Conference (RusAutoCon).

Yadav, D. N., T. Anand, M. Sharma, and R. K. Gupta. 2014. "Microwave Technology for Disinfestation of Cereals and Pulses: An Overview." *Journal of Food Science and Technology* 51 (12): 3568–3576. doi:10.1007/s13197-012-0912-8.

Yadav, D. N., P. E. Patki, G. K. Sharma, and A. S. Bawa. 2008. "Effect of Microwave Heating of Wheat Grains on the Browning of Dough and Quality of Chapattis." *International Journal of Food Science & Technology* 43 (7): 1217–1225. doi:10.1111/j.1365-2621.2007.01593.x.

Ye, J., T. Hong, Y. Wu, L. Wu, Y. Liao, H. Zhu, Y. Yang, and K. Huang. 2017. "Model Stirrer Based on a Multi-Material Turntable for Microwave Processing Materials." *Materials* 10 (2): 95. doi:10.3390/ma10020095.

Yu, D. U., B. L. Shrestha, and O. D. Baik. 2015. "Radio Frequency Dielectric Properties of Bulk Canola Seeds under Different Temperatures, Moisture Contents, and Frequencies for Feasibility of Radio Frequency Disinfestation." *International Journal of Food Properties* 18 (12): 2746–2763. doi:10.1080/10942912.2015.1013630.

12 Cold Plasma for the Disinfestation of Stored Grains

R. Anbarasan and R. Mahendran
National Institute of Food Technology Entrepreneurship and Management, Thanjavur, Tamil Nadu, India

12.1 INTRODUCTION

According to the FAO, approximately 17% of global food production is wasted during storage, with 10% attributed to insect infestations and 7% to mites, rodents, and diseases. In this regard, about 80% of the loss occurs in plant-based cereal grains like wheat, rice, maize, millet, sorghum, and barley, as well as in legumes such as beans, soybeans, and peanuts. Insect infestations can occur at various stages, starting on the farm as pests of the plants or during transportation, processing, or storage due to favorable conditions and nutrient-rich stored foods. In addition to causing quantitative damage, insects contaminate stored products with their body parts, rendering the food unsuitable for consumption. Furthermore, their metabolic activities alter the chemical composition of stored products, affecting amino acids, proteins, carbohydrates, fats, and organoleptic characteristics. These infestations result in reduced commercial and nutritional value of raw materials and foods, potentially posing health risks to consumers (Stathas et al. 2023). Stored product insects can be categorized as primary and secondary insects. Primary insects, such as grain weevils and flour beetles, directly harm stored food products by feeding on them, causing significant damage. They are the most economically dangerous pests in stored product management, capable of infesting and reproducing within stored food, compounding their destructive potential. On the other hand, secondary insects like mites and dermestid beetles infest damaged grains and processed grain products. They feed on mold, fungi, and damaged food particles rather than intact grains or products, often serving as indicators of existing primary pest infestations in stored food management (Ahmad et al. 2022).

In contemporary pest management practices, chemical disinfectants such as malathion, chlorpyrifos-methyl, and deltamethrin function as contact pesticides, while methyl bromide (MeBr) and phosphine (PH_3) serve as fumigants. Notably, MeBr has been prohibited in developed countries since 2005 and in developing countries since 2015, due to the harmful health consequences it poses to humans and the ecosystem. Equally significant is the reduced efficacy observed in certain chemical disinfectants, specifically PH_3 and malathion, attributable to insect resistance mechanisms, particularly with regard to unhatched eggs concealed beneath grain kernels. This highlights the need to replace chemical disinfectants with more efficient, non-chemical methods that have fewer negative impacts on the grain's physiochemical attributes, personal health, and ecosystem. Therefore, the focus shifted to non-chemical disinfestation techniques such as modified atmospheric storage, irradiation, cold plasma (CP), radio/microwave heating, e-beam technology, etc. (Moirangthem and Baik 2021). Among these, CP technology is gaining more attention as it offers the opportunity to reduce microbial loads, pesticide residues, seed dormancy periods, and soaking times during post-harvest processing, in addition to its disinfestation capabilities (Rajan et al. 2023).

Plasma is often referred to as the fourth state of matter, in addition to the more familiar states of solid, liquid, and gas. It is a state of matter that exists at extremely high temperatures, such as

DOI: 10.1201/9781003609834-12

those found in stars, lightning, and certain high-energy environments. In a plasma, the atoms and molecules are so energized that electrons are stripped from their nuclei, resulting in a mixture of free electrons and positively charged ions. Therefore, the plasma state comprises photons, electrons, ions, atoms, free radicals, and excited and non-excited molecules. The concentration of these elements varies depending on the energy input into the system. Broadly, these constituents can be classified into reactive species, including reactive nitrogen species (RNS) and reactive oxygen species (ROS), ultraviolet (UV) photons, and accelerated electrons/atoms. These components exhibit strong oxidizing properties and can induce DNA damage in living cells. When it comes to plasma types, it is categorized into nonthermal plasma and thermal plasma, among which the nonthermal plasma, otherwise known as CP is preferred for food applications due to its atmospheric operational temperature. Maintaining sub-atmospheric pressure inside the plasma chamber allows the generation of plasma at low temperature (Rajan and Radhakrishnan 2023). This chapter will focus on the nonthermal plasma disinfestation mechanism, its efficacy variations for internal and external feeders, and its treatment impact on stored grain quality.

12.2 MECHANISM INVOLVED IN COLD PLASMA DISINFESTATION

Insect mortality resulting from the nerve toxin effect induced by plasma-generated ROS on their neuromuscular systems is a multifaceted process. This phenomenon predominantly occurs when these ROS penetrate the insect's body through the respiratory pathway. Once inside, they exert influence on various aspects of the insect's physiology, encompassing its behavior, immune responses, and circulatory system. It is worth emphasizing that the specific mechanisms driving insect mortality can vary significantly depending on the insect's type and current life stage, whether in the egg, larval, pupal, or adult stage. The distinct mechanisms responsible for inducing mortality in insects throughout these diverse life stages are explained below (Rajan et al. 2023).

12.2.1 Eggs

The primary factors contributing to egg mortality include disruptions in embryogenesis caused by ROS, cell membrane damage of the eggshell due to the accumulation of electrostatic charges, and hatching delays or failures resulting from oxygen-deprived (anoxic) or low-oxygen (hypoxic) conditions.

12.2.2 Larva

CP treatment leads to larval mortality by initiating oxidative stress within the larvae. This stress causes the damage to the insect's cuticle and epidermis. Additionally, oxidative reactions induce melanization of the hemolymph and physical damage, ultimately causing the death of the larvae. Nevertheless, these impacts are expected to be less pronounced in internal feeders than in external feeders due to the limited mobility and reduced penetration of plasma reactive species into the grains.

12.2.3 Pupa

Physical damage to insects in the pupal stage occurs due to the impact of accelerated electrons and ions on the insect's surface. However, in the case of internal feeders, only gaseous RS interact with the insects because the seed coat prevents direct exposure of the pupal-stage insects to accelerated electrons. It's believed that these reactive species oxidize the pupa's cell-wall lipids and proteins, leading to their mortality, as there are fewer chances for accelerated electrons and ions-induced surface etching when the pupa is located inside the seed. For external feeders, a synergistic effect is observed, as all plasma reactive species (including accelerated electrons and ions) can induce damage to the pupal insects.

12.2.4 Adult

High-voltage plasma discharge induces electrostatic excitation, disturbing the insect's cell membrane. Simultaneously, the low-oxygen environment (anoxia condition) leads to damage in the nerves and neuromuscular system. These effects collectively serve as the primary mode of action in causing insect death. Moreover, secondary to these primary effects, ROS cause damage to the insect's surface cuticular lipid layer by breaking C-H bonds. This secondary mechanism accelerates the insect's morality by inducing excessive dehydration.

12.3 COLD PLASMA DISINFESTATION OF EXTERNAL FEEDER, PRIMARY AND SECONDARY INSECTS

Stored product insect categories, such as external feeders, primary, and secondary insects, are commonly present on the sample surface rather than feeding inside the food grains. Therefore, plasma treatment directly interacts with these insects and induces mortality (Figure 12.1a). Nevertheless, the direct indirection of plasma species with food grains is not always possible, as food grains are often packed in retail packaging. Therefore, to identify the effectiveness of plasma treatment on

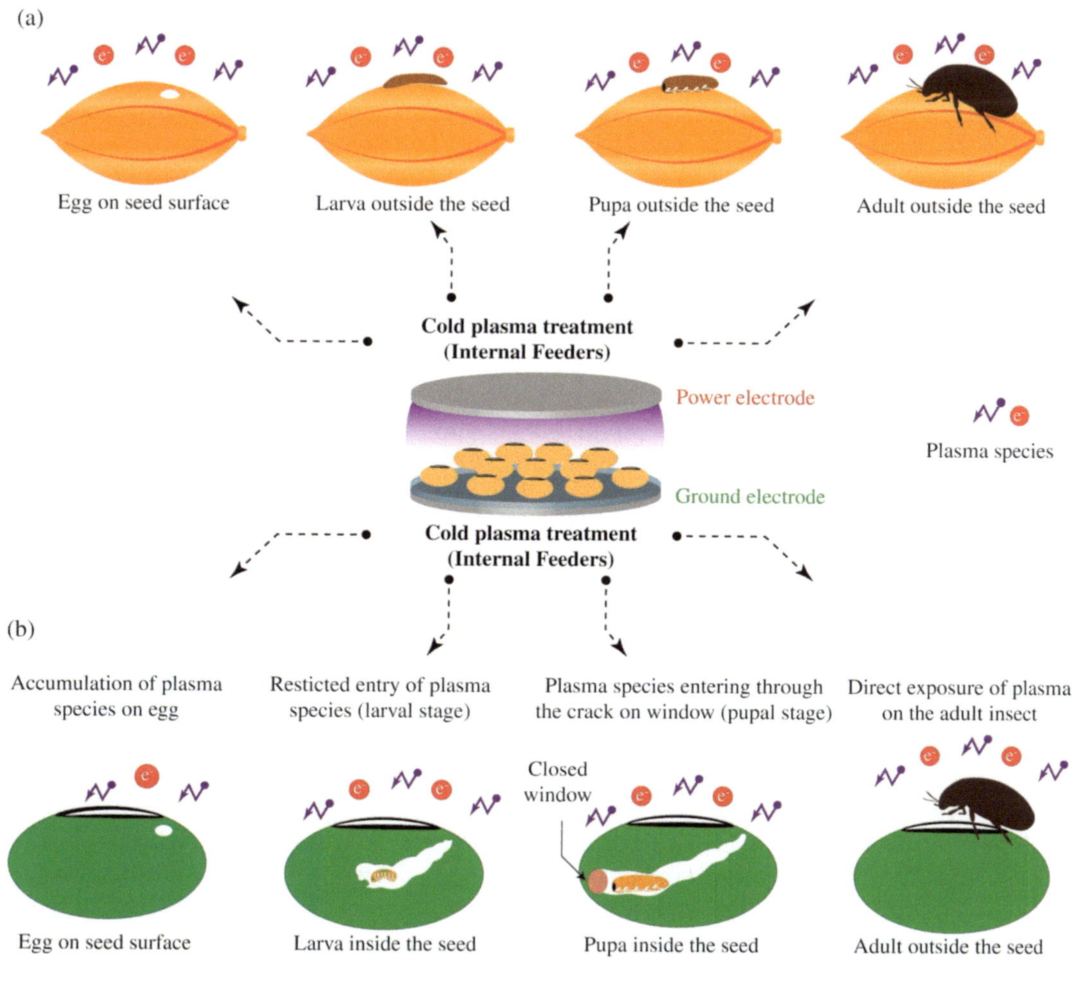

FIGURE 12.1 Interaction of plasma species with (a) external and (b) internal feeders.

in-package insect control, Madathil et al. (2021) used a continuous atmospheric pressure nonthermal jet plasma system to disinfect *Rhyzopertha dominica* adults in whole wheat kernels. The wheat kernels infested with *R. dominica* adults were enclosed in polypropylene packs and subjected to plasma treatment at 44–47 kV voltages for 4–7 minutes. The results revealed a significant correlation between plasma voltage, treatment time, and insect mortality. For instance, the mortality of *R. dominica* increased from 3.33% at 44 kV for 4 minutes to 88.33% at 47 kV for 7 minutes. Furthermore, the study elucidated the insecticidal plasma species, including ROS such as O_3, H_2O_2, O_2^-, OH, and HO_2. These species were observed to reduce the insects' respiration rate, induce asphyxiation, and damage cellular structures and DNA, ultimately leading to insect mortality. Additionally, charged particles generated during plasma formation accumulate on the membrane surface, creating electrostatic forces and resulting in cell rupture. Moreover, the intense bombardment of free radicals on insects also damaged the insect's outer skin, hastening its mortality. Besides its effectiveness in disinfestation, plasma treatment significantly improved the milling yield, protein, and fiber content of wheat. In addition to immediate disinfestation, plasma treatment also assists in controlling the infestation during the prolonged storage of food grains. In order to store wheat flour for 90 days, Sutar et al. (2021) treated the samples with 60 W plasma for 30 minutes and found no infestation issues till the 90th day of storage. Similarly, plasma treatment can also disinfect various life stages of stored product insects, including eggs, larvae, pupa, and adults. However, the efficiency of plasma disinfestation varies across insect life stages. For example, *Tribolium castaneum* pre-adult stages exhibit high susceptibility, with mortality rates reaching 95.0%–100% within seconds of treatment. Meanwhile, adult insects sustained up to 10 minutes of plasma treatment before attaining 100% mortality. In this regard, the time required for the disinfestation of different life stages of *T. castaneum* is as follows: adult (10 minutes) > pupae (5 minutes) > egg (2 minutes) > young (30 seconds) > and old larvae (1 minute) (Ziuzina et al. 2021).

The impact of plasma feed gases on insect mortality is a significant factor, prompting Sayed et al. (2021) to explore the disinfestation potential of Radio Frequency Plasma (RFP) utilizing argon (Ar) and helium (He) gases against *T. castaneum*. The study was conducted at 100 W power with varied treatment times (0, 20, 40, 60, and 90 seconds) and revealed that He plasma exhibited greater efficacy in inducing mortality in *T. castaneum* than in Ar plasma. The optimal exposure time for achieving 100% mortality across all insect life stages was determined to be 90 seconds with He plasma. In contrast, Ar plasma treatment for 90 seconds resulted in only 71.4%, 65.3%, and 36.7% mortality for pupae, larvae, and the adult stage, respectively. In contrast, introducing gases such as oxygen and nitrogen in the feed gas will negatively impact insect mortality, as they reduce the anoxic stress induced by an oxygen-free environment Table 12.1. This effect was observed when O_2 (mortality reduced from 88% to 1.3% and 96% to 0% for *T. castaneum* and *Tribolium confusum*, respectively,) or N_2 (mortality reduced from 90% to 46% and 93% to 57% for *T. castaneum* and *T. confusum*, respectively,) gas was mixed with Ar gas during the disinfestation study (Carpen et al. 2019).

Nasr et al. (2020) study examined the effectiveness of CP generated through corona discharge in disinfestating the *Oryzaephilus surinamensis* and the immature stages of the *Plodia interpunctella* in stored products. For the experiment, three voltage levels (150, 200, and 250 V) and six exposure times (1–25 minutes) were selected, and it was found that CP induced a more pronounced effect on *O. surinamensis* stages compared to the *P. interpunctella*, with the larval stage being more susceptible. For *O. surinamensis*, at 150 V, mortality percentages for adults, eggs, larvae, and pupae were 53.3%, 44.4%, 61.1%, and 25.4%, respectively, after 5 minutes and it increased to 81.1%, 69.6%, 95.6%, and 64.4% for the respective stages after 15 minutes. However, at maximum voltage of 250 V, complete mortality was achieved in all the life stages of the *O. surinamensis* at all the given treatment times. Meanwhile, *P. interpunctella* mortality at 150 V was 27.8%, 60.0%, and 22.2% for eggs, larvae, and pupae, respectively, after 5 minutes, and the values increased to 81.1%, 96.7%, and 76.7% after 20 minutes of plasma treatment. Further increase in the mortality (86.7%, 96.7%, and 71.1%) was observed at 250 V treatment after 10 minutes treatment time. Therefore, it was found

TABLE 12.1
Cold Plasma Treatment Condition and Its Impact on External Feeder, Primary and Secondary Insects

Grain Infestation	Treatment Condition	Results	Reference
Rhyzopertha dominica in wheat sample	V: 44–47 kV T: 4–7 min d: 5 cm P: ambient pressure Gas: atmospheric air	After 24 h, 88.33% mortality was recorded in the 47 kV – 7 min plasma-treated sample.	Madathil et al. (2021)
Tribolium castaneum, Tribolium confusum and *Plodia interpunctella* in wheat flour	Power: 40–60 W T: 0–30 min d: 3 cm P: 0.15 mbar Gas: atmospheric air	60 W plasma treatment for 25 min completely eradicated *T. castaneum* and *T. confusum* during the 90 days' storage. An extended 30 min treatment prevented the *P. interpunctella* presence during the 3 months storage.	Sutar et al. (2021)
T. castaneum	V: 80 kV T: 0–5 min d: 50 mm P: ambient pressure Gas: atmospheric air Mode: direct and indirect	After 0 and 24 h post-treatment, the mortality rates for eggs, young larvae, old larvae, and pupae were as follows: 84.6% and 95.0%, 97.1% and 100%, 80.0% and 96.7%, and 76.7% and 96.7%, respectively.	Ziuzina et al. (2021)
T. castaneum	Power: 100 W T: 0–90 sec F: 13.65 MHz P: 150 kPa Gas: He and Ar	Mortality of 71.4, 65.3, and 36.7% were recorded in pupae, larvae, and adult, respectively, after 90 sec plasma exposure.	Sayed et al. (2021)
Oryzaephilus surinamensis (L.) in	V: 150, 200, and 250 V T: 1–25 min	Egg, larva, and adult stages attained 100% mortality after 25 min of 250 V plasma treatment. Meanwhile a maximum of 99% mortality was achieved in pupae stage insects.	Nasr (2020)
P. interpunctella (Hübner)		Egg and larva attained 100% mortality. Around 98.9 ± 1.1% mortality was observed for pupae stage insects after 25 min plasma treatment at 250 V input voltage.	
T. castaneum and *T. confusum* in wheat	Power: 90–130 W T: 10–15 min F: 13.65 MHz d: 8 cm	*T. castaneum* mortality reduced from 88% to 65% and 1.3% when 3000 sccm Ar was mixed with 2.8 sccm and 8.4 sccm O_2 gas, respectively. Similarly, *T. confusum* mortality reduced from 96% to 51%, and 0% after 15 min treatment.	Carpen et al. (2019)
T. castaneum in maida flour	V: 1000–2500 V T: 1–5 min d: 3–5 cm P: 1 mbar Gas: atmospheric air	1750 V plasma for 7 min at 4 cm electrode distance induced 100% mortality. While higher voltage of 3000 V took only 3 min to achieve the same.	Mahendran et al. (2016)
T. confusum and *Ephestia kuehniella* in wheat	V: 10 kV T: 0–30 sec P: ambient pressure Gas: atmospheric air	100% mortality achieved within 20 sec of treatment.	Mohammadi et al. (2015)

(Continued)

TABLE 12.1 *(Continued)*
Cold Plasma Treatment Condition and Its Impact on External Feeder, Primary and Secondary Insects

Grain Infestation	Treatment Condition	Results	Reference
P. interpunctella	V: 10–20 kV Pulses: 1–20 d: plasma jet nozzle distance varied from 11 to 15 cm. P: atmospheric pressure Gas: atmospheric air	Larval mortality: maximum of 86.6%, 63.0%, and 53.3% was recorded after 20 pulses at 11, 13, and 15 cm nozzle distances. Pupal mortality: maximum of 53.3%, 40.0%, and 33.3% was recorded after 20 pulses at 11, 13, and 15 cm nozzle distances	Abd El-Aziz, Mahmoud, and Elaragi (2014)

d – distance between electrodes; F – frequency; h – hour(s); min – minute(s); P – pressure; sec – second(s); T – time; V – voltage

that the pupae stage of *O. surinamensis* and *P. interpunctella* were more resistant to plasma treatment than other life stages of these insects.

Insects do not necessarily need to be subjected to plasma for longer durations; even an exposure period of 30 seconds can cause 100% mortality, depending on the type of insect, its life stage, plasma type, and the operating conditions (Mohammadi et al. 2015; Mahendran et al. 2016). Furthermore, plasma exposure as a pulse treatment was also effective in disinfesting insects such as *P. interpunctella* (Abd El-Aziz, Mahmoud, and Elaragi 2014).

12.4 COLD PLASMA DISINFESTATION OF INTERNAL FEEDER

Internal feeders constitute a distinct category of insects that reside within grains and come out only after emerging as adults (Figure 12.1). Consequently, conventional surface treatments often fail to impact insects within the grains. In order to evaluate the plasma treatment impact on these internal feeders, Anbarasan et al. (2023) treated four stages (egg, larva, pupa, and adult) of *Callosobruchus maculatus* infested soybeans at different plasma voltages and exposure time Table 12.2. Among the life stages, adults proved most susceptible to plasma treatment (100% mortality at 2.0 kV within 12 minutes), given their external habitat. Conversely, larvae, sheltered inside the soybean seeds, exhibited greater protection from plasma reactive species (maximum mortality 11.11%). However, with improved permeability due to seed windows (grain holes covered only with seed coat), pupae displayed higher sensitivity (maximum mortality 98.89%) than larvae. Freshly laid eggs, protected by a wet outer layer, demonstrated lower mortality rates (maximum 37.33%). Eggs laid 36 hours prior exhibited slightly greater resistance to plasma treatment compared to those laid 18 hours ago. Additionally, the study noted significantly higher mortality in adult males than females up to the 1.0 kV – 12 minutes plasma treatment. The orientation of seeds and insects toward plasma electrodes was observed to influence mortality, with insects on seeds facing the electrode succumbing faster than those positioned away from it. While the earlier study on *C. maculatus* demonstrated higher resistance of eggs and larvae to plasma treatment, increasing the voltage to 70 kV resulted in their mortality exceeding 90% within 3 minutes of treatment (Kirk-Bradley et al. 2023). In another study on internal feeders, four varieties of chickpea (Kripa, Virat, Vishal, and Rajas) were treated using plasma to evaluate their disinfestation efficacy during lone-time storage. Among the four varieties, Kripa, Virat, and Rajas were not observed with any eggs up to a storage period of 90 days after the 40, 50, and 60 W plasma treatments. Meanwhile, eggs were spotted on chickpeas at

TABLE 12.2
Cold Plasma Treatment Condition and Its Impact on Internal Feeder

Grain Infestation	Treatment Condition	Results	Reference
Callosobruchus maculatus in cowpea seeds	V: 20, 50, and 70 kV T: 0–3 min d: 50 mm P: atmospheric pressure Gas: O_2 (65%), CO_2 (30%), and N_2 (5%)	Egg, first instar larvae, and adult mortality were found to be 91%, 93%, and 100% at 70 kV – 3 min treatment.	Kirk-Bradley et al. (2023)
C. maculatus in soybean seeds	V: 1.0–2.0 kV T: 2–24 min P: <2 mbar d: 3 cm	100%, 98.89%, and 37.33% mortality were observed in adult, pupae, and egg stage insects, respectively, following 24 min of plasma treatment at 2.0 kV. Maximum of 11.11% mortality was achieved in 1.5 kV – 18 min plasma treatment.	Rajan et al. (2022)
Callosobruchus chinensis in chickpea	V: 40, 50, and 60 kV T: 10, 15, and 20 min d: 3 cm P: 0.05 mbar Gas: atmospheric air	Grain damages were not observed in Kripa, Virat, and Rajas after plasma treatments. In the Vishal variety, 40 W treatments for 15 and 20 min did not eradicate the insect damage in the grain.	Pathan, Deshmukh, and Annapure (2021)

V – voltage; T – time; d – distance between electrodes; P – pressure; F – frequency; h – hour(s); min – minute(s); sec – second(s)

low-intensity plasma treatment (40 W); however, higher power levels did not let any eggs appear on the food grains (Pathan, Deshmukh, and Annapure 2021).

12.5 BENEFITS OF COLD PLASMA DISINFESTATION OF FOOD GRAINS

Besides serving as a disinfestation technique, CP can also potentially improve the quality of food grains. Studies on CP disinfestation were conducted for grains such as wheat, soybean, chickpea, and cowpea. For these food grains, CP treatment was found to be effective in either reducing pesticide contamination and microbial load or increasing germination, hydration rate, flour properties, or plant-based milk yield.

A study conducted by Anbarasan et al. (2023) demonstrated that CP treatment could reduce more than 50% of chlorpyrifos pesticide residues from soybean seeds within 6 minutes of a 2 kV plasma treatment. Similarly, the authors also showed the ability of plasma treatment to improve the soymilk extraction yield (92.44%–97.22%) when the seeds were subjected to plasma treatment to reduce infestation and later used for soymilk extraction. Additionally, soymilk extracted from the plasma-treated seeds had a higher protein content than that extracted without plasma treatment. The higher protein content in the soymilk was due to the increased water hydration of plasma-treated seeds (Anbarasan et al. 2023). Similarly, wheat seed hydration increased after plasma treatment, leading to a 21% increase in the germination rate of wheat and a reduction in microbial load by up to two logs (Zahoranová et al. 2016). Furthermore, the ability of CP to inhibit enzyme activity, such as lipase and lipoxygenase, played a crucial role in maintaining the quality of wheat germ during storage (Tolouie et al. 2018). Furthermore, Menkovska, Mangova, and Dimitrov (2014) observed an increase in farinograph and alveograph values of wheat dough prepared from plasma-treated flour over time, along with improvements in the shape formation ratio and loaf volumes. These enhancements were attributed to enhanced gas retention and increased crumb cell formation. These results serve as evident proof of the multifaceted applications of plasma treatments.

12.6 CHALLENGES AND FUTURE SCOPE

Commercializing CP as a disinfectant for stored food grain poses several challenges that need to be addressed before its implementation (Kirty, Mamta, and Vikas 2021; Dharini, Jaspin, and Mahendran 2023; Rajan et al. 2023). These challenges include as follows:

- *Regulatory Approval:* CP application on food grain disinfestation may need regulatory approvals and safety assessments to ensure that the process meets food safety standards. However, navigating the regulatory landscape can be time-consuming and costly.
- *Cost-Effectiveness:* Developing and implementing CP systems can be expensive. Finding ways to make the technology cost-effective for manufacturers and consumers is crucial for its adoption.
- *Scalability:* Scaling up CP technology for large-scale food grain processing can be challenging. The technology needs to be adaptable to different production volumes and configurations.
- *Energy Efficiency:* CP systems may require a significant amount of energy. Improving energy efficiency and minimizing environmental impacts are essential considerations.
- *Equipment Design and Maintenance:* Designing robust, reliable, and low-maintenance equipment is critical. Maintenance costs and downtime can impact the economic viability of the technology.
- *Safety Concerns*: Ensuring the safety of workers and consumers when using CP technology is paramount. Adequate training and safety protocols must be established.
- *Quality Assurance:* Maintaining the quality and nutritional value of food grains during CP treatment is essential. Furthermore, minimizing any negative effects on taste, texture, and shelf life is crucial.
- *Consumer Acceptance:* Convincing consumers that CP -treated food grains are safe and equivalent in quality to traditional methods may require education and marketing efforts.
- *Supply Chain Integration:* Integrating CP technology into food grain supply chains can be complex. Compatibility with current processing and storage methods must be ensured.
- *Market Competition:* Competition from other disinfestation methods and technologies may pose a challenge. Demonstrating the superiority of CP in terms of effectiveness and safety is necessary.
- *Awareness and Education:* Raising awareness among stakeholders, including farmers, processors, and policymakers, about the benefits and potential of CP is crucial.
- *Sustainability:* Demonstrating the sustainability benefits of CP, such as reduced chemical use and food waste, can enhance its appeal.

Overcoming these challenges will require collaboration among researchers, industry stakeholders, and policymakers to drive innovation, address safety concerns, and develop effective strategies for the successful commercialization of CP applications in food grain disinfestation.

12.7 CONCLUSION

In conclusion, applying CP for the disinfestation of stored grains presents a promising avenue with several advantages. Notably, it can potentially enhance seed germination, mitigate microbial contaminations, and reduce the presence of toxins and pesticides. However, optimizing the CP process is crucial for future studies and practical applications, as prolonged exposure may lead to seed damage. Practical implementation of CP requires a well-designed system for effectively disinfecting grains entering silos. As we move forward, it is imperative to conduct more comprehensive studies on CP disinfestation, specifically focusing on the sensitivity of insects at different life stages to plasma treatment. This targeted evaluation will enable future researchers to refine their approaches, concentrating on specific insect life stages and minimizing excessive plasma exposure to food grains. The potential benefits of CP offer a valuable contribution to enriching the safety and quality of stored grains.

REFERENCES

Abd El-Aziz, M. F., E. A. Mahmoud, and G. M. Elaragi. 2014. "Non Thermal Plasma for Control of the Indian Meal Moth, *Plodia interpunctella* (Lepidoptera: Pyralidae)." *Journal of Stored Products Research* 59: 215–221. doi:10.1016/j.jspr.2014.03.002.

Ahmad, R., S. Hassan, S. Ahmad, S. Nighat, Y. K. Devi, K. Javeed, S. Usmani, et al. 2022. "Stored Grain Pests and Current Advances for Their Management." In *Postharvest Technology – Recent Advances, New Perspectives and Applications.* doi:10.5772/intechopen.101503; https://www.intechopen.com/chapters/79822.

Anbarasan, R., B. Boopathy, J. Stephen, and M. Radhakrishnan. 2023. "Cold Plasma Disinfestation of *Callosobruchus maculatus* Infested Soybeans: Its Subsequent Impact on Soymilk Extraction Yield and Quality." *Journal of Food Process Engineering* 46 (2). doi:10.1111/jfpe.14246.

Carpen, L. G., C. Chireceanu, M. Teodorescu, A. Chiriloaie, A. Teodoru, and G. Dinescu. 2019. "The Effect of Argon/Oxygen and Argon/Nitrogen Atmospheric Plasma Jet on Stored Products Pests." *Romanian Journal of Physics* 64: 3–4. https://rjp.nipne.ro/2019_64_3-4/RomJPhys.64.503.pdf. Accessed on: 18 December, 2024.

Dharini, M., S. Jaspin, and R. Mahendran. 2023. "Cold Plasma Reactive Species: Generation, Properties, and Interaction with Food Biomolecules." *Food Chemistry* 405 (Part A). Elsevier Ltd: 134746. doi:10.1016/j.foodchem.2022.134746.

Kirk-Bradley, N. T., T. G. Salau, K. Z. Salzman, and M. C. Moore. 2023. "Atmospheric Cold Plasma (ACP) Treatment for Efficient Disinfestation of the Cowpea Weevil, *Callosobruchus maculatus*." *Journal of the ASABE* 66 (4): 921–927. doi:10.13031/ja.15449.

Kirty, P., T. Mamta, and N. Vikas. 2021. "Applications of Cold Plasma Technology in Grain Processing." In *Non-Thermal Processing Technologies for the Grain Industry*, 95–120. doi:10.1201/9781003109501-4.

Madathil, R. V., R. G. Thirugnanasambandan Kalaivendan, A. Paul, and M. Radhakrishnan. 2021, May. "In Package Control of *Rhyzopertha dominica* in Wheat Using a Continuous Atmospheric Jet Cold Plasma System." *Frontiers in Advanced Materials Research* 10–25. doi:10.34256/famr2112.

Mahendran, R., K. R. Ramanan, R. Sargunam, and R. Sarumathi. 2016. "Effect of Cold Plasma on Mortality of *Tribolium castaneum* on Maida Flour." *Agricultural Engineering* 37–44.

Menkovska, M., M. Mangova, and K. Dimitrov. 2014. "Effect of Cold Plasma on Wheat Flour and Bread Making Quality." *Macedonian Journal of Animal Science* 4 (1): 27–30. doi:10.54865/mjas1441027m.

Mohammadi, S., D. Dorranian, S. Tirgari, and M. Shojaee. 2015. "The Effect of Non-Thermal Plasma to Control of Stored Product Pests and Changes in Some Characters of Wheat Materials." *Journal of Biodiversity and Environmental Sciences* 150 (5): 150–156.

Moirangthem, T. T., and O.-D. Baik. 2021, January. "Disinfestation of Stored Grains Using Non-Chemical Technologies – A Review." *Trends in Food Science & Technology* 107: 299–308. doi:10.1016/j.tifs.2020.11.002.

Nasr, M. 2020. "Susceptibility of the Saw-Toothed Grain Beetle, *Oryzaephlius surinamensis* (L.) and The Indian Meal Moth, *Plodia interpunctella* (Hübner) Infested Stored Products to Cold Plasma." *Journal of Plant Protection and Pathology* 11 (6): 299–302. doi:10.21608/jppp.2020.111727.

Pathan, F. L., R. R. Deshmukh, and U. S. Annapure. 2021. "Potential of Cold Plasma to Control *Callosobruchus chinensis* (Chrysomelidae: Bruchinae) in Chickpea Cultivars during Four Year Storage." *Scientific Reports* 11 (1): 13425. doi:10.1038/s41598-021-92792-x.

Rajan, A., B. Boopathy, M. Radhakrishnan, L. Rao, O. K. Schlüter, and B. K. Tiwari. 2023. "Plasma Processing: A Sustainable Technology in Agri-Food Processing." *Sustainable Food Technology* 1 (1): 9–49. doi:10.1039/D2FB00014H.

Rajan, A., B. Boopathy, J. Stephen, and M. Radhakrishnan. 2022. "Cold Plasma Disinfestation of *Callosobruchus maculatus* Infested Soybeans: Its Subsequent Impact on Soymilk Extraction Yield and Quality." *Journal of Food Process Engineering* 46 (2). doi:10.1111/jfpe.14246.

Rajan, A., and M. Radhakrishnan. 2023. "Computational Plasma Dynamics for Wastewater Treatment Systems." In *Advanced Computational Approaches for Water Treatment*, 141–168. New York: CRC Press. doi:10.1201/9781003325147-8.

Sayed, W. A. A., R. S. Hassan, T. M. Sileem, and B. A. Rumpold. 2021. "Impact of Plasma Irradiation on *Tribolium castaneum*." *Journal of Pest Science* 94 (4): 1405–1414. doi:10.1007/s10340-021-01360-9.

Stathas, I. G., A. C. Sakellaridis, M. Papadelli, J. Kapolos, K. Papadimitriou, and G. J. Stathas. 2023. "The Effects of Insect Infestation on Stored Agricultural Products and the Quality of Food." *Foods* 12 (10). doi:10.3390/foods12102046.

Sutar, S. A., R. Thirumdas, B. B. Chaudhari, R. R. Deshmukh, and U. S. Annapure. 2021, May. "Effect of Cold Plasma on Insect Infestation and Keeping Quality of Stored Wheat Flour." *Journal of Stored Products Research* 92: 101774. doi:10.1016/j.jspr.2021.101774.

Tolouie, H., M. A. Mohammadifar, H. Ghomi, and M. Hashemi. 2018. "Cold Atmospheric Plasma Manipulation of Proteins in Food Systems." *Critical Reviews in Food Science and Nutrition* 58 (15): 2583–2597. doi:10.1080/10408398.2017.1335689.

Zahoranová, A., M. Henselová, D. Hudecová, B. Kaliňáková, D. Kováčik, V. Medvecká, and M. Černák. 2016. "Effect of Cold Atmospheric Pressure Plasma on the Wheat Seedlings Vigor and on the Inactivation of Microorganisms on the Seeds Surface." *Plasma Chemistry and Plasma Processing* 36 (2): 397–414. doi:10.1007/s11090-015-9684-z.

Ziuzina, D., R. van Cleynenbreugel, C. Tersaruolo, and P. Bourke. 2021. "Cold Plasma for Insect Pest Control: *Tribolium castaneum* Mortality and Defense Mechanisms in Response to Treatment." *Plasma Processes and Polymers* 18 (10). doi:10.1002/ppap.202000178.

13 Ionizing Radiations

Cristiano Dietrich Ferreira and Aldrey Nathália Ribeiro Corrêa
Technological Institute in Food for Health, University of Vale do Rio dos Sinos, São Leopoldo, Brazil

Nairiane Bilhalva and Paulo Carteri Coradi
Department of Agricultural Engineering, Rural Sciences Center, Federal University of Santa Maria, Santa Maria, Rio Grande do Sul, Brazil
Department of Agricultural Engineering, Laboratory of Postharvest (LAPOS), Campus Cachoeira do Sul, Federal University of Santa Maria, Cachoeira do Sul, Rio Grande do Sul, Brazil

13.1 INTRODUCTION

The use of radiation to decontaminate foods such as fruit, meat, cereals, or processed products has been used for a long time. The different radiations used originate and can be identified in the electromagnetic spectrum. Electromagnetic waves, or rays, are classified according to their wavelength. Electromagnetic radiation comprises self-propagating waves, which have an electric field component and a magnetic field component that oscillates in phase and directions perpendicular to each other. The typical electromagnetic spectrum of wavelengths ranges from 10^{-15} m (cosmic radiation) to 10^{7} m (10,000 km; high-, medium-, and low-frequency alternating currents). Only a small fraction of all electromagnetic waves are visible to the human eye and are located between 450 and 700 nm.

Table 13.1 summarizes the characteristics of ionizing and non-ionizing radiation in the electromagnetic spectrum. Radiations in the electromagnetic spectrum can be divided into two groups based on their ionization potential. Ionizing radiation is that which produces ionization, that is, it is the process through which an atom or molecule loses or gains electrons to form ions. Non-ionizing radiation cannot ionize and may have thermal (radio frequency, microwave, and infrared) or non-thermal (visible and ultraviolet light) action.

13.2 IONIZING RADIATIONS

The atom is formed by a nucleus with a positive electrical charge, with protons and neutrons (called nucleons), and around it are electrons with a negative electrical charge, as proposed by the Rutherford-Bohr atomic-molecular model "planetary model." The nucleus is very compact, measuring 10^{-15} m, and is 100,000 times smaller than the atom, which has a diameter of 10^{-10} m. When in equilibrium, the nucleus and electrons of an atom are in stationary orbitals. Electrons gravitate in orbits or layers with different energy levels, and when they move from a level closer to the center to layers further away, the atom is said to be in an excited (unstable) state. The process then occurs in which the atom tends to return to a condition of equilibrium, that is, to its fundamental state. This process can generate the emission of photons with different radiant energies. With sufficient energy, the electron escapes from the nucleus, and the atom becomes positively charged, a process known as ionization.

Ionizing radiation represents an emerging technology characterized by extremely short wavelengths and intensity high enough to alter atoms through the extraction of an electron, resulting in the formation of an ion. However, it does possess the requisite intensity to induce atom splitting, preventing the sources subjected to it from becoming radioactive.

DOI: 10.1201/9781003609834-13

TABLE 13.1
Effect of Gamma Rays on Insects, Functional Properties, and Fungi from Different Food Sources

Study Object	Food Source	Dose Used (kGy)	Main Results	Reference
		Insects		
Sitophilus oryzae (L.)	Wheat	1.00	100% mortality after 96h	Nasr et al. (2022)
Tribolium castaneum	Wheat	1.00	100% mortality	Sileem, Mohamed, and Mehany (2019)
Callosobruchus maculatus	Cowpea	1.50	100% mortality	Ashraf et al. (2021)
C. maculatus	Cowpea	0.65	100% mortality	Hammad, Gabarty, and Zinhoum (2020)
Rhyzopertha dominica	Wheat	1.00	100% mortality	Dehkordi, Ahmadi, and Hatami (2018)
R. dominica	Wheat	0.28	100% mortality and maintained grain viability	Abotaleb, Zinhoum, and Gabarty (2021)
Sitophilus zeamais and *C. maculatus*	Maize and cowpea	0.20–0.50	100% mortality	Enu and Enu (2014)
H. hampei	Green coffee	3.20	100% adult mortality	Kiran, Shenoy, and Venkatesha (2019)
		Functional Properties		
Anti-nutritional factors	Sorghum	30.00	Chemical composition (↔) Digestibility of crude starch protein (↑) Phyto and tannins (↓) Total phenols (↑) Mineral content and bioavailability (↔)	Rousta et al. (2014)
Proximate composition, pasting properties, lipoxygenase activity, and antioxidant properties	Millet	15.00	Moisture contents (↓) Protein contents (↑) Fat content (↔) Pasting properties (↓)	Reddy and Viswanath (2019)
Physicochemical, functional, and antioxidant properties	Brown rice flour	5.00	Apparent amylose content (↓) Taste properties (↓) Water and oil absorption (↑) Emulsifying capacity (↑) Emulsion stability and foam formation (↓) Antioxidant activity (↑)	Bhat, Wani, and Sultan (2023)
Fungal growth, free fatty acids, in vitro protein digestibility (IVPD), protein solubility, and functional properties	Sorghum	5.00	Storage properties (↑) Fungal incidence (↓) Free fatty acid content (↓) IVPD (↑) Soluble protein (↑)	Ahmed et al. (2018)
Gel consistency, water absorption, total anthocyanin content, phenolic compounds, protein, and antioxidant activity	Pigmented rice	1.00	Gel consistency (↔) Total phenolic content (↑) Antioxidant capacity (↔)	Singh et al. (2022)
Phenolic compounds and antioxidant activity	Brown rice	10.00	Bound phenolics (↑) Free, bound, and total antioxidant activity (↑) Color parameters (↔) Gel viscosity and hardness (↓)	Shao et al. (2013)

(Continued)

TABLE 13.1 *(Continued)*

Effect of Gamma Rays on Insects, Functional Properties, and Fungi from Different Food Sources

Study Object	Food Source	Dose Used (kGy)	Main Results	Reference
Baking properties and safety indicators	Wheat flour	47.52	Aerobic and anaerobic microorganisms (↓) Mold (↓) Total deformation, elastic properties, and elasticity index (↓) Safety indicators (↑)	Romanov et al. (2022)
Free and bound total phenolic compounds, antioxidant activity, physicochemical, and microbiological properties	Black rice flour	3.00	Phenolic compounds (↑) Antioxidant activity (↑) Total anthocyanins, cyanidin-3-glucoside, and flavonoids (↑)	
Protein modifications	Soybean	2.00	Turbidity (↓) Hydrophobicity (↓) Oxidation (↓) Solubility (↑) Sulfhydryl contents (↑) Free phenolics (↑) DPPH and FRAP activities (↑)	Kumari et al. (2020)
Physicochemical, functional, and paste properties	Rice	1.5	pH (↓) Swelling power (↓) Solubility (↑) X-ray digestion pattern (↔) Paste properties (↓)	Ocloo et al. (2017)
Storage stability	Sorghum	10.0	Number of insects (↓)	Abdel Rahman, Abdalla, and Ahmed (2020)
Nutritional quality and functional properties	Soybean flour	1.00	Shelf life (↑) Protein content (↑) Oil absorption capacity (↑) Antioxidant activity (↑) Anti-nutritional facts (↓)	Jabeen, Kola, and Jalarama (2015)
Protein characteristics and functional properties	Sesame	1.00	Globulin and albumin contents (↑) Digestibility (↑) Solubility (↑) Emulsifying and foaming properties (↑)	Hassan et al. (2018)
Microbial, chemical, and sensory properties	Sesame	9.00	Moisture, ash, and fat content (↔) Acidity (↓) TVBN (↑) Flavor, color, and texture (↑)	Al-Bachir (2016)
Microbial, chemical, and sensory properties	Tunisian millet	5.00	pH (↔) Peroxide (↓) Phenylalanine (↓) Total phenolic content (↓)	Mustapha et al. (2014)
Physicochemical properties	Wheat flour	1.00	Solubility and foaming capacity (↔)	Manupriya et al. (2020)
Physicochemical and functional properties	Whole wheat flour	10.00	Proximate composition (↔) Pasting properties of flour and starch (↓) Amylase content (↑) Bulk density (↔)	Bashir et al. (2017)
		Fungi and Mycotoxin		
AFB1	Peanut	9.00	43% of degradation	
AFB1	Rice	6.00	95% of degradation	
AFB1	Soybean	10.00	62.20% of degradation	Zhang, Xie, and Che (2018)

(Continued)

TABLE 13.1 *(Continued)*

Effect of Gamma Rays on Insects, Functional Properties, and Fungi from Different Food Sources

Study Object	Food Source	Dose Used (kGy)	Main Results	Reference
OTA	Wheat flour	30.50	24% of degradation	Calado et al. (2018)
AFB1	Wheat	20.00	90% of degradation	Hathout et al. (2020)
AFB1	Maize, wheat, and rice	8.00	60.29, 64.68, and 6.29% of degradation	Mohamed et al. (2015)
AFB_1 and AFB_2	Wheat	10.00	65.83% of degradation	Tatar, Fadavi, and Koohsari (2021)
ZEA	Wheat and maize	20.00	69.3% of degradation	Sebaei et al. (2022)
AFB1, AFB2 E OTA	Maize	20.00	40.1, 33.3, and 61.1% of degradation	
Aspergillus flavus	Peanut	8.00	Inactivates or eliminates the fungus	Borges et al. (2020)
OTA and AFB1	Sorghum	10.00	59 and 32% reduction	Ben Amara et al. (2022)

The general Codex standard for irradiated foods allows three forms of ionizing radiation Table 13.2 for food irradiation purposes: high-energy gamma rays (^{60}Co and ^{137}Cs), X-rays (energies up to 5 MeV, and in some countries up to 7, 5 MeV), and electron beam (energies up to 10 MeV) Fan and Niemira 2020; Khan et al., 2019).

The effect of ionizing radiation is similar, whether on human pathogens or product quality, whether directly or indirectly. In the direct effect, cellular components such as carbohydrates, DNA,

TABLE 13.2

Characterization of Ionizing and Non-Ionizing Radiation Present in the Electromagnetic Spectrum

	Non-ionizing	
Types		General characteristics
Radio-frequency (RF)	Frequency (HZ): 3×10^3	Low frequency/low energy
Microwave (MW)	Wavelength (m): 10^5–10^{-8}	Long wavelength
Infrared (IR): near, mid, far	Photon energy (eV)	Low penetration force
Visible light (Vis): 400–750 nm		Interaction with matter through
Ultraviolet (UV-A): 315–400 nm		Absorption and scattering
Ultraviolet (UV-B): 280–315 nm		Primary effect: excitation
Ultraviolet (UV-C): <280 nm		No molecular bond breakages
	Ionizing	
X-rays	Frequency (HZ): 3×10^{17}–3×10^{22}	High frequency/high energy
Y-rays	Wavelength (m): 10^{-9}–10^{-14}	Short wavelength
	Photon energy (eV): 1.24 × 103–1.24 × 108	High penetration depth
		Interaction with matter through
		Absorption and scattering
		Primary effect: ionization and excitation
		Chemical bond breakages

and lipids are directly damaged by ionizing radiation, while in the indirect effect, free radicals and reactive species such as hydro electrons, hydrogen atoms, or hydroxyl radicals) from water, radiolysis reacts with cells or food components. The main ingredient of fresh produce is water, and therefore, the indirect effect is the main mechanism for the transition of ionizing irradiation (Fan and Niemira 2020; Bisht et al. 2021). The result of indirect and direct effects is the occurrence of physiological, genetic, epigenetic, and biological changes (Desouky, Ding, and Zhou 2015).

Several factors play an essential role in the elimination of different microorganisms by irradiation technology. These factors involve the composition and physical state of the food, type, and dose of radiation used, exposure time, species, number, age, type and location of microorganisms, inoculation time, the regeneration capacity of microorganisms, and environmental factors (Delorme et al. 2020; Pi et al. 2022). Therefore, a proper understanding of these factors and their influence on the longevity or death of microorganisms is crucial in deciding acceptable radiation so that beneficial inactivation can be achieved (Fan, Huang, and Chen 2017).

13.2.1 Gamma

Gamma rays are radiation that has a high frequency and energy, which enables them to traverse long distances through the air, making them highly pervasive (Erramli and El Asri 2019). Gamma-ray photons possess a higher recurrence and energy compared to UV and X-ray photons, as they can penetrate a target food to varying depths (Pathak et al. 2018). Gamma irradiation is widely acknowledged as an effective sterilization or decontamination method, significantly impacting the destruction of microbes and disrupting the covalent bonds in the DNA of viruses and bacteria.

This radiation is produced from the radioisotopes ^{60}Co and ^{137}Cs, which use electricity to produce electron beams (up to 10 MeV). However, ^{137}Cs are rarely used because large sources of ^{137}Cs are not readily available. For food irradiation purposes, ^{60}Co gamma radiation sources are frequently preferred due to their insolubility in water, which results in a lower risk of environmental contamination. On the other hand, ^{137}Cs are soluble in water, making them a higher risk for potential environmental contamination (Ashraf et al. 2019).

Commercially, gamma radiation has been used for more than 35 years in industry to ensure food safety and is currently the most widely used in food processing (Byun et al. 2019). The dose of gamma radiation is measured in kGy (Kilogray), where 1 Gy is equivalent to 1 joule/kg. The radioresistance of a microorganism or toxin is the dose necessary to inactivate 90% of the population or concentration (Urbain, 1986).

13.2.1.1 Physical Principle

Gamma radiation works by exciting electrons in food molecules, producing charged particles known as ions. However, gamma rays do not induce radioactivity in food as they lack the necessary energy to impact neutrons. The reduction in the concentration of mycotoxin present is what determines the efficiency of this technology (Guo et al. 2021). The advantages of applying this irradiation are the few changes that are caused to food components and the depth of penetration in liquid and solid media.

^{60}Co can be produced by particle accelerators (Ozan, 2017); however, it is usually produced by bombarding isotope 59 in a nuclear power reactor in which cobalt-59 absorbs neutrons (Howard et al. 2021). Activated metal (Co-60) is double encapsulated with rods or discs in stainless steel casings before being released to radiation facilities. When the radiation source is not in use, it is enclosed in a shielded container or submerged in a pool of water that absorbs all the emitted radiation. It does not generate any waste, as the deteriorated source is returned to the supplier for replacement or retention.

Gamma rays exhibit quick penetration through various materials and are commonly used in large-scale food where electron beams are utilized for surface irradiation (Pathak et al. 2018).

Gamma radiation is non-thermal processing extensively applied in various aspects of the food industry, including disinfestation, microbial inactivation, sprout inhibition, and delayed ripening (Mir et al. 2015; Ravindran and Jaiswal 2019). Notably, this technology is both energy-efficient and environmentally sustainable, generating no waste. The utilization of irradiation technology helps in increasing the shelf life of diverse food products without affecting their nutritional properties.

13.2.1.2 Search Results

Radiation in food processing aims to increase storage stability, eradicate parasites or insects by impeding enzymatic activity and reduce bacteria, fungi, viruses, and foodborne mycotoxins. In Table 13.3, it is possible to analyze some of the most recent studies using gamma radiation on grains and by-products.

13.2.1.2.1 Insect Control

Gamma, which in the adult insect includes the gonads and midgut. Exposure to a cumulative dose of gamma rays totaling 0.20 kGy (0.66 rad/s dose rate) from the 137Cs source induces cellular perturbations in the midgut epithelium of F1 progeny of the *Blaps polycresta* beetle (Kheirallah and El-Samad 2020). At minimal doses of gamma irradiation, which disrupts the functionality of the digestive system, observations indicate that the midgut lacks a well-defined membrane, and the intercellular connections with adjacent cells are decreased (Moamen and El Bakary 2020). As a consequence, insects cease reproduction and feeding due to the incapacity of the midgut epithelium to process food.

Nasr et al. (2022), when studying the influence of gamma radiation at doses of 0.10, 0.25, 0.50, and 1.00 kGy on the mortality rates of *Sitophilus oryzae* adults and its effects on weevil sterility and seed germination, it was determined that a dose of 1.00 kGy resulted in 100% mortality within 96 h of irradiation. Furthermore, the application of irradiation at a 1.00 kGy dose against *S. oryzae* adults exhibited a strong impact on histological changes and a negative impact on the percentage of germination.

According to Salim, Ahmadi, and Moharramipour (2018), a dose of 0.50 kGy of gamma radiation can effectively eliminate 100% of both *Tribolium confusum* adults and larvae in 22 days. On the other hand, infestation by *Tribolium castaneum*, according to (sileem, Mohamed, and Mehany 2019), is effectively controlled with a gamma irradiation dosage of 1.00 kGy.

Gamma irradiation was employed as a method to manage *Callosobruchus maculatus* infestation in cowpea seeds during storage (Ashraf et al. 2021). Hammad, Gabarty, and Zinhoum (2020), in

TABLE 13.3
Comparison of Radiation Sources Used for Food Sterilization

Radioactive isotope		**General Characteristic**
	Gamma-ray	Temperature/chemical independent
	Most popular	applicable to packaged/fully sealed
	Simple, easily controlled, widely applicable	product safe and effective
	E-beam	
Elector beam accelerator	Short treatment time + high-dose rate	
	No waste products	
	X-ray	
	High depth penetrations	
	Good dose uniformity radio	

turn, recommend a dose of 0.65 kGy for the phytosanitary control of this insect, as this dose does not significantly alter the main constituents of cowpea.

Dehkordi, Ahmadi, and Hatami (2018) employed gamma irradiation to assess its impact on the mortality and sterility of *Rhyzopertha dominica*. Irradiation dosage above 1.00 kGy resulted in 100% sterility in the *R. dominica* population. Abotaleb, Zinhoum, and Gabarty (2021), in turn, when investigating the effect of low doses of gamma irradiation in the range of 0.02–0.30 kGy on the different stages of development of *R. dominica* in wheat grains, found that no adults emerged after irradiating eggs, larvae, and pupae at 0.14, 0.24 and 0.26 kGy, respectively. Furthermore, they found that the viability of wheat grains, in terms of germination, was not affected by the dose of 0.28 kGy, which is a recommended irradiation dose for effective phytosanitary control.

Kiran, Shenoy, and Venkatesha (2019) evaluated the efficacy of gamma radiation as a post-harvest disinfestation method against the life stages of the coffee borer (*Hypothenemus hampei*), aiming to determine the effective doses of irradiation to inhibit the development of various stages of the coffee borer. For eggs, irradiation was conducted at five levels ranging from 0.01 to 0.16 kGy. Regarding larvae and prepupa, seven levels are tested, ranging between 0.10 and 2.00 kGy. The dose of 2.00 kGy resulted in 100% mortality in first- and second-instar larvae, as well as 98.99% mortality in prepupa. For pupa, six irradiation levels were examined, spanning from 0.10 to 1.60 kGy. The dose of 1.60 kGy effectively prevented adult hatching from the irradiated pupa. Finally, when it came to adult coffee borers, ten irradiation levels were assessed, ranging from 0.10 to 3.20 kGy. At a dose of 3.20 kGy, 100% mortality was achieved in the adult stage. Therefore, a dose of 3.20 kGy can ensure the complete safety of all developmental stages of *H. hampei* and prevent yield loss in green coffee as well as the spread of the pest.

Insect mortality induced by irradiation is attributed to the formation of highly reactive free radicals. As a result, this method is often regarded as a favorable option due to its absence of chemical residues, cheaper processing, and ease of use. Although irradiation is quite effective against insects at low doses (less than 1.00 kGy), it does not provide a guarantee of effectiveness against viruses and any toxins that may be generated in the grain.

13.2.1.2.2 Changes in Functional Properties

The impact of gamma irradiation on the nutritional quality and functional properties of both fresh soy flour and germinated soy flour. The irradiation did not result in any discernible organoleptic or chemical changes in fresh flour or germinated soy flour, in addition to improving shelf life. At the irradiation dose of 1.00 kGy, positive effects were observed, including an increase in protein content, oil absorption capacity, and total antioxidant activity. Furthermore, a decrease in the levels of anti-nutritional factors was observed.

Hassan et al. (2018) conducted a study on the effects of gamma irradiation (0.5, 1.0, 1.5, and 2 kGy) on the protein characteristics and functional properties of sesame seeds (*Sesamum indicum*). The authors reported an increase in the globulin and albumin contents of sesame protein, as well as an improvement in the digestibility and solubility of the sesame flour protein. Furthermore, the emulsifying and foaming properties were improved after gamma irradiation.

Al-Bachir (2016) observed that gamma irradiation of sesame seeds led to an increase in protein content, particularly up to a dose of 9.00 kGy, when compared to the control sample. Protein content also decreased over a 12-month storage period. Furthermore, the free fatty acid content in the irradiated samples decreased, which was attributed to either the inactivation of the lipase enzyme or the loss of volatile free fatty acids.

The microbiological and biochemical properties of Tunisian millet were examined after gamma radiation by Ben Mustapha et al. (2014). At a gamma irradiation dose of 1.00 kGy, the concentration of ochratoxin decreased by 13%, while at a dose of 3.00 kGy, it reduced by 44%. When the dose was increased to 5.00 kGy, the pH remained unchanged, and the peroxide value increased at the initial concentration from 26.16 meq O_2/kg to 34.43 O_2/kg from the non-irradiated sample to the irradiated sample (3.00 kGy), respectively, but with increasing irradiation dose (5.00 kGy), the peroxide value decreased. The authors also reported a dose-dependent decrease in phenylalanine concentration from 32 to 23 mg/g in millet flour irradiated with 5.00 kGy and a non-significant decrease in total phenolic content (TPC).

Manupriya et al. (2020) studied the effect of gamma irradiation (0, 0.25, 0.50, 1, 500, and 10.00 kGy) on the physicochemical properties of wheat flour. Their findings indicate that there were no significant changes in the properties of the flour of wheat, including water absorption capacity, water absorption index, swelling property, loose apparent density, compacted apparent density, total carbohydrate content, total antioxidant capacity, lipids, and protein profile. Solubility and foaming capacity remained unchanged up to 1.00 kGy. However, reductions were observed at 5.00 and 10.00 kGy. Most of the properties of wheat flour irradiated at 1.00 kGy are below the level of significance with minimal surface cracks, indicating that this dose of irradiation could be recommended for use in the preparation of bakery products to enhance their quality and flavor.

The physicochemical and functional attributes of whole wheat flour and starch irradiated with gamma rays (0.50, 1.00, 2.50, 5.00, and 10.00 kGy) were studied by Bashir et al. (2017). The authors verified that there were no notable changes in the proximate composition of the samples across different dosage levels. However, the pasting properties of both flour and starch showed a significant decrease in peak viscosity, final viscosity, and setback with increasing dosage. The amylose content exhibited a significant increase with escalating dosage, rising from 25.33 to 36.03%. The apparent density of the flour remained relatively stable throughout the irradiation process. Additionally, swelling, solubility, syneresis, and freeze-thaw stability were improved with dosing. The water and oil absorption capacity of flour increased significantly with dosage and was found in the range of 0.85–0.91 and 1.10–1.91 g/g flour, respectively.

Abdel Rahman, Abdalla, and Ahmed (2020) sought to extend the storage stability of sorghum grains using a gamma radiation source (Co-60) as a means of preservation by applying doses of 10.00 and 30.00 kGy to the grains. Irradiated sorghum grains and a control sample were stored at room temperature (30°C) and 30% relative humidity and examined during storage for changes in physical properties, presence of insects, broken grains, and organoleptic properties. The number of insects in grains irradiated with 10.00 kGy was only one insect per kg compared to 140 insects per kg in the control (non-irradiated grains).

The effect of gamma irradiation on some physicochemical, functional, and paste properties of rice grown in Ghana was studied by Ocloo et al. (2017). Four locally grown rice cultivars and one commercially imported type were acquired, cleaned, and irradiated at doses of 0.0, 0.25, 0.50, 0.75, 1.00, and 1.50 kGy. The irradiated rice cultivars were then ground, and their physicochemical, functional, and paste properties were examined. Following gamma irradiation, there were reductions in pH and swelling power, accompanied by an increase in the solubility of rice cultivars. The X-ray diffraction pattern of rice cultivars remained unaffected by the gamma irradiation process. Moreover, gamma irradiation significantly decreased the peak time for two rice cultivars. Gamma irradiation significantly decreased peak viscosity, trough/hot slurry viscosity, breakthrough viscosity, final viscosity, and setback viscosity for all rice cultivars.

Kumari et al. (2020) investigated the modifications of proteins extracted from irradiated soybeans. The irradiation dose of 2.00 kGy resulted in a reduction in turbidity, surface hydrophobicity, and protein oxidation but caused an increase in protein solubility and sulfhydryl contents in both genotypes. Free phenolics (FPs), DPPH, and FRAP activities in all genotypes significantly increased at 2.0 kGy and decreased at 5.0 kGy, except DPPH. However, lipoxygenase and bound phenolics decreased with increasing doses of gamma irradiation in both tested genotypes.

Black rice has been categorized as a functional food due to its substantial content of bioactive compounds. Therefore, Ito et al. (2019) evaluated the effects of gamma radiation (0, 1.00, 2.00, and 3.00 kGy) on total free and bound phenolic compounds, antioxidant activity, physicochemical, and microbiological properties of black rice flour samples during storage. The results showed an increase in phenolic compounds and the antioxidant activity of the free and total phenolic fraction, as well as in the content of total anthocyanins, cyanidin-3-glucoside, and total flavonoids, particularly up to a dose of 3.00 kGy. However, after 120 days of storage, a decline in these levels was observed for all samples. Nevertheless, the irradiated samples displayed greater stability by the end of storage. In terms of color attributes, it was observed that gamma irradiation slightly modified all measured parameters. For microbiological analysis, there was no growth of microorganisms in the samples during the 120-day storage period.

Romanov et al. (2022) conducted a study to assess the effect of irradiation on the safety indicators of wheat flour. The investigation included wheat flour subjected to gamma irradiation (0–47.52 kGy). The flour was shown to be safe in terms of residual gamma radiation after 24 and 72 h. At the maximum dose of gamma radiation, the total viable count of aerobic and facultative anaerobic mesophilic microorganisms decreased 15-fold, while the amount of mold decreased 5-fold. The total strain energy, elastic properties, and elasticity index of the dough decreased by more than 50%. The microbiological safety indicators of wheat flour increased at the maximum irradiation dose.

A study conducted by Shao et al. (2013) analyzed three rice genotypes with different colors. These rice samples were subjected to gamma irradiation at doses of 2.00, 4.00, 6.00, 8.00, and 10.00 kGy, aiming to investigate the effect of gamma irradiation on the content of phenolic compounds and antioxidant activity, as well as the physicochemical properties of brown rice. The content of bound phenolics in all genotypes increased significantly with increasing irradiation dose. High-dose gamma irradiation resulted in an increase in the free, bound, and total antioxidant activities of three rice genotypes, except for the free antioxidant activities of red rice. Although the changes in color parameters they were not visually discernible. However, gel viscosities (RVA) and hardness decreased continuously with increasing irradiation doses.

Singh et al. (2022) studied the impact of varying doses of gamma radiation (0.25–1.00 kGy) on quality attributes from different prominent ethnic pigmented rice cultivars. These cultivars, consisting of five different types, were sourced from the state of Manipur, India. The study aimed to assess changes in quality attributes such as gel consistency, water absorption capacity, total anthocyanin content, phenolic compounds, protein content, and antioxidant activity in response to the different irradiation doses. The cultivars showed significant variation in their physical quality attributes, including hulled grain color, grain weight, and length/width ratio. No significant difference was observed in the gel consistency profile of these rice cultivars after processing by gamma radiation (0.25–1.00 kGy). A wide range of total anthocyanin content (0.17–4.90 mg cyanidin-3-glu/g) was observed in non-irradiated samples among pigmented rice cultivars. Following the radiation processing, the impact on anthocyanin content varied across the different cultivars, with values ranging from 0.24 to 3.96 mg cyanidin-3-glu/g in the irradiated samples. TPC (0.22–2.46 mg GAE/g in non-irradiated samples) increased marginally (0.32–2.47 mg GAE/g in irradiated samples) in all rice cultivars after radiation processing. Rice cultivars also exhibited prominent antioxidant capacity (DPPH elimination: 20.5–80.5% in non-irradiated samples), and this health-protective functional property was maintained (DPPH elimination: 20.8–80.7% in non-irradiated samples) after gamma irradiation.

Bhat, Wani, and Sultan (2023) evaluated the effect of gamma irradiation on the physicochemical, functional, and antioxidant properties of brown rice flour conditioned to two different moisture contents of 10 and 12% and treated with gamma irradiation doses of 0 (native or control), 2.50 and 5.00 kGy. The moisture, protein, fat, ash, and carbohydrate content of native flour was 10/100 g, 6.54/100 g, 1.54/100 g, 1.0/100 g, and 81.48/100 g, respectively. It was observed that the apparent amylose content decreased as the gamma irradiation dose and moisture content increased, with values decreasing from 29.97 to 20.30/100 g. The pasting properties, including peak viscosity, minimum viscosity, and final viscosity of all flour samples, decreased as the irradiation dose increased.

Functional properties such as water and oil absorption and emulsifying capabilities were increased, while emulsion stability and foaming stability decreased with irradiation. Irradiation led to a general increase in the antioxidant activity of rice flour. In general, Fourier transform infrared spectroscopy (FTIR) spectra revealed a decrease in the absorption intensities of the functional groups.

Rousta et al. (2014) studied the effects induced by gamma radiation on the nutritional quality and anti-nutritional factors of sorghum grain. In their research, sorghum grains were subjected to gamma radiation at doses of 10.00, 20.00, and 30.00 kGy. The study encompassed an examination of the chemical composition, mineral content, and bioavailability, digestibility in vitro of protein and starch, total phenolic compounds, phytate, and tannin contents of sorghum grain. The results showed that gamma irradiation did not induce changes in the chemical composition of sorghum grains; it remained consistent. Furthermore, the results demonstrated an increase in the digestibility of starch by 3.8 and by 6.6% in the digestibility of crude protein from sorghum grain after irradiation. Phytic acid and tannin levels were significantly reduced, while total phenols were increased relative to unprocessed control samples. The results indicated no substantial changes in mineral content and bioavailability among the sample, except significant increases in extractable calcium, phosphorus, zinc, and iron.

Ahmed et al. (2018) evaluated the effect of gamma irradiation at dose levels of 0.50, 1.00, 2.00, 3.00, 4.00, and 5.00 kGy on fungal growth, free fatty acids, protein digestibility in vitro (in vitro protein digestibility (IVPD)), protein solubility, and functional properties of sorghum grains. The results indicated that the radiation process improved storage properties and eliminated fungal incidence mainly at the highest irradiation doses of 4.00 and 5.00 kGy. Additionally, gamma irradiation resulted in a significant reduction in the free fatty acid content of sorghum grains, lowering it to the range of 3.4–3.2 mg/g. Furthermore, the study revealed that there was a significant increase in IVPD and soluble protein in the irradiated grains compared to the non-irradiated sample, and this increased with increasing dose; however, maximum IVPD (17.6%) and protein solubility (11.7%) were observed in sorghum grains irradiated at 2.00 kGy. On the other hand, a significant reduction in emulsion capacity was noted after grain irradiation. However, emulsion activity and stability remained stable up to 1.0 kGy. As for the foaming properties, there were no significant changes observed in the irradiated samples, particularly at lower doses of up to 2.0 kGy, when compared with untreated samples subjected to millet grains.

Reddy and Viswanath (2019) conducted a study in which millet flour was exposed to gamma radiation at different dose levels: 2.0, 5.0, 10.00, and 15.00 kGy. The research aimed to investigate the impact of gamma irradiation on several aspects of millet flour, including its proximate composition, pasting properties, lipoxygenase activity, and antioxidant properties. This study sought to assess how gamma irradiation influenced these specific properties in millet flour. The moisture, protein, fat, ash, and carbohydrate content of native flour was 12.96, 6.85, 2.73, 1.57, and 73.69/100 g, respectively. Gamma irradiation significantly reduced moisture content and increased protein content; however, the fat content of the irradiated flour was either equal to or lower than that of the native flour. The pasting properties of millet flour were reduced by gamma irradiation. Compared with native flour, lipoxygenase activity, and malondialdehyde content decreased, and radical scavenging, catalase, and superoxide dismutase activity increased in irradiated flour with increasing gamma radiation doses.

According to Timakova, Iliukhina, and Iliukhin (2023), the application of inactivating (1.00–5.00 kGy) and sterilizing (25.00 kGy) doses of gamma radiation should guarantee the biosafety of grains treated with radiation. The authors report that no changes were observed in the primary indicators of freshness (color, odor) following treatment with different doses of ionizing radiation and during storage for up to 30 days. The actors also reported reductions in moisture content in the irradiated grains, both initially and after 30 days of storage. The study identified a significant decrease in grain microflora measured by the quantitative parameters of QMAFAnM (Quantity of mesophilic aerobic and facultative anaerobic microorganisms) and fungal microbiota, which makes it possible to establish the positive effect of using gamma radiation to improve the storage capacity of food grains in food systems.

13.2.1.2.3 Fungi and Mycotoxins

Once mycotoxins contaminate cereal grains in the field or during storage, they are likely to be found on the consumer's plate (Luo, Liu, and Li 2018). Mycotoxins commonly detected and controlled worldwide in cereal grains are aflatoxins (AFB1 and AFB2), ochratoxins (OTA), fumonisins (FBs), zearalenone (ZEA), and deoxynivalenol (DON). Ingestion can lead to acute or chronic effects in consumers (Ostry et al. 2017). Exposure to high levels of aflatoxin has been associated with edema, hemorrhage, liver damage, cancer, nutritional disorders, and, in severe cases, fatal outcomes (Guo et al. 2021). Grains are more susceptible to aflatoxins and can contribute to serious health effects; therefore, (Khalil, Hammad, and Sebaei 2021) studied the inhibition and reduction of aflatoxins and ochratoxin A in corn by gamma irradiation. The authors report that 40.1% degradation of aflatoxin B1 33.3% degradation of AFB2, and 61.1% degradation of OTA were achieved in grains after the application of gamma radiation with a dose of 20.00 kGy. This treatment also increased storage time.

Aspergillus flavus is indeed one of the most important fungal species known for its ability to produce toxins during improper grain storage. Therefore, (Patil et al. 2019), when analyzing gamma irradiation to reduce AFB1 and microbiological contamination in peanuts artificially enriched at a concentration of 300 μg/kg of AFB1, they found that gamma irradiation (5.00, 7.00, and 9.00 kGy) reduced 20–43% of AFB1.

Ahsan et al. (2013) investigated an effective extraction, cleaning, and analysis methodology for aflatoxins, amino acids, and fatty acid methyl esters in rice samples. After treatment with gamma irradiation (0.00, 2.00, 4.00, and 6.00 kGy), a reduction of more than 95% in AFB1 was observed in the samples. The efficacy of gamma radiation on rice samples contaminated with high concentrations of AFB1 and total aflatoxins produced by *A. flavus* and *Aspergillus parasiticus* was significantly dependent on the radiation dose. The initial fungal population in the rice samples was determined to be 2.4×10^6 CFU/g. The study found that the maximum percentage reduction in the fungal population was achieved at a radiation dose of 6 kGy, resulting in a count of 6.4 CFU/g, as compared to the control sample. Amino acid profiles generally remained unaltered, except for an increase in leucine content observed at 4.00 kGy. Rice fat, after irradiation, showed small changes in the composition of fatty acid methyl esters. Total biomass (CFU/g) showed linear behavior with increasing gamma irradiation dose.

The effects of gamma irradiation on AFB1 levels in soybeans and the properties of soybeans and soybean oil were investigated by Zhang, Xie, and Che (2018). Gamma irradiation eliminated fungi and significantly reduced AFB1 content in soybeans at 10.00 kGy. For soybeans incubated for 30 days, there were no significant changes in quality attributes, tocopherol content, and oil oxidation stability when the gamma irradiation dose was lower than 20.00 kGy.

The effect of gamma irradiation on OTA in wheat flour was studied by (Calado et al. 2018). It was observed that despite increasing the moisture content of wheat flour, the maximum elimination achieved was 24% with a radiation dose of 30.00 kGy. Hathout et al. (2020) evaluated the impact of gamma irradiation on AFB1 in enriched wheat samples (20 μg/kg). They found a reduction in AFB1 concentration to 1.22 and 0.94 μg/kg for samples irradiated with gamma at doses of 10.00 and 20.00 kGy, respectively. Greater reductions (2.35, 1.93 μg/kg) were observed for AFB1-enriched samples (10 μg/kg) exposed to 10.00 and 20.00 kGy of gamma irradiation, respectively. Similar results were found by Mohamed et al. (2015), who found a 69.29% reduction in AFB1 content in wheat samples following gamma irradiation at a dose of 8.00 kGy.

Sebaei et al. (2022) analyzed the occurrence of ZEA in grains and its reduction by gamma radiation, aiming to minimize the risks of exposure and economic and health consequences. Wheat grains, yellow corn, and white corn were analyzed. To investigate influence of gamma rays, 12 uncontaminated samples were enriched with ZEA at the maximum allowable limit (100 μg/kg) and subsequently exposed to varying gamma-ray energies (5.00, 10.00, 15.00, and 20.00 kGy). The ZEA reduction varied between 49.0 and 97.0% in wheat samples, between 24.8 and 51.1% in yellow corn samples, and between 20.3 and 59.9% in white corn samples. Finally, at a dose of 20.00 kGy, the reductions reached 97.0, 51.1 and 59.9% in wheat, yellow corn, and white corn, respectively.

Tatar, Fadavi, and Koohsari (2021) analyzed various quantities of calcium oxide (CaO) (0, 0.5, and 1%) mixed with wheat samples containing mold spores. After a 20-day period, these samples were subjected to gamma radiation at different levels (0.00, 5.00, 10.00, 15.00, and 20.00 kGy). The presence of *A. flavus*, AFB1, and AFB2 was noted in the samples. The findings revealed that calcium oxide, gamma irradiation, and their combined effects significantly influenced *A. flavus*, AFB1, and AFB2 contamination. Notably, an additional reduction in AFB1 and AFB2 occurred when irradiation was coupled with CaO, with the most substantial inhibition of aflatoxin production achieved at a CaO concentration of 0.5%.

Abdel-Rahman et al. (2019) aimed to identify the natural levels of mycotoxins in roasted Egyptian peanuts and the destructive effect of gamma radiation on AFB1, tested doses of 5.00, 10.00, and 20.00 kGy. The levels of AFB1 reduction varied according to the gamma radiation dose levels. The dose of 5.00 kGy was ineffective, while a significant reduction in AFB1 was observed using 10.00 and 20.00 kGy, recording a reduction of >26.7% notably in samples enriched with 20 μg/kg AFB1 and treated with 20.00 kGy (59% reduction).

Alivandi Farkhad and Hosseini (2020) investigated the effects of irradiation on the phytochemical content of soybeans. Soybean seeds underwent irradiation at varying doses of 0.00, 0.50, 1.00, and 2.00 kGy, with a focus on assessing alterations in isoflavone aglycone concentrations. Additionally, the study evaluated the antioxidant potential through DPPH free radical scavenging activity. Genistein and daidzein concentrations, as well as antioxidant activity and total phenolic and flavonoid content in soy extracts, were found to be more pronounced at a radiation dose of 0.5 kGy. Lower levels of irradiation enhanced the concentration and antioxidant activity of soy isoflavones, along with the overall flavonoid and phenolic content.

Ben Amara et al. (2022) evaluated the effects of different doses of gamma irradiation on the survival of natural mycoflora and the reduction of contamination by OTA and AFB1 mycotoxins in naturally contaminated sorghum samples. The results demonstrated that a 3.00 kGy dose effectively eliminated 90% of the natural fungal population in sorghum. Increasing the dose of gamma irradiation showed significant effects on the initial mycotoxin content. Specifically, at a radiation dose of 10.00 kGy, there was a maximum reduction of 59% for AFB1 and 32% for OTA in naturally contaminated sorghum.

13.2.2 Electron Beam (E-Beam)

The electron beam is an advanced ionizing irradiation technology used in food decontamination. It is generated from non-nuclear sources and offers significant control over machine-generated radiation, making it a more convenient option compared to gamma radiation (Lung et al. 2015). This technology finds applications in various sectors, including healthcare (radiotherapy and brachytherapy), environmental applications (sludge disinfection and pollution monitoring and control), and the chemical industries (crosslinking and polymerization of polyethylene and polypropylene. However, the low penetration capacity compared to other ionizing radiation is the main problem associated with this type of radiation.

Electron beam irradiation (EBI) moves through space in the form of electromagnetic waves, such as gamma radiation. However, there are differences in the penetration power and the form of application (Mustapha et al. 2014). EBI stands out in some characteristics, such as high effectiveness, little heat required, and short processing time.

13.2.2.1 Physical Principles

In EBI, electrons are emitted by an electrical machine, enabling precise control over the characteristics of the electron flow, including energy, intensity, and geometry. While various principles of acceleration may be employed, the fundamental components of these machines remain consistent: an electron source (cathode), an accelerator tube, and a beam-shaping system. The cathode, often composed of materials like tungsten, is stimulated electrically within a vacuum system, causing it to release electrons. These emitted electrons are captured and exposed to a potential difference in an electron gun. As a result, the electrons are accelerated to an energy level proportional to the applied voltage and are then shaped into a controlled beam directed toward the target product for treatment (Bouzarjomehri et al. 2020).

The beam current (I) is expressed in milliamps (mA), which determines the dose rate and potential yield. The delivered dose (D) is directly proportional to I and inversely proportional to the product speed (v), making the relationship $D = k\ (I/v)$. The coefficient k, however, varies with different factors, including the distance between the product and the beam emergence point, the scanning width, and the geometry of the conveyor system. The accelerator power is the product of I and the electron energy; for example, the power of a 10 MeV accelerator at 20 mA is 200 kW.

In a general context, the items intended for treatment are passed beneath the electron beam through a transport system designed in accordance with radiological protection principles. When the electrical excitation of the cathode is discontinued, it is evident that the electron flow also ceases. This on/off operation may allow the installation of this equipment in production plants, as the radiation emission is not permanent. The utilization of E-beam technology is also contingent on the energy level of these radiations. Low-energy (0.1–1 MeV) E-beam technology finds application in surface sterilization, aseptic packaging, and modifications in food packaging. Medium energy E-beams (1–5 MeV) are suitable for various applications, including surface sterilization, food pasteurization in custom packaging, and modifications of packaging material. While, applications of high-energy E-beam technology (5–10 MeV) include food sterilization and phytosanitary treatment, pasteurization, and food industry waste treatment (Gryczka et al. 2020).

In the direct decontamination mechanism, the generated electrons make direct contact with the mycotoxin, leading to the destruction of its structure. Another method is the indirect method caused by the reactive species of H•, OH•, and other reactive species produced by the radiolysis of the water molecule in food (Mousavi Khaneghah et al. 2020).

13.2.2.2 Search Results

EBI plays a significant role in the destruction of microorganisms, as the high-energy electron breaks the molecular bonds of the microorganism, resulting in the degradation of the DNA structure and denaturation of proteins and enzymes. This disruption of the cell's overall physiological metabolism and interference with chromosomal replication result in injury or death of the microorganism, thus ensuring food safety and quality through effective decontamination of microorganisms and mycotoxins (Nunes et al. 2021). The effectiveness of this radiation depends on the size, thickness, density, and uniformity of the food packaging (Mousavi Khaneghah et al. 2020). In Table 13.4, it is possible to view some of the recent studies using X-ray irradiation on grains and by-products.

13.2.2.2.1 Insects

Sang et al. (2016) evaluated the effects of EBI exposure in a range of doses (0.03–0.12 kGy) on the development of cowpea bruchid (*C. maculatus*) at various stages of its life cycle. Differential radiosensitivity was detected during egg development. Early- and intermediate-stage eggs did not hatch following exposure to a 0.03 kGy dose, while a significant portion of late-stage eggs survived irradiation even at 0.12 kGy. However, the further development of the hatched larvae was hindered. Although midgut protein digestion remained unaffected, irradiated larvae (0.06 kGy or

TABLE 13.4
Effect of Electron Beams on Insects, Functional Properties, and Fungi from Different Food Sources

	Study Object	Food Source	Dose Used (kGy)	Main Results	Reference
Insects	*C. maculatus* (F.) *and Bactrocera zonata* (Saunders)	Cowpea	0.41–0.30	100% mortality	Gabarty et al. (2022)
	C. maculatus	Cowpea	0.06	Prevented population expansion	Sang et al. (2016)
	Cryptolestes pusillus, *Cryptolestes ferrugineus*, *and Cryptolestes turcicus*	Wheat	2.00	100% mortality	
Functional properties	Phytochemical composition, total bacterial count, lipase, and fatty acid activity	Quinoa	8.00	Saponin content (↔) Phytic acid levels (↓) Total phenolic content (↑) Bacteria (↓)	Luo et al. (2021)
	Nutritional quality and anti-nutritional factors	Sorghum	30.00	Chemical composition (↔) Digestibility of friend and crude protein (↑) Phytic acid and tannin levels (↓) Total phenols (↑) Mineral content (↔)	Rousta et al. (2014)
	Physicochemical properties	Maize starch	30.00	Molecular weight (↓) Branched-chain length, melting temperature, enthalpy of melting, and relative crystallinity (↓)	Zhou et al. (2020)
	Storage and physicochemical properties	Rice	10.00	Lipase activity (↓) Fatty acids (↔) Granular structure of isolated starches and molecular structures of amylopectin and amylose (↓) Relative crystallinity (↓)	Pan et al. (2020)
	Freshness and quality	Brown and milled rice	5.00	Viscosities and amylose content (↓) Crystalline forms (↔) Crystallinity (↓) Microstructure (↔) Free fatty acids (↑) Lipase activity (↓)	Luo et al. (2019)
	Gelatinization and physicochemical properties	Maize flour	5.40	Total starch and crude fiber contents (↓) Moisture and reducing sugar contents (↑) Viscosity peak (↓)	Xue et al. (2017)
	Water distribution	Rice	4.00	Few effects on rice properties	Pan et al. (2020)
	Parameters used in the food industry	Wheat	10.00	Hydration properties from 50 to 70°C (↑) Nutrient content (↔) TPC, TFC, and antioxidant activity (↑)	
	Expansion, color, paste, and sensory properties	Rice	<4.00	Expansion power and viscosity of the paste (↓) Sensory properties (↑)	Du et al. (2019)

(Continued)

TABLE 13.4 *(Continued)*

Effect of Electron Beams on Insects, Functional Properties, and Fungi from Different Food Sources

	Study Object	Food Source	Dose Used (kGy)	Main Results	Reference
Fungi and Mycotoxins	*Aspergillus* spp. *Penicillium* spp., and *Fusarium* spp.	Maize	6.20	88% destruction	Nemţanu et al. (2014)
	ZEA and OTA	Maize	10.00–50.00	Reduces ZEA and OTA toxicity by 71.1% and 67.9	Luo et al. (2017)
	AFB1	Maize	20.00	Significant reduction in aflatoxins	
	Aspergillus niger, *A. flavus*, and *Penicillium chrysogenum*	Bean (*Canavalia maritima*)	10.00	Improved shelf life of up to six months without deterioration under normal laboratory conditions	Supriya, Sridhar, and Ganesh (2014)
	AFB1	Pistachio	7.00	77.17% destruction	Hashemi et al. (2020)

higher) failed to progress into normal, healthy adults; instead, they either died as pupae or hatched abnormally, suggesting a detrimental impact of EBI on the metamorphosis process. The irradiated pupae that did emerge had shorter lifespans and were incapable of producing any eggs at doses of 0.06 kGy or higher. Within this dose range, eggs laid by irradiated adults were nonviable. EBI treatment led to a dose-dependent reduction in adult longevity. Reciprocal crosses indicated that females were more sensitive to EBI exposure than males. Dissection of the female reproductive system revealed that EBI treatment prevented oocyte formation.

Gabarty et al. (2022) evaluated the effect of EBI on the different stages of development of *C. maculatus* (F.) and *Bactrocera zonata* (Saunders). No adults emerged from three-day-old *C. maculatus* eggs irradiated with 0.30 kGy of EBI, and no adults emerged from the F1 generation of larvae and pupae irradiated with 0.10 kGy. The adult stage of *C. maculatus* exhibited the highest resistance to EBI, with the administration of high-dose irradiation (0.41 kGy) to the adult stage, resulting in the complete suppression of F1 generation adults' emergence. Extensive trials corroborated that a dose of 0.41 kGy proved to be efficacious for phytosanitary and safety treatment in the case of *C. maculatus*. Meanwhile, an EBI dosage of 0.41 kGy effectively hindered the hatching of *B. zonata* eggs. Additionally, EBI doses of 0.30, 0.41, and 0.65 kGy effectively arrested the development of the first, second, and third larval instars, respectively, until the pupation stage. Using the adult emergence criteria from the F1 generation, targeting the third larval stage, considered the most tolerant stage to EBI, the dose of 0.30 kGy was effective for the quarantine and safety treatment of *B. zonata*. The lethal effects of electronic beam irradiation from 0 to 1.00 kGy on *Cryptolestes pusillus*, *Cryptolestes ferrugineus*, and *Cryptolestes turcicus* were studied, the susceptibilities to irradiation of the three were compared, and the effects of electronic beam irradiation from 0 to 5.00 kGy on the qualities of wheat processing were investigated. The results showed that the higher the irradiation dose, the faster the death of three types of adult Cucujidae insects. Wet gluten content and flour yield were not significantly changed after EBI, but development time and dough stability decreased significantly after doses greater than 2.00 kGy. Thus, an irradiation dose of 0.10–2.00 kGy can be considered an adequate dose for controlling the development of three species of Cucujidae insects in stored grains.

Chizh et al. (2018) analyzed the effect of a low-energy (up to 160 keV) repetitive pulsed electron beam extracted in ambient air on disease incidence and sowing quality of spring barley seeds and on the viability of insect pests during barley storage. Culture irradiation of barley seeds five days before sowing greatly reduces the incidence of the disease in its sprouts: for *Drechslera teres*, the incidence decreases by 2.1–3.2 times depending on the dose and dosage rate, and for *Penicillium* sp., it decreases by 3.5–13.7 times at all doses and rates used. The incidence of *Fusarium* sp. is not affected by radiation. Disease spread decreases 1.5–3.2 times for *D. teres* and 4–8 times for *Penicillium* sp. Electron beam irradiation 28 days before seeding shows similar suppression (2.2–3.1-fold) only for *D. teres* at a dose of 1.00 kGy. The seed sowing quality improves when irradiation is carried out 5 days before planting, but it diminishes when irradiation is performed 28 days in advance. The viability of *Oryzaephilus surinamensis* in infested grains is completely inhibited after irradiation of 0.20–1.00 kGy.

13.2.2.2.2 Functional Properties

The alterations in the phytochemical composition, total bacterial count, lipase activity, and fatty acid content of quinoa were analyzed by (Luo et al. 2021), after exposure to EBI at different doses (0, 1.00, 2.00, 4.00, and 8.00 kGy). The results demonstrated that it effectively suppresses the growth of total bacterial count, lipase activity, and the accumulation of free fatty acids during storage. The saponin content in quinoa remained unchanged following irradiation, while the phytic acid content decreased by 19% at a dose of 8.00 kGy. Notably, the TPC of FPs exhibited a significant increase of 33% at 8.00 kGy. The total flavonoid content (TFC) of hydrolysable phenolic acids (HPAs) significantly decreased by 24% at 4.00 kGy. TFC in AHP and TFC in base hydrolysable phenolics

(BHPs) showed no significant changes, while TPC in BHP and TFC in FP improved. The content of polyunsaturated fatty acids increased by 3% at 1.00 kGy. The value of ω3/ω6 (0.21) increased by 5% after radiation.

Rousta et al. (2014) studied the effects induced by EBI on the nutritional quality and anti-nutritional factors of sorghum grains. EBI was applied at doses of 10, 20, and 30 kGy, and the chemical composition, mineral content, bioavailability, in vitro digestibility of protein and starch, total phenolic compounds, phytate, and tannin contents of sorghum grain were investigated. The results showed that the chemical composition remained unchanged by irradiation, starch digestibility increased by 5.4%, and the digestibility of crude protein from sorghum grain increased by 5%. Additionally, irradiation resulted in a significant reduction in phytic acid and tannin levels, while total phenolic compounds increased in comparison to unprocessed control samples. The results indicated no substantial changes in mineral content and bioavailability among the sample, except significant increases in extractable calcium, phosphorus, zinc, and iron.

Waxy corn starch was irradiated under different radiation doses (2.00–30.00 kGy), and the physicochemical properties of the starch were analyzed by Zhou et al. (2020). Films were subsequently produced from native and irradiated waxy corn starches, and their properties were assessed. The molecular weight of starch decreased markedly with increasing irradiation dose. The branched-chain length, melting temperature, melting enthalpy, and relative crystallinity decreased slightly, particularly at irradiation doses below 15.00 kGy. This observation suggested that a lower dose of irradiation primarily cleaved more α-1,6-glycosidic bonds as opposed to α-1,4-glycosidic bonds, resulting in the release of additional linear starch chains. Films prepared from waxy corn starch irradiated with 10.00 kGy exhibited improved mechanical properties and increased solubility due to a moderate increase in linear starch chains and a decrease in starch molecular weight, respectively. Fast-dissolving films resulting from irradiated waxy corn starch have the potential for use in instant food packaging.

Rice grains were treated with EBI and the storage properties of irradiated rice, as well as the physicochemical properties of isolated starches, were studied by Pan et al. (2020). With the elevation of the irradiation dose from 0 to 10.00 kGy, the lipase activity in irradiated rice displayed a decline, decreasing from 7.82 mg KOH/100 g to 5.15 mg KOH/100 g. EBI treatment did not induce significant alterations in the fatty acid values. The granular structure of the isolated starches was partially destroyed after severe irradiation, and EBI treatment also caused the degradation of the molecular structures of amylopectin and amylose. Both irradiated and non-irradiated starches displayed type A crystalline structures. However, a dose of 10.00 kGy disrupted the double helix structures and consequently reduced the relative crystallinity of the starches. The formation of carboxylic groups had an adverse impact on starch digestibility, whereas the disruption of crystalline structures facilitated the access of digestive enzymes to the degraded starch chains. In summary, the findings indicated that a low dose of irradiation had minimal effects on the quality of rice grains and the associated starches.

The effects of EBI treatment on the freshness and quality of brown and milled rice at different irradiation doses were investigated by Luo et al. (2019). The colors of both brown rice and milled rice exhibited slight changes after irradiation at 5.00 kGy. Post-irradiation, the viscosities and amylose contents of the samples decreased, while the crystalline forms remained unaltered, although there was a reduction in crystallinity. The microstructures of the samples remained unaltered, as indicated by the results of scanning electron microscopy. Throughout the 120-day storage period, the irradiated samples displayed a slower increase in free fatty acids compared to the non-irradiated samples, signifying effective inhibition of lipase activity.

Xue et al. (2017) investigated the effects of EBI on the gelatinization and physicochemical properties of corn flour irradiated by 0–5.40 kGy electron beam. After EBI treatment, a substantial decrease in both total starch and crude fiber contents in corn flour was observed, while moisture and reducing sugar contents increased significantly. EBI-induced perforations on the surfaces of corn flour particles, with the irradiated regions gradually shedding to reveal smoother surfaces, adopting spherical structures and reduced particle sizes. The molecular chains in the corn flour broke due to EBI. After irradiation, the peak viscosity of the paste decreased drastically from 1251.74 to 7.16 Pa.s, showing that the gelatinization of corn flour was completely inhibited. Thus, EBI can be used to inhibit the gelatinization of corn flour, which can be beneficial for industrial and food formulations.

Pan et al. (2020) used low-dose EBI (0–4.00 kGy) to study water distribution in freshly harvested rice (15.03%) and dry rice (11.97%) using low-field nuclear magnetic resonance (LF-NMR). The gelatinization, texture, and rheological properties of rice and the thermal and digestion properties of rice starch were determined. The findings indicated that low-dose EBI had an impact on water distribution in rice, primarily affecting free water in low-humidity conditions and both free water and bound water in high-humidity conditions. Rice with high moisture content exhibited gradual changes in gelatinization, resulting in smoother rheological and texture properties. The swelling power and solubility index suggested that irradiation promoted the depolymerization of starch chains. Overall, low-dose EBI had minimal effects on rice properties. High-moisture rice demonstrated superior quality and flavor, while rice with low moisture content presented superior nutritional quality.

Du et al. (2019) analyzed the effects of EBI on the swelling power, color, paste, and sensory properties of white rice after short-term storage. Samples were electron beam irradiated at 0, 2.00, 4.00, 6.00, or 8.00 kGy and stored at 25 or 37°C for up to 75 days. The results showed that the expansion power and main viscosities of the paste (including peak, break, and recoil viscosities) at both storage temperatures decreased with increasing irradiation dose. Negative correlations were also observed between the main paste properties' viscosities and the irradiation dose at both storage temperatures. In the sensory evaluation, rice subjected to high-dose EBI (>4 kGy) received exceptionally low scores in terms of hardness, appearance, flavor, and overall acceptability. However, rice stored at 37°C performed worse than rice at 25°C about the aforementioned parameters. On the other hand, the sensory properties at irradiation doses between 2.00 and 4.00 kGy were better than those of the control group at both storage temperatures.

Huang et al. (2022) conducted an analysis of the effects of EBI processing on the characteristics of whole tartar buckwheat flour, focusing on parameters commonly used in the food industry. The study revealed that EBI processing resulted in the appearance of cross-section, small granules, and fragment reaggregation. Additionally, it led to higher values for L*, a*, and whiteness index. Hydration properties were predominantly positively affected by EBI processing in the temperature range of 50–70°C, with a contrasting reduction at higher temperatures. Importantly, EBI processing did not have a significant impact on the nutrient content, and TPC, TFC, and antioxidant activity exhibiting notable increases following EBI processing at doses of up to 10 kGy, with the generation of only a few free radicals.

13.2.2.2.3 Fungi and Mycotoxins

Nemţanu et al. (2014) studied the effect of accelerated electrons (0.1–6.2 kGy) on fungi (*Aspergillus* spp., *Penicillium* spp., and *Fusarium* spp.) naturally attached to corn seeds and found that fungal contamination of corn seeds decreased significantly with increasing irradiation dose. The sensitivity of the most prevalent fungi to electron beam treatment exhibited the following order: *Penicillium* spp. > *Fusarium* spp. > *Aspergillus* sp., with complete inactivation occurring at irradiation doses of 1.7, 2.5, and 4.8 kGy, respectively. The efficacy of electron beam treatment in combating fungi in naturally contaminated corn seeds is contingent on the irradiation dose, thus offering a means to manage and reduce the fungal load in corn.

The influence of EBI on the physical and chemical properties of corn contaminated with ZEA and OTA was studied by Luo et al. (2020). The amylose content and starch crystallinity in corn decreased significantly at an irradiation dose greater than 20.00 kGy. Scanning electron microscopy results revealed that corn starch particles began to be crushed at 10.00 kGy. The contents of essential and total amino acids in corn decreased significantly with increasing EBI dose. Therefore, EBI treatment can influence the quality of corn to a certain extent and effectively reduce the toxicity of ZEA and OTA in contaminated corn.

Luo et al. (2017) investigated the effects of EBI on ZEA and OTA in corn. The results indicated that the electron beam significantly affected ZEA and OTA. In a solution, the degradation rates of ZEA and OTA at a dose of 10.00 kGy were 65.6 and 75.2%, respectively. The initial quantities of these mycotoxins significantly influenced the degradation rate. When irradiated in corn, the levels of ZEA and OTA decreased with the irradiation dose, and at 50.00 kGy, their degradation rates were 71.1 and 67.9%, respectively. Significantly, ZEA and OTA were more readily degraded in corn grain compared to corn flour.

A. B. Mohamed et al. (2022) evaluated the effectiveness of EBI doses in reducing viable fungal populations and destroying aflatoxins and fumonisins in naturally highly contaminated maize samples from eastern Kenya. In this study, 24 samples with >100 ng/g total aflatoxins and > 1000 ng/g of total fumonisins were selected and processed by pulping with sterile deionized water. These paste samples were exposed to target doses of 5.00, 10.00, and 20.00 kGy, and control. In the control samples, a significant positive correlation was observed between total and fumonisin levels ($r = 0.68$; $p < 0.001$). Exposure to EBI doses of 5.00 kGy effectively reduced fungal loads to below detection limits by plating (<1.9 log (CFU/g)). Furthermore, a significant ($p = 0.03$) mean reduction of 0.3 log (ng/g) in aflatoxin content was achieved at 20.00 kGy (range −0.9–1.4 log (ng/g)). However, there was no significant reduction in fumonisin levels, even at the 20.00 kGy dose. Importantly, EBI doses below 20.00 kGy did not result in a reduction of mycotoxins.

Hashemi et al. (2020) evaluated AFB1 contamination of pistachio kernels after irradiation with EBI. Pistachios were inoculated with known concentrations of *A. flavus*. Then, pistachio samples were exposed to EBI at five different doses (0, 1.00, 3.00, 5.00, and 7.00 kGy) and at various storage times (0, 30, 60, 90, and 180 days) at room temperature. Subsequently, the sensory attributes, fungal contamination, and AFB1 levels in pistachio kernels were examined. On different storage days and with two replications, the average percentage reduction of aflatoxin in samples treated at doses of 1.00, 3.00, 5.00, and 7.00 kGy was 38.84, 48.79, 53.50, and 77.17%, respectively. The 1.00 kGy dose was found to be effective in reducing mold levels in pistachios without altering their sensory characteristics. However, organoleptic properties such as color, texture, and overall palatability of pistachios underwent significant changes after exposure to a 5.00 kGy dose.

Supriya, Sridhar, and Ganesh (2014) effectively controlled the main fungi on dry beans *Canavalia maritima*, by EBI. Although nine fungi were isolated from non-irradiated dried beans, the incidence of *Aspergillus niger* was highest (33–50%), followed by *A. flavus* (14–20%), and *Penicillium chrysogenum* (7–13%). Both non-irradiated beans and beans subjected to electron beam irradiation at doses of 2.50, 5.00, 10.00, and 15.00 kGy were examined for the presence of fungal species and their prevalence during a storage period of zero, three, and six months under laboratory conditions. The authors suggest that an EBI dose of 10.00 kGy is advisable for fungal decontamination and enhancing the shelf life of dry beans.

13.2.3 X-Ray

X-rays are produced through the bombardment of dense target materials by high-energy accelerated electrons, resulting in a continuous energy spectrum (Ashraf et al. 2019). These X-rays exhibit the ability to traverse thick materials, approximately 30–40 cm, which makes them a favorable option for the treatment of pre-packaged foods, thus reducing the risk of recontamination

(Ricciardi et al. 2019). For food irradiation purposes, the energy of the impacting electrons must be 5 MeV or less, which generates X-rays with penetrating power similar to that of Co-60 (Prakash and Ornelas-Paz 2019).

X-ray irradiation at a dose of 10 kGy has been demonstrated as a safe method to effectively reduce the load of pathogenic microorganisms (Cho and Ha 2019; Indiarto and Qonit 2020; Bisht et al. 2021). These rays offer the advantage of treating large quantities of materials without the need for radioactive substances. They can penetrate thick materials of approximately 30–40 cm, making them suitable for use in packaged products (Ashraf et al. 2019). According to Bisht et al. (2021), these types of radiation are chosen because (a) they produce the desired effects on food preservation, (b) in food and product packaging do not cause radioactivity, and (c) are available in quantities and costs that allow the irradiation method to be used commercially.

13.2.3.1 Physical Principles

Exposing an atom to X-rays leads to the conversion of energy from photons and electrons directed into the atom's orbit. This high-energy interaction results in the excitation of an electron, causing it to transition from a low-energy orbit to a high-energy orbit. When the atoms within a material are subjected to X-rays, the high-energy photons originating from the source ionize the atom by displacing an electron from the outer orbital. The level of ionization is dependent on both the energy of the photons and the material's physical properties (Pillai and Shayanfar 2017). This interaction with water molecules triggers the radiolysis of water, producing free radicals. In recent years, X-rays and EBs generated from accelerated ionizing radiation sources have gained increased attention compared to gamma radiation due to their controlled radiation generation, enabled by on-and-off switching mechanisms (Zehi et al. 2020).

13.2.3.2 Search Results

In Table 13.5, it is possible to view some of the recent studies using X-ray irradiation on grains and by-products.

13.2.3.2.1 Insects

Amubieya et al. (2021) conducted a laboratory investigation to assess the efficacy of X-ray irradiation at three energy levels (60, 70, and 80 KeV) both independently and in conjunction with oven drying for controlling *Acanthoscelides obtectus* infestations in stored bean seeds. The experiments were conducted in controlled laboratory conditions with a temperature of 29.7°C and a relative humidity of 76.5%. The results indicated that female *A. obtectus* deposited a maximum of 2.20 eggs per day, a relatively low number compared to other bruchinae females. This reduced reproductive capacity in female *A. obtectus* was primarily attributed to the impact of X-ray irradiation. Furthermore, the study revealed that X-ray irradiation caused a maximum mortality rate of 37% among *A. obtectus* adults within 24 h. After seven days of exposure, mortality became highly significant, reaching a minimum of 86.67%. When combined with oven drying, there were significant reductions in the daily and cumulative emergence of new *A. obtectus* offspring, as well as a decrease in seed damage and dust production. Significantly, X-ray irradiation did not negatively impact the viability of the bean seeds following the pest control process. Jiang et al. (2022) evaluated the effects of X-ray irradiation on the growth, development, survival, reproduction, and flight of *Spodoptera frugiperda*. The results indicated that irradiation with 0.05–0.40 kGy did not have significant effect on pupal hatching. However, it was observed that the females were more sensitive than males in terms of reproductive parameters, especially when radiation doses were > 0.35 kGy. After irradiation with a sub-sterilizing dose of 0.25 kGy, the sterility rate of parents was >85%, and sterility traits could be passed on to offspring, resulting in a continuous decrease in the population of the F1 and F2 generations. The authors confirm the feasibility of sterile insect technology to control *S. frugiperda* in the field using X-ray radiation.

TABLE 13.5
Effect of X-rays on Insects, Functional Properties, and Fungi from Different Food Sources

	Study Object	Food Source	Dose Used	Main Results	Reference
Insects	*Acanthoscelides obtectus*	Bean	80 keV	86.67% mortality after 7 days	Amubieya et al. (2021)
	Spodoptera frugiperda	Maize	0.25 kGy	85% parental sterility	Jiang et al. (2022)
	Hypothenemus hampei	Coffee	0.10 kGy	Prevents subsequent reproduction	Follett (2018)
	Sitophilus zeamais Motschulsky	Maize	>80 keV	Did not significantly restrict the emergence of *S. zeamais* and did not significantly protect grains from damage and flour production	Bell-Gam et al. (2021)
	Sitophilus granarius (L.)	Maize	0.20 kGy	97.5% mortality after 14 days	Keszthelyi, Lavák, and Ám (2015)
Functional properties	Vegetative and productive parameters	Cowpea	10–15 mGy	Induced variability and increased vegetative growth and productivity	Victor et al. (2021)
Fungi and mycotoxins	*Aspergillus niger*	Rice	Dosage rate of 0.76 kGy/h	D 10 Value: 571.9 Gy	Shankar et al. (2020)
	Fusarium sp., *Bipolaris sorokiniana*, and *Alternaria* sp.	Wheat	1.50 kGy	Total disinfection of fungi and reduction of germination from 79.0 to 12.7%	Isemberlinova et al. (2021)

Follett (2018) conducted dose-response tests with adult coffee borer beetles (CBB) (*H. hampei*) (the most serious coffee insect worldwide), aiming to identify a dose of sterilizing ionizing radiation between 0.05, 0.075, 0.10 and 0.125 kGy, followed by large-scale confirmatory tests with adults infesting coffee berries at 0.10 kGy X-rays. In total, 6,598 adult CBBs naturally infesting dried coffee berries were irradiated with 0.10 kGy and did not produce viable offspring, while 1,033 non-irradiated controls produced 327 eggs, 411 larvae, and 58 pupae three weeks after treatment. Irradiation of CBB adults infesting coffee fruits at 0.10 kGy prevented further reproduction. Therefore, irradiation could be used as a post-harvest disinfestation treatment to prevent the movement of CBB during the shipment of green coffee.

The use of X-ray irradiation to control the corn weevil, *S. zeamais* Motschulsky, in stored corn, was tested by Bell-Gam et al. (2021) at doses 60, 70, and 80 KeV. The results showed that there was a 10% mortality rate among adult weevils after 24 h, and this increased to 40% mortality after seven days. The result indicates a moderate effect on mortality. Furthermore, there was a progressive increase in mortality was recorded with increasing doses from 60 to 80 KeV. X-ray irradiation at the doses tested did not significantly restrict the emergence of *S. zeamais* from corn grains and also did not significantly protect the grains from damage and flour production. There was no difference in the daily emergence pattern of new progenies and the accumulated emergence pattern. The absence of a significant difference between the viability of irradiated grains and non-irradiated grains strongly suggests that X-ray irradiation did not adversely affect seed viability, and this gives the irradiation technique an opportunity for greater acceptability in agriculture. The authors recommend >80 KeV X-ray irradiation or its incorporation into the integrated pest management system to achieve effective post-harvest control.

The effectiveness of X-ray radiation as a method of protecting corn grains against the granary weevil *Sitophilus granarius* (L.) was evaluated by Keszthelyi, Lavák, and Ám (2015) in laboratory tests. Irradiation was applied in five doses: 0.01, 0.05, 0.07, 0.10, and 0.20 kGy. After each irradiation treatment, the samples were divided into two parts. In one group (n = 50), the samples remained unaltered. In the other group, 25 treated adults were substituted with 25 untreated adults in an attempt to assess the impact of sterility. Mortality among *S. granarius* adults in the unchanged group was assessed after 48 h, as well as 7, 14, and 21 days of exposure. After counting for 21 days, all exposed adults were removed, and progeny production was evaluated 45 days later. The mortality consequences of X-ray radiation on barn weevil were highly influenced by the time after exposure and the dose applied. At doses of 0.05 and 0.07 kGy, the accepted efficacy in plant protection (insect mortality: 95%) was recorded only in 21 days. During the application of doses of 0.10 and 0.02 kGy, high mortality rates (83.5 and 97.5%) were observed in 14 days. Furthermore, the use of multiple doses at a minimum treatment of 0.05 kGy was effective against *S. granarius* in corn under operational conditions. The authors confirmed the complete sterilizing effect of doses of 0.07 kGy and above. The autocidal effect of doses of 0.05 and 0.07 kGy was demonstrated when irradiated specimens were mixed in groups of fertile specimens. Consequently, these doses can be successfully applied to put the sterile insect technique into practice.

13.2.3.2.2 Functional Properties

Andrew et al. (2021) evaluated the effects of varying X-ray irradiation doses on the vegetative and productive traits of four cowpeas (*Vigna unguiculata* (L.) Walp) genotypes (black eye, light brown eye, dark brown eye, and ife brown). The seeds were exposed to different X-ray doses (0, 5, 10, 15, and 20 mGy). The results demonstrated that X-ray irradiation doses of 10 and 15 mGy induced significant variations in all vegetative parameters (such as days to 50% emergence, plant height, number of leaves, leaf area, stem circumference, and number of branches). Additionally, field parameters, including days to 50% flowering, pod length, weight of 100 seeds, seed yield per pod (g), and grain yield per hectare (t/ha), were significantly enhanced by the various X-ray doses compared to the control plants. Among the four varieties and treatments, light brown eyes exposed to 15 mGy displayed superior performance compared to the other varieties studied. Therefore, the overall findings suggest that low X-ray irradiation doses of 10 and 15 mGy can introduce variability and enhance the vegetative growth and productivity of cowpea plants.

13.2.3.2.3 Fungi and Mycotoxins

X-ray irradiation can reduce microbial populations in various food products, including fruits, milk, dairy items, fish, meat, and cereals. In a recent study, rice samples artificially inoculated *A. niger* were exposed to X-ray treatment at a rate of 0.76 Gy/h and showed a D10 value of 571.9. When combining treatment with essential oil, X-ray increased the radiosensitization of *A. niger* by 1.63 times (Shankar et al. 2020).

Isemberlinova et al. (2021) analyzed pulsed X-ray treatment of wheat against pathogenic fungi. Grains of the Irene wheat variety naturally infected by *Fusarium* spp., *Bipolaris sorokiniana*, and *Alternaria* sp. were treated with pulsed (0.5 μs) X-ray irradiation (280 keV) at a high-dose rate (up to 40 kGy/s). The results indicated a dose-dependent effect of irradiation on fungal contamination in the grains. The authors show that a dose of 1.5 kGy led to complete disinfection of wheat from fungi, while wheat seed germination was reduced from 79.0 to 12.7%.

13.3 CONCLUSION

Ionizing radiation is an emerging technology known for its ability to alter atoms and create ions without inducing radioactivity. Its effects on pathogens and products are evident both directly, by damaging cellular components such as DNA and lipids and indirectly, through free radicals. Gamma radiation, in particular, is effective in inhibiting the cell division of insects, impairing their reproduction and feeding by damaging the midgut epithelium. Additionally, ionizing radiation has shown benefits in improving functional properties and reducing anti-nutritional factors in various foods. Electron beam irradiation (EBI) also stands out for its precise control over the beam's intensity and energy, proving effective in surface sterilization and phytosanitary treatment. EBI has been effective in inhibiting insect development and enhancing food quality. Both gamma radiation and EBI present themselves as effective methods for reducing contaminants and improving food safety, with gamma radiation showing efficacy in aflatoxin reduction and EBI demonstrating versatility and precise control in industrial applications.

REFERENCES

Abdel-Rahman, G. N., Y. Y. Sultan, S. H. Salem, and M. M. Amer. 2019. "Identify The Natural Levels of Mycotoxins in Egyptian Roasted Peanuts and the Destructive Effect of Gamma Radiation." *Journal of Microbiology, Biotechnology and Food Sciences* 8 (5): 1174–77. https://doi.org/10.15414/jmbfs.2019.8.5.1174-1177.

Abotaleb, A. O., R. A. Zinhoum, and A. Gabarty. 2021. "Evaluation of Irradiation Quarantine Dose – Induced DNA Damage in *Rhyzopertha dominica* and Its Effect on Wheat Grains Using Comet Assay." *International Journal of Tropical Insect Science* 41 (1): 693–701. https://doi.org/10.1007/s42690-020-00258-w.

Ahmed, M. M., I. G. Abdalla, A. M. Salih, and A. B. Hassan. 2018. "Effect of Gamma Radiation on Storability and Functional Properties of Sorghum Grains (*Sorghum bicolor* L.)." *Food Science & Nutrition* 6 (7): 1933–1939. https://doi.org/10.1002/fsn3.752.

Ahsan, S., Z. Hussain, S. A. Raza Naqvi, and M. R. Asi. 2013. "Effect of Gamma Radiation on Aflatoxin Load, Amino Acid and Fatty Acid Composition of *Oryza sativa* L." *Pakistan Journal of Botany* 45 (5): 1577–1580.

Al-Bachir, M. 2016. "Some Microbial, Chemical and Sensorial Properties of Gamma Irradiated Sesame (*Sesamum indicum* L.) Seeds." *Food Chemistry* 197 (April): 191–197. https://doi.org/10.1016/j.foodchem.2015.10.094.

Alivandi Farkhad, S., and A. Hosseini. 2020. "Effect of Gamma Irradiation on Antioxidant Potential, Isoflavone Aglycone and Phytochemical Content of Soybean (*Glycine max* L. Merrill) Cultivar Williams." *Journal of Radioanalytical and Nuclear Chemistry* 324 (2): 497–505. https://doi.org/10.1007/s10967-020-07100-0.

Amara, A. B., A. Mehrez, C. Ragoubi, R. Romero-González, A. G. Frenich, A. Landoulsi, and I. Maatouk. 2022. "Fungal Mycotoxins Reduction by Gamma Irradiation in Naturally Contaminated Sorghum." *Journal of Food Processing and Preservation* 46 (3). https://doi.org/10.1111/jfpp.16345.

Amubieya, S., L. C. Nwosu, U. Zakka, O. M. Azeez, V. C. Okereke, A. N. Eluwa, G. M. Ugagu, et al. 2021. "Effect of X-Ray Irradiation and Oven-Drying on the Bionomics of *Acanthoscelides obtectus* Say (Coleoptera : Chrysomelidae) Infesting Common Bean in Storage: Can X-Ray Irradiation Affect Seed Viability after Pest Control Process" *Journal of Entomology and Zoology Studies* 9 (3): 13239.

Andrew, O. V., K. C. Egbucha, G. Omosun, F. E. Akanwa, and S. I. Akpanudo. 2021. "Effect of X-Ray Irradiation on the Growth and Yield Parameters of Four Cowpea (*Vigna Unguiculata* (L.) Walp) Genotypes." *NIPES-Journal of Science and Technology Research* 3 (4): 1–8.

Ashraf, A., B. Pathrose, M. Chellappaappan, and B. V. Indirabai. 2021. "Efficacy of Gamma Radiation against Pulse Beetle *Callosobruchus maculatus* (F.)." *Indian Journal of Entomology*, September, 1–5. https://doi.org/10.55446/IJE.2021.1.

Ashraf, S., M. Sood, J. D. Bandral, M. Trilokia., and M. Manzoor. 2019. "Food Irradiation: A Review." *International Journal of Chemical Studies* 7 (2): 131–136.

Bashir, K., T. L. Swer, K. S. Prakash, and M. Aggarwal. 2017. "Physico-Chemical and Functional Properties of Gamma Irradiated Whole Wheat Flour and Starch." *LWT – Food Science and Technology* 76 (March): 131–139. https://doi.org/10.1016/j.lwt.2016.10.050.

Bell-Gam, S. P., L. C. Nwosu, K. D. Ileke, and U. O. Aguwa. 2021. "Effect of X-Ray Irradiation on the F1 Generation of *Sitophilus zeamais* Motschulsky and the Germination Rate of Maize Grain." *The Journal of Basic and Applied Zoology* 82 (1): 25. https://doi.org/10.1186/s41936-021-00225-y.

Bhat, N. A., I. A. Wani, and N. Sultan. 2023. "Effect of Gamma-Irradiation on the Physicochemical, Functional, and Antioxidant Properties of Unpigmented Brown Whole Rice Flour." *Food Science and Technology International* 29 (3): 228–242. https://doi.org/10.1177/10820132211069244.

Bisht, B., P. Bhatnagar, P. Gururani, V. Kumar, M. S. Tomar, R. Sinhmar, N. Rathi, and S. Kumar. 2021. "Food Irradiation: Effect of Ionizing and Non-Ionizing Radiations on Preservation of Fruits and Vegetables—A Review." *Trends in Food Science & Technology* 114 (August): 372–385. https://doi.org/10.1016/j.tifs.2021.06.002.

Bouzarjomehri, F., V. Dad, B. Hajimohammadi, S. P. Shirmardi, and A. Yousefi-Ghaleh Salimi. 2020. "The Effect of Electron-Beam Irradiation on Microbiological Properties and Sensory Characteristics of Sausages." *Radiation Physics and Chemistry* 168 (March): 108524. https://doi.org/10.1016/j.radphyschem.2019.108524.

Byun, K.-H., M.-J. Cho, S.-Y. Park, H. S. Chun, and S.-D. Ha. 2019. "Effects of Gamma Ray, Electron Beam, and X-Ray on the Reduction of *Aspergillus flavus* on Red Pepper Powder (*Capsicum annuum* L.) and Gochujang (Red Pepper Paste)." *Food Science and Technology International* 25 (8): 649–658. https://doi.org/10.1177/1082013219857019.

Calado, T., M. L. Fernández-Cruz, S. C. Verde, A. Venâncio, and L. Abrunhosa. 2018. "Gamma Irradiation Effects on Ochratoxin A: Degradation, Cytotoxicity and Application in Food." *Food Chemistry* 240 (February): 463–471. https://doi.org/10.1016/j.foodchem.2017.07.136.

Chizh, T. V., N. N. Loy, A. N. Pavlov, M. S. Vorobyov, and S. Y. Doroshkevich. 2018. "Low-Energy Electron Beams for Protection of Grain Crops from Insect Pests and Diseases." *Journal of Physics: Conference Series* 1115 (November): 022025. https://doi.org/10.1088/1742-6596/1115/2/022025.

Cho, G.-L., and J.-W. Ha. 2019. "Application of X-Ray for Inactivation of Foodborne Pathogens in Ready-to-Eat Sliced Ham and Mechanism of the Bactericidal Action." *Food Control* 96 (February): 343–350. https://doi.org/10.1016/j.foodcont.2018.09.034.

Dehkordi, S. A., M. Ahmadi, and B. Hatami. 2018. "The Investigation of the Combined Effects of Gamma Irradiation and Environmental Manipulation on Mortality and Sterility of *Rhyzopertha dominica* (Fabricius) (Coleoptera: Bostrichidae)." *Journal of Asia-Pacific Entomology* 21 (4): 1337–41. https://doi.org/10.1016/j.aspen.2018.10.013.

Delorme, M. M., J. T. Guimarães, N. M. Coutinho, C. F. Balthazar, R. S. Rocha, R. Silva, L. P. Margalho, et al. 2020. "Ultraviolet Radiation: An Interesting Technology to Preserve Quality and Safety of Milk and Dairy Foods." *Trends in Food Science & Technology* 102 (August): 146–154. https://doi.org/10.1016/j.tifs.2020.06.001.

Desouky, O., N. Ding, and G. Zhou. 2015. "Targeted and Non-Targeted Effects of Ionizing Radiation." *Journal of Radiation Research and Applied Sciences* 8 (2): 247–254. https://doi.org/10.1016/j.jrras.2015.03.003.

Du, Z., J. Xing, X. Luo, L. Wang, L. Pan, Y. Li, R. Wang, Y. Liu, X. Li, and Z. Chen. 2019. "Characterization of the Physical Properties of Electron-Beam-Irradiated White Rice and Starch during Short-Term Storage." Edited by Walid Elfalleh. *PLOS ONE* 14 (12): e0226633. https://doi.org/10.1371/journal.pone.0226633.

Erramli, H., and J. El Asri. 2019. "Gamma Rays: Applications in Environmental Gamma Dosimetry and Determination Samples Gamma-Activities Induced by Neutrons." In *Use of Gamma Radiation Techniques in Peaceful Applications*. IntechOpen. https://doi.org/10.5772/intechopen.85503.

Fan, X., R. Huang, and H. Chen. 2017. "Application of Ultraviolet C Technology for Surface Decontamination of Fresh Produce." *Trends in Food Science & Technology* 70 (December): 9–19. https://doi.org/10.1016/j.tifs.2017.10.004

Fan, X., and B. A. Niemira. 2020. "Gamma Ray, Electron Beam, and X-Ray Irradiation." In *Food Safety Engineering*, edited by A. Demirci, H. Feng, and K. Krishnamurthy, 1–13. *Food Engineering Series*. Cham: Springer.

Follett, P. A. 2018. "Irradiation for Quarantine Control of Coffee Berry Borer, *Hypothenemus hampei* (Coleoptera: Curculionidae: Scolytinae) in Coffee and a Proposed Generic Dose for Snout Beetles (Coleoptera: Curculionoidea)." *Journal of Economic Entomology* 111 (4): 1633–1637. https://doi.org/10.1093/jee/toy123.

Gabarty, A., A. Hammad, R. Zinhoum, and A. Negm. 2022. "Impact of Electron Beam Irradiation on the Developmental Stages of *Callosobruchus maculatus* (Coleoptera: Bruchidae) and *Bactrocera zonata* (Diptera: Tephritidae)." *Journal of Entomological Science* 57 (4): 460–476. https://doi.org/10.18474/JES21-72.

Gryczka, U., H. Kameya, K. Kimura, S. Todoriki, W. Migdał, and S. Bułka. 2020. "Efficacy of Low Energy Electron Beam on Microbial Decontamination of Spices." *Radiation Physics and Chemistry* 170 (May): 108662. https://doi.org/10.1016/j.radphyschem.2019.108662.

Guo, Y., L. Zhao, Q. Ma, and C. Ji. 2021. "Novel Strategies for Degradation of Aflatoxins in Food and Feed: A Review." Food Research International 140 (February): 109878. https://doi.org/10.1016/j.foodres.2020.109878.

Hammad, A., A. Gabarty, and R. A. Zinhoum. 2020. "Assessment Irradiation Effects on Different Development Stages of *Callosobruchus maculatus* and on Chemical, Physical and Microbiological Quality of Cowpea Seeds." *Bulletin of Entomological Research* 110 (4): 497–505. https://doi.org/10.1017/S0007485319000865.

Hashemi, S., M. H. Ehrampoush, M. Jalili, S. K. Limaki, and B. Hajimohammadi. 2020. "Discrimination of Sensorial Characteristics, Fungal, and Aflatoxin B_1 Contamination of Pistachio Kernels after E-Beam Irradiation," *International Journal of Environmental Health Engineering* 9 (1): 19. https://journals.lww.com/ijeh/fulltext/2020/09000/discrimination_of_sensorial_characteristics,.19.aspx

Hassan, A. B., N. S. Mahmoud, K. Elmamoun, O. Q. Adiamo, and I. A. M. Ahmed. 2018. "Effects of Gamma Irradiation on the Protein Characteristics and Functional Properties of Sesame (*Sesamum indicum* L.) Seeds." *Radiation Physics and Chemistry* 144 (March): 85–91. https://doi.org/10.1016/j.radphyschem.2017.11.020.

Hathout, A. S., S. M. Abel-Fattah, Y. H. Abou-Sree, and A. S. M. Fouzy. 2020. "Incidence and Exposure Assessment of Aflatoxins and Ochratoxin A in Egyptian Wheat." *Toxicology Reports* 7: 867–73. https://doi.org/10.1016/j.toxrep.2020.07.003.

Howard, R. H., M. A. Reichenberger, J. M. Urban-Klaehn, J. L. Peterson-Droogh, J. V. Brookman, B. Gross, C. Tyler, J. Navarro, M. Lillo, and A. Zillmer. 2021. "Overcoming Challenges to Support Us Resumption of High Specific Activity Cobalt-60." *Applied Radiation and Isotopes* 169 (March): 109494. https://doi.org/10.1016/j.apradiso.2020.109494.

Huang, Y., H. Wang, Q. Liu, J. Zhou, S. Gao, J. Yu, X. Zhang, and M. Wang. 2022. "Comparing Impacts of Dielectric Barrier Discharge Plasma and Electron Beam Irradiation Processing on Characteristics of Tartary Buckwheat Whole Flour." *Innovative Food Science & Emerging Technologies* 77 (May): 102986. https://doi.org/10.1016/j.ifset.2022.102986.

Indiarto, R., and M. A. H. Qonit. 2020. "A Review of Irradiation Technologies on Food and Agricultural Products." *International Journal of Scientific and Technology Research* 9 (1): 4411–4414.

Isemberlinova, A. A., I. S. Egorov, S. A. Nuzhnyh, A. V. Poloskov, E. A. Pokrovskaya, A. V. Vertinskiy, F. V. Konusov, E. S. Sukhikh, M. V. Chubik, and G. E. Remnev. 2021. "The Pulsed X-Ray Treatment of Wheat against Pathogenic Fungi." *Nuclear Instruments and Methods in Physics Research Section B: Beam Interactions with Materials and Atoms* 503 (September): 75–78. https://doi.org/10.1016/j.nimb.2021.07.011.

Ito, V. C., A. A. F. Zielinski, I. M. Demiate, M. Spoto, A. Nogueira, and L. G. Lacerda. 2019. "Gamma Radiation Effects on Physicochemical, Microbiological and Antioxidant Properties of Black Rice (*Oryza sativa* L.) Flour during Storage." *Carpathian Journal of Food Science & Technology* 11 (3): 163–74.

Jabeen, N. M., M. Kola, and K. Jalarama. 2015. "Impact of Irradiation on Nutritional Quality and Functional Properties of Soy Flour and Sprouted Soy Flour." *International Journal of Advanced Research* 3 (3): 1120–1129

Jiang, S., L. M. He, W. He, H. Y. Zhao, X. M. Yang, X. Q. Yang, and K. M. Wu. 2022. "Effects of X-Ray Irradiation on the Fitness of the Established Invasive Pest Fall Armyworm *Spodoptera frugiperda*." *Pest Management Science* 78 (7): 2806–2815. https://doi.org/10.1002/ps.6903.

Keszthelyi, S., C. Glavák, and F. Pál-FÁm. 2015. "Mortality and Sterility Consequences of X-Ray Radiation on the Granary Weevil, *Sitophilus granarius*." *Bulletin of Insectology* 68 (1): 7–11.

Khalil, O. A. A., A. A. Hammad, and A. S. Sebaei. 2021. "*Aspergillus flavus* and *Aspergillus ochraceus* Inhibition and Reduction of Aflatoxins and Ochratoxin A in Maize by Irradiation." *Toxicon* 198 (July): 111–120. https://doi.org/10.1016/j.toxicon.2021.04.029.

Khan, S., M. Sayed, M. Sohail, L. A. Shah, and M. A. Raja. 2019. "Advanced oxidation and reduction processes." In Advances in water purification techniques, ed. Ahuja, S. 135–164. https://doi.org/10.1016/b978-0-12-814790-0.00006-5

Kheirallah, D., and L. El-Samad. 2020. "Biochemical changes induced by gamma irradiation in the ground beetle." *Journal of Advances in Biology* 9 (3): 1937–1947.

Kiran, R., K. B. Shenoy, and M. G. Venkatesha. 2019. "Effect of Gamma Radiation as a Post-Harvest Disinfestation Treatment against Life Stages of the Coffee Berry Borer, *Hypothenemus hampei* (Ferrari) (Coleoptera: Curculionidae)." *International Journal of Radiation Biology* 95 (9): 1301–1308. https://doi.org/10.1080/09553002.2019.1619950.

Kumari, S., O. P. Gupta, C. B. Mishra, V. Thimmegowda, V. Krishnan, B. Singh, A. Sachdev, and A. Dahuja. 2020. "Gamma Irradiation, an Effective Strategy to Control the Oxidative Damage of Soy Proteins during Storage and Processing." *Radiation Physics and Chemistry* 177 (December): 109134. https://doi.org/10.1016/j.radphyschem.2020.109134.

Lung, H.-M., Y.-C. Cheng, Y.-H. Chang, H.-W. Huang, B. B. Yang, and C.-Y. Wang. 2015. "Microbial Decontamination of Food by Electron Beam Irradiation." *Trends in Food Science & Technology* 44 (1): 66–78. https://doi.org/10.1016/j.tifs.2015.03.005.

Luo, X., L. Qi, Y. Li, R. Wang, D. Yang, L. Wang, Y. Li, W. Fang, and Z. Chen. 2020. "Influences of Electron Beam Irradiation on the Physical and Chemical Properties of Zearalenone- and Ochratoxin A-Contaminated Corn and In Vivo Toxicity Assessment." *Toxins* 12 (3): 1–18. https://doi.org/10.3390/toxins12030198

Luo, X., Z. Du, K. Yang, J. Wang, J. Zhou, J. Liu, and Z. Chen. 2021. "Effect of Electron Beam Irradiation on Phytochemical Composition, Lipase Activity and Fatty Acid of Quinoa." *Journal of Cereal Science* 98 (March): 103161. https://doi.org/10.1016/j.jcs.2021.103161.

Luo, Y., X. Liu, and J. Li. 2018. "Updating Techniques on Controlling Mycotoxins – A Review." *Food Control* 89 (July): 123–132. https://doi.org/10.1016/j.foodcont.2018.01.016.

Luo, X., L. Qi, Y. Liu, R. Wang, D. Yang, K. Li, L. Wang, Y. Li, Y. Zhang, and Z. Chen. 2017. "Effects of Electron Beam Irradiation on Zearalenone and Ochratoxin A in Naturally Contaminated Corn and Corn Quality Parameters." *Toxins* 9 (3): 84. https://doi.org/10.3390/toxins9030084.

Luo, X., L. Yulin, D. Yang, J. Xing, K. Li, M. Yang, R. Wang, L. Wang, Y. Zhang, and Z. Chen. 2019. "Effects of Electron Beam Irradiation on Storability of Brown and Milled Rice." *Journal of Stored Products Research* 81 (March): 22–30. https://doi.org/10.1016/j.jspr.2018.12.003.

Manupriya, B. R., S. Lathika, H. M. Somashekarappa, S. L. Patil, and K. B. Shenoy. 2020. "Study of Gamma Irradiation Effects on the Physico-Chemical Properties of Wheat Flour (*Triticum aestivum*, L.)." *Radiation Physics and Chemistry* 172 (July): 108693. https://doi.org/10.1016/j.radphyschem.2020.108693.

Mir, S. A., S. J. D. Bosco, M. A. Shah, M. M. Mir, and K. V. Sunooj. 2015. "Effect of Gamma Irradiation on Physicochemical Properties of Brown Rice." *International Journal of Food Engineering* 11 (4): 563–571. https://doi.org/10.1515/ijfe-2015-0038.

Moamen, S. A. A., and N. El Bakary. 2020. "Histological Variants Induced by Gamma Irradiation in the Alimentary Tract of *Callosbruchus maculatus* (Coleoptera: Bruchidae)." *Arab Journal of Nuclear Sciences and Applications* 53 (2): 118–124. https://doi.org/10.21608/ajnsa.2020.16705.1267.

Mohamed, A. B., R. A. Chavez, M. J. Wagacha, C. K. Mutegi, J. W. Muthomi, S. D. Pillai, and M. J. Stasiewicz. 2022. "Efficacy of Electron Beam Irradiation in Reduction of Mycotoxin-Producing Fungi, Aflatoxin, and Fumonisin, in Naturally Contaminated Maize Slurry." *Toxicon: X* 16 (December): 100141. https://doi.org/10.1016/j.toxcx.2022.100141.

Mohamed, N. F., R. S. S. El-Dine, M. A. M. Kotb, and A. Saber. 2015. "Assessing the Possible Effect of Gamma Irradiation on the Reduction of Aflatoxin B1, and on the Moisture Content in Some Cereal Grains." *American Journal of Biomedical Sciences* (January): 33–39. https://doi.org/10.5099/aj150100033.

Mousavi Khaneghah, A., M. H. Moosavi, C. A. F. Oliveira, F. Vanin, and A. S. Sant'Ana. 2020. "Electron Beam Irradiation to Reduce the Mycotoxin and Microbial Contaminations of Cereal-Based Products: An Overview." *Food and Chemical Toxicology* 143 (September): 111557. https://doi.org/10.1016/j.fct.2020.111557.

Mustapha, M. B., M. Bousselmi, T. Jerbi, N. B. Bettaïeb, and S. Fattouch. 2014. "Gamma Radiation Effects on Microbiological, Physico-Chemical and Antioxidant Properties of Tunisian Millet (*Pennisetum glaucum* L.R.Br.)." *Food Chemistry* 154 (July): 230–37. https://doi.org/10.1016/j.foodchem.2014.01.015.

Nasr, G. M., E.-K. A. Taha, A. M. Hamza, E. A. Negm, N. L. Eryan, A. Noureldeen, H. Darwish, M. S. Zayed, and E.-S. M. Elnabawy. 2022. "Gamma Radiation: An Eco-Friendly Control Method for the Rice Weevil, *Sitophilus oryzae* (L.) (Coleoptera: Curculionidae)." *Biology* 11 (9): 1295. https://doi.org/10.3390/biology11091295.

Nemţanu, M. R., M. Braşoveanu, G. Karaca, and İ. Erper. 2014. "Inactivation Effect of Electron Beam Irradiation on Fungal Load of Naturally Contaminated Maize Seeds." *Journal of the Science of Food and Agriculture* 94 (13): 2668–2673. https://doi.org/10.1002/jsfa.6607.

Nunes, V. M. R., M. Moosavi, A. M. Khaneghah, and C. A. F. Oliveira. 2021. "Innovative Modifications in Food Processing to Reduce the Levels of Mycotoxins." *Current Opinion in Food Science* 38 (April): 155–161. https://doi.org/10.1016/j.cofs.2020.11.010.

Ocloo, F. C. K., M. Owureku-Asare, J. Agyei-Amponsah, W. S. K. Agbemavor, M. N. Y. H. Egblewogbe, F. B. Apea-Bah, A. Sarfo, et al. 2017. "Effect of Gamma Irradiation on Physicochemical, Functional and Pasting Properties of Some Locally-Produced Rice (*Oryza* spp.) Cultivars in Ghana." *Radiation Physics and Chemistry* 130 (January): 196–201. https://doi.org/10.1016/j.radphyschem.2016.08.025.

Ostry, V., F. Malir, J. Toman, and Y. Grosse. 2017. "Mycotoxins as Human Carcinogens—The IARC Monographs Classification." *Mycotoxin Research* 33 (1): 65–73. https://doi.org/10.1007/s12550-016-0265-7.

Ozan, A. 2017. "Investigation of the Production of Cobalt-60 via Particle Accelerator." *Nuclear Technology & Radiation Protection* 32 (4): 327–333. https://doi.org/10.2298/NTRP1704327A

Pan, L., J. Xing, X. Luo, Y. Li, D. Sun, Y. Zhai, K. Yang, and Z. Chen. 2020. "Influence of Electron Beam Irradiation on the Moisture and Properties of Freshly Harvested and Sun-Dried Rice." *Foods* 9 (9): 1139. https://doi.org/10.3390/foods9091139.

Pan, L., J. Xing, H. Zhang, X. Luo, and Z. Chen. 2020. "Electron Beam Irradiation as a Tool for Rice Grain Storage and Its Effects on the Physicochemical Properties of Rice Starch." *International Journal of Biological Macromolecules* 164 (December): 2915–2921. https://doi.org/10.1016/j.ijbiomac.2020.07.211.

Pathak, B., P. K. Omre, B. Bisht, and D. Saini. 2018. "Effect of Thermal and Non-Thermal Processing Methods on Food Allergens." *Progressive Research-An International Journal* 13 (4): 314–319.

Patil, H., N. G. Shah, S. N. Hajare, S. Gautam, and G. Kumar. 2019. "Combination of Microwave and Gamma Irradiation for Reduction of Aflatoxin B1 and Microbiological Contamination in Peanuts (*Arachis hypogaea* L.)." *World Mycotoxin Journal* 12 (3): 269–280. https://doi.org/10.3920/WMJ2018.2384.

Pillai, S. D., and S. Shayanfar. 2017. "Electron Beam Technology and Other Irradiation Technology Applications in the Food Industry." *Topical Reviews in Current Chemistry* 375 (6): 1–20. https://doi.org/10.1007/s41061-016-0093-4

Pi, X., Y. Yang, Y. Sun, X. Wang, Y. Wan, G. Fu, X. Li, and J. Cheng. 2022. "Food Irradiation: A Promising Technology to Produce Hypoallergenic Food with High Quality." *Critical Reviews in Food Science and Nutrition* 62 (24): 6698–6713. https://doi.org/10.1080/10408398.2021.1904822.

Prakash, A., and J. de Jesús Ornelas-Paz. 2019. "Irradiation of Fruits and Vegetables." In *Postharvest Technology of Perishable Horticultural Commodities*, 563–589. Elsevier. https://doi.org/10.1016/B978-0-12-813276-0.00017-1.

Rahman, R. Y., M. I. Abdalla, and A. H. R. Ahmed. 2020. "Effect of Using Gamma Radiation on Storability and Sensory Acceptability of Sudanese Sorghum." *International Journal for Research in Applied Sciences and Biotechnology* 7 (4): 5–11.

Ravindran, R., and A. K. Jaiswal. 2019. "Wholesomeness and Safety Aspects of Irradiated Foods." *Food Chemistry* 285 (July): 363–368. https://doi.org/10.1016/j.foodchem.2019.02.002.

Reddy, C. K., and K. K. Viswanath. 2019. "Impact of γ-Irradiation on Physicochemical Characteristics, Lipoxygenase Activity and Antioxidant Properties of Finger Millet." *Journal of Food Science and Technology* 56 (5): 2651–2659. https://doi.org/10.1007/s13197-019-03753-2.

Ricciardi, E. F., V. Lacivita, A. Conte, E. Chiaravalle, A. Vittorio Zambrini, and M. A. Del Nobile. 2019. "X-Ray Irradiation as a Valid Technique to Prolong Food Shelf Life: The Case of Ricotta Cheese." *International Dairy Journal* 99 (December): 104547. https://doi.org/10.1016/j.idairyj.2019.104547.

Romanov, A., A. Markov, I. Sergeeva, L. Kozubaeva, and D. Protopopov. 2022. "Effect of Ionising Irradiation on Wheat Flour." *Food Processing: Techniques and Technology* 52 (4): 729–738. https://doi.org/10.21603/2074-9414-2022-4-2401.

Rousta, M., A. A. Sadeghi, P. Shawrang, M. A. Afshar, and M. Chamani. 2014. "Effect of Gamma, Electron Beam and Infrared Radiation Treatment on the Nutritional Value and Anti-Nutritional Factors of Sorghum Grain" *Iranian Journal of Applied Animal Science* 4 (4): 661–664.

Salim, S. M., M. Ahmadi, and S. Moharramipour. 2018. "Evaluating the Interaction between Gamma Radiation and Carum Copticum Essential Oil for the Control of *Tribolium confusum* (Coleoptera: Tenebrionidae)." *Journal of Crop Protection* 7 (2): 231–242.

Sang, W., M. Speakmon, L. Zhou, Y. Wang, C. Lei, S. D. Pillai, and K. Zhu-Salzman. 2016. "Detrimental Effects of Electron Beam Irradiation on the Cowpea Bruchid *Callosobruchus maculatus*." *Pest Management Science* 72 (4): 787–795. https://doi.org/10.1002/ps.4053.

Sebaei, A. S., H. M. Sobhy, A. S. M. Fouzy, and O. A. Hussain. 2022. "Occurrence of Zearalenone in Grains and Its Reduction by Gamma Radiation." *International Journal of Environmental Analytical Chemistry* 102 (11): 2503–2511. https://doi.org/10.1080/03067319.2020.1756282.

Shankar, S., P. Follett, S. Ayari, F. Hossain, S. Salmieri, and M. Lacroix. 2020. "Microbial Radiosensitization Using Combined Treatments of Essential Oils and Irradiation-Part B: Comparison between Gamma-Ray and X-Ray at Different Dose Rates." *Microbial Pathogenesis* 143 (June): 104118. https://doi.org/10.1016/j.micpath.2020.104118.

Shao, Y., F. Tang, F. Xu, Y. Wang, and J. Bao. 2013. "Effects of γ-Irradiation on Phenolics Content, Antioxidant Activity and Physicochemical Properties of Whole Grainrice." *Radiation Physics and Chemistry* 85 (April): 227–233. https://doi.org/10.1016/j.radphyschem.2013.01.022.

Sileem, T., S. Mohamed, and A. Mehany. 2019. "First Trial to Assess Gamma-Irradiation and Diatomaceous Earth Combination on Mortality of Some Stored Product Insects." *Arab Journal of Nuclear Sciences and Applications* 52 (3): 115–120. https://doi.org/10.21608/ajnsa.2019.6159.1134.

Singh, K. S., S. Saxena, Y. Sinam, S. Gautam, and G. A. S. Devi. 2022. "Effect of Gamma Radiation Processing on the Quality Characteristics of Anthocyanin Rich Ethnic Rice Cultivars." *Applied Food Research* 2 (1): 100081. https://doi.org/10.1016/j.afres.2022.100081.

Supriya, P., K. R. Sridhar, and S. Ganesh. 2014. "Fungal Decontamination and Enhancement of Shelf Life of Edible Split Beans of Wild Legume *Canavalia maritima* by the Electron Beam Irradiation." *Radiation Physics and Chemistry* 96 (March): 5–11. https://doi.org/10.1016/j.radphyschem.2013.08.007.

Tatar, Y., A. Fadavi, and H. Koohsari. 2021. "Mitigation of *Aspergillus flavus* and Its Aflatoxins in Wheat Grains by Gamma Irradiation and Calcium Oxide." *Journal of Crop Protection* 10 (2): 261–270.

Timakova, R., I. Iliukhina, and R. Iliukhin. 2023. "Practical Applicability of Radiation Technologies in Food Systems." Edited by I. S. Fadeeva and O. Yu. Patlasov. *BIO Web of Conferences* 57 (January): 02007. https://doi.org/10.1051/bioconf/20235702007.

Urbain, W. M. (1986). Fruits, vegetables and nuts. *Food irradiation*, 170-216. https://doi.org/10.1016/B978-0-12-709370-3.50014-2

Xue, P., Y. Zhao, C. Wen, S. Cheng, and S. Lin. 2017. "Effects of Electron Beam Irradiation on Physicochemical Properties of Corn Flour and Improvement of the Gelatinization Inhibition." *Food Chemistry* 233 (October): 467–475. https://doi.org/10.1016/j.foodchem.2017.04.152.

Zehi, Z. B., A. Afshari, S. M. A. Noori, B. Jannat, and M. Hashemi. 2020. "The Effects of X-Ray Irradiation on Safety and Nutritional Value of Food: A Systematic Review Article." *Current Pharmaceutical Biotechnology* 21 (10): 919–926. https://doi.org/10.2174/1389201021666200219093834.

Zhang, Z. S., Q. F. Xie, and L. M. Che. 2018. "Effects of Gamma Irradiation on Aflatoxin B1 Levels in Soybean and on the Properties of Soybean and Soybean Oil." *Applied Radiation and Isotopes* 139 (September): 224–230. https://doi.org/10.1016/j.apradiso.2018.05.003.

Zhou, X., X. Ye, J. He, R. Wang, and Z. Jin. 2020. "Effects of Electron Beam Irradiation on the Properties of Waxy Maize Starch and Its Films." *International Journal of Biological Macromolecules* 151 (May): 239–246. https://doi.org/10.1016/j.ijbiomac.2020.01.287.

14 Use of Other Gases

P. Paulin Patricia and R. Meenatchi
National Institute of Food Technology Entrepreneurship and Management, Thanjavur, Tamil Nadu, India

14.1 INTRODUCTION

The International Grains Council estimates global grain production to be 2.29 billion tons for the marketing year 2023–2024. Population is growing at a rate of around 0.88% per year during 2023, and it is expected to reach 9 billion in the year 2037. World hunger statistics says that about one in nine people globally do not have enough food to lead a healthy life. As the global population grows and the demand for food increases, there is pressure to expand agricultural areas, often at the expense of natural ecosystems.

To meet increasing food demands, farmers started to intensify crop production through the use of fertilizers, pesticides, and water. While these practices can boost yields, they also have negative environmental impacts. Excessive use of fertilizers and pesticides like phosphine and methyl bromide can lead to water pollution, harm non-target species, and contribute to the development of pesticide-resistant insect populations. Stored pest causes both quantitative and qualitative losses by consuming or damaging stored grains. This can result in economic losses for farmers and food security issues.

Growing demand for organically produced foods, restored eco-friendly alternative methods, and use of natural gas as a fumigant is gaining importance. Hence, it is of utmost important to develop gradually some alternate strategies like modified atmosphere (MA) or controlled atmosphere (CA). Different gas concentrations such as oxygen (O_2), carbon dioxide (CO_2), ozone (O_3) as well as nitric oxide (NO) proved to be economical and effective against stored insects.

14.2 IMPORTANCE OF GAS FUMIGATION

Fumigation is a pest control method that involves the application of gases as a fumigant to eliminate pests and pathogens from stored grains and other commodities. To ensure the success of fumigation, it is essential to expose all insects within a grain mass to the specific concentration of the fumigant for a sufficient duration to kill all stages of the insects present. A non-toxic gas fumigant diffuses rapidly and is capable of penetrating deeply into bulk grains. Some beneficial aspects of using fumigants are as follows:

- The use of non-toxic fumigants aims to effectively control and manage pests without causing uninterrupted harm to the environment and human health
- No chemical residues
- No resistance insect or pest population
- Eliminates insects in all stages of development
- Requires less investment and is safe for people and the environment

14.3 FACTORS TO BE CONSIDERED DURING GAS FUMIGATION

Non-chemical gas fumigation, often referred to as controlled atmosphere storage (CAS) or modified atmosphere storage (MAS), is a method used to control pests and to maintain the quality of agricultural products and stored commodities without the use of chemical fumigants. This approach involves controlling the atmospheric conditions within storage facilities to inhibit the growth of pests and pathogens. Some key factors to consider during non-chemical gas fumigation are as follows.

DOI: 10.1201/9781003609834-14

14.3.1 Gas Composition

The gas composition required for each gas concerning the commodity varies. Common gases used in non-chemical gas fumigation include carbon dioxide (CO_2), nitrogen (N_2), and oxygen (O_2).

14.3.2 Gas Concentration

It is required to maintain precise control over the concentration of gases within the storage facility. The optimal gas concentration varies depending on the commodity, temperature, humidity, and targeted pests.

14.3.3 Temperature and Humidity Control

Monitor and control temperature and humidity levels within the storage facility. These factors may influence the effectiveness of non-chemical gas fumigation and the rate of pest mortality.

14.3.4 Gas Monitoring and Control Systems

Implement gas monitoring and control systems to continuously measure and adjust gas concentrations. These systems ensure that the desired conditions are maintained throughout the fumigation period.

14.3.5 Sealing and Airtightness

Ensure that the storage facility is well-sealed and airtight to prevent the exchange of gases with the external environment. Proper sealing is crucial for maintaining the desired gas composition.

14.3.6 Commodity Compatibility

Verify that the commodity being stored is compatible with the selected gas composition and storage conditions. Some commodities may require specific gas concentrations or exposure durations.

14.3.7 Pest Monitoring

Regularly monitor the facility for signs of pest activity. Pest traps and visual inspections can help assess the effectiveness of non-chemical gas fumigation and determine if any adjustments are needed.

14.3.8 Exposure Duration

Determine the required exposure duration based on the type of pests present and the gas composition used. Extended exposure periods may be necessary to ensure pest mortality.

14.4 CATEGORIES OF GRAIN STORAGE STRUCTURES

Grain storage structures are essential for preserving the harvested grains, preventing losses due to spoilage and pests, and also ensuring a stable food supply. These structures can vary in size, design, and materials used. Some common categories of grain storage structures are as follows:

1. ***Silos*:** Silos are tall, cylindrical structures typically made of concrete, steel, or even wood. They are designed to store large quantities of grain vertically. Silos can be either tower silos, which are freestanding structures, or bunker silos, which are partially below ground.

2. ***Grain bins*:** Grain bins are smaller, often cylindrical, storage structures made of steel or sometimes corrugated metal. They are commonly used on farms and can store moderate to large quantities of grain. Grain bins are available in various sizes and configurations.
3. ***Bag storage*:** Grain bags, also known as silage bags or grain pouches, are flexible containers made of heavy-duty plastic. They are filled with grain and sealed, creating a temporary storage solution. Grain bags are commonly used in situations where traditional storage structures are not available or practical.
4. ***Tarpaulin storage*:** Tarpaulins are used to cover and protect harvested grains stored in the open. They are secured over the grain piles or stacks to provide temporary protection against weather and pests.
5. ***Traditional storage structures*:** In many regions, traditional storage methods such as mud-brick silos, thatched huts, or woven baskets are still used for grain storage, especially in rural and subsistence farming communities.
6. The choice of grain storage structure depends on factors like the quantity of grain to be stored, local climate conditions, available resources, and the desired duration of storage. Proper management and maintenance of these structures are crucial to ensure grain quality and minimize losses.

14.5 MODIFIED ATMOSPHERE STORAGE

MAS alters the gas composition, such as oxygen, carbon dioxide, and nitrogen present in the stored environment, causing pest-lethal effects (Odjo et al. 2020). Modified atmosphere can be achieved normally by allowing metabolic process in an airtight storage condition to alter the gas composition through respiration of the produce is a passive method of storage which is also known to be hermetic storage. In the active method of storage, the gas composition in the storage atmosphere is modified purposely by the addition or removal of particular gases. Reduced O_2 or increased CO_2 levels can diminish metabolic activity, rate of respiration, ethylene production, and microbial growth of products.

The effectiveness of a modified atmosphere totally depends on biotic and abiotic factors such as temperature, moisture content, species and stage of stored pest, gas composition and uniformity of gas distribution, exposure time, and permeability of packaging material. Modified atmosphere packaging (MAP) not only serves as a means to control existing pests within the packaged product but can also act as a physical barrier to prevent further immigration of insects and other pests (Riudavets et al. 2010). The role and responsibility of each gas used in MAP are indeed related to their specific properties.

14.5.1 Effect of Modified Atmosphere on Grain Quality

- Grain moisture content and temperature are important environmental factors that influence grain quality, germination loss, and relative humidity
- Ageing of seeds and germination loss cannot be stopped but can be minimized by modified atmosphere storage
- When the seed moisture content is between 14 and 5%, the life of the seed gets doubled for every 1% decrease in seed moisture content
- MAP helps to regulate the metabolic processes of seeds, including those related to cellular respiration and energy production

14.6 CONTROLLED ATMOSPHERE STORAGE

CA storage is distinct from airtight storage. In airtight storage, changes in gas ratios (such as O_2 depletion and CO_2 accumulation) occur naturally due to the metabolic processes of insects, microflora, and the stored grain itself. This natural change in gas composition can sometimes lead to pest infestations if not managed properly. In CA storage, gases are introduced intentionally to create a controlled and stable atmosphere (Odjo et al. 2020). This proactive approach can prevent damage to the stored grain and minimize the need for chemical treatments or fumigation.

14.7 HERMETIC STORAGE

Hermetic storage refers to a method of storing agricultural commodities, particularly grains and seeds, in airtight storage systems. Hermetic storage is an eco-friendly and non-chemical method of pest control. It reduces the reliance on chemical pesticides and environmental health implications. It is classified into three different types, namely,

- ***Organic hermetic storage*:** In this method, reduced oxygen and elevated carbon dioxide are achieved by the metabolic process of grains and insects in the airtight storage atmosphere.
- ***Vacuum hermetic fumigation*:** By using a vacuum pump, oxygen present in the stored environment is sucked out thoroughly to create partial pressure of air inside the storage structure, which leads to the depletion of oxygen.
- ***Gas hermetic fumigation*:** Non-toxic gases such as carbon dioxide or nitrogen are artificially flushed into the storage environment using pumps or generators to kill stored pests.

Thus, depleted oxygen level in the stored environment achieves 100% mortality of all life stages of insects and prevent mold development. Hermetic storage doubles the lifespan of stored seeds.

14.8 VARIOUS GASES AS A FUMIGANT

14.8.1 Reduced Oxygen (Hypoxia Effects)

Oxygen is essential for the growth of aerobic (oxygen-dependent) microorganisms. In the presence of oxygen, bacteria, yeasts, and molds can thrive and proliferate. Oxygen promotes the oxidation of fats and oils, leading to the development of rancid flavors and odors in food products. Because of its harmful effect, reduced oxygen is preferred in modified storage. Reduced oxygen (hypoxia) is generally achieved by the respiration of insects and grains at airtight storage conditions. Table 14.1

TABLE 14.1
Effect of Reduced Oxygen Level against Mortality of Adult *Sitophilus oryzae* L.

Oxygen Level (%)	Exposure Time (days)	Mortality (%)
1	3	46.67
	4	89.33
	5	100.00
3	7	30.33
	9	75.33
	11	100.00
5	8	07.67
	10	14.67
	12	21.00

Source: Kandel et al. (2021).

shows the effect of reduced oxygen level against mortality of adult *Sitophilus oryzae* L. The insect population growth and survival are suppressed at sub-optimal levels of oxygen.

14.8.2 Increased Carbon Dioxide (Hypercapnia Effect)

Carbon dioxide is effective in controlling stored pests for long-term storage in sealed structures. Its ability to create a low-oxygen atmosphere helps to slow down insect metabolism, respiration, and reproduction, making it suitable for extended storage periods (Husain et al. 2017). CO_2 is soluble both in water and lipids (fats). Its solubility increases as the temperature decreases. This property allows CO_2 to penetrate and distribute throughout stored grains, ensuring effective pest control. CO_2 can have a profound effect on insect respiration. Molecules of CO_2 can enter an insect's body through tiny openings called spiracles, which are used for gas exchange. Exposure to elevated levels of CO_2 can cause the spiracles to remain open, making it difficult for insects to regulate their respiration effectively. This disrupts their normal respiratory function, resulting in insect mortality.

14.8.3 Vacuum Hermetic Fumigation

This process typically begins with placing the grains or commodities to be treated in a specialized vacuum-sealed storage container or bag. These containers are designed to be airtight and impermeable to external gases. A vacuum pump is used to remove air and oxygen inside the container. The vacuum pump creates a low-pressure environment within the container, effectively eliminating atmospheric gases. At this juncture, a known concentration of inert gasses such as nitrogen or carbon dioxide is purged into the airtight container to eliminate stored pests. After the fumigation period is complete, the container is typically degassed and aerated to remove any remaining fumigant residues and stored at normal atmospheric conditions (Lawrence, Bicksler, and Duncan 2017).

It is highly effective in controlling a wide range of stored grain pests, including insects and mites, without the need for chemical pesticides. The installation, operation, and maintenance costs of machinery are high.

14.8.4 Nitrogen

Nitrogen is indeed an inert gas, which means it does not react chemically with the products it surrounds. It is colorless, odorless, and tasteless, making it an ideal choice for preserving food without altering its sensory properties. Nitrogen itself does not possess antimicrobial properties, meaning it doesn't actively kill or inhibit the growth of microorganisms like bacteria, molds, or yeast. However, its use in MAP can indirectly contribute to microbial control by creating an environment where oxygen-dependent microorganisms cannot thrive.

Thus, it is used as a filler gas to replace oxygen within the packaging. This helps to extend the shelf life of products by reducing the oxidative reactions that can cause spoilage.

14.8.5 Nitric Oxide

Nitric oxide (NO) as a fumigant for pest control and postharvest quality improvement is an interesting and promising development in the field of agriculture and food preservation. It spontaneously reacts with oxygen (O_2) to form nitrogen dioxide (NO_2). Therefore, NO fumigation needs to be carried out under ultralow oxygen (ULO) conditions in airtight chambers to minimize the oxidation of NO within the fumigation chamber. NO fumigation must be terminated by flushing the chamber with an inert gas like nitrogen (N_2) to dilute NO and prevent the formation of NO_2. NO_2 can potentially cause harm to fresh products (Bahadoran et al. 2016). Thus, Table 14.2 shows the effect of nitric oxide fumigation against mortality of different life stages of stored pests.

TABLE 14.2
Effect of Nitric Oxide Fumigation against Mortality of Different Life Stages of Stored Pest

Targeted Species	Nitric Oxide (%)	Temperature (°C)	Developmental Stage	Exposure Time (h)	Mortality (%)
Tribolium confusum	0.5	20	Larva	8	83.80
				12	92.00
				24	100.00
	0.5	20	Pupa	8	35.60
				12	93.60
				24	100.00
	2.0	10	Egg	8	77.00
				12	83.60
				24	100.00
Sitophilus oryzae	0.5	25	Adult	24	93.00
	1.0	25	Adult		100.00
	0.5	25	Egg	48	99.60
	1.0	25	Egg		100.00

Source: Liu (2013).

Nitric oxide has shown effectiveness against wide range of insect pests, including both external and internal feeders, at various life stages. This makes it a potentially valuable tool for pest control in both fresh and stored products (Liu 2013).

14.8.6 Ozone (O_3)

Ozone is a triatomic molecule composed of three oxygen atoms (O_3) and is characterized by its distinctive pungent odor. It is generated by splitting diatomic oxygen molecules (O_2) into free radical oxygen (O), which then reacts with another O_2 molecule to form ozone. This process typically requires a significant amount of energy. Ultraviolet (UV) radiation with a wavelength of 188 nm and corona discharge methods are commonly used to initiate the production of free radical oxygen and generate ozone. The mode of action against insects involves causing oxidative stress to their cells, leading to the destruction of cell membranes and the breakdown of DNA strands, ultimately causing the death of the insects.

Several factors, including seedbed thickness, moisture content, ozone concentration, exposure time, and mycobiota contamination of seeds, should be considered when using ozone for pest control and grain preservation (Pandiselvam et al. 2017). One of the drawbacks of using ozone is its potential to cause rusting in storage bins, and it affects rubber couplings used in fumigation chambers. Therefore, careful selection of containers and accessories is necessary when using ozone (Hardin et al. 2009). Table 14.3 shows the effect of ozone treatment against mortality of different life stages of stored pests.

14.8.7 Biogas (Methane+Carbon Dioxide)

Biogas mainly contains two gases which include methane (60%) and carbon dioxide (40%), while carbon dioxide showed efficacy against stored insect pests. Although carbon dioxide can be used directly to combat stored grain insect pests, the cost and delivery to the actual site made it a non-viable option. On the other hand, biogas from cow dung can be produced within the farm; only some

TABLE 14.3
Effect of Ozone Treatment against Mortality of Different Life Stages of Stored Pests

Targeted Species	Ozone Concentration (ppm)	Exposure Time	Developmental Stage	Mortality (%)
Tribolium confusum	5	5 days	Adult	100.00
	125	6 h	Adult	100.00
	35–42	6 days	Larva	100.00
			Adult	100.00
Tribolium castaneum	50	3 days	Adult	92.20
	1800	1 h	Larva	92.10
			Pupa	79.20
			Egg	88.30
			Adult	84.30
Plodia interpunctella	50	3 days	Adult	94.50
	170	4 h	Larva	10.00
	300	4 h	Pupa	98.00
	1800	1 h	Larva	66.90
			Adult	84.30
Sitophilus zeamais	50	3 days	Adult	100.00
Sitophilus oryzae	125	6 h	Adult	100.00
Rhyzopertha dominica				
Oryzaephilus surinamensis	5	3 days	Adult	100.00

Source: Isikber and Athanassiou 2015).

pipes and other arrangements need diversion as and when needed through the airtight grain storage structures. This is a very cheap and convenient method. It is used as a fumigant in stored seeds and grains of pigeon peas in a specially designed airtight storage bin having one quintal capacity to manage *Callosobruchus chinensis.*

14.8.8 High-Pressure Carbon Dioxide

The application of high-pressure CO_2 could rapidly control infestations compared to other commercial methods. This approach can be highly effective but necessitates specialized equipment, which requires a significant capital investment. The effectiveness of high-pressure CO_2 treatment depends on several factors, including the species of the pest, the pressure applied, exposure time, and temperature. Different pests may have varying sensitivities to these factors. The mortality of stored pests against high-pressure carbon dioxide at 15 bar for 30 min was found to be effective against stored pests Table 14.4.

14.9 INSECT RESPIRATORY BEHAVIOR TO TOXIC GASES

1. Insects have a system of tiny tubes called tracheae that branch throughout their body. These tracheae open to the outside through small openings called spiracles, which are typically located along the abdomen and thorax.
2. Insects open and close their spiracles to regulate the flow of air in and out of their tracheal system. This allows them to conserve water in arid environments and minimize exposure to toxic gases.

TABLE 14.4
Mortality of Stored Pest Treated with High-Pressure CO_2 at 15 Bars for 30 Min

Target Species	Developmental Stage	Mortality (%)
Oryzaephilus surinamensis	Adult	94.3
	Larvae	82.4
Lasioderma serricorne	Adult	59.3
	Larvae	92.6
Sitophilus oryzae	Adult	75.2
	Larvae	98.4
Rhyzopertha dominica	Adult	6.1
	Larvae	97.1

Source: Riudavets et al. 2010).

3. Generally, insects can survive in hypoxic conditions (Bailey 1955). There are few insects that may survive by enlarging the tracheal diameter, reducing respiration as well as the metabolic rates (Zhou et al. 2007). Pest resurgence condition may also arise if the insect survived the conditions of short hypoxia and then normoxia. In such cases, it is very important to monitor the insect activity throughout the storage period.

14.10 CONCLUSION

It is crucial to understand the properties of different gases used as a fumigant for stored pests and create the right storage environment to preserve product quality and prevent pest infestation is essential. Each gas has unique characteristics that make it suitable for specific applications. Non-toxic fumigants offer exciting potential benefits in terms of pest control and postharvest quality improvement. The use of gas fumigants may help to reduce reliance on chemical pesticides, minimize residues, and extend the shelf life of products.

There are challenges to overcome when implementing non-toxic fumigants on a larger scale. This includes the need for specialized equipment, dosage, and time required to control all the stages of stored pests, the potential impact on product quality, and regulatory approval. Continued research and collaboration among researchers, industry stakeholders, and regulatory agencies are essential for realizing the full potential of non-toxic fumigation in agriculture and food preservation. This includes studying the efficacy and safety of these methods, as well as developing best practices and guidelines for their use.

REFERENCES

Bahadoran, Z., P. Mirmiran, S. Jeddi, F. Azizi, A. Ghasemi, and F. Hadaegh. 2016. "Nitrate and Nitrite Content of Vegetables, Fruits, Grains, Legumes, Dairy Products, Meats and Processed Meats." *Journal of Food Composition and Analysis* 51: 93–105. doi:10.1016/j.jfca.2016.06.006.

Bailey, S. T. 1955. "Air-Tight Storage of Grain; Its Effects on Insect Pests. I. *Calandra granaria* L. (Coleoptera, Curculionidae). "*Australian Journal of Agricultural Research* 6 (1): 33–51. doi:10.1071/AR9550033.

Hardin, J. A., C. L. Jones, E. L. Bonjour, R. T. Noyes, R. L. Beeby, D. A. Eltiste, and S. Decker. 2009. "Ozone Fumigation of Stored Grain; Closed-Loop Recirculation and Rate of Ozone Consumption." In *2009 Reno, Nevada, June 21–June 24, 2009*. St. Joseph, MI: American Society of Agricultural and Biological Engineers. doi:10.13031/2013.27354.

Husain, M., S. Sukirno, K. Mehmood, M. Tufail, K. G. Rasool, W. S. Alwaneen, and A. S. Aldawood. 2017. "Effectiveness of Carbon Dioxide against Different Developmental Stages of *Cadra cautella* and *Tribolium castaneum*." *Environmental Science and Pollution Research* 24 (14): 12787–12795. doi:10.1007/s11356-017-8860-5.

Isikber, A. A., and C. G. Athanassiou. 2015. "The Use of Ozone Gas for the Control of Insects and Micro-Organisms in Stored Products." *Journal of Stored Products Research* 64 (October): 139–145. doi:10.1016/j.jspr.2014.06.006.

Kandel, P., M. E. Scharf, L. J. Mason, and D. Baributsa. 2021. "Effect of Hypoxia on the Lethal Mortality Time of Adult *Sitophilus oryzae* L." *Insects* 12 (10): 952. doi:10.3390/insects12100952.

Lawrence, B., A. J. Bicksler, and K. Duncan. 2017. "Local Treatments and Vacuum Sealing as Novel Control Strategies for Stored Seed Pests in the Tropics." *Agronomy for Sustainable Development* 37 (2): 6. doi:10.1007/s13593-017-0415-0.

Liu, Y.-B. 2013. "Nitric Oxide as a Potent Fumigant for Postharvest Pest Control." *Journal of Economic Entomology* 106 (6): 2267–2274. doi:10.1603/EC13249.

Odjo, S., J. Burgueño, A. Rivers, and N. Verhulst. 2020. "Hermetic Storage Technologies Reduce Maize Pest Damage in Smallholder Farming Systems in Mexico." *Journal of Stored Products Research* 88 (September): 101664. doi:10.1016/j.jspr.2020.101664.

Pandiselvam, R., S. Sunoj, M. R. Manikantan, A. Kothakota, and K. B. Hebbar. 2017. "Application and Kinetics of Ozone in Food Preservation." *Ozone: Science & Engineering* 39 (2): 115–126. doi:10.1080/01919512.2016.1268947.

Riudavets, J., C. Castañé, O. Alomar, M. J. Pons, and R. Gabarra. 2010. "The Use of Carbon Dioxide at High Pressure to Control Nine Stored-Product Pests." *Journal of Stored Products Research* 46 (4): 228–233. doi:10.1016/j.jspr.2010.05.005.

Zhou, D., J. Xue, J. Chen, P. Morcillo, J. D. Lambert, K. P. White, and G. G. Haddad. 2007. "Experimental Selection for *Drosophila* Survival in Extremely Low O_2 Environment." *PLoS ONE* 2 (5): e490. doi:10.1371/journal.pone.0000490.

15 Modifying Gas Concentrations in Storage Environments

Cristiano Dietrich Ferreira
Technological Institute in Food for Health, University of Vale do Rio dos Sinos, São Leopoldo, Brazil

Larissa Alves Rodrigues and Newiton da Silva Timm
Department of Agroindustry Science and Technology, Federal University of Pelotas, Pelotas, Rio Grande do Sul, Brazil

15.1 INTRODUCTION

After harvesting, grains go through post-harvest processes, which include reception, cleaning, drying, and storage. Recognizing that even during storage, grains continue to have an active metabolism, storage aims to maintain the quality of the grain to prevent deterioration and contamination by external agents, which is one of the challenges that the industry seeks to address (Cañizares et al. 2021). Despite the importance of the post-harvest stage being extensively studied and acknowledged, losses in this process are still high, estimated at 20% per year.

The presence of insects in stored products is of great economic importance due to the numerous damages caused and also the contamination of the product with their fragments, resulting in devaluation or even rejection of the product for commercialization. During grain deterioration, there is a significant release of carbon dioxide (CO_2), moisture, and heat produced by the metabolism of the grain itself and associated organisms, such as fungi, insects, and mites, and the release of distinct odors (Neethirajan et al. 2010).

Currently, the most common method for protecting against infestations of stored grains is chemical control, using pesticides, primarily the fumigant phosphine. It is the most widely used technique because it is efficient in controlling all stages of insects, requires a short exposure time, and has low cost and easy management. However, the toxicity of these products has negative aspects, as well as the resistance that insects can develop when the application is done incorrectly. These two points are reasons to limit their recurring use in pest control (Nayak et al. 2020).

15.2 METHODS FOR MODIFYING GAS CONCENTRATIONS IN STORAGE ENVIRONMENTS

Modified atmosphere storage is an alternative technique for protecting stored grain mass and controlling infesting pests while maintaining grain quality without using pesticides. In summary, this grain preservation technique involves replacing the atmosphere within the grain mass by introducing a gas or a mixture of inert gases (nitrogen and carbon dioxide) until a significant reduction in oxygen concentration is achieved. The goal of this method is to restrict oxygen to the insect's respiratory system and, at the same time, take advantage of the harmful effect of high concentrations of other gases, primarily carbon dioxide.

15.2.1 Hermetic Storage System

One of the various methods for managing and controlling insects in stored grains is the use of physical methods as a modifier of the storage atmosphere (Viebrantz, Radunz, and Dionello 2016). Among these

DOI: 10.1201/9781003609834-15

physical methods, hermetic storage is considered the most traditional because storing grains in hermetically sealed bags has proven to be an effective method for controlling moisture and insects in grains, limiting gas exchange between the internal and external environments and the grain mass, maintaining initial moisture levels, and controlling pests through a lack of oxygen (Quezada et al. 2006).

It acts by reducing the concentration of O_2 due to the respiration of grains and associated organisms, making the atmosphere rich in CO_2 and poor in O_2. These changes inhibit the development of aerobic fungi and mycotoxins, as well as reduce insect incidence and grain metabolic activity, leading to a reduction in oxidation and product preservation. In summary, the atmosphere in the hermetic grain mass storage is characterized by a high concentration of CO_2 (Guenha et al. 2014). This storage method is based on reducing the available O_2 in the storage environment to lethal or limiting levels for living organisms. This reduction of O_2 can occur naturally through the gas exchange related to grain and organism respiration, ensuring that hermetic grain storage is a way to reduce insect and fungi attacks (Rupollo et al. 2006).

15.2.2 Modified Atmosphere System with Carbon Dioxide (CO_2)

Unlike more traditional chemical fumigants like phosphine and methyl bromide, which are toxic to insects at low concentrations, the required concentration of carbon dioxide (CO_2) to create a lethal environment for insects is about 10,000 times higher. Therefore, for the effective fumigation of stored grains using CO_2, the storage structure needs to be hermetically sealed, or CO_2 needs to be regularly or continuously added to maintain lethal concentration levels for insects. Additionally, as CO_2 is about 50% heavier than air, after some time, there may be a high concentration in the lower layer of the storage cell. In this case, recirculation of the gas mixture is necessary to balance the concentration throughout the cell, aiming for effective pest control (Navarro, Jay, and Leesch 1986). In summary, CO_2 treatment is advantageous because no toxic chemical residues remain in the grains after fumigation.

CO_2 is highly permeable to animal tissue and acts by increasing the respiration of insects, accelerating their metabolism. Concentrations greater than 5% of CO_2 cause spiracle opening, leading to water loss, desiccation, and insect death. In addition to these effects, researchers report that insects exposed to a high concentration of CO_2 for a short period exhibit initial immobilization due to the anesthetic effect of the gas (Choi, Waller, and Dennehy 1997).

The toxicity of the modified atmosphere with carbon dioxide for stored grain pests is indifferent when the concentration exceeds 60% CO_2, considering that concentrations between 60% and 35% allow some species to survive or be tolerant to levels below 35%. Furthermore, research has shown that an atmosphere with 60% CO_2 is more effective against primary pests, such as *Sitophilus* sp., *Rhyzopertha dominica*, and *Trogoderma granarium*, while an atmosphere containing 100% CO_2 is more effective against secondary pests, such as *Tribolium castaneum* and *Oryzaephilus surinamensis*, possibly due to the adsorption and desorption of CO_2 on the grain (Semple 1992).

15.2.3 Modified Atmosphere System with Nitrogen (N)

Historically, the use of nitrogen gas as an insecticide in stored grains is divided into two phases: in the initial phase, called purging, the required atmosphere is obtained in the cell by the continuous injection of nitrogen gas to replace intergranular air until the oxygen concentration in the upper space of the silo is reduced to 1%. The second phase is called maintenance due to the frequent injection of nitrogen to maintain the required concentration in the storage cell (Banks and Annis 1977).

In addition to being effective in non-polluting insect control programs, nitrogen gas has the benefit of initially occupying 78% of the intergranular space. However, it has been normally considered a diluent since its action is similar to that of inert gases, such as helium. Thus, it has some disadvantages, such as longer exposure time and higher concentrations, resulting in higher costs than traditional fumigants. However, research in the 1970s demonstrated that nitrogen storage in a modified atmosphere reduces aflatoxin formation by storing fungi *Aspergillus flavus*.

15.2.4 Modification by Organic Acids

The use of organic acids for grain preservation during storage has a recent history. Researchers report an inhibitory effect on storage fungi, primarily *Aspergillus* spp. and *Penicillium* spp (Rangel and Villela 2004). Initially, the use of organic acids spread due to problems caused by high humidity storage, which occurs due to adverse weather conditions or a lack of resources, knowledge, and/or technologies compatible with the producer's conditions (Elias, Oliveira, and Vanier 2018).

Therefore, it became necessary to develop technologies that would allow for the delay or even the replacement of the drying step to promote grain preservation, the release of land for other crops, or other necessary factors. One way to delay or even replace the drying process is the use of short-chain organic acids, such as acetic and propionic, which can also be associated with hermetic storage. The use of organic acids represents an effective alternative for small- and medium-sized producers who do not have the structures or resources to install more elaborate drying and storage systems. Delaying drying allows for the rationalization of system sizing and usage without increasing losses or even reducing them by reducing the idle time of drying structures, transportation, and its impact on freight (Elias, Oliveira, and Vanier 2018).

15.2.5 Absorption

To preserve a wide variety of foods, new technologies have been developed to eliminate or reduce oxygen levels within packaging. These technologies include oxygen absorbers. Oxygen absorbers were first used and commercialized in Japan in 1970 (Mu et al. 2013). Since then, they have been widely used to maintain the characteristics and quality and inhibit the development of microorganisms in the products in which they are used (Aday and Caner 2013). Furthermore, some studies on pest control in stored products using oxygen absorbers have been conducted, and lethal effects on certain pest insects were found (Kucerova et al. 2014). In general, the use of oxygen absorbers has the advantages of being easy to use, approved by the Food and Drug Administration (FDA), preventing the growth of aerobic microorganisms, and delaying lipid oxidation. Disadvantages include the possibility of package collapse (which can be avoided by using a CO_2 generator), accidental ingestion of the contents, potential sachet rupture and product contamination, and triggering metal detectors when using sachets based on reactions involving metals (Braga and Peres 2010).

15.2.6 Adsorption

Adsorption is defined as a physical-chemical phenomenon in which a component in a gas or liquid phase is transferred to the surface of a solid phase. The components that adhere to the surface are called adsorbates, while the solid phase that retains the adsorbate is called the adsorbent. The removal of these components from one phase to another is driven by the difference in concentrations between the fluid's interior and the adsorbent's surface (Borba 2006).

The goal of the adsorbent is to provide the necessary surface area for the selective adsorption of the preferred adsorbed species. Ideally, the adsorbent should not only have high selectivity for the component to be separated but also a high adsorption capacity. The most commonly used adsorbents are activated carbon and zeolites. Activated carbon is used as an adsorbent, catalyst, or catalyst support. In effluent treatment, it is used in liquid-phase adsorption, such as the adsorption of organic molecules that cause taste, odor, and toxicity. The best characteristic of activated carbon is its large internal surface area located within a network of narrow pores, where most of the adsorption process will take place and where the size and shape of the pores can influence adsorption selectivity. The high adsorption capacity of activated carbon is related to its large surface area, structure, and distribution of micropores, as well as possible surface reactivity properties arising from the complexity of surface chemical groups.

15.2.7 Membrane Separation

A membrane is defined as a barrier that separates two phases and completely or partially restricts the transfer of chemical species. The phase that passes through the permeable membrane is called permeate, and the one retained is called concentrate. The use of membrane separation is common in the food, pharmaceutical, and chemical industries because it is energetically favorable, as it promotes separation without changing the phase and is selective, allowing the separation of heat-labile compounds and having a simpler operation.

Separation through membranes occurs due to the existence of a potential gradient that acts on the membrane. Most membrane separation methods use the chemical potential gradient as the driving force, which can be in terms of pressure, concentration, and temperature differences. Another driving force is the electric potential gradient. One of the most common limiting factors in using membrane separation is fouling, which is a variation in the permeate flow due to chemical species in the solution, resulting in a total or partially irreversible case. The main factors that cause fouling are the adsorption of solute molecules on the membrane's surface or in the pores, pore blockage by suspended molecules, precipitation, and the deposit of suspended material on the membrane (Habert, Borges, and Nobrega 2006).

15.3 APPLICATIONS OF GAS CONCENTRATION MODIFICATION IN STORAGE ENVIRONMENTS

Modified or controlled atmospheres are a safe and sustainable alternative for insect pest control in many types of stored grains (Cui et al. 2017). Their success is tied to both their residue-free preservation capacity and their insecticidal action, as the reduction of O_2 and the increase of CO_2 create a microclimate that renders the respiration of pest insects insufficient (Makino et al. 2020).

Gas concentration modification is widely used not only in grain storage but also in the preservation of fruits, vegetables, and other agricultural products (Qu et al. 2022). Research has reported that storage with higher concentrations of nitrogen can effectively delay the deterioration of grain quality and protect its internal enzyme elimination system (Wang et al. 2018). Related studies have also found that vitamin E and polyphenols in grains can be preserved during storage with nitrogen (Moncini et al. 2020). With the same objective, current studies have also proven that nitrogen storage in a modified atmosphere can reduce the degradation of phenolic compounds in black rice during storage (Lang et al. 2019).

Studies conducted on three weevil species (*Callosobruchus maculatus*, *Zabrotes subfasciatus*, and *Acanthoscelides obtectus*) have confirmed the effectiveness of atmospheres with low O_2 concentrations and high CO_2 concentrations in the mortality of these insects (Iturralde-García et al. 2016). The most environmentally tolerant stages are the egg and pupa stages, with adults being the least tolerant (Wong-Corral, Castañé, and Riudavets 2013). In addition to mortality, Dawson (1995) demonstrated that modified atmospheres can also affect the reproductive potential of surviving *C. maculatus* females.

In conclusion, the gas concentration modification method has evolved over the years and has proven to be quite efficient in both preservation and insect disinfestation. However, like any other method, it requires specialized labor and quality equipment to achieve successful results.

15.4 CONCLUSION

As seen throughout this chapter, decontamination and prevention of insect infestation in grain storage facilities go beyond chemical and traditional methods, which are now less recommended due to the high risk of consumer contamination and environmental damage. Gas concentration modification is a method that started as a preservation technique, and with advances in research, its insecticidal effect was observed in pests resistant even to chemical methods of disinfestation.

It is also considered an "environmentally friendly" method by authorities, as it poses no risks to human health, delays grain degradation, maintains bioactive compounds, and affects the reproductive capacity of female insects, leading to a significant reduction in pest insect populations.

REFERENCES

Aday, M. S., and C. Caner. 2013. "The Shelf Life Extension of Fresh Strawberries Using an Oxygen Absorber in the Biobased Package." *LWT – Food Science and Technology* 52 (2): 102–109. doi:10.1016/j.lwt.2012.06.006.

Banks, H. J., and P. C. Annis. 1977. *Suggested Procedures for Controlled Atmosphere Storage of Dry Grain*. Melbourne: CSIRO. doi: 10.25919/5n23-fh15.

Borba, C. 2006. "Modeling the Removal of Heavy Metals in a Fixed Bed Adsorption Column." State University of Campinas.

Braga, L. R., and L. Peres. 2010. "Novas Tendências Em Embalagens Para Alimentos: Revisão." *Boletim Centro de Pesquisa de Processamento de Alimentos* 28 (1): 69–84.

Cañizares, L. d. C. C., N. d. S. Timm, G. H. Lang, C. A. Gaioso, C. D. Ferreira, and M. d. Oliveira. 2021. "Effects of Using Wind Exhausters on the Quality and Cost of Soybean Storage on a Real Scale." *Journal of Stored Products Research* 93 (September): 101834. doi:10.1016/j.jspr.2021.101834.

Choi, C. Y., P. M. Waller, and T. M. Dennehy. 1997. "Insect Control with Carbon Dioxide Foam." *Transactions of the ASAE* 40 (5): 1475–1480. doi:10.13031/2013.21362.

Cui, S., L. Wang, J. Qiu, Z. Liu, and X. Geng. 2017. "Comparative Metabolomics Analysis of *Callosobruchus chinensis* Larvae under Hypoxia, Hypoxia/Hypercapnia and Normoxia." *Pest Management Science* 73 (6): 1267–1276. doi:10.1002/ps.4455.

Dawson, C. 1995. "The Effect of Carbon Dioxide-Induced Anesthesia on the Fecundity of *Callosobruchus maculatus (Coleoptera: Bruchidae)*." *Journal of Stored Products Research*, 31: 49–54. doi: 10.1016/0022-474X (94)00035-R.

Elias, M. C., M. Oliveira, and N. L. Vanier. 2018. Grain pre-storage, storage and conservation technologies. UFPel. https://labgraos.com.br/manager/uploads/arquivo/apostila---unidade-i---prova-i.pdf

Guenha, R., B. d. V. Salvador, J. Rickman, L. F. Goulao, I. M. Muocha, and M. O. Carvalho. 2014. "Hermetic Storage with Plastic Sealing to Reduce Insect Infestation and Secure Paddy Seed Quality: A Powerful Strategy for Rice Farmers in Mozambique." *Journal of Stored Products Research* 59 (October): 275–281. doi:10.1016/j.jspr.2014.06.007.

Habert, A.C., C.P. Borges, and Nobrega, R. 2006. Membrane Separation Processes. Série Escola Piloto de Engenharia-Química COPPE/UFRJ. Rio de Janeiro: E-papers ServicosEditoriais Ltda.

Iturralde-García, R. D., J. Borboa-Flores, F. J. Cinco-Moroyoqui, J. Riudavets, C. L. Del Toro-Sánchez, E. O. Rueda-Puente, O. Martínez-Cruz, and F. J. Wong-Corral. 2016. "Effect of Controlled Atmospheres on the Insect *Callosobruchus maculatus* Fab. in Stored Chickpea." *Journal of Stored Products Research* 69 (October): 78–85. doi:10.1016/j.jspr.2016.06.004.

Kucerova, Z., K. Kyhos, R. Aulicky, J. Lukas, and V. Stejskal. 2014. "Laboratory Experiments of Vacuum Treatment in Combination with an O_2 Absorber for the Suppression of *Sitophilus granarius* Infestations in Stored Grain Samples." *Crop Protection* 61 (July): 79–83. doi:10.1016/j.cropro.2014.03.018.

Lang, G. H., I. d. S. Lindemann, C. D. Ferreira, J. F. Hoffmann, N. L. Vanier, and M. de Oliveira. 2019. "Effects of Drying Temperature and Long-Term Storage Conditions on Black Rice Phenolic Compounds." *Food Chemistry* 287 (July): 197–204. doi:10.1016/j.foodchem.2019.02.028.

Makino, Y., A. Nishizaka, M. Yoshimura, I. Sotome, K. Kawai, and T. Akihiro. 2020. "Influence of Low O_2 and High CO_2 Environment on Changes in Metabolite Concentrations in Harvested Vegetable Soybeans." *Food Chemistry* 317 (July): 126380. doi:10.1016/j.foodchem.2020.126380.

Moncini, L., G. Simone, M. Romi, G. Cai, G. Guerriero, A. Whittaker, S. Benedettelli, and R. Berni. 2020. "Controlled Nitrogen Atmosphere for the Preservation of Functional Molecules during Silos Storage: A Case Study Using Old Italian Wheat Cultivars." *Journal of Stored Products Research* 88 (September): 101638. doi:10.1016/j.jspr.2020.101638.

Mu, H., H. Gao, H. Chen, F. Tao, X. Fang, and L. Ge. 2013. "A Nanosised Oxygen Scavenger: Preparation and Antioxidant Application to Roasted Sunflower Seeds and Walnuts." *Food Chemistry* 136 (1). Elsevier Ltd: 245–250. doi:10.1016/j.foodchem.2012.07.121.

Navarro, S., E. G. Jay, and J. G. Leesch. 1986. "Recirculation Rate Requirements for Adequate Distribution of Carbon Dioxide in Grain Bins." *Transactions of the ASAE* 29 (5): 1348–1354. doi:10.13031/2013.30321.

Nayak, M. K., G. J. Daglish, T. W. Phillips, and P. R. Ebert. 2020. "Resistance to the Fumigant Phosphine and Its Management in Insect Pests of Stored Products: A Global Perspective." *Annual Review of Entomology* 65: 333–350. doi:10.1146/annurev-ento-011019-025047.

Neethirajan, S., M. S. Freund, D. S. Jayas, C. Shafai, D. J. Thomson, and N. D. G. White. 2010. "Development of Carbon Dioxide (CO_2) Sensor for Grain Quality Monitoring." *Biosystems Engineering* 106 (4): 395–404. doi:10.1016/j.biosystemseng.2010.05.002.

Quezada, M. Y., J. Moreno, M. E. Vázquez, M. Mendoza, A. Méndez-Albores, and E. Moreno-Martínez. 2006. "Hermetic Storage System Preventing the Proliferation of *Prostephanus truncatus* Horn and Storage Fungi in Maize with Different Moisture Contents." *Postharvest Biology and Technology* 39 (3): 321–326. doi:10.1016/j.postharvbio.2005.10.004.

Qu, C., W. Li, Q. Yang, Y. Xia, P. Lu, and M. Hu. 2022. "Metabolic Mechanism of Nitrogen Modified Atmosphere Storage on Delaying Quality Deterioration of Rice Grains." *Food Chemistry: X* 16 (December): 100519. doi:10.1016/j.fochx.2022.100519.

Rangel, M. A. S., and F. A. Villela. 2004. "Conservação de Sementes de Soja Pela Aplicação de Produtos à Base de Ácido Propiônico." *Revista Brasileira de Sementes* 26 (1): 57–61. doi:10.1590/S0101-3122200 4000100009.

Rupollo, G., L. C. Gutkoski, I. R. Martins, and M. C. Elias. 2006. "Efeito Da Umidade e Do Período de Armazenamento Hermético Na Contaminação Natural Por Fungos e a Produção de Micotoxinas Em Grãos de Aveia." *Ciência e Agrotecnologia* 30 (1): 118–125. doi:10.1590/S1413-70542006000100017.

Semple, R. L. 1992. *Towards Integrated Commodity and Pest Management in Grain Storage*. In: REGNET Publication – NAPHIRE/FAO, 526.

Viebrantz, P. C., L. L. Radunz, and R. G. Dionello. 2016. "Mortality of Insects and Quality of Maize Grains in Hermetic and Non-Hermetic Storage." *Revista Brasileira de Engenharia Agrícola e Ambiental* 20 (5): 487–492. doi:10.1590/1807-1929/agriambi.v20n5p487-492.

Wang, W., A. He, S. Peng, J. Huang, K. Cui, and L. Nie. 2018. "The Effect of Storage Condition and Duration on the Deterioration of Primed Rice Seeds." *Frontiers in Plant Science* 9 (February). doi:10.3389/fpls.2018.00172.

Wong-Corral, F. J., C. Castañé, and J. Riudavets. 2013. "Lethal Effects of CO_2-Modified Atmospheres for the Control of Three Bruchidae Species." *Journal of Stored Products Research* 55 (October): 62–67. doi:10.1016/j.jspr.2013.08.005.

16 Role of Active Packaging in Preventing Insect Infestation in Food Products

L. Kavitha, J. Thendral, M. D. Tejeshwini, and S. Anandakumar
National Institute of Food Technology Entrepreneurship and Management, Thanjavur, Tamil Nadu, India

16.1 INTRODUCTION

Insects can lead to food degradation by either creating openings in packaging with their developed mouthparts or by being present in packages, resulting in the food product being initially infested with insects. Packaging materials like polymers, paper, paperboard, textiles, and metal foils are commonly used for food packaging (Navarro et al. 2007). However, a significant issue with these materials is their vulnerability to pests, which can lead to food contamination. The extent of pest infestation in packaged foods is influenced by factors such as the type of pests, the duration of exposure to them, and the environmental conditions (Scheff 2016).

Over the past decade, there has been a significant emergence of new advancements in food packaging technologies, with "active packaging" standing out as one of the most promising techniques. Active packaging refers to a packaging system that has been adapted to uphold or enhance the sensory, safety, and quality characteristics of food. This approach involves integrating active compounds into the packaging materials, which serve the purpose of safeguarding food by providing antioxidant and/or antimicrobial benefits (Licciardello et al. 2013). Numerous studies have been conducted and published on the use of active packaging techniques aimed at preventing insects in food products (Carvalho et al. 2012; Licciardello et al. 2013; Herrera et al. 2018; Divija, Kumar, and Subramanya 2022).

This chapter delves into the critical issue of insect infestations in food products and the role of active packaging in mitigating this problem. It explores various types of active packaging, including oxygen scavengers, vacuum packaging, modified atmosphere packaging (MAP), essential oil (EO), and anti-insect packaging, highlighting their effectiveness in controlling insects and preserving food quality. Additionally, it discusses recent advancements in active packaging technology, such as ozone treatment and biopesticide incorporation, to provide innovative solutions for safeguarding food products from infestation and deterioration.

16.2 INSECT INFESTATION

Insect infestations can result in food deterioration, either due to common insects or stored-product insects that infiltrate packaged items by creating openings in the packaging with their developed mouthparts. In other scenarios, insects can be found in packages, particularly in dry foodstuffs and cereal-based products where larvae are present and not removed during processing, resulting in the food product being initially infested with insects (Singh and Kaur 2018). Furthermore, insect infestations can occur during the transportation and storage of these products in damp storage facilities (Oladejo et al. 2017). Insects can damage food products through various means, such as contaminating, feeding, adding moisture, and introducing fungi and bacteria (Kłyś, Malejky, and Nowak-Chmura 2017). Furthermore, the presence of proteins found in insects' bodies, eggs, excrement, and

DOI: 10.1201/9781003609834-16

secretions can potentially trigger allergic reactions in some individuals. Extensive insect infestation in food products can result in significant losses, including both food damage and financial setbacks for suppliers due to reduced customer satisfaction with their offerings (Arlian 2002).

Insects that pose a threat to packaged food products can be categorized into two groups: "penetrators" and "invaders." Penetrators are insects capable of creating holes in the packaging materials, while invaders are those that enter through pre-existing holes, which can result from mechanical damage, holes made by other penetrating insects, or improper sealing (Navarro et al. 2007). Table 16.1 presents examples of insects capable of infiltrating food packaging. For instance, Riudavets, Salas, and Pons (2007) found that penetrators such as adult *Sitophilus oryzae*, *Rhyzopertha dominica*, and *Lasioderma serricorne* could penetrate polypropylene, polyethylene, and polyester packaging

TABLE 16.1

Example of Penetrators and Invaders in Packaging Material

Penetrators	**Invaders**
Warehouse beetle (*Trogoderma glabrum*)	Merchant grain beetle (*Oryzaephilus mercator*)

Rice weevil (*Sitophilus oryzae*)	Sawtoothed grain beetle (*Oryzaephilus surinamensis*)

Lesser grain borer (*Rhyzopertha dominica*)	Squarenecked grain beetle (*Cathartus quadricollis*)

Cadelle (*Tenebroides mauritanicus*)	Flat grain beetle (*Cryptolestes pusillus*)

Drugstore beetle (*Stegobium paniceum*)	Rice moth (*Corcyra cephalonica*)

Source: Highland (1991).

materials. *L. serricorne* and *R. dominica* adults were more effective at penetrating plastic films than *S. oryzae* adults. In addition, the larvae of *L. serricorne* couldn't produce openings in polyester, and as the material thickness increased, the ability of the mentioned species to penetrate it diminished. Chung et al. (2011) reported that the *Plodia interpunctella* larvae created pinholes in all the tested films, while *Tribolium castaneum* adults were incapable of puncturing any of the film samples. The sharper mandibles of *P. interpunctella* larvae played a significant role in their ability to penetrate films with pinholes.

On the other hand, invaders such as *Oryzaephilus surinamensis* adults typically do not gnaw on plastic to make openings but instead, they access it through pre-existing holes (Adler 2004). Adler (2008) reported that most species inhabiting packaged products are generally incapable of passing through holes smaller than 0.71 mm in width, except for neonate larvae (first instars), which can typically infiltrate polyethylene, polypropylene, and polyvinyl chloride (PVC) through holes with diameters often less than 0.5 mm. This holds particular significance for packages intentionally punctured to facilitate gas exchange during fumigation or the controlled/modification of atmospheric conditions within storage areas. Indeed, all packaged goods incorporate openings to facilitate pressure variations, whether due to changes in elevation or during transportation. Furthermore, openings in packages are designed to enable efficient stacking on pallets or store shelves. Stitched packages offer accessible entry points for stored-product insects, while insects can exploit areas lacking continuous glue seals as pathways for infestation (Athanassiou, Riudavets, and Kavallieratos 2011).

Beyond packaging concerns, it's crucial to eradicate infestations in transportation vehicles and retail stores to prevent the spread of infestations into packaged products. Numerous instances of infestation in packaged goods can be traced back to infestations in transportation vehicles and retail establishments.

16.3 OVERVIEW OF ACTIVE PACKAGING

Packaging is a technique employed to safeguard food products, shielding them from external factors, including physical, chemical, and biological elements, with a particular focus on preventing insect infestation. Effective packaging is essential to maintain the quality of processed food items, ensuring they remain undamaged until they reach the end consumer. The primary aim of packaging is to protect food materials from factors that can lead to spoilage, such as oxygen, water vapor, and various biological and chemical elements and insect eradication, which can be achieved through the utilization of active packaging technology (Prasad and Kochhar 2014). Active packaging refers to "packaging in which subsidiary components are intentionally integrated into either the packaging material itself or the headspace within the package to improve the overall performance of the packaging system" (Janjarasskul and Suppakul 2018). Active packaging plays a significant role in managing the presence of spoilage-inducing insects. This can be achieved through two main approaches: altering the atmospheric conditions around the product by introducing gases like carbon dioxide or employing physical and chemical methods (Navarro 2012). A key benefit of active packaging is the reduction in the need for potent and potentially harmful pesticides or insecticides that can pose risks to human health when ingested or inhaled. To address such concerns, innovative active packaging methods have been developed. These methods involve incorporating insect repellents directly into the packaging materials, utilizing sachets, or incorporating natural repellents obtained from natural sources (Riudavets et al. 2007).

16.4 TYPES OF ACTIVE PACKAGING

Active packaging enhances product quality and safety by interacting with the product or environment. Figure 16.1 describes the types of active packaging used for insect infestation. Active packaging systems can be broadly categorized into two main groups: emitters and absorbers. Emitters are designed to release specific substances into the packaging in order to maintain a safe environment

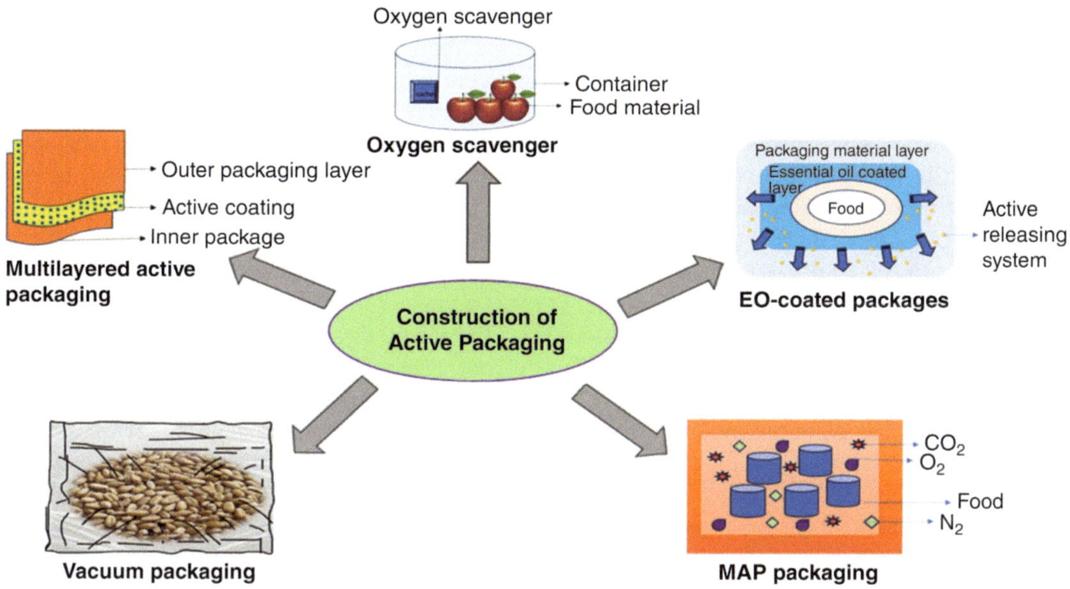

FIGURE 16.1 Construction of active packaging.

for the product being stored. Absorbers, on the other hand, are responsible for removing undesirable components like oxygen, moisture, carbon dioxide, ethylene, antioxidants, and moisture from the packaging (Chauhan, Chauhan, and Chauhan 2022).

Furthermore, active packaging systems can also be classified based on their physical form within the packaging. They can be implemented as sachets or pouches, or they can take the form of films integrated into the packaging (Prasad and Kochhar 2014). The types of active packaging involved in preventing insect infestation from food packaging are described in the following sections.

16.4.1 Oxygen Scavengers

Packaged food products commonly include a certain volume of headspace gases, alongside entrapped oxygen. Concerns arise regarding the permeation of oxygen through plastic containers. Molecular oxygen (O_2) has the capacity to undergo reduction through the acquisition of electrons, yielding various intermediate forms, namely hydroxyl radicals, superoxide, and hydrogen peroxide. These intermediates, with the exception of water, exhibit high reactivity. Their free radical nature imparts an autocatalytic quality to the oxidative reactions in which they actively participate (Prasad and Kochhar 2014).

Undesirable oxygen can enter food packaging due to inadequate evacuation, presence in food or packaging materials, poor sealing, or micro-perforations. Excess oxygen diminishes food quality and shelf life by promoting the deterioration of food products. Additionally, it encourages microbial growth and insect infestation. To address this, oxygen in packaging can be eliminated through vacuum sealing or by using inert gases like nitrogen (N_2) or carbon dioxide (CO_2), or a combination of both (Mohan, Ravishankar, and Srinivasagopal 2008; Prasad and Kochhar 2014). Removing oxygen during packaging is crucial, and managing the remaining oxygen levels with the help of oxygen-absorbing materials is essential to slow down the degradation and spoilage of food products. Ascorbic acid, iron, catechol, and low-molecular-weight ingredients are the oxygen scavengers that are commercially available (Tarr and Clingeleffer 2005). These techniques can bring down oxygen levels to less than 0.01%, significantly lower than the usual range of 0.3%–3.0% achieved in MAP packaging with residual oxygen (Sen, Mishra, and Srivastav 2012).

In a study conducted by Divija et al. (2022), it was found that lowering oxygen content can increase the mortality rate of *T. castaneum* beetles in small millet rice. The study used four different pouch types (HDPE, LDPE, Ecotact, and Swiss Pac multi-layered pouches), along with oxygen absorbers and beetle specimens. The results showed that Ecotact and Swiss Pac multi-layered pouches with oxygen scavengers maintained zero oxygen levels, leading to 100% beetle mortality. These multi-layered pouches were identified as the most effective packaging for safely storing foxtail millet grains by creating an oxygen-free environment that killed the beetles. Similarly, Tarr and Clingeleffer (2005) evaluated the impact of an oxygen absorber (Ageless®) on the mortality of red flour beetles in sealed packs of golden raisins at different temperatures and durations. The results showed that when O_2 absorbers were used, there was complete mortality of beetles, preventing a second-generation population from being stored at 27°C for 60 days. In hermetically sealed bags, 100% adult mortality occurred at 22.5°C and 30°C, but eggs and pupae survived. The study also examined the effect of a low O_2 atmosphere on fruit color during storage. At 15°C and with a low O_2 atmosphere, fruit color remained stable over 84 days of storage. However, at higher temperatures (30°C) and longer storage durations without a modified atmosphere, there were significant negative effects on fruit color.

Kucerova et al. (2014) compared vacuum treatment (V) and vacuum treatment with an oxygen absorber (V + A) to control *Sitophilus granarius* L. infestations in grains at different temperatures (15°C and 25°C) and durations. V + A was generally more effective, particularly at lower temperatures. After three weeks at 15°C, V + A achieved a 99.8% reduction in hidden infestations, while V alone took 4 weeks to achieve a 95.8% reduction.

16.4.2 Vacuum Packaging

Vacuum packaging is a process where air is removed from a product pouch and then hermetically sealed. This method helps extend the product's storage or shelf life by preventing insect infestation and the growth of microorganisms. It also enhances hygiene by reducing the risk of cross-contamination and preserves the product's flavor while protecting it from dehydration and weight loss (Meena 2022). There is a limited number of practical studies on the impact of vacuum on major cereal seed pests. A laboratory study conducted by Kučerová et al. (2013) focused on *S. granarius* and found that applying a high vacuum of 7.6 mm Hg (equivalent to 1 kPa of absolute pressure) to wheat stored in aluminized multi-layer barrier pouches resulted in a lethal exposure time to achieve 99% mortality (LT_{99}) of 160 hours at 25°C and 275 hours at 15°C. However, this vacuum level seems impractical for common use, except for very long-term seed storage banks aimed at preserving seed viability. Kumar et al. (2020) found that insects generally exhibit reduced metabolic rates when subjected to negative pressures of 50–100 mm Hg (6.7–13.4 kPa) during vacuum hermetic fumigation.

Sarazin et al. (2023) examined the vacuum storage's impact on cereal pests (*S. oryzae*, *S. granarius*, and *R. dominica*). Infested wheat seeds were vacuum-sealed in bags at 30, 50, and 70 kPa depression levels. These bags were stored at an average temperature of 25.8°C, and insect survival and offspring production were monitored at regular intervals after opening. The time required for a 99% reduction in the survival rate and offspring production (LT_{99}) was determined for each pest species. *S. oryzae* and *S. granarius* achieved LT_{99} at 70 kPa after approximately 37.5 and 31 days of exposure, respectively. In contrast, *R. dominica* was notably more resistant, even producing offspring under vacuum conditions.

16.4.3 Modified Atmospheric Packaging (MAP)

Modified atmosphere (MA) provides a safe and environmentally friendly alternative to traditional chemical fumigants for managing insect pests in stored grains, processed goods, oilseeds, and packaged foods without producing residues. MA storage involves maintaining a specific gas composition within a storage enclosure, typically comprising elevated levels of nitrogen (N_2) carbon dioxide (CO_2) and reduced levels of oxygen (O_2), either under standard atmospheric pressure conditions or under altered pressure conditions as required (Navarro 2012). This controlled atmospheric environment is sustained for a specified duration to effectively manage storage pests and uphold the overall quality of the stored commodity. When grains are exposed to these MA conditions, it induces significant changes in the physiological conditions of insects, leading to their mortality. This is because storage grain pests, being aerobic organisms dependent on oxygen, face restricted access to oxygen due to the altered gas composition (Navarro 2006). Environments rich in CO_2, comprising more than 40% of the air composition, have demonstrated a swifter capacity to manage pests compared to atmospheres having higher nitrogen content. CO_2 primarily influences insects by causing a constant opening of their spiracles, which govern their breathing, leading to substantial water loss and desiccation. Furthermore, CO_2 induces acidification of internal insect fluids, alters various metabolic pathways, and creates an anesthetic-like effect by changing the permeability of neuron cell membranes (Riudavets et al. 2009; Riudavets et al. 2018). The utilization of modified atmosphere packaging (MAP) with elevated CO_2 levels (exceeding 50%) and minimal remaining O_2 concentrations (below 3%) has been proven to be efficient in managing *Sitophilus zeamais* (Riudavets et al. 2009). Table 16.2 provides information about the MAP application on insect control in food products.

Pascua et al. (2021) conducted a study on the impact of MA conditions on *Callosobruchus maculatus* egg development in mung beans. Eggs were exposed to different gases (air, 100% CO_2, 100% N_2, vacuum) for 48 hours in sealed bags. The results indicated that the low oxygen level in MA led to delayed egg-to-adult emergence, reduced adult emergence rates (especially in vacuum), and shorter adult lifespans. Furthermore, scanning electron microscopy (SEM) showed that low-pressure conditions disrupted the egg structure of *C. maculatus*. MA also affected biomolecules in the eggs, impacting insect growth, with vacuum having the most significant effect, followed by 100% N_2 and 100% CO_2.

TABLE 16.2
MAP on Insect Infestation on Food Products

MAP + Other Combinations	Gas Composition	Insect	Food Products	Mortality Rate (%)	Reference
MAP + Temperature (4°C)	85% CO_2 + 3% O_2 + 12% N_2 75% CO_2 + 12% O_2	Sawtoothed grain beetle (*Oryzaephilus surinamensis*)	Sayer date fruits	100	Dehghan-Shoar, Hamidi-Esfahani, and Abbasi (2010)
MAP	1. 80% CO_2 + 20% O_2 2. 60%CO_2 + 20% O_2 + 20% N_2	Rice weevil (*Sitophilus* spieces)	Maize grain	98	Carli et al. (2010)
MAP + Low dose radiation	1. H34M Bags – 7.8% O_2 + 8.7% CO_2 2. MAP (H34M) + Irrad (150Gy)	Fruit fly (*Bactrocera dorsalis*)	Mango	21 100	Srimartpirom et al. (2018)
MAP	90% CO_2, 5% O_2, and 0	Maize weevil (*Sitophilus zeamais*)	Maize	87	Riudavets et al. (2018)
MAP	1. 50% CO_2 + 10% O_2 + 40% N_2 2. 70% CO_2 +6% O_2 + 24% N_2 3. 90% CO_2 +3% O_2 + 7% N_2	Lesser grain borer (*Rhyzopertha dominica*)	Chickpeas	96.4 100 100	Iturralde-García et al. (2020)

16.4.4 Anti-Insect Packaging

Infestations can occur not only during the processing and warehousing stages but also at various points along the supply chain, including during transportation, while products are on store shelves, and even in consumers' homes (Kim et al. 2013). One effective strategy for preventing infestations and reducing quality-related risks is to develop packaging that is resistant to insect attacks, commonly referred to as anti-insect packaging. Such packaging needs to effectively address two primary types of insect threats: penetrators and invaders (Vachon et al. 2020). Insect resistance varies among different materials, and a study conducted a ranking of various packaging materials based on their susceptibility to insect penetration. Polyethylene terephthalate, polyester, and polycarbonate nylon plastics were found to be highly effective in preventing insect infestations. On the other hand, polyvinyl chloride, thick polyethylene, polyamide, and cellulose acetate displayed some level of resistance to insects. Whereas, polylactic acid, acrylonitrile, and thinner polyethylene (~125 μm) were identified as materials that were vulnerable to insect penetration. Ethyl vinyl acetate, kraft paper, and thin PE films (less than 100 μm) exhibited no resistance to insects (VanRyckeghem 2011; Mullen, Vardeman, and Bagwell 2012; Hassan et al. 2016). Table 16.3 describes the application of anti-insect packaging in insect invasion.

Debnath et al. (2011) conducted a study to determine the effectiveness of surface-functionalized silica nanoparticles (SNP) were at controlling the rice weevil *S. oryzae* when compared to bulk-sized silica particles (particles >1 μm). The results showed that SNP was highly successful in controlling this insect pest, causing more than 90% mortality. The study aimed to develop biodegradable coatings incorporating propionic acid as a repellent to control insect pests in cereal product packaging. Results showed that mono and multi-layer coatings of zein, corn, and polycaprolactone containing

TABLE 16.3
Application of Anti-Insect Packaging in Prevention of Insects

Prevention of Insect	Compound Used in Packaging Material	Concentration and Formulation	Application Mode and Type of Package	References
Sitophilus oryzae, *Plodia interpunctella*	Lignin	HDPE and lignin (2%–5%) powders were dry blended and let dry at 70°C in vacuum oven + extruded	High density polyethylene (HDPE)	Vachon et al. (2020)
P. interpunctella, *Trogoderma variabile*	Methoprene	0.1%, 0.25%, and 0.5%	Impregnated on foil type packaging material	Scheff (2016)
Tribolium confusum	Nano silica	6–12 ppm	Incorporate along with polyethylene and cellophane bags with thickness 4 mm was treated with 3 cm^3 of nano silica	El sayed, Rizk, and Sayed (2023)
Callosobruchus chinensis	–	Micro bags (having 14 × 19 cm dimensions) with a capacity of 200 g	Used plastic material like Polyethylene	Hanif et al. (2023)
Tribolium castaneum, *T. variabile*	Methoprene (Outer later polypropylene, inner layer woven fabric)	0.1% or 1000 ppm per area	Methoprene impregnated and untreated woven packaging materials	Scheff (2016)

propionic acid significantly reduced insect infestation compared to uncoated cartons, while coating materials without propionic acid had no effect on insect entry (Germinara et al. 2010).

16.4.5 Active FIBC Bags for Grain Storage

Flexible Intermediate Bulk Containers (FIBC) bags are now used in different food and agro-processing industries for different packaging and storage applications. The type of FIBC bag will be selected in the industries based on the shipment, storage condition, and shelf life of products. FIBC bags are available in the form of coated and non-coated. The coated FIBC bags are manufactured with different laminated layers of food grade 100% polypropylene and polyethylene liner to provide a moisture barrier, insect, and dust protection and good storage stability. There are four types of FIBC bags available such as type A, B, C, and D. The coated FIBC bags are used in transporting and storage of dry and granular materials such as flour, salt, sugar, pulses, cereals, and milled products. Similarly, uncoated FIBC bags are made without any laminated layers. The uncoated FIBC bags have more air passage micro vents used for breathable and ventilated fresh commodities. These bags are mainly used in the packaging of fresh fruits and vegetable products such as onion, potatoes, sugar beets, nuts, garlic, coffee beans, and others, which prevent moisture accumulations in grain bags and support quality maintenance. The different sizes of FIBC bags with different shapes, loop construction, top loading, and bottom distributions are available, as shown in Figure 16.2.

Active FIBC bags such as flame-retardant bags, tamper-proof bags, and UV-resistant bags are used for different purposes in the processing industries. These bags have higher strength and durability,

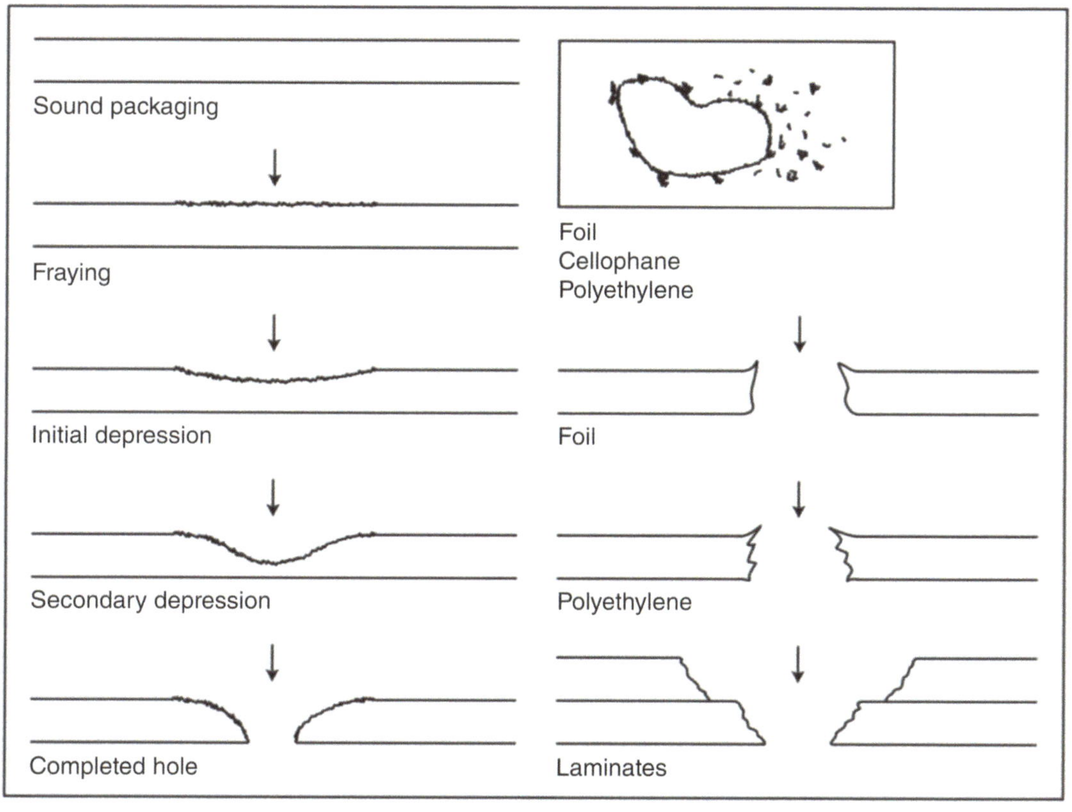

FIGURE 16.2 Penetration of insects into food packaging materials. (Adapted from Mullen and Mowery [2006].)

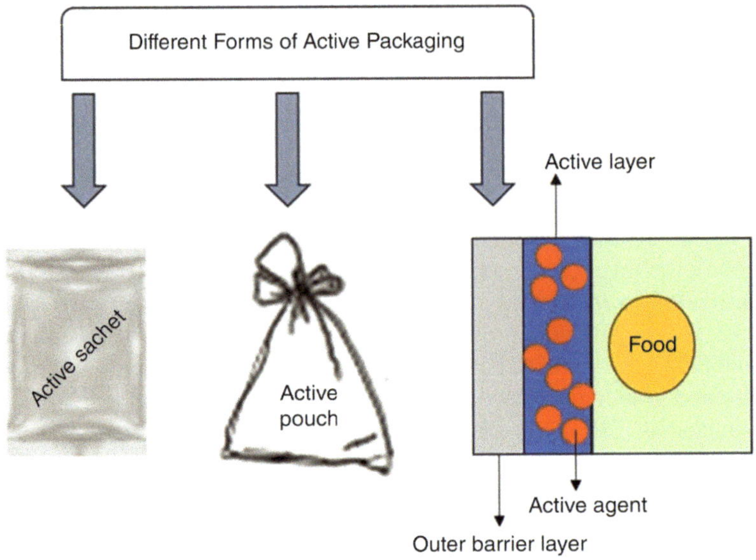

FIGURE 16.3 Different forms of active packaging.

can be reusable multiple times, and are made by recyclable plastic materials as environmentally friendly. The capacity of the bags varies from 500 to 2000 kg (Bradna et al. 2019) reported that the biological activity of the stored grains mainly depends on the storage temperature of grains, relative humidity, and moisture content. It was observed that no significant changes in the quality of wheat grain during two years of storage at 1-tonne capacity FIBC bags (0.9 × 0.9 × 1.1 m). However, the stable temperature was not maintained during the storage period at different grain bed depths. Hence, thermal insulated and shaded warehouses or a combination can be used for controlling quality losses.

16.4.6 Essential Oil-Based Packaging

Essential oils (EOs) are a complex blend of volatile organic compounds produced by plants as secondary metabolites. These compounds encompass hydrocarbons like terpenes and sesquiterpenes alongside oxygenated substances such as esters, aldehydes, phenols, alcohols, lactones, ketones, ethers, and phenol esters (Tavares et al. 2018). These volatile compounds are released by plants when they experience stress or attacks. These EOs are derived from different plant parts that contain oil glands, like the pulp, leaves, stems, and roots. EOs, such as natural insect repellents, are preferred for integration into food packaging due to their ease of extraction from plants (Batish et al. 2008). EOs, known for their anti-insect properties, act as insecticides by targeting the nervous system. Onion (*Allium cepa* L.) and garlic (*Allium sativum* L.) are among the most popular herbs globally and have Eos that effectively act against *P. interpunctella* larvae (Lee et al. 2017).

Na, Hong, and Ryoo (2008) studied the incorporation of cinnamon extract into the packaging materials of chocolate and the adhesive that is used to attach stickers to chocolate packages. The result showed that this approach effectively repelled *P. interpunctella*. Chocolate packages treated with an adhesive containing either 0.02% or 0.025% of cinnamon extract remained free from infestations by the Indian meal moth for a period of 60 days when stored at 28.1°C and 70%–75% relative humidity. Similarly, Nazurah et al. (2022) reported that the chitosan films containing 0.5% curry leaves Eos effectively deterred around 60%–70% of *Musca domestica* in a preference test conducted in the area. Furthermore, films enhanced with 1.0% curry leaves EOs, and 0.3% Tween 80 exhibited complete repellency, ensuring a 100% deterrence rate. Table 16.4 provides information about the role of EO packaging in insect infestation.

TABLE 16.4
Essential Oil (EO) Based Packaging on Insect Infestation on Food Products

Source of EOs	Repellent to Insect	Concentration and Formulation	Mode of Application	Efficiency (%)	Reference
Thyme oil	*Plodia interpunctella*	37% formaldehyde + melamine (formaldehyde/melamine molar ratio: 3.7) + 100 mL H_2O + 7 g/100 mL thyme oil + 0.15% PVA	Microcapsule film	90	Chung et al. (2013)
Cinnamon	*P. interpunctella*	2 g PVA + 98 g distilled H_2O + 5.5g CO + 0.07 (Tween 80)	Microcapsule film	95	Jo et al. (2015)
Citronella	*Tribolium castaneum*	(Kaolin clay + 14% latex binder + calcium stearate + polyacrylate) – precoat + 90% ethanol, citronella	Coating on carton board	50	Wong et al. (2005)
Rosemary Citronella Oregano	*T. castaneum*	Rosemary oil 4% on polypropylene (30 µm thickness)	Coating on film	80 50 60	Licciardello et al. (2013)
Ginger Garlic Onion Black Pepper Fennel	*P. interpunctella*	5% (w/w) rice flour + 2% glycerol (plasticizer) + 1% Tween 20 (emulsifier) + distilled water + (2% onion/5% garlic/other mentioned EO)	Multi-layered film	88 77 8171 45	Lee et al. (2017)
Neem	*Sitophilus oryzae* *T. castaneum* *Cryptolestes ferrugineus*	Defatted pea protein extract or DEET (>95%) + 3% neem EO	Coated on film	86–90	Hou, Fields, and Taylor (2004)
Sweet basil and lemon grass	*T. castaneum* *Rhyzopertha dominica*	50:50 ratio of sweet basil + lemon grass (1% w/w) + 10 g starch/100 mL	Coated and covered by another layer	96	Utono, Coote, and Gibson (2014)

In another study, encapsulated *Eucalyptus globulus* Labill EO, rich in terpenic compounds like 1,8-cineol and p-cymene, was found to be toxic to larvae of the Mediterranean flour moth (*Ephestia kuehniella* Zeller). This encapsulation method improved both its insecticidal activity and its persistence (Ebadollahi et al. 2022). Kavallieratos et al. (2021) evaluated the effectiveness of Tansy (*Tanacetum vulgare* L.) EOs as a wheat protectant against four harmful stored-product insect species at two concentrations, 500 and 1000 ppm. The findings revealed that *T. vulgare* EO caused complete mortality (100%) in *T. castaneum* larvae but only 25.6% mortality in adults at 1000 ppm after six and seven days. *Tribolium confusum* experienced 56.7% larval and 8.9% adult mortality at 1000 ppm after seven days. *Tenebrio molitor* adults showed a 52.2% mortality rate, while larvae had a lower 8.9% mortality at 1000 ppm. *O. surinamensis* larvae had a high mortality rate of 93.3% at 1000 ppm after seven days, with only 13.3% of exposed adults affected at the same concentration and duration.

16.5 RECENT ADVANCEMENTS IN ACTIVE PACKAGING AGAINST INSECTS' INFESTATION

New developments in active packaging technology have yielded innovative solutions to combat insect infestations in various products. These advancements leverage various strategies to deter or eliminate insects, ensuring the quality and safety of stored goods. Ozone gas is a suitable choice for disinfection and decontamination because it leaves no residual byproduct and is recognized as generally recognized as safe (GRAS). Muniswamy et al. (2023) examined the effectiveness of ozone treatment on proso millets, which are known to be susceptible to storage pests such as *T. castaneum*. Ozone treatment was tested on proso millet with varying moisture levels and ozone concentrations against different life stages of *T. castaneum*. The most effective treatment was 750 ppm ozone at 8% moisture, with lethal times of 268 minutes for adults, 475 minutes for pupae, 160 minutes for larvae, and 679 minutes for eggs. Ozone had no significant impact on millet's protein, fat, and ash content.

Mishra et al. (2019) explored the use of ozone (O_3) treatment to control the common grain pest *R. dominica* in stored wheat. The optimal ozone treatment conditions were found to be 12% grain moisture, 2.5 g/m^3 ozone concentration, and an 8-hour treatment duration. This treatment resulted in a high mortality rate of 97% for adult *R. dominica*, with 100% mortality for pupae, 99% for larvae, and 100% for eggs. Ozone treatment reduced the moisture and protein content in wheat compared to infested wheat. SEM revealed visible microstructural changes in treated samples, while molecular-level changes were minimal. Ozone was found to be an effective disinfection method against all life stages of *R. dominica* in stored wheat, leaving no residue. Rotation strategies could enhance mortality while maintaining grain usability for various purposes.

Biopesticides are often inherent to the natural defense mechanisms of various plant species. They typically exhibit strong selectivity against specific pests, pose minimal toxicity, and are biodegradable. They are employed to safeguard seeds and crops, as insect infestations during transportation and storage can lead to significant economic losses (Goñi et al. 2017). Biopesticides can be incorporated into packaging materials to establish an insect-repelling effect and prevent substantial alterations in the grain composition. While this approach can effectively eliminate adult insects, it's important to acknowledge that the deceased insects will remain within the packaging. The insecticide is gradually discharged from the packaging over a specified period.

Soujanya (2018) suggested an innovative approach involving the placement of a biopesticide between two layers of plastic material within a double-layered bag. Importantly, the biopesticide, which consisted of *Tinospora cordifolia* leaf powder, was positioned in such a way that it did not come into direct contact with the rice. The research findings indicated favorable results in terms of efficacy, thereby endorsing the idea of adopting a more diverse and expanded use of biopesticides as a viable method for pest control. Similarly, de Fátima Silva et al. (2022) documented some fungicidal effects resulting from the use of chitosan-coated paperboard. However, their most effective outcome in preventing insect infestation was found to be less than 80%. The development of active packaging that incorporates biopesticides represents an innovative technology for preserving food, as it offers protection against various threats like fungi, insects, repellents, and herbicides. Herrera et al. (2018) demonstrated a bioactive material designed for safeguarding stored grains was created by adding 1-octen-3-ol to low-density polyethylene films. This incorporation of the biopesticide was achieved using a high-pressure cell, magnetic stirring, and a high-pressure impregnation system assisted by supercritical CO_2. The films demonstrated strong insecticidal activity against *S. zeamais*, resulting in 100% mortality within 24 hours.

16.6 CONCLUSION

In conclusion, active packaging plays a crucial role in safeguarding the quality and safety of food products by addressing the threat of insect infestations. Oxygen scavengers, vacuum packaging, and MAP are effective methods for controlling insect populations and extending the shelf life of various food items. Recent advancements, such as ozone treatment and biopesticide incorporation into packaging materials, offer innovative solutions to combat insect infestations without leaving harmful residues. These developments emphasize the importance of ongoing research and innovation in active packaging to protect both food quality and economic interests. Overall, active packaging emerges as a promising technique to mitigate food deterioration and insect-related losses in the food industry.

REFERENCES

Adler, C. 2004. "Significance of Hermetic Seals, Controlled Ventilation, and Wire-Mesh Screens to Prevent the Immigration of Stored Product Pests." *IOBC/WPRS Bulletin* 27 (9). Citeseer: 13–16.

Adler, C. 2008. "Insect-Proof Packaging to Avoid Stored Product Insects." *IOBC/WPRS Bulletin* 40: 363–369.

Arlian, L. G. 2002. "Arthropod Allergens and Human Health." *Annual Review of Entomology* 47 (1). Annual Reviews: 395–433. doi:10.1146/annurev.ento.47.091201.145224.

Athanassiou, C. G., J. Riudavets, and N. G. Kavallieratos. 2011. "Preventing Stored-Product Insect Infestations in Packaged-Food Products." *Stewart Postharvest Review* 3 (8). Stewart Postharvest Review: 1–5.

Batish, D. R., H. P. Singh, R. K. Kohli, and S. Kaur. 2008. "Eucalyptus Essential Oil as a Natural Pesticide." *Forest Ecology and Management* 256 (12): 2166–2174. doi: 10.1016/j.foreco.2008.08.008.

Bradna, J., J. Šimon, D. Hájek, D. Vejchar, I. Polišenská, and I. Sedláčková. 2019. "Comparison of a 1 t and a 55 t Container When Storing Spelt Grain in Mild Climate of the Czech Republic." *Agronomy Research* 17 (6): 2203–2210. doi:10.15159/AR.19.200.

Carli, M. D., B. Bresolin, C. P. Z. Noreña, I. Lorini, and A. Brandelli. 2010. "Efficacy of Modified Atmosphere Packaging to Control Sitophilus spp. in Organic Maize Grain." *Brazilian Archives of Biology and Technology* 53 (6): 1469–1476. doi:10.1590/S1516-89132010000600024.

Carvalho, M. O., I. Pires, A. Barbosa, G. Barros, J. Riudavets, A. C. Garcia, C. Brites, and S. Navarro. 2012. "The Use of Modified Atmospheres to Control *Sitophilus zeamais* and *Sitophilus oryzae* on Stored Rice in Portugal." *Journal of Stored Products Research* 50. Elsevier: 49–56.

Chauhan, U., E. S. Chauhan, and P. Chauhan. 2022. "Role of Active Packaging in the Food Industry: A Review." *Journal of Pharmaceutical Research International* 34 (50B): 20–32.

Chung, S. K., J. Y. Seo, J. H. Lim, H. H. Park, Y. T. Kim, K. H. Song, S. J. Park, S. S. Han, Y. S. Park, and H. J. Park. 2011. "Barrier Property and Penetration Traces in Packaging Films against *Plodia interpunctella* (Hübner) Larvae and *Tribolium castaneum* (Herbst) Adults." *Journal of Stored Products Research* 47 (2): 101–105. doi: 10.1016/j.jspr.2011.01.005.

Chung, S. K., J. Y. Seo, J. H. Lim, H. H. Park, M. J. Yea, and H. J. Park. 2013. "Microencapsulation of Essential Oil for Insect Repellent in Food Packaging System." *Journal of Food Science* 78 (5). Wiley Online Library: E709–E714.

de Fátima Silva, M., V. B. V. Maciel, A. P. R. Noletto, A. C. Venturini, R. A. de Carvalho, and C. M. P. Yoshida. 2022. "Chitosan Active Coating on Paperboard Surface Forming an Anti-Insect Grain-Based Food Packaging." *Packaging Technology and Science* 35 (4). Wiley Online Library: 361–372.

Debnath, N., S. Das, D. Seth, R. Chandra, S. C. Bhattacharya, and A. Goswami. 2011. "Entomotoxic Effect of Silica Nanoparticles against *Sitophilus oryzae* (L.)." *Journal of Pest Science* 84. Springer: 99–105.

Dehghan-Shoar, Z., Z. Hamidi-Esfahani, and S. Abbasi. 2010. "Effect of Temperature and Modified Atmosphere on Quality Preservation of Sayer Date Fruits (Phoenix Dactylifera L.)." *Journal of Food Processing and Preservation* 34 (2). Wiley Online Library: 323–334.

Divija, S. D., H. D. K. Kumar, and S. Subramanya. 2022. "Oxygen Scavenger: Promising Tool for the Management of *Tribolium castaneum* (Herbst) in Small Millet Rice." *International Journal of Tropical Insect Science* Springer: 16.

Ebadollahi, A., J. J. Sendi, W. N. Setzer, and T. Changbunjong. 2022. "Encapsulation of *Eucalyptus largiflorens* Essential Oil by Mesoporous Silicates for Effective Control of the Cowpea Weevil, *Callosobruchus maculatus* (Fabricius) (Coleoptera: Chrysomelidae)." *Molecules* 27 (11). MDPI: 3531.

El sayed, T. S., S. A. Rizk, and R. M. Sayed. 2023. "Cellophane Packaging Treated with Nano Silica Is Superior to Polyethylene in Reducing Stored Irradiated Flour from *Tribolium confusum* Infestation." *International Journal of Tropical Insect Science* Springer: 17.

Germinara, G S, A Conte, L Lecce, A Di Palma, and M A Del Nobile. 2010. "Propionic Acid in Bio-Based Packaging to Prevent *Sitophilus granarius* (L.) (Coleoptera, Dryophthoridae) Infestation in Cereal Products." *Innovative Food Science & Emerging Technologies* 11 (3): 498–502. doi: 10.1016/j.ifset.2010.03.001.

Goñi, M. L., N. A. Gañán, J. M. Herrera, M. C. Strumia, A. E. Andreatta, and R. E. Martini. 2017. "Supercritical CO_2 Iof LDPE Films with Terpene Ketones as Biopesticides against Corn Weevil (*Sitophilus zeamais*)." *The Journal of Supercritical Fluids* 122: 18–26. doi: 10.1016/j.supflu.2016.11.017.

Hanif, M., S. Saeed, M. Ali, and M. Younas. 2023. "Efficacy of Packaging Materials against Pulse Beetle *Callosobruchus chinensis*." *Indian Journal of Entomology*: 16.

Hassan, M. W., Gulraize, U. Ali, F. U. Rehman, H. Najeeb, M. Sohail, B. Irsa, Z. Muzaffar, M. S. Chaudhry. 2016. "Evaluation of Standard Loose Plastic Packaging for the Management of *Rhyzopertha dominica* (F.) (Coleoptera: Bostrichidae) and *Tribolium castaneum* (Herbst) (Coleoptera: Tenebriondiae)." *Journal of Insect Science* 16 (1). Entomological Society of America: 91.

Herrera, J. M., N. A. Gañán, M. L. Goñi, J. A. Zygadlo, and R. E. Martini. 2018. "Active LDPE Films Loaded with Biopesticides by Supercritical CO_2-Assisted Impregnation for Stored Grain Protection." *Food Packaging and Shelf Life* 18: 80–86. doi: 10.1016/j.fpsl.2018.03.001.

Highland, H. A. 1991. "Post-Irradiation Protection from Infestation by Insect Resistant Packaging." In *Insect Disinfestation of Food and Agricultural Products by Irradiation: Proceedings of the Final Research Co-Ordination Meeting, Beijing, China*, 51–57. Citeseer.

Hou, X., P. Fields, and W. Taylor. 2004. "The Effect of Repellents on Penetration into Packaging by Stored-Product Insects." *Journal of Stored Products Research* 40 (1). Elsevier: 47–54.

Iturralde-García, R. D., F. J. Wong-Corral, C. Castañé, and J. Riudavets. 2020. "Susceptibility of *Rhyzopertha dominica* to High CO_2 Modified Atmospheres in Packaged Chickpeas." *Journal of Stored Products Research* 85: 101537. doi: 10.1016/j.jspr.2019.101537.

Janjarasskul, T., and P. Suppakul. 2018. "Active and Intelligent Packaging: The Indication of Quality and Safety." *Critical Reviews in Food Science and Nutrition* 58 (5). Taylor & Francis: 808–831.

Jo, H.-J., K.-M. Park, J. H. Na, S. C. Min, K. H. Park, P.-S. Chang, and J. Han. 2015. "Development of Anti-Insect Food Packaging Film Containing a Polyvinyl Alcohol and Cinnamon Oil Emulsion at a Pilot Plant Scale." *Journal of Stored Products Research* 61. Elsevier: 114–118.

Kavallieratos, N. G., A. Skourti, E. P. Nika, P. Mártonfi, E. Spinozzi, and F. Maggi. 2021. "*Tanacetum vulgare* Essential Oil as Grain Protectant against Adults and Larvae of Four Major Stored-Product Insect Pests." *Journal of Stored Products Research* 94: 101882. doi: 10.1016/j.jspr.2021.101882.

Kim, I. H., J. Han, J. H. Na, P. S. Chang, M. S. Chung, K. H. Park, and S. C. Min. 2013. "Insect-resistant Food Packaging Film Development Using Cinnamon Oil and Microencapsulation Technologies." *Journal of Food Science* 78 (2). Wiley Online Library: E229–E237.

Kłyś, M., N. Malejky, and M. Nowak-Chmura. 2017. "The Repellent Effect of Plants and Their Active Substances against the Beetle Storage Pests." *Journal of Stored Products Research* 74. Elsevier: 66–77.

Kucerova, Z., K. Kyhos, R. Aulicky, J. Lukas, and V. Stejskal. 2014, July. "Laboratory Experiments of Vacuum Treatment in Combination with an O_2 Absorber for the Suppression of *Sitophilus granarius* Infestations in Stored Grain Samples." *Crop Protection* 61: 79–83. doi:10.1016/j.cropro.2014.03.018.

Kučerová, Z., K. Kýhos, R. Aulický, and V. Stejskal. 2013. "Low-Pressure Treatment to Control Food-Infesting Pests (*Tribolium castaneum*, *Sitophilus granarius*) Using a Vacuum Packing Machine." *Czech Journal of Food Sciences* 31 (1): 94–98.

Kumar, S., D. Mohapatra, N. Kotwaliwale, and K. K. Singh. 2020. "Efficacy of Sensor Assisted Vacuum Hermetic Storage against Chemical Fumigated Wheat." *Journal of Stored Products Research* 88. Elsevier: 101640.

Lee, S.-H., Y. Chang, J. H. Na, and J. Han. 2017. "Development of Anti-Insect Multilayered Films for Brown Rice Packaging That Prevent *Plodia interpunctella* Infestation." *Journal of Stored Products Research* 72. Elsevier: 153–160.

Licciardello, F., G. Muratore, P. Suma, A. Russo, and C. Nerín. 2013, July. "Effectiveness of a Novel Insect-Repellent Food Packaging Incorporating Essential Oils against the Red Flour Beetle (*Tribolium castaneum*)." *Innovative Food Science & Emerging Technologies* 19. Elsevier: 173–180. doi:10.1016/J.IFSET.2013.05.002.

Meena, M. K. 2022. "Vacuum Packaging as Novel Approach to Extend Storability and Quality of Various Crop Seeds."

Mishra, G., A. A. Palle, S. Srivastava, and H. N. Mishra. 2019. "Disinfestation of Stored Wheat Grain Infested with *Rhyzopertha dominica* by Ozone Treatment: Process Optimization and Impact on Grain Properties." *Journal of the Science of Food and Agriculture* 99 (11). Wiley Online Library: 5008–5018.

Mohan, C. O., C. N. Ravishankar, and T. K. Srinivasagopal. 2008. "Effect of O_2 Scavenger on the Shelf-life of Catfish (Pangasius Sutchi) Steaks during Chilled Storage." *Journal of the Science of Food and Agriculture* 88 (3). Wiley Online Library: 442–448.

Mullen, M. A., and S. V. Mowery. 2006. "Insect-Resistant Packaging." In *Insect Management for Food Storage and Processing*, 35–38. Elsevier. doi:10.1016/B978-1-891127-46-5.50011-8.

Mullen, M. A., J. M. Vardeman, and J. Bagwell. 2012. "12 Insect-Resistant Packaging." *Stored Product Protection* 135.

Muniswamy, K., A. Sugumar, S. Yarrakula, L. Manickam, and D. V. Chidanand. 2023. "Effect of Ozone Fumigation on Controlling Common Storage Pest *Tribolium castaneum* (Herbst) in Proso Millet during Storage." *Ozone: Science & Engineering*. Taylor & Francis" 117.

Na, J.-H., E.-I. Hong, and M.-I. Ryoo. 2008. "Protection of Chocolate Products from Indian Meal Moth by Adding Cinnamon Extract to the Adhesive on the Wrapping" 47 (4): 491495.

Navarro, S. 2006. "Modified Atmospheres for the Control of Stored-Product Insects and Mites." In *Insect Management for Food Storage and Processing*, 105–146. Minnesota: AACC International St. Paul.

Navarro, S. 2012. "The Use of Modified and Controlled Atmospheres for the Disinfestation of Stored Products." *Journal of Pest Science* 85 (3). Springer: 301–322.

Navarro, S., D. Zehavi, S. Angel, and S. Finkelman. 2007. "Natural Nontoxic Insect Repellent Packaging Materials." In *Intelligent and Active Packaging for Fruits and Vegetables*, 201–236. New York: CRC Press.

Nazurah, R. N. F., M. A. Noranizan, M. A. R. Nor-Khaizura, and Z. A. Nur Hanani. 2022. "The Potential of Chitosan-Based Film with Curry Leaf Essential Oil as Natural Insect-Repellent Food Packaging." *Food Packaging and Shelf Life* 34: 100993. doi: 10.1016/j.fpsl.2022.100993.

Oladejo, A. O., H. Ma, W. Qu, C. Zhou, B. Wu, and X. Yang. 2017. "Influence of Ultrasound Pretreatments on Diffusion Coefficients, Texture and Colour of Osmodehydrated Sweet Potato (Ipomea Batatas)." *International Journal of Food Science and Technology* 52 (4): 888–896. doi:10.1111/ijfs.13352.

Pascua, G. F. S, E. R. V. Bayogan, L. Salaipeth, and S. Photchanachai. 2021. "Pretreating *Callosobruchus maculatus* (F.) Eggs in Mung Bean with Modified Atmosphere Conditions Influence Its Adult Emergence and Survival." *Journal of Stored Products Research* 91: 101771. doi: 10.1016/j.jspr.2021.101771.

Prasad, P., and A. Kochhar. 2014. "Active Packaging in Food Industry: A Review." *Journal of Environmental Science, Toxicology and Food Technology* 8 (5): 1–7.

Riudavets, J., C. Castañé, O. Alomar, M. J. Pons, and R. Gabarra. 2009. "Modified Atmosphere Packaging (MAP) as an Alternative Measure for Controlling Ten Pests That Attack Processed Food Products." *Journal of Stored Products Research* 45 (2): 91–96. doi: 10.1016/j.jspr.2008.10.001.

Riudavets, J., M. J. Pons, J. Messeguer, and R. Gabarra. 2018. "Effect of CO_2 Modified Atmosphere Packaging on Aflatoxin Production in Maize Infested with *Sitophilus zeamais*." *Journal of Stored Products Research* 77: 89–91. doi: 10.1016/j.jspr.2018.03.005.

Riudavets, J., I. Salas, and M. J. Pons. 2007. "Damage Characteristics Produced by Insect Pests in Packaging Film." *Journal of Stored Products Research* 43 (4): 564–570. doi: 10.1016/j.jspr.2007.03.006.

Sarazin, J. B., C. Cathala, and F. Fleurat-Lessard. 2023. "Disinfestation of Wheat Seed from *Sitophilus granarius* (L.), *S. oryzae* (L.), and *Rhyzopertha dominica* (F.) by Packing under Vacuum in High-Barrier Multi-Layered Plastic Bags." *Journal of Stored Products Research* 100: 102045. doi: 10.1016/j.jspr.2022.102045.

Scheff, D. S. 2016. *Packaging Technologies for the Control of Stored-Product Insects*. Kansas State University.

Sen, C., H. N. Mishra, and P. P. Srivastav. 2012. "Modified Atmosphere Packaging and Active Packaging of Banana (Musa spp.): A Review on Control of Ripening and Extension of Shelf Life." *Journal of Stored Products and Postharvest Research* 3 (9). Academic Journals: 122–132.

Singh, B., and A. Kaur. 2018. "Control of Insect Pests in Crop Plants and Stored Food Grains Using Plant Saponins: A Review." *LWT* 87. Elsevier: 93–101.

Soujanya, P. L. L. 2018. "Biopesticide Treated Double Layered Bags: Novel Method of Application of Botanicals for *Sitophilus oryzae* L. Management in Stored Maize." *Maydica* 63 (1): 6.

Srimartpirom, M., I. Burikam, W. Limohpasmanee, T. Kongratarporn, T. Thannarin, A. Bunsiri, and P. A. Follett. 2018. "Low-Dose Irradiation with Modified Atmosphere Packaging for Mango against the Oriental Fruit Fly (Diptera: Tephritidae)." *Journal of Economic Entomology* 111 (1). Oxford University Press US: 135–140.

Tarr, C. R., and P. R. Clingeleffer. 2005. "Use of an Oxygen Absorber for Disinfestation of Consumer Packages of Dried Vine Fruit and Its Effect on Fruit Colour." *Journal of Stored Products Research* 41 (1): 77–89. doi: 10.1016/j.jspr.2004.01.003.

Tavares, M., M. R. Ma. da Silva, L. B. de Oliveira de Siqueira, R. A. S. Rodrigues, L. Bodjolle-d'Almeida, E. P. dos Santos, and E. Ricci-Júnior. 2018. "Trends in Insect Repellent Formulations: A Review." *International Journal of Pharmaceutics* 539 (1): 190–209. doi: 10.1016/j.ijpharm.2018.01.046.

Utono, I. M., C. Coote, and G. Gibson. 2014. "Field Study of the Repellent Activity of 'Lem-Ocimum'-Treated Double Bags against the Insect Pests of Stored Sorghum, *Tribolium castaneum* and *Rhyzopertha dominica*, in Northern Nigeria." *Journal of Stored Products Research* 59. Elsevier: 222–230.

Vachon, J., D. Assad-Alkhateb, S. Baumberger, J. Van Haveren, R. J. A. Gosselink, M. Monedero, and J. M. Bermudez. 2020. "Use of Lignin as Additive in Polyethylene for Food Protection: Insect Repelling Effect of an Ethyl Acetate Phenolic Extract." *Composites Part C: Open Access* 2. Elsevier: 100044.

VanRyckeghem, A. 2011. "Insect Resistant Packaging: The Last Line of Defense." *Fumigants Pheromones* 99 (2).

Wong, K. K. Y., F. A. Signal, S. H. Campion, and R. L. Motion. 2005. "Citronella as an Insect Repellent in Food Packaging." *Journal of Agricultural and Food Chemistry* 53 (11). ACS Publications: 4633–4636.

17 Insect Traps for Monitoring and Management of Stored Product Insects

Suganthy Mariappan, Senguttuvan Kannan, and Mohan Sarma
Department of Agricultural Entomology, Tamil Nadu Agricultural University, Coimbatore, India

17.1 INTRODUCTION

The issue of food security has become a matter of great significance considering the rapid expansion of the global populace, projected to surpass 9.1 billion individuals by the year 2050 (Parfitt, Barthel, and Macnaughton 2010; FAO 2014). Approximately 20–40% of post-harvest losses manifest themselves both in the field and during subsequent post-harvest operations. As per the findings of the World Bank in 2011, it was determined that a significant proportion of these losses, amounting to 55%, transpire during the transportation phase. As per the scholarly findings of Asrar et al. (2016), it has been elucidated that the annual global food grain losses attributable to the pernicious influence of insect infestation exhibit a considerable range, spanning from 10–40%. In accordance with the findings of Sharon, Abirami, and Alagusundaram (2014), it has been determined that a significant proportion, precisely 50%, of post-harvest storage losses can be attributed to the activities of insects and fungi. The presence of insects within stored commodities can give rise to significant detriments in terms of both nutritional value and economic implications. According to Sarwar et al. (2013), it has been observed that whole pulses exhibit a higher degree of vulnerability to pest infestation, with oilseeds and cereal grains following suit in terms of susceptibility. The ability to objectively quantify the quantity and dispersion of the insect populace is of utmost importance in order to ascertain the degree of susceptibility of the agricultural yield. This comprehension would facilitate precise temporal coordination of pest management strategies, thereby circumventing premature and belated implementation. Traps have traditionally served as a means for conducting qualitative assessments or as an initial indicator of insect presence within the context of food grain storage.

To unveil the presence of insects within the confines of stored food cereals, a plethora of detection technologies, both internal and external in nature, have been meticulously devised and refined. Various methodologies employed in the field encompass detection probes, kernel staining, the Berlese funnel method, acoustic approaches, the uric acid method, X-ray imaging, nuclear magnetic resonance imaging, thermal imaging, and solid-phase micro-extraction, as elucidated by Neethirajan et al. (2007). Several of these methodologies are characterized by their protracted nature and exorbitant costs, while also potentially engendering health risks and exhibiting diminished levels of effectiveness. On agricultural premises, it is customary to utilize manual sample traps and equipment, while in the context of grain storage and processing facilities, the prevailing practices involve manual inspection, sifting, and the employment of the Berlese funnel method, as expounded upon by Neethirajan et al. (2007). However, it is important to note that these methodologies are characterized by their inherent sluggishness and limited ability to discern incipient infestations of significant pest species belonging to the Curculionidae, Bostrichidae, Gelechiidae, and Bruchidae lineages. The populations of these species exhibit a magnitude that is tenfold greater in comparison to the independently existing insects, as stated by Fleurat-Lessard (2006).

DOI: 10.1201/9781003609834-17

The primary objective of monitoring the preservation of food grains is to assess fluctuations in the population of insects, their various stages of development, and the extent of infestation as these parameters evolve over a given duration. This methodology not only elucidates the way insects react to various environmental factors but also evaluates the efficacy of pest control strategies. The timely identification of insect infestations in stored commodities is of utmost importance to mitigate potential losses and ensure the preservation of secure storage facilities, thereby facilitating the continuity of agricultural production.

Insect traps have long been introduced into the market as a means of mitigating insect populations in stored commodities, primarily employed for the purpose of detecting insects within wheat stocks situated on agricultural premises. The scholarly works (Loschiavo and Atkinson 1967, 1973; Loschiavo and Smith 1986; Hagstrum, Flinn, and Subramanyam 1998; Wakefield and Cogan 1999) have made significant contributions to the field of study. As per the work conducted by Barak, Burkholder, and Faustini (1990), the development of insect traps for the purpose of effectively managing insect pests in stored food commodities and food processing facilities considered fundamental insect behaviors, including their patterns of movement and flight. Consequently, the efficacy of capturing techniques predominantly hinges upon the dynamic and transient developmental phases of insects.

Traps have proven to be efficacious tools for the purpose of insect sampling due to their inherent convenience, limited temporal and labor requirements, enhanced capacity to discern adult insects (Barak and Harein 1982; Lippert and Hagstrum 1987), and the capability to conduct uninterrupted sampling endeavors over an extended duration. The utilization of trapping to detect novel infestations in large quantities of stored grain, grain repositories, and warehouses is deemed to be a more commendable approach compared to sampling and sieving techniques. This is substantiated by the fact that trapping has the capacity to identify insects at an earlier stage, as supported by the studies conducted by Lippert and Hagstrum (1987); Vela-Coiffier et al. (1997); Hagstrum, Flinn, and Subramanyam (1998); Cogan and Wakefield (1987); Reed et al. (1991).

Numerous classifications of entomological contraptions have been devised subsequent to the year 1990, with the primary objective of surveilling and managing the proliferation of vermin that inflict detrimental consequences upon preserved commodities (Mohan 1993). Traps obviate the necessity for recurrent grain sampling, a task that is characterized by its laborious nature and the consumption of valuable time. This strategy is commonly associated with temporal control measures, such as the implementation of fumigation or the application of residual pesticides. Consequently, the implementation of traps serves as a preventive measure against the contamination of food by pesticides, as evidenced by Wright's scholarly work in 1989. Insects can be ensnared through the construction of traps, which may be designed with or without the inclusion of attractants. Probe traps, pitfall traps, pitfall cone (PC) traps, two-in-one model traps, stack probe traps, sticky traps, refuge traps, electronic grain probe traps, flour traps, and automatic insect elimination containers exemplify a range of physical variations devoid of attractants.

Contrarily, traps that incorporate allurements consist of light traps, traps featuring enticing food, and pheromone traps. In the context of physical traps, the inherent proclivities of insects to seek out sustenance, companionship, or evade threats and adverse circumstances have been ingeniously exploited. As per the scholarly works of Mohan (2004); Rajendran (2005), a diverse array of physical apparatuses have been employed for the purpose of detecting, monitoring, and effectively capturing insects that have been identified in stored commodities.

17.2 INSECT TRAPS WITHOUT ATTRACTANTS

17.2.1 Probe Traps

The pioneering work of Loschiavo and Atkinson (1967) led to the inception of the probe trap, which subsequently underwent various modifications and redesigns by subsequent researchers. A probe

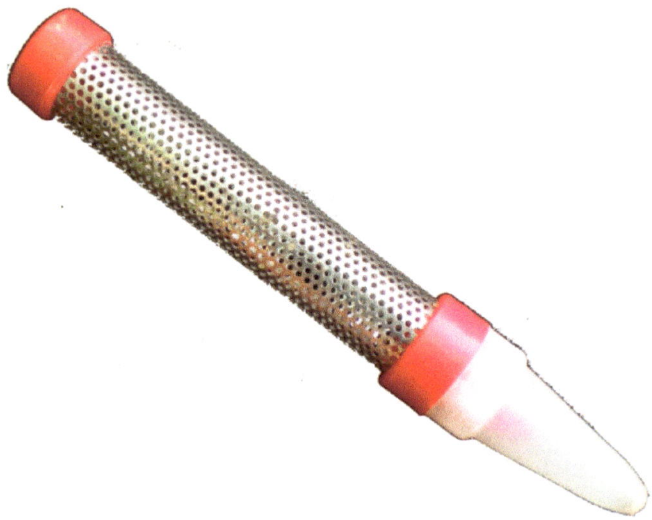

FIGURE 17.1 Probe trap.

trap (Figure 17.1) comprises a plastic cylinder that is punctured with holes of varying diameters, typically ranging from 2.8 to 4.0 mm. The specific hole size is contingent upon the type of grains in which the trap is intended to be employed. These holes are strategically angled downward into the main body of the trap. Within this body, a funnel is present to guide the captured insects toward a removable collecting tube. The trap is strategically positioned within the grain mass in a vertical orientation, where it remains undisturbed for a duration of one week or potentially longer.

The arthropods residing within the lower strata of the grains proceed to traverse the apertures and subsequently descend into the receptacles designed for their capture. The arthropods persist ensnared within the confines of the trap until such time as the trap is extracted from the grain and subjected to inspection. The application of a luon coating, specifically liquid Teflon, encircling the neck of the collecting tube within the probe trap serves as an effective measure to impede the egress of ensnared insects from the said trap, as elucidated by Cogan and Wakefield (1987). The probe trap exhibits a heightened sensitivity toward arthropods of the order Coleoptera, specifically the species *Tribolium* spp., *Cryptolestes* spp., and *Oryzaephilus surinamensis*. These organisms, known for their remarkable agility and propensity to thrive in granaries, pose a significant threat to stored agricultural commodities. Physical or mechanical techniques, such as the utilization of rotational forces, tumbling motions, and the application of impactful forces, have proven to be efficacious in the eradication of stored product insects. The implementation of a mechanical approach for the extraction of insects from grain during the storage phase can be classified as a pivotal strategy within the realm of pest control, as elucidated by Bailey (1962) in his seminal work in 1962.

The propensity for locomotion exhibited by stored product insects has been effectively harnessed by entomologists in their endeavors to create innovative contraptions aimed at capturing these insects (Barak and Harein 1982). Probe traps were employed in the context of bulk-stored grains to effectively detect and ensnare stored product insects, as elucidated by the scholarly works of Loschiavo and Atkinson (1967), as well as Mohan et al. (1994). The efficacy of probe traps in capturing and monitoring *Sitophilus oryzae*, *Rhyzopertha dominica*, and *Tribolium castaneum* has been demonstrated by Mohan (1993).

According to the findings of Mohan et al. (2006), it was observed that the efficacy of the probe traps exhibited an enhancement when the wheat grains were subjected to treatment with protein-enriched pea flour. The utilization of the protein-enriched pea flour's repellent attribute is harnessed to augment the efficacy of the probe trap through the application of flour onto the grain. It has been observed that a greater quantity of *Cryptolestes ferrugineus*, commonly known as rusty grain beetles, were discovered through the utilization of probe traps in wheat that had been treated with protein-enriched pea flour. Additionally, it has been documented that the incorporation of protein-enriched pea flour at various concentrations of 1, 5, and 10% *(w/w)* exhibits a disruptive effect on the upward locomotion propelled by positive geotaxis in *C. ferrugineus* insects. The observed disturbance results in heightened apprehensions in the superior stratum (measuring 6 in.) of processed cereals within the confines of the Tamil Nadu Agricultural University (TNAU) investigative apparatus, as opposed to exceedingly negligible apprehensions (less than 1) in unprocessed receptacles. The utilization of protein-enriched pea flour, specifically within the uppermost 6 in. of the grain stratum, greatly augments the discerning capabilities of probe pitfall traps in the identification of *C. ferrugineus*. The efficacy of this enhancement is heightened in direct correlation with elevated concentrations of protein-enriched pea flour.

17.2.2 Pitfall Traps

Pitfall traps serve as efficacious mechanisms for trapping insects that exhibit activity upon the grain surface and within various strata of grain, thereby functioning as a valuable tool for both monitoring and engaging in mass trapping endeavors. The TNAU model pitfall trap is equipped with a perforated lid and a cone-shaped bottom that gradually narrows into a funnel-shaped trapping tube. In their seminal work, Mohan, Devadoss, and Mahendran (2001) conducted a comprehensive investigation into the efficacy of pitfall traps as a means of mitigating the presence of bruchids in green gram. Their findings revealed that the implementation of pitfall traps yielded a noteworthy reduction in bruchid adults, with a substantial number of 491 insects being removed after a period of five months of storage. Consequently, this intervention led to a discernible decrease in the extent of damage inflicted upon pulses during the storage phase. As the *Callosobruchus* spp., *O. surinamensis*, and *Cryprolestes* spp. traverse the grain, they inadvertently traverse the mesh and enter the trap. In certain instances, it is conceivable that the foundation of a pitfall trap could incorporate a pad that has been permeated with oil, thereby impeding the insects' ability to extricate themselves from the said trap.

17.2.3 Two-in-one Model Trap

The cone-shaped pitfall trap (Figure 17.2) with a perforated lid and a bottom tapering cone, the probe trap with the perforated tube, the pitfall mechanism, and a collection tube were all integrated into one device. Insect trapping efficiency is increased when a pitfall and probe are combined. For pulse beetles, it works best. It does not involve laborious steps like applying adhesive to the inside of the PC before trapping pulse beetles. Since beetles are caught alive in this trap, other insects may be drawn in by the pheromone that is released (Mohan 2019). The PC trap, which combines features from pitfall and probe traps, makes it easier to catch insects that are both on the surface and in the deeper layers of bulk grain. It has been discovered that although this trap is less costly than the probe trap, it is nonetheless equally effective in identifying *O. surinamensis*, *C. ferrugineus*, and *R. dominica* in grain kept in bins.

The effectiveness of the new model was evaluated in a lab setting by Mohan et al. (2008), and it was contrasted with the TNAU pitfall trap of the PC type as well as the conventional TNAU probe trap. A funnel-shaped trapping tube is formed by the bottom cone of a PC-type trap narrowing. With this design, the PC-type trap's requirement to apply a specific sticky coating in order to retain trapped insects is eliminated.

Insects enter a pitfall probe trap through apertures, descend through a void, and are collected in a reservoir beneath the trap from which they cannot escape. The advantages of using traps for insect sampling include their user-friendliness, limited time and personnel requirements, and ability to

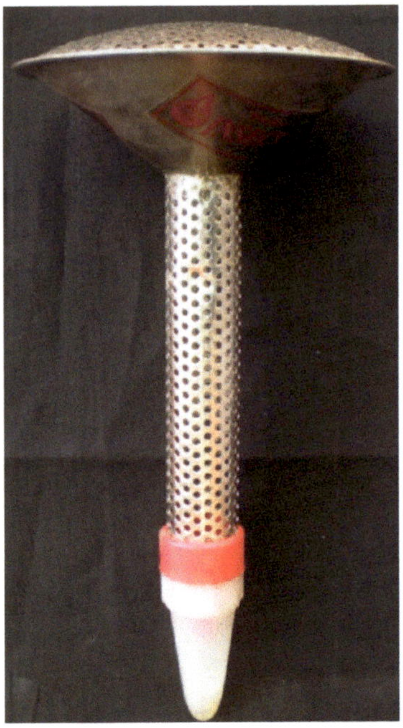

FIGURE 17.2 Two-in-one trap.

detect infestations earlier than grain sampling. The monitoring of *C. ferrugineus* populations using pitfall probe traps is well documented (Loschiavo 1974; White and Loschiavo 1986; White et al. 1990).

17.2.4 Stack Probe Trap

The device known as the TNAU stack probe trap (Indian Patent Number 284727) is constructed of stainless steel and has a main hollow tube with a diameter of between 1.8 and 2.0 cm, equispaced perforations on its upper portion measuring between 1.8 and 2.0 mm, and a bend at one end that ends in a transparent collection unit to catch insects that fall from the bend. The other end of the main tube is closed. When a standard probe trap cannot be utilized to monitor product insects kept in bags, a device called a stack probe trap (Figure 17.3) is utilized. It facilitates insect detection in bag storage without endangering the bags (Mohan et al. 2008).

According to research by Hategekimana, Mohan, and Thirupathi (2013), the TNAU stack probe trap can be used to identify insects in bag stacks of stored grain, verify the efficacy of fumigation, and map the spread of these pests. During the week before to fumigation, the detection rates of *T. castaneum* using trapping and spear sampling were 90 and 17.7%, respectively. The non-baited

FIGURE 17.3 Stack robe trap.

stack probe trap outperformed spear sampling methods before and after fumigation in terms of the number of insects collected and the diversity of those insects. The distribution of stored grain insects showed that *R. dominica* and *T. castaneum* were most common in the uppermost layer (more than 3.5 m above ground) of the bag stacks, while *S. oryzae* and *O. surinamensis* were most common in thc lowest layer (less than 0.5 m above ground). According to a study conducted by Rajesh and Mohan (2016), the most cost-effective way to monitor insect pests in stored turmeric is through the deployment of 16 traps per bag stack of 5 metric tons.

17.2.5 Sticky Traps

It is standard practice to use an adhesive surface to entangle airborne arthropods. In order to identify and capture flying arthropods, such as *R. dominica*, *Ephestia* spp., *Sitotroga cerealella*, and *Plodia interpunctella*, sticky traps that are infused with adhesive materials like petroleum jelly and polybutene gel are utilized (Hagstrum, Dowdy, and Lippert 1994). It is imperative to recognize the intrinsic drawback of sticky traps, namely their impracticability in environments marked by a high concentration of particle matter or widespread infestations. This limitation results from the probability that the sticky surface will be impeded by dust or insect deposition (Phillips, Cogan, and Fadamiro 2000). The application of commercially available adhesive traps resulted in a lower number of *T. castaneum* arthropods in situations when housing and food were available at the same time. Pheromone-equipped sticky traps or baits that appeal to insects have demonstrated a notable degree of effectiveness in identifying and tracking insect infestations in a variety of settings, including food establishments, marketing networks, warehouses, and grain storage facilities (Vick et al. 1990).

17.2.6 Refuge Traps

Many crawling beetle pests and the pupating larvae of *Ephestia* spp. frequently use corrugated paper as a hiding place or shelter. The design of refuge traps has taken advantage of insects' tendency to seek cover. When grain oils or pheromones are added to these traps as attractants, their efficacy increases (Burkholder 1984).

17.2.7 Indicator Device

Mohan, Devadoss, and Mahendran (2001) found that an indication device is helpful in spotting the appearance of field carryover beetles in retailers at the right time. It's a perforated cup in the shape of a cone, with a 3 mm hole spacing and a cover on top. The cup's base is attached to a container and a circular dish, both of which can be coated with adhesives. Place 200 g of sample in the cup before storing pulses. The beetles that have been transported from the field enter the perforations, get slipped off, and fall into the trapping sections as soon as they begin emerging. Farmers can readily find beetles thanks to their adherence to the adhesive materials and the remove of the bulk-stored pulses to dry in the sun. Cereals can be processed with the same machine with 2 mm holes. This will aid in reducing the starting population, which is the primary driver of expansion. As a result, prompt identification will aid farmers in protecting their precious pulses while they are stored.

17.2.8 Automatic Insect Removal Bin

The success rate of probe traps for mass catching of stored product insects was low, and they were mainly useful for early detection. As a result, the "container concept" has replaced the "trap concept" with the intention of removing the insects as early as possible so that a major effect may be noticed in the stored produce in terms of damage and offspring production (Mohan 2000). Therefore, Mohan et al. (1994) developed the TNAU automatic insect removal bin (Figure 17.4),

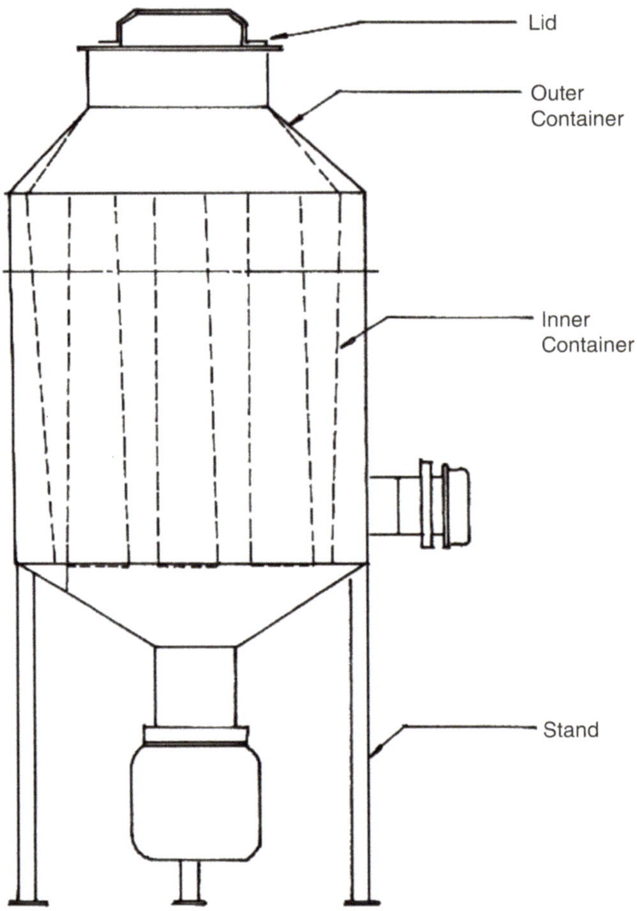

FIGURE 17.4 Automatic insect removal bin.

which takes advantage of the insects' tendency to roam and their preference for areas with good ventilation. Insects discovered in stored products were eliminated at a rate of 83–94% within ten days, leading to a dramatic drop in both damage and offspring generation. Insects that wander into the perforation of an automatic insect remover bin on their way to the aerated section are slid off and into a collection vessel through a pitfall mechanism (Mohan 1995). Mohan (1996) discovered that more than 90% of the time, stored grains that were attacked by *S. oryzae*, *R. dominica*, *T. castaneum*, and *O. surinamensis* were eliminated mechanically within 20 days.

According to research conducted by Mohan (1997), when experimentally infested milled rice was stored in the automatic insect removal bin, 94.0% of the adults of *S. oryzae* were removed during a period of ten days. In addition, after ten days in the insect removal bin, 75.34% of the adult *R. dominica* had been removed from the contaminated paddy, resulting in a significant drop in damage and offspring generation among the grains kept there.

17.2.9 Insect Egg Removal Machine

Insects are more vulnerable to impact than acceleration; hence, several authors have recorded the use of mechanical devices for the control of stored grain insects. Acceleration followed by deceleration, however, is particularly dangerous for insects (Plarre and Reichmuth 2000). Stratil and Wohlgemuth (1989) demonstrated that an impact machine was able to eradicate adult stages of

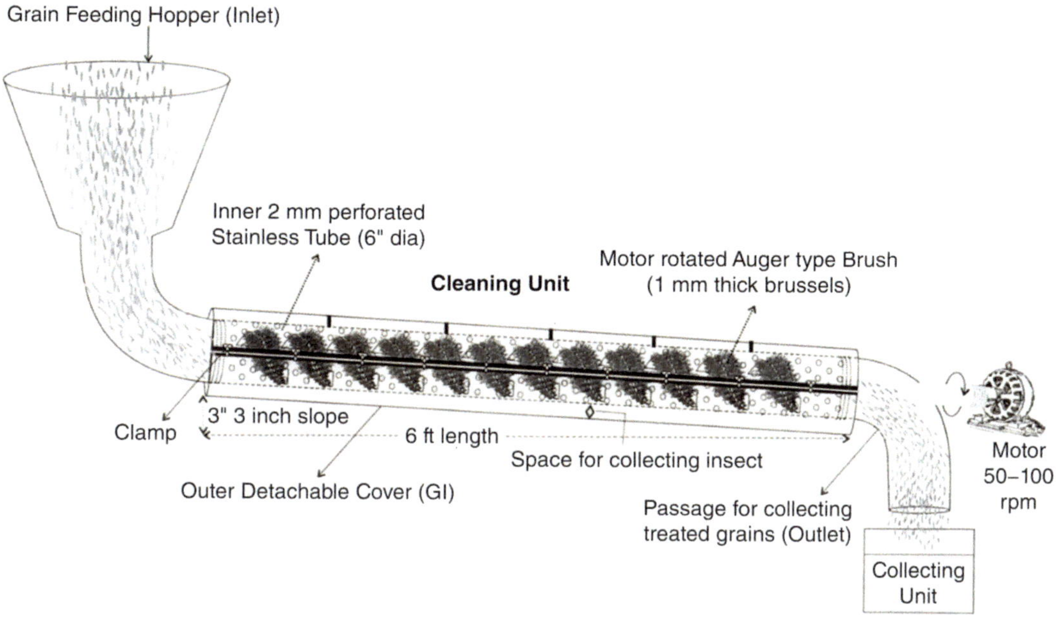

FIGURE 17.5 Insect egg removal machine.

O. surinamensis, *T. castaneum*, and *Sitophilus granarius* from grains. The use of non-chemical strategies to reduce insect populations becomes more urgent in this setting. Their lack of chemical residues, safety to applicators, and low risk of insect resistance make them preferable to fumigation and insecticide application. Banks and Fields (1995) and Hollingsworth and Armstrong (2005) highlight the importance of physically removing insects from stored grains as a method of pest control. There is a variety of fast equipment that can be used to remove adults, in addition to the standard methods of sieving and entoleters. Therefore, the invention of the insect egg remover (Mohan 2010) is a major step forward in the direction of more sustainably minded approaches to controlling stored grain insects (Mohan 2004). It takes advantage of the free-ranging and oxygen-hungry habits of the stored product insects. Adult insects and their eggs can be removed from stored products with the help of insect egg removal prototype models (Figure 17.5) that have undergone rigorous testing (Divya and Mohan 2009; Jayaprakash, Mohan, and Ramaraju 2010). As commercial versions of the tool have just emerged (Mohan 2011) there is a pressing need to explore how effectively they can eliminate adult insects from stored foods.

17.2.10 Electronic Grain Probe

The detection process can now be automated using probe trap techniques. Both Chinese and American studies (Vick, Webb, and Litzkow 1991) confirm that probe traps with infrared beams or piezoelectric sensors can be used to detect and monitor insects within stored grain. The electronic grain probe insect counter (EGPIC) was created by Shuman, Coffelt, and Weaver (1996). Insects may be tracked and recorded as they enter the trap thanks to the infrared beam sensor housed within the probe of this gadget. Computers are used to reliably record the sensor's output signals in the probe. In Canada, for instance, commercial applications of this technology, like the StorMax Insector, are now possible. It's useful for keeping tabs on pest populations in places like grain silos, elevators, and other large, sealed containers. It is an integral part of a grain storage management system. The authors suggested that the EGPIC may be used to identify and differentiate between captured insect species.

17.2.11 Multiple Traps

In order to catch or identify beetle and moth pests, Trematerra, Rotundo, and De Cristofaro (1994) created a trap with a series of vertically aligned funnels stacked on top of one another. This multi-funnel trap successfully caught flying adult beetles, including *R. dominica*, *T. castaneum*, and *C. ferrugineus*, during a six-week trial in an Italian cereal warehouse. The trap caught all four species of moths commonly found in flour mills during field trials: *Ephestia kuehniella*, *Pyralis farinalis*, *Nemapogon granella*, and *S. cerealella*.

17.2.12 Flour Trap

T. castaneum, commonly known as the red flour beetle, is one of the most damaging secondary insect pests that can damage flour products (RW 1965). Damage to the feedstock and the introduction of contaminants reduces the quality and quantity of the final product. Webbing, dead skins, feces, shed exoskeletons, bodily smells, and frass are all examples of this kind of contamination. Insects that feed on flour and spices can be lured into a trap (Mohan 2018). The trap consists of a cylindrical tumbler-shaped unit with a sieve at the bottom with a size of 0.8–0.9 mm, and a lid with 2 mm perforations. Attaching the gadget to the top of jars of dry goods like flour or spices is a breeze thanks to the included carry handle. Insects enter the tumbler part through the 2 mm perforations at the top and fall through to the 0.8–0.9 mm sieve at the bottom, where they are trapped. Flour that has settled to the bottom of the trap can be easily removed by giving it a light shake. This method prevents clogs by removing the smallest possible amount of flour, along with any insect eggs that may have been caught in the process. Using these procedure, insects can be found and removed from flours and spice powders without the use of any chemicals.

17.2.13 Other Traps

Using a computerized automated monitoring system and flying traps, Ho and Boon (1995) documented the detection of *T. castaneum* in a Singaporean rice mill. In this configuration, flying traps were hung from the ceiling of the mill's storage space, 6 m above the ground. The insects were placed in a 5-m PVC tube with a specimen vial at the bottom; the vial was balanced on a top pan. The increase in mass attributable to the ensnared insects was noted mechanically. Using an automated detection and monitoring system, this strategy is expected to drastically reduce labor and time costs.

17.3 INSECT TRAPS WITH ATTRACTANTS

17.3.1 Bait Trap

Food volatiles generated by stored foods are a major attractant for insect infestations. This effect has been utilized to keep tabs on the presence of beetles and other insect pests within stored foods and warehouses (Pierce et al. 1990; Phillips et al. 1993). In comparison to synthetic pheromones, the price of food attractants is typically lower, and they are not species-specific. When food products are already present in the facility, however, the range of attraction given by food baits is limited, unlike that of pheromones. Insect larvae and adults can be monitored effectively with the help of food attractant traps (Pinniger 1990). There are three distinct varieties of food baits: those made of whole grains, those made of crushed grains, and those made of liquid ingredients such as cereal and vegetable oils and carob (locust bean) or other grain extracts. In the first group, bags of fabric, jute, or plastic encapsulate broken grains that are used either alone or in mixtures. These baited sacks are placed on the warehouse floor and around grain stacks. The bags are checked after a week or two, and the number of insects caught is recorded. Insects of many types are attracted to the bait bags, and many of them choose to stay within the bags for long periods of time. However, the bait loses its appeal after two weeks and needs to be replaced. Although bait bags are effective, their usage is frequently restricted in food processing and storage facilities because of the possibility of product contamination.

The presence of insects in stored grains can be lured with several types of food bait bags (Strong 1970). Broken grains of barley, wheat, sorghum, oats, corn, and coffee beans were used in previous tests, along with sacks composed of metal mesh, cheesecloth, jute, or hessian. Insect pests in food storage facilities, cargo holds, and other similar environments were the focus of these investigations, which sought to employ these food baits for detection and surveillance purposes. Research conducted after this found that the most enticing bait was either brown rice alone or a mixture of wheat, peanuts, and kibbled carob (Hodges et al. 1985). Monitoring infestations in warehouses storing milled rice in Indonesia was successful when using plastic bait bags with 2 mm apertures containing 60 g of brown rice (Hodges et al. 1985). When compared to Pinniger (1975) mixed grain bait bags, those made with brown rice performed similarly well. Both *S. zeamais* and *R. dominica* were more attracted to the former than to the latter. There was a significant improvement in detection rates when using brown rice bait bags instead of the standard approach of randomly sampling a certain area. Experiments carried out in Jakarta, Indonesia, warehouses further confirmed this effectiveness. *T. castaneum*, *O. surinamensis*, *S. zeamais*, and *R. dominica* were just some of the beetle pests that the brown rice bait bags attracted and killed (Haines et al. 1990). Adult *S. oryzae* were found to be drawn to oat oil and corn oil in the lab, while adults of *T. castaneum* were attracted to oils derived from rice, soybeans, wheat germ, and corn (Phillips et al. 1993). *T. granarium* larvae were particularly drawn to oat seeds, pumpkin seeds, and sesame oils (Barak 1989). When added to pheromones or physical insect traps, these oils have been demonstrated to have a synergistic impact (Barak 1989).

Insects like *R. dominica* (Dowdy et al. 1993) and *O. surinamensis* (Pierce et al. 1990), but not *P. truncatus* (Fadamiro, Wyatt, and Birch 1998), have been shown to be drawn to distillates or volatile compounds generated from natural food sources. Insects, including *S. oryzae*, *S. zeamais*, *S. granarius*, *Aedes advena*, and *C. ferrugineus* are drawn to hexanoic acid in the volatiles released by carob tree pods (Obeng-Ofori, Torto, and Hassanali 1993; Wakefield and Cogan 1999).

While there are several advantages to using food as an attractant, commercially manufactured food bait lures are still unavailable. While pheromones have been studied extensively, the potential of food attractants for use in insect detection and monitoring has only just begun to be explored.

17.3.2 Light Traps

Different wavelengths of light attract a wide variety of Insect species. While certain species react differently to ranges of the visible and infrared spectrum, fluorescent bulbs or ultraviolet-emitting bulbs (black lights) are commonly used in insect traps and other devices that employ light to draw insects. Artificial light attracts hundreds of species of insects, most of which are not pests, such as flies, beetles, and moths. They might simply fly to lights for a few hours at night, or they might do so all night long.

Physical methods, like light, can be used to both attract and identify insects found in stored products. Insect pests of products that are kept in storage are attracted to light with wavelengths between 280 and 600 nm. Among these, green light with a wavelength between 500 and 560 nm and long wavelength UV light at 365 nm are particularly attractive (Rees 1985). UV light can come from "black light" tubes or bulbs, which release mercury vapor and radiate in the 300–500 nm range, peaking at 365 nm for tubes and 400 nm for bulbs. In food manufacturing and storage facilities, for example, *Lasioderma serricorne* are commonly found near windows, suggesting a positive response to light (phototaxis). Numerous investigations, such as those conducted by Soderstrom (1970) Kirkpatrick, Yancey, and Marzke (1970), showed positive phototaxis in response to ultraviolet light. Rees (1985) proposed that green light is more attractive to moth pests, which are stored product insects, than UV light. The way an insect responds to an alluring light source depends on several variables, including the type of insect, its age, gender, and temperature. Other environmental considerations include the amount of light in the area and the duration of the photoperiod.

For the control of *R. dominica* and *T. castaneum* in grain storage warehouses, a TNAU UV-light trap has been suggested (Mohan et al. 1994). Additionally, they proposed that the rice warehouse's alleyways and corners, where a 4 W UV light trap with a peak emission of 250 nm was placed 1.5 m above the ground, was the location where *R. dominica* was most attracted.

Using blue (470 nm), green (520 nm), yellow (590 nm), and red (625 nm) light-emitting diodes (LEDs), Kim et al. (2013) assessed the repellent effects of various light wavelengths against cigarette beetles and found that the blue LED was roughly 1.3 times more repellent to adult *L. serricorne* than a luring lamp black light. The findings imply that environmentally friendly insect management could benefit from the application of blue LEDs. Katsuki et al. (2013) examined the attraction efficiency of LEDs with varying colors (wavelengths) and examined the spectrum sensitivity of the compound eye of *L. serricorne*. Regardless of mating or oviposition status, their findings indicated that both blue (470 nm) and ultraviolet (UV, 375 nm) LEDs attracted the greatest number of cigarette beetles of both sexes. However, the UV LED consistently tended to attract the greatest number of beetles. The green LED (520 nm) seldom attracted insects, even though the compound eye's principal sensitivity peak was at that wavelength. Additionally, Brindha et al. (2020) demonstrated that the adult sesame seed beetle, *Elasmolomus sordidus*, could be effectively mass-trapped using a TNAU UV-light trap.

17.3.3 UV Light Trap in Phosphine Resistance Management

Phosphine-resistant phenotypes of *T. castaneum* were found at high rates of 100% in India, according to a thorough statewide investigation of phosphine resistance in stored grain insect pests (Rajendran 1999). According to Raghavendra (2009), *T. castaneum* samples from 14 godowns of the Tamil Nadu Civil Supplies Corporation demonstrated phosphine resistance ranging from 6.67 to 71.67%. Phosphine resistance is a challenge to the pest's successful management. When *T. castaneum* was tested at low concentration (0.03 mg/L), the frequency of phosphine resistance ranged from 40.29 to 100% in Tamil Nadu, 71.40 to 93.40% in Kerala, and 73.99 to 100% in Andhra Pradesh populations, according to Rajan (2015). When the test was conducted at a high concentration (0.25 mg/L), the frequency of phosphine resistance ranged from 25.84 to 100%, 67.79 to 85.39, and 45.55 to 100% in Tamil Nadu, Kerala, and Andhra Pradesh, sequentially.

Effective management of phosphine resistance may be aided by the use of UV light traps to capture resistant beetles that emerge following fumigation. The use of light traps in the phosphine resistance management strategy for *T. castaneum* in Indian grain storage warehouses was investigated by Rajesh and Mohan (2016). They found that TNAU UV-light traps were effective in mass trapping *T. castaneum* both before and after fumigation. Furthermore, compared to populations obtained prior to fumigation, *T. castaneum* populations taken from UV light traps after fumigation had a higher frequency of phosphine resistance. The findings showed that during the post-fumigation phase in grain storage warehouses, UV light traps can effectively catch populations of *T. castaneum* that are resistant to phosphine. As a result, one instrument to employ in a phosphine resistance management plan for *T. castaneum* in grain storage warehouses is the TNAU UV-light trap.

17.3.4 Pheromone Traps

One of the most promising methods for managing stored product insects is the use of pheromones. Utilizing these materials could result in a significant decrease in chemical treatments, which would have significant financial benefits and improve the quality of the final product. Considerable progress has been achieved in recent years in monitoring and controlling insects found in stored products through the use of mass trapping, mating disruption, and attracticide (lure and kill) techniques.

The best outcomes have been obtained by integrated programs that employ many treatments. All preventative measures that successfully keep contaminated items out of mills, storehouses, and food processing facilities should be given a significant role (Burkholder 1985).

Because pheromones are unique to target species, a pheromone trap can identify the presence of an insect where many traditional sample techniques would not find any. This makes pheromones extremely sensitive tools for insect identification. They can be employed for pheromone-based mass capturing of insects that are stored in products, in addition to monitoring (Phillips, Cogan, and Fadamiro 2000).

Male *L. serricorne* was found to exhibit considerable affinity to the sex pheromone released by females, according to Coffelt and Burkholder (1972). 4,6-Dimethyl-7-hydroxy-nonan-3-one, also known as serricornin, was discovered by Chuman et al. (1979) to be a sex pheromone of *L. serricorne. L. serricorne* was used in trapping research by Watanabe et al. (1982) in flat adhesive traps that contained serricornin and methyl phenyl acetate, a food attractant. The trap worked well for both capturing beetles and figuring out their seasonal emergence patterns. The usage of the traps to locate the sites of infestation helped with beetle management.

According to Burkholder (1985), *L. serricorne*'s short adult life span and reliance on fat body stores for adult survival and egg development allow them to profit from a highly synchronized manner of pheromone transmission. Unlike lepidopterous insects, these beetles were known to conceal in nooks and crannies in their native habitats and to emerge during sexually active periods. Lower pheromone release rates, greater pheromone response thresholds, and limited mobility are characteristics of *L. serrulata*. A laboratory experiment by Vick et al. (1990) and a long-term study on cigarette beetles in a Hawaiian food storage facility showed that widespread and continuous use of pheromone-baited traps for cigarette beetle populations was beneficial for long-term suppression as well as monitoring.

Chin and Lin (1990) demonstrated the effectiveness of using a "serrico" sticky trap to capture cigarette beetles, or *L. serricorne*, at a tobacco storage warehouse and cigarette manufacturing factory located in Taiwan. Only males were drawn to the flavor that an adult female *L. serricorne* had previously possessed in the lab, but larvae and adult females avoided it, indicating the presence of a sex pheromone, according to Ambadkar and Howlader (1992). Gilberg and Roach (1991) used a commercial sex pheromone trap to track the *L. serricorne* infestation in an Australian museum. The Fuji Flavor Company in Japan produced the novel Serico Tobacco insect trap, which included the attractant serricornin.

The most successful results were obtained with multisurface traps when baited with 10 mg anhydroserricornin, according to Buchelos and Levinson (1993) comparison of the effectiveness of pheromone-baited multisurface traps and Lasiotraps for mass trapping the males of *L. serricorne* in two insecticide-free tobacco stores at Piraeus (Southern Greece) for 15 months. In April and May of 1996, investigated the use of pheromone traps to monitor *L. serricorne* in a tobacco warehouse in Sao Paulo, Brazil. Testing was done on sex pheromone-baited traps, food-baited traps, and food-baited traps. While traps with food alone proved ineffectual, those baited with sex pheromone alone did not differ significantly from those placed with food alone.

Trematerra and Zanetti (1998) examined four distinct commercial pheromone dispensers at a tobacco shop in central Italy that sold tobacco kinds from Paraguay and Maryland. They discovered that pheromone age had no bearing on the effectiveness of the trapping process. Studies on the effectiveness of five distinct pheromone traps—Loline, Borgwaldt, Mini Delta, FLIT-TRAK M2, and New Serico with or without food lures were carried out by. Throughout the entire trap, (4S, 6S, 7S)-4,-di methyl-7 Hyydoxynonan-3-one was the main active pheromone component. Compared to other traps, the Mini Delta trap captured more *L. serricorne*. Nonetheless, there was no discernible variation in the average catches of this kind of trap and the recently developed Serrico trap. While the bait in the tiny delta trap lasted for at least six weeks, the pheromone lure in the new serco only lasted for a month. The least amount of *L. serricorne* captures was recorded with the FLIT-TRAK M2 trap.

17.4 COMBINATION TRAPPING

The study conducted by Buchelos and Levinson (1993) unveiled the efficacy of utilizing multisurface traps, which were baited with 10 mg of anhydroserricornin, as a highly advantageous method for the mass trapping of male *L. serricorne*. Furthermore, this approach was found to facilitate insectistasis, thereby effectively controlling the population of this species in tobacco storage facilities. Lasiotraps, when baited with a precise quantity of 10 mg of anhydroserricornin, exhibit a significantly reduced adhesive surface area compared to multisurface traps. Nevertheless, these traps possess the remarkable ability to detect tobacco beetle infestations with equal swiftness as their multisurface counterparts. Consequently, they can be effectively utilized for the purpose of monitoring populations of *L. serricorne*. The efficacy of pheromone-baited multisurface traps in comparison to pheromone-baited Lasiotraps is primarily attributed to their augmented adhesive area, which is vertically suspended in proximity to a central pheromone source. The heightened allure of pheromone-infused multisurface traps likely stems from the amalgamation of sensory stimulation experienced by male tobacco beetles when they are concurrently exposed to figural and pheromone stimuli, which act in a synergistic optical-olfactory interplay.

In their study, Papadopoulou and Buchelos (2002b) employed a set of adhesive traps, each possessing an equivalent adhesive area. These traps were categorized into four distinct types: electric (black light), pheromone (serricornin and anhydroserricornin), food attractant, and a control group. The purpose of this experiment was to capture specimens of *L. serricorne* (F.) (Coleoptera: Anobiidae). The electric trap successfully ensnared a total of 8912 *L. serricorne* adults, while the pheromone trap proved effective in capturing 6608 individuals. Additionally, the food attractant trap managed to attract and trap 1974 specimens. In comparison, the control group yielded a relatively lower count of 1231 *L. serricorne* adults. The efficacy of the electric adhesive trap surpassed that of the alternative traps and the control group by a notable margin. The electric trap exhibited a diminished count of adult males in comparison to the pheromone trap, while conversely, it successfully ensnared a significantly greater quantity of adult females. The proportion of female to male adults in the electric trap was 60.15–39.84%; in the pheromone trap, it was 8.75–91.25%; in the food attractant trap, it was 51.57–48.42%, and in the control trap it was 43.41–56.58%.

Papadopoulou and Buchelos (2002a) elucidated that the process of discerning the female adult *L. serricorne* can be accomplished through a straightforward examination of the abdomen using a stereomicroscope after its immersion in a solution of 70% alcohol or benzene for approximately 5 min. Consequently, the urosternites underwent a transformation, acquiring a state of translucency, thereby enabling the discernment of the V-shaped apodeme, a pivotal sexual characteristic of the female, through the straightforward means of external observation, obviating the need for abdominal dissection. The identification of the V-shaped apodeme via rudimentary external observation, devoid of the need for dissection, has streamlined and expedited the assessment of a substantial population of *L. serricorne* adult specimens.

In their seminal study, Mahroof and Phillips (2008) delved into the intricate realm of adult *Stegobium paniceum and L. serricorne*, both belonging to the esteemed order Coleoptera and the distinguished family Anobiidae. Their noble pursuit involved examining the reactions of these esteemed insects to a myriad of distinct fabrications, ranging from commercially available to prototype, of the alluring sex pheromones produced by their female counterparts. The findings indicated that the quantity of *S. paniceum* specimens captured in traps utilizing the commercially accessible lures exhibited a notable increase in comparison to those captured in traps utilizing the prototype stegobinone lures. The efficacy of capturing *L. serricorne* was found to be equitably distributed among the three commercially available brands of serricornin lures that were subjected to investigation. Moreover, it has been documented that the number of beetles apprehended in traps infused with a blend of serricornin and chili volatiles exhibited a noteworthy increase compared to traps baited solely with pheromone or chili volatiles. This observation suggests the possibility of enhancing the monitoring or mass trapping of *L. serricorne* by synergistically combining pheromone compounds with naturally occurring plant-derived volatiles found in *Capsicum* spp.

In their study, Saeed, Khan, and Shahid (2007) conducted a comparative analysis of five distinct monitoring traps utilized in tobacco warehouses. These traps included STORGARD® THINLINE™ CB and COMBI™, STORGARD DOME®CB, New Serrico trap baited with sex pheromone 4S,6S,7S-serricornin, and the Diamond trap, which was traditionally employed as a Light Beetle trap. The researchers sought to evaluate the effectiveness of these traps in their intended monitoring purposes. The examination of population dynamics pertaining to beetles confined within various trap types over the course of a year has revealed that the STORGARD® THINLINE™ CB trap exhibits the utmost efficacy in facilitating meticulous surveillance of *L. serricorne*, as it has yielded the highest number of beetles (3.35). Subsequently, the New Serrico trap has demonstrated a commendable performance with a value of 2.47 beetles. A meager quantity of beetles, specifically 0.58, were meticulously recorded in the context of the Light Beetle trap. Additionally, it has been noted that a significant proportion of beetles were captured during the month of August.

In their seminal work, Arbogast, Kendra., and Chini (2003) conducted a rigorous investigation to assess the veracity of contour mapping as a means of monitoring pest population in warehousing and retail environments. Their study focused specifically on the utilization of *L. serricorne*, a test insect, to evaluate the efficacy of this technique. The findings indicated that as the beetles underwent dispersion and the overall trap catch escalated, the peripheral traps exhibited a progressively higher capture of insects. However, the cumulative trap catch remained most pronounced near the release points. The initial period following release exhibited the utmost rate of capture, which subsequently diminished over time, initially at an accelerated pace and subsequently at a more gradual rate until reaching a state of near constancy. The numerical aggregates obtained from each trap following time intervals of 6, 24, and 48 h exhibited a discernible exponential decline as the distance from the initial release point increased. The empirical examination of spatial patterns in trap catch in relation to infestation sources and the inverse correlation between trap catch and distance from said sources lend credence to the efficacy of contour mapping as a method for monitoring stored product insects and identifying areas of infestation.

17.5 FACTORS AFFECTING TRAP CAPTURE OF STORED GRAIN INSECT PESTS

The efficacy of all insect traps is contingent upon the locomotive behavior exhibited by insects. Any variable that exerts an influence on the locomotion of insects will inevitably have a corresponding impact on the efficacy of trap capture. The magnitude of this phenomenon is contingent upon a multitude of factors, namely the taxonomic classification of the insect, ambient temperature, the specific variety of grain, and the overall quality and state of the grain. The quantification of insects ensnared within traps exhibits a correlation with the behavioral tendencies of the species (such as aggregation and mobility), the composition of the grain, ambient temperature and moisture levels, the duration of the trapping endeavor, the geographical placement and quantity of the traps, and ultimately, the density of the insect population (Hagstrum et al. 1990).

17.5.1 Behavior of Species Relation to Trap Capture

The prevailing trend among stored grain insects is the manifestation of aggregated distributions within the grain mass. The phenomenon of insects colonizing the surface of the grain and subsequently dispersing into the grain mass also plays a significant role in the formation of aggregated distributions (Hagstrum 1989). *T. castaneum* and *C. ferrugineus* exhibit notable levels of locomotor activity, whereas *R. dominica*, in contrast, displays comparatively diminished mobility among its counterparts. Consequently, the latter species is less prone to becoming ensnared, as substantiated by Subramanyam and Harein's research in 1989.

17.5.2 Insect Density

Research conducted on *C. ferrugineus*, colloquially known as the rusty grain beetle, has shed light on its propensity to infest grain externally, making it a prevalent pest in stored wheat across the globe. Notably, the findings of Toews and Phillips (2002) indicate a significant correlation between the capture rates of these beetles and the densities of their presence within wheat.

17.5.3 Grain Temperature

Multiple researchers have observed a notable increase in trap catch when grain temperatures are elevated, as evidenced by the studies conducted by Loschiavo and Smith (1986); N. D. G. White and Loschiavo (1986); Fargo et al. (1989). The findings of these studies indicate a notable escalation in the capture rates of *Cryptolestes* spp. as temperatures rise to higher levels. The presence of temperature and moisture fluctuations within a grain mass, as elucidated by Hagstrum in 1987, engenders disparate conditions that render certain regions more conducive to insect habitation than others. The locomotion of *C. ferrugineus* is influenced by the variables of moisture, temperature, and gravity, as elucidated by the scholarly works of Loschiavo (1983); Watters (1969).

17.5.4 Trapping Duration

Fargo et al. (1989) demonstrated a positive correlation between the duration of trapping and the observed abundance of insects belonging to a specific species. He alluded to the fact that the reaction to the duration of trapping exhibits variability across different species, and therefore, it ought to be considered when analyzing the results of trap catch. The selection of the trapping duration may present a complex predicament, as it is influenced by the dynamic nature of trapping efficiency. This is attributed to the phenomenon wherein insects that have already been captured release an aggregation or sex pheromone, thereby augmenting the allure of the trap as time progresses (Loschiavo 1974; Barak and Harein 1982). The notable presence of the probe trap within the grain for a certain duration significantly contributes to the substantial quantity of captured entities.

Additional variables that influence the effectiveness of trapping methods include the composition and quality of the grain (Wright and Mills 1984), the presence of moisture and fungi, which can impact the movement patterns of insects (Watters 1969) as well as the design, placement, and quantity of traps, all of which play a significant role in the success of the probe trap (Fargo et al. 1989; Wakefield and Cogan 1999).

17.6 CONCLUSION

The undeniable significance of utilizing insect traps in the realm of pest control for stored grains cannot be overstated. These traps provide a myriad of advantages to both the agricultural and food industries. The expeditiousness with which they can discern and address pest infestations facilitates the implementation of timely measures aimed at mitigating the peril of substantial agricultural harm and economic setbacks. Furthermore, the utilization of these substances serves to diminish dependence on chemical pesticides, thereby fostering a shift toward more sustainable and ecologically aware methodologies for the control of pests. The integration of diverse trap mechanisms and technological advancements serves to augment the precision and efficacy of strategies employed in the surveillance and control of pest populations. With the relentless progression of technology, it is only reasonable to anticipate the emergence of increasingly sophisticated trap configurations that augment the accuracy of detection mechanisms and elevate the efficacy of trap captures. The incorporation of data-centric methodologies, such as remote monitoring and real-time data analysis, harbors the capacity to fundamentally transform the way pest infestations are approached. Furthermore, the exploration of the symbiotic relationship between insect traps and biological control techniques has the potential to unveil novel approaches to mitigating pest infestations while maintaining ecological equilibrium and sustainability. Therefore, it can be deduced that the utilization of insect traps is regarded as a pivotal component of efficacious stored grain pest control in contemporary times. Furthermore, the continuous advancement and refinement of these traps offer intriguing prospects for the days to come. Through the adoption of innovative practices and the cultivation of collaborative efforts, we have the potential to harness the inherent capabilities of these mechanisms. This will not only serve to safeguard the integrity of stored grain and guarantee the stability of our food supply but also position us as pioneers in the establishment of ecologically sound pest management methodologies.

REFERENCES

Ambadkar, P. M., and A. J. Howlader. 1992. *Responses of Adults and Larvae of Lasioderma serricorne F. (Coleoptera: Anobiidae) to Medium Conditioned by Con-Specific Adults and Larvae.*

Arbogast, R. T., P. E. Kendra., and S. R. Chini. 2003. "*Lasioderma serricorne* (Coleoptera: Anobiidae): Spatial Relationship between Trap Catch and Distance from an Infested Product." *Florida Entomologist* 86 (4): 437–444.

Asrar, M., N. Ashraf, M. Dildar Gogi, S. M. Hussain, K. Zia, and R. Bilal. 2016. "Toxicity and Repellence of Plant Oils Against *Tribolium castaneum* (Herbst), *Rhyzopertha dominica* (F.) and *Trogoderma granarium* (E.)" *Pakistan Entomologist* 38 (1): 55–63.

Bailey, S. W. 1962. "The Effects of Percussion on Insect Pests of Grain1." *Journal of Economic Entomology* 55 (3): 301–304. https://doi.org/10.1093/jee/55.3.301.

Banks, J., and P. Fields. 1995. *Physical Methods for Insect Control in Stored-Grain Ecosystem.* New York: Marcel Dekker.

Barak, A. V., W. E. Burkholder, and D. L. Faustini. 1990. "Factors Affecting the Design of Traps for Stored-Product Insects." *Journal of the Kansas Entomological Society*, 466–485.

Barak, A. V., and P. K. Harein. 1982. "Trap Detection of Stored-Grain Insects in Farm-Stored, Shelled Corn12." *Journal of Economic Entomology* 75 (1): 108–111. https://doi.org/10.1093/jee/75.1.108.

Barak, A. V. 1989. "Development of a New Trap to Detect and Monitor Khapra Beetle (Coleoptera: Dermestidae)." *Journal of Economic Entomology* 82 (5): 1470–1477. https://doi.org/10.1093/jee/82.5.1470.

Brindha, K., C. A. Mahalingam, P. A. Saravanan, and K. Raja, and Mohan. 2020. "TNAU – UV Light Trap: A Device for Mass Trapping Sesame Seed Bug-*Elasmolomus sordidus* (F.) (Hemiptera: Lygaeidae) in Sesame Storage Warehouses." *Madras Agricultural Journal* 107 (7–9). https://doi.org/10.29321/MAJ.2020.000380.

Buchelos, C. T., and A. R. Levinson. 1993. "Efficacy of Multisurface Traps and Lasiotraps With and Without Pheromone Addition, for Monitoring and Mass-trapping of *Lasioderma serricorne* F. (Col., Anobiidae) in Insecticide-free Tobacco Stores." *Journal of Applied Entomology* 116 (1–5): 440–448. https://doi.org/10.1111/j.1439-0418.1993.tb01220.x.

Burkholder, W. E. 1984. "Use of Pheromones and Food Attractants for Monitoring and Trapping Stored-Product Insects." In: F. J. Baur (Ed). St. Paul, Minnesota: Insect Management for Food Storage and Processing, American Association of Cereal Chemists.

Burkholder, W. E. 1985. *The Use of Pheromones and Food Attractants for Monitoring and Trapping Stored-Product Insects.*

Chin, W. T., and H. C. Lin. 1990. "Studies on the Occurrence of Cigarette Beetle, *Lasioderma serricorne* (F.) with 'serrico' trap in Fen-Yuan Tobacco Factory." *Bulletin of the Tobacco Research Institute, Taiwan Tobacco & Wine Monopoly Bureau* 32: 53–58.

Coffelt, J. A., and W. E. Burkholder. 1972. "Reproductive Biology of the Cigarette Beetle, *Lasioderma serricorne.* 1.1 Quantitative Laboratory Bioassay of the Female Sex Pheromone from Females of Different Ages 2,3." *Annals of the Entomological Society of America* 65 (2): 447–450. https://doi.org/10.1093/aesa/65.2.447.

Cogan, P. M., and M. E. Wakefield. 1987. *Further Developments in Traps Used to Detect Low-Level Infestations of Beetle Pests in Bulk Stored Grain.*

Shuman, D., J. A. Coffelt, and D. K. Weaver. 1996. "A Computer-Based Electronic Fall-Through Probe Insect Counter for Monitoring Infestation in Stored Products." *Transactions of the ASAE* 39 (5): 1773–1780. https://doi.org/10.13031/2013.27653.

Divya, S., and S. Mohan. 2009. *TNAU Egg Removal Device in the Management of Insect Pests of Stored Siddha and Ayurvedic Ingredients.*

Dowdy, A. K., R. W. Howard, L. M. Seitz, and W. H. McGaughey. 1993. "Response of *Rhyzopertha dominica* (Coleoptera: Bostrichidae) to Its Aggregation Pheromone and Wheat Volatiles." *Environmental Entomology* 22 (5): 965–970. https://doi.org/10.1093/ee/22.5.965.

Fadamiro, H. Y., T. D. Wyatt, and M. C. Birch. 1998. "Flying Beetles Respond as Moths Predict: Optomotor Anemotaxis to Pheromone Plumes at Different Heights." *Journal of Insect Behavior* 11: 549–557. https://doi.org/10.1023/A:1022367430354.

FAO. 2014. "Global Initiative on Food Losses and Waste Reduction. Definitional Framework of Food Loss." Rome, Italy.

Fargo, W. S., D. Epperly, G. W. Cuperus, B. C. Clary, and R. Noyes. 1989. "Effect of Temperature and Duration of Trapping on Four Stored Grain Insect Species." *Journal of Economic Entomology* 82 (3): 970–973. https://doi.org/10.1093/jee/82.3.970.

Fleurat-Lessard, F. 2006. *Advanced Methods for the Monitoring of Fungal Spoilage and Insect Pest Infestation Risks During Cereal Storage*. Industries Des Céréales.

Gilberg, M., and A. Roach. 1991. "The Use of a Commercial Pheromone Trap for Monitoring *Lasioderma serricorne* (F.) Infestations in Museum Collections." *Studies in Conservation* 36 (4): 243–247.

Hagstrum, D. W., P. W. Flin, B. Subramanyam, D. W. Keever, and G. W. Cuperus. 1990. "Interpretation of Trap Catch for Detection and Estimation of Stored-Product Insect Populations." *Journal of the Kansas Entomological Society* 63: 500–505.

Hagstrum, D. W. 1989. "Infestation by *Cryptolestes ferrugineus* (Coleoptera: Cucujidae) of Newly Harvested Wheat Stored on Three Kansas Farms." *Journal of Economic Entomology* 82 (2): 655–659. https://doi.org/10.1093/jee/82.2.655.

Hagstrum, D. W., A. K. Dowdy, and G. E. Lippert. 1994. "Early Detection of Insects in Stored Wheat Using Sticky Traps in Bin Headspace and Prediction of Infestation Level." *Environmental Entomology* 23 (5): 1241–1244. https://doi.org/10.1093/ee/23.5.1241.

Hagstrum, D. W., P. W. Flinn, and B. Subramanyam. 1998. "Predicting Insect Density from Probe Trap Catch in Farm-Stored Wheat." *Journal of Stored Products Research* 34 (4): 251–262. https://doi.org/10.1016/S0022-474X(98)00006-X.

Haines, C., D. Rees, K. Ryder, S. Sistyanto, and Y. Cahyana. 1990. "Cahyana. "Brown-Rice Bait-Bags for Monitoring Insect Pest Populations in Bag Stacks of Milled Rice as an Aid to Pest Control Decision-Making."

Hategekimana, A., S., K. Ramaraju. Mohan, and V. Thirupathi. 2013. "Efficiency of Stack Probe Trap for Detection, Validation of Fumigation and Distribution of Stored Grain Insects in Bag Stacks." *Journal of Renewable Agriculture* 1 (3).

Ho, S. H., and K. S. Boon. 1995. "Spatial Distribution of Flying *Tribolium castaneum* (Coleoptera: Tenebrionidae) in a Rice Warehouse." *Bulletin of Entomological Research* 85 (3): 355–359.

Hodges, R. J., H. Halid, D. P. Rees, J. Meik, and J. Sarjono. 1985. "Insect Traps Tested as an Aid to Pest Management in Milled Rice Stores." *Journal of Stored Products Research* 21 (4): 215–229. https://doi.org/10.1016/0022-474X(85)90018-9.

Hollingsworth, R. G., and J. W. Armstrong. 2005. "Potential of Temperature, Controlled Atmospheres, and Ozone Fumigation to Control Thrips and Mealybugs on Ornamental Plants for Export." *Journal of Economic Entomology* 98 (2): 289–298.

Jayaprakash, S. A., S. Mohan, and K. Ramaraju. 2010. "Egg Removal Device for the Management of Three Stored Product Pests." *Julius-Kühn-Archiv* 425: 673.

Katsuki, M., K. Arikawa, M. Wakakuwa, Y. Omae, K. Okada, R. Sasaki, K. Shinoda, and T. Miyatake. 2013. "Which Wavelength Does the Cigarette Beetle, *Lasioderma serricorne* (Coleoptera: Anobiidae), Prefer? Electrophysiological and Behavioral Studies Using Light-Emitting Diodes (LEDs)." *Applied Entomology and Zoology* 48 (4): 547–551. https://doi.org/10.1007/s13355-013-0205-3.

Kirkpatrick, R. L., D. L. Yancey, and F. O. Marzke. 1970. "Effectiveness of Green and Ultraviolet Light in Attracting Stored-Product Insects to Traps." *Journal of Economic Entomology* 63 (6): 1853–1855. https://doi.org/10.1093/jee/63.6.1853.

Lippert, G. E., and D. W. Hagstrum. 1987. "Detection or Estimation of Insect Populations in Bulk-Stored Wheat With Probe Traps." *Journal of Economic Entomology* 80 (3): 601–604. https://doi.org/10.1093/jee/80.3.601.

Loschiavo, S. R. 1974. "Laboratory Studies of a Device to Detect Insects in Grain, and of The Distribution of Adults of The Rusty Grain Beetle, *Cryptolestes ferrugineus* (Coleoptera: Cucujidae), In Wheat-Filled Containers." *The Canadian Entomologist* 106 (12): 1309–1318. https://doi.org/10.4039/Ent1061309-12.

Loschiavo, S. R. 1983. "Distribution of the Rusty Grain Beetle (Coleoptera: Cucujidae) in Columns of Wheat Stored Dry or With Localized High Moisture Content." *Journal of Economic Entomology* 76 (4): 881–884. https://doi.org/10.1093/jee/76.4.881.

Loschiavo, S. R., and J. M. Atkinson. 1967. "A Trap for the Detection and Recovery of Insects in Stored Grain." *The Canadian Entomologist* 99 (11): 1160–1163. https://doi.org/10.4039/Ent991160-11.

Loschiavo, S. R., and J. M. Atkinson. 1973. "An Improved Trap to Detect Beetles (Coleoptera) in Stored Grain." *The Canadian Entomologist* 105 (3): 437–440. https://doi.org/10.4039/Ent105437-3

Loschiavo, S. R., and L. B. Smith. 1986. "Population Fluctuations of The Rusty Grain Beetle, *Cryptolestes ferrugineus* (Coleoptera: Cucujidae), Monitored With Insect Traps in Wheat Stored in a Steel Granary." *The Canadian Entomologist* 118 (7): 641–647. https://doi.org/10.4039/Ent118641-7.

Mahroof, R. M., and T. W. Phillips. 2008. "Responses of Stored-product Anobiidae to Pheromone Lures and Plant-derived Volatiles." *Journal of Applied Entomology* 132 (2): 161–167. https://doi.org/10.1111/j.1439-0418.2007.01251.x.

Mohan, S. 2019. *A Grain Saved Is a Grain Produced - A Journey Towards Eco-Friendly Grain Storage.* Jodhpur, India: Scientific Publishers.

Mohan, S. 2011. "A Device to Remove Adult Insects and Crush Their Eggs in Stored Grains." *Pestology* 35 (9).

Mohan, S. 1993. *Studies on Detection & Management of Stored Product Insect Pests of Rice.* Coimbatore: Tamil Nadu Agricultural University.

Mohan, S. 1995. "A Storage Container for Automatic Removal of Insects from Grains." *Indian Food Industry* 14 (3): 45.

Mohan, S. 1996. "A New Storage Bin." *TNAU Newsletter* 26 (1): 2–3.

Mohan, S. 1997. "Laboratory Studies of a New Storage Bin to Remove *Sitophilus oryzae* L. and *Rhyzopertha dominica* F. from Stored Paddy." *Pestology* 22 (12): 30–35.

Mohan, S. 2000. "Insect Removal Bin Could Help Farmers in Developing Countries." *Magazine of American Association of Agrl. Engineers*, 2000.

Mohan, S. 2004. "TNAU Gadgets for the Management of Stored Product Insects in Food Grains." *Pestology* 28 (9): 10–12.

Mohan, S. 2010. A device to monitor stored grain insects in bag stacks. Indian Patent Application Number 1773/CHE/2008, issued 2010.

Mohan, S. 2018. "TNAU Devices a 'Pit Fall' for Flour, Millet Pests. Business Line." *The Hindu*, 2018.

Mohan, S., C. T. Devadoss, and D. Mahendran. 2001. *New Devices for Pulse Beetle Management.*

Mohan, S., M. Gopalan, P. C. Sundarababu, and V. V. Sreenarayanan. 1994. "Practical Studies on the Use of Light Traps and Bait Traps in the Management of *Rhyzopertha dominica* (F.) in Rice Warehouses." *International Journal of Pest Management* 40 (2): 148–152. https://doi.org/10.1080/09670879409371873.

Mohan, S., S. S. Sivakumar, Z. Kavitharaghavan, S. Venkatesh, and G. S. V. Raghavan. 2008. "A New Trap Model to Increase the Trapping of *Cryptolestes ferrugineus* (Coleoptera: Laemophloeidae) in Wheat Filled Containers." *Madras Agricultural Journal* 95 (7–12): 390–393.

Mohan, S., S. S. Sivakumar, S. R. Venkatesh, and G. S. V. Raghavan. 2006. "Simple Technique to Increase the Sensitivity of Probe Traps in Detecting *Cryptolestes ferrugineus* (Coleoptera: Laemophloeidae) in Stored Wheat." *The Canadian Entomologist* 198 (5): 733–735.

Neethirajan, S., C. Karunakaran, D. S. Jayas, and N. D. G. White. 2007. "Detection Techniques for Stored-Product Insects in Grain." *Food Control* 18 (2): 157–162. https://doi.org/10.1016/j.foodcont.2005.09.008.

Obeng-Ofori, D., B. Torto, and A. Hassanali. 1993. "Evidence for Mediation of Two Releaser Pheromones in the Aggregation Behavior of the Gregarious Desert Locust, *Schistocerca gregaria* (Forskal) (Orthoptera: Acrididae)." *Journal of Chemical Ecology* 19 (8): 1665–1676. https://doi.org/10.1007/BF00982299.

Papadopoulou, S. C., and C. T. Buchelos. 2002a. "Identification of Female Adult *Lasioderma serricorne* (F.) by Simple External Observation of the Abdomen." *Journal of Stored Products Research* 38 (3): 315–318. https://doi.org/10.1016/S0022-474X(01)00029-7.

Papadopoulou, S. C., and C. T. Buchelos. 2002b. "Comparison of Trapping Efficacy for *Lasioderma serricorne* (F.) Adults With Electric, Pheromone, Food Attractant and Control-Adhesive Traps." *Journal of Stored Products Research* 38 (4): 375–383. https://doi.org/10.1016/S0022-474X(01)00039-X.

Parfitt, J., M. Barthel, and S. Macnaughton. 2010. "Food Waste Within Food Supply Chains: Quantification and Potential for Change to 2050." *Philosophical Transactions of the Royal Society B: Biological Sciences* 365 (1554): 3065–3081. https://doi.org/10.1098/rstb.2010.0126.

Phillips, T. W., X.-L. Jiang, W. E. Burkholder, J. K. Phillips, and H. Q. Tran. 1993. "Behavioral Responses to Food Volatiles by Two Species of Stored-Product Coleoptera, *Sitophilus oryzae* (Curculionidae) and *Tribolium castaneum* (Tenebrionidae)." *Journal of Chemical Ecology* 19 (4): 723–734. https://doi.org/10.1007/BF00985004.

Phillips, T. W., P. M. Cogan, and H. Y. Fadamiro. 2000. "Pheromones." In *Alternatives to Pesticides in Stored-Product IPM*, 273–302. Boston, MA: Springer US. https://doi.org/10.1007/978-1-4615-4353-4_10.

Pierce, A. M., H. D. Pierce, A. C. Oehlschlager, and J. H. Borden. 1990. "Attraction Of *Oryzaephilus surinamensis* (L.) and *Oryzaephilus mercator* (Fauvel) (Coleoptera: Cucujidae) to Some Common Volatiles of Food." *Journal of Chemical Ecology* 16 (2): 465–475. https://doi.org/10.1007/BF01021778.

Pinniger, D. B. 1975. *The Use of Bait Traps for Assessment of Stored-Product Insect Populations.*

Pinniger, D. B. 1990. "Food-Baited Traps; Past, Present and Future." *Journal of the Kansas Entomological Society* 63 (4): 533–538.

Plarre, R., and F. Reichmuth. 2000. "Impact." In: *Alternatives to Pesticides in Stored Product IPM.* Edited by Bhadriraju Subramanyam. Hagstrum and D. W. Hagstrum. Dordrecht, Netherlands: Kluwer Academic Publishers.

Raghavendra, D. 2009. *Insecticide resistance in stored product pests.* M.Sc. (Agriculture) Thesis, Tamil Nadu Agricultural University, Coimbatore, India.

Rajan, S. 2015. *Studies on frequency, distribution and molecular analysis of phosphine resistance, field ecology and population genetics of Tribolium castaneum (Herbst) and Sitophilus oryzae (L.) in three major southern states of India.* M.Sc. (Agriculture) Thesis, Tamil Nadu Agricultural University, Coimbatore, India.

Rajendran, S. 1999. "Resistência à Fosfina Em Pragas de Insetos de Grãos Armazenados Na Índia." In *In Proceedings of the 7th International Working Conference on Stored Product Protection*, 635–641.

Rajendran, S. 2005. "Detection of Insect Infestation in Stored Foods." *Advances in Food and Nutrition Research* 49 (4): 163–232.

Rajesh, A., and S. Mohan. 2016. "Studies on Use of TNAU- UV Light Trap for Management of Phosphine Resistance in *Lasioderma serricorne* (F.) (Coleoptera: Anobiidae) in Turmeric Warehouses." *Madras Agricultural Journal* 103 (June): 137–140. https://doi.org/10.29321/maj.10.001006.

Reed, C. R., V. F. Wright, T. W. Mize, J. R. Pedersen, and B. J. Evans. 1991. "Pitfall Traps and Grain Samples as Indicators of Insects in Farm-Stored Wheat." *Journal of Economic Entomology* 84 (4): 1381–1387. https://doi.org/10.1093/jee/84.4.1381.

Rees, D. P. 1985. *Review of the Response of Stored Product Insects to Light of Various Wavelengths, with Particular Reference to the Design and Use of Light Traps for Population Monitoring.*

RW, H. 1965. "Losses Caused by Insects and Mites in Stored Foods and Foodstuffs." *Nutrition Abstracts and Reviews* 35: 285–302.

Saeed, M., S. M. Khan, and M. Shahid. 2007. "Effective Monitoring of Cigarette Beetle *Lasioderma serricorne* (F.) Coleoptera : Anobiidae in Tobacco Warehouses of NWFP Pakistan." *Sarhad Journal of Agriculture.* 23 (1): 123–28.

Sarwar, M., M. Hamed, M. Yousaf, and M. Hussain. 2013. "Identification of Resistance to Insect Pests Infestations in Cotton (*Gossypium hirsutum* L.) Varieties Evaluated in the Field Experiment." *International Journal of Scientific Research in Environmental Sciences*, no. November 2013: 317–323. https://doi.org/10.12983/ijsres-2013-p317-323.

Sharon, M. E. M., C. V. K. Abirami, and K. Alagusundaram. 2014. "Review Article JPHT Grain Storage Management in India." *Journal of Postharvest Technology* 2 (01): 12–24.

Soderstrom, E. L. 1970. "Phototactic Response of Stored-Product Insects to Various Intensities of Ultra-Violet Light." *Journal of Stored Products Research* 6 (3): 275–277. https://doi.org/10.1016/0022-474X(70)90018-4.

Stratil, H., and R. Wohlgemuth. 1989. "Investigations into the Mechanism of Action of Impact Machines on Insects That Damage Stored Products." *Anzeiger Für Pestkunde, Pflanzenschutz, Umweltschutz* 62 (3): 41–47.

Strong, R. G. 1970. "Distribution and Relative Abundance of Stored-Product Insects in California: A Method of Obtaining Sample Populations." *Journal of Economic Entomology* 63 (2): 591–596. https://doi.org/10.1093/jee/63.2.591.

Toews, M. D., and T. W. Phillips. 2002. "Factors Affecting Capture of *Cryptolestes ferrugineus* (Coleoptera: Laemophloeidae) in Traps Placed in Stored Wheat." *Journal of Economic Entomology* 95 (1): 200–207. https://doi.org/10.1603/0022-0493-95.1.200.

Trematerra, P., G. Rotundo, and A. De Cristofaro. 1994. "Trapping Stored-Product Insects Using an Unbaited Multifunnel Trap." In *In Proceedings of the 6th International Working Conference on Stored-Product Protection, Canberra, Australia*, 451–454.

Trematerra, P., and P. Zanetti. 1998. *Activity of Commercial Pheromone Dispensers in the Capture of the Tobacco Beetle Lasioderma serricorne F. (Coleoptera, Anobiidae).*

Vela-Coiffier, E. L., W. S. Fargo, E. L. Bonjour, G. W. Cuperus, and W. D. Warde. 1997. "Immigration of Insects into On-Farm Stored Wheat and Relationships Among Trapping Methods." *Journal of Stored Products Research* 33 (2): 157–166. https://doi.org/10.1016/S0022-474X(96)00043-4.

Vick, K. W., R. W. Mankin, R. R. Cogburn, V. F. Mullen, J. E. Throne, V. F. Wright., and L. D. Cline. 1990. "Review of Pheromone-Baited Sticky Traps for Detection of Stored-Product Insects." *Journal of the Kansas Entomological Society* 63: 526–532.

Vick, K.W., J.C. Webb, and Litzkow. 1991. Insect detection using a pitfall probe trap having vibration detection. U.S. Patent 5,005,416, issued 1991.

Wakefield, M. E., and P. M. Cogan. 1999. "The Use of a Managed Bulk of Grain for the Evaluation of PC, Pitfall Beaker, Insect Probe and WBII Probe Traps for Monitoring Sitophilus Granarius During the Winter and Summer in the UK." *Journal of Stored Products Research* 35 (4): 329–338. https://doi.org/10.1016/S0022-474X(99)00016-8.

Watanabe, T., K. Arai, J. I. Kato, M. Kohno, and T. Chuman. 1982. "Integrated Control of Cigarette Beetle (*Lasioderma serricorne* F.) with a Synthetic Sex Pheromone and Food Attractant in the Cigarette Manufacturing Factory." *Kenkyu Hokoku Scientific Papers of the Central Research Institute, Japan Tobacco & Salt Public Corporation.*

Watters, F. L. 1969. "The Locomotor Activity of *Cryptolestes ferrugineus* (Stephens) (Coleoptera: Cucujidae) in Wheat." *Canadian Journal of Zoology* 47 (6): 1177–1182. https://doi.org/10.1139/z69-182.

White, N. D. G., and S. R. Loschiavo. 1986. "Effects of Insect Density, Trap Depth, and Attractants on the Capture of *Tribolium castaneum* (Coleoptera: Tenebrionidae) and *Cryptolestes ferrugineus* (Coleoptera: Cucujidae) in Stored Wheat." *Journal of Economic Entomology* 79 (4): 1111–1117. https://doi.org/10.1093/jee/79.4.1111.

White, N. D. G. R. T., P. G. F. Arbogast, S. R. R. C. Hillmann, B. Loschiavo, J. E. Subramanyam, J. E. Throne, and V. F. Wright. 1990. "The Development and Use of Pitfall and Probe Traps for Capturing Insects in Stored Grain." *Journal of the Kansas Entomological Society 63* 4: 506–525. http://www.jstor.org/stable/25085217.

Wright, V. F., and R. B. Mills. 1984. *Estimation of Stored-Product Insect Populations in Small Bins Using Two Sampling Techniques.* Proceedings of the Third International Working Conference on Stored Product Entomology October 23-28, 1983, Kansas State University, Manhattan, Kansas USA: 672-679

18 Mechanical Approaches for Insect Management in Grain Storage

Detection, Prevention, and Control

Vinass Jamali Pitchai Mydeen, Vismaya K. Sachithanandhan, and Eyarkai Nambi Vijayaram
Department of Food Plant Operations, Incubation and Entrepreneurship, National Institute of Food Technology Entrepreneurship and Management, Thanjavur, Tamil Nadu, India

Loganathan Manickam
Department of Food Packaging and Storage Technology, National Institute of Food Technology Entrepreneurship and Management, Thanjavur, Tamil Nadu, India

Shanmugasundaram Saravanan
Department of Food Plant Operations, Incubation and Entrepreneurship, National Institute of Food Technology Entrepreneurship and Management, Thanjavur, Tamil Nadu, India

Chandrasekar Veerapandian
Department of Food Process Technology, National Institute of Food Technology Entrepreneurship and Management, Thanjavur, Tamil Nadu, India

18.1 INTRODUCTION

The global dependence on stored grains necessitates rigorous protection against insect infestation. At the same time, chemical insecticides have traditionally held sway, concerns over residues, resistance development, and environmental impact fuel the search for safer alternatives. This chapter delves into the exciting realm of mechanical and physical approaches, offering research scholars an in-depth exploration of their potential and limitations. This chapter encompasses a diverse palette of techniques, including:

- Acoustic detection,
- Ozone,
- Low-pressure treatment of grains,
- Irradiation,
- Use of storage structures and packaging system,
- Thermal methods, and
- Modified atmosphere packaging (MAP)

DOI: 10.1201/9781003609834-18

Each section will be meticulously researched, presenting the latest scientific findings alongside insightful commentary from a research scholar's perspective. This chapter critically evaluates the effectiveness of each approach, highlighting its limitations and potential trade-offs. Finally, the chapter will culminate in a comprehensive conclusion, outlining future research directions and emphasizing the potential of these methods to reshape the landscape of stored grain insect control.

This chapter aims to serve as a valuable resource for scholars seeking to understand the intricacies of mechanical and physical approaches in stored grain insect control. By offering a balanced and critical evaluation, we hope to inspire further research and development in this rapidly evolving field, ultimately contributing to a safer and more sustainable future for global food security.

18.2 DETECTION METHODS OF INSECT INFESTATION IN STORED GRAINS

18.2.1 Grain Probes and Insect Traps

Insect populations in stored grain can be monitored using traps such as grain triers, deep bin cups, or vacuum probes (Hagstrum, Flinn, and Shuman 1996). An alternative trap option is the brass probe trap, initially developed by Atkinson (1971) and later redesigned with plastic materials. Pitfall and probe traps, available commercially (Treite Inc., Salinas, CA, USA), are commonly used for detecting adult insects in stored grain (White et al. 1990) are cylindrical tubes with perforations in the upper section that allows insects to drop into the trap but prevent their escape. These traps have a pointed tip for easy insertion into the grain. To assess captured insects, the traps need to be periodically removed from the grain bin and inspected to determine their quantity and type.

Estimating insect numbers is typically necessary for research on insect ecology and assessment of the efficacy of pest management. To measure the density of insects, probe traps can be utilized. A stacked variant of the probe trap was created by White et al. (1990) to look at insect behavior at various grain depths. Instrumentation for electrically counting probe trap catches was invented in 1996 and it is currently available for purchase (Stormax Insector, OPI systems, Calgary, AB, Canada). Because they stay in the grain for extended periods of time, probe traps find insects when typical methods of grain sampling, including the grain trier method (Barak and Harein 1982), fail to detect them.

The labor-intensive nature of this insect detection technique restricts the temporal availability of data and places the probe traps in inaccessible areas. The fact that so many factors affect trap catch makes it very difficult to interpret. When the length of the trapping process increases, the catch also rises proportionately (Fargo et al. 1989). The type of grain, temperature, and insect species all affect the rate of capture (Reed et al. 1991).

TNAU, Coimbatore (India), has created a variety of traps (Figure 18.1) (Mohan et al. 1994). Insects that stray toward the air are used as design inspiration. Because the two-in-one probe trap (Figure 18.1) combines a pitfall trap with a probe, it has a high efficiency. Pulse beetles are ideal for capturing since they tend to stray on the grain surface. An indicator device is made up of a circular dish and container covered in a sticky substance that is fastened at the bottom of a perforated, cone-shaped cup with a lid. The bug is removed, and its eggs are crushed by the TNAU automatic insect removal bin (Figure 18.2). Given that the majority of the insects can be eliminated in ten days, its efficiency is extremely high (almost 90%). After ten months of storage, it caused 1–4% damage to grains, compared to 33–65% damage in the regular bin for rice and sorghum grains. Storage godowns located 1.5 m above ground utilize UV light traps (Figure 18.2) that are incorporated with a 4-W germicidal lamp that emits ultraviolet light with a wavelength of 250 nm. Many stored grain insects can be captured with these traps, such as rice weevil (*Sitophilus oryzae* L.), red flour beetle (*Tribolium castaneum* Herbst), saw-toothed grain beetle (*Oryzaephilus surinamensis*), and lesser grain borer (LGB) (*Rhyzopertha dominica* F.) (Mohan and Rajesh 2016).

Comparative experiments have shown that physical traps were more efficient in detecting insect infestation than conventional sampling methods (Reed et al. 1991; White et al., 1990). The trap

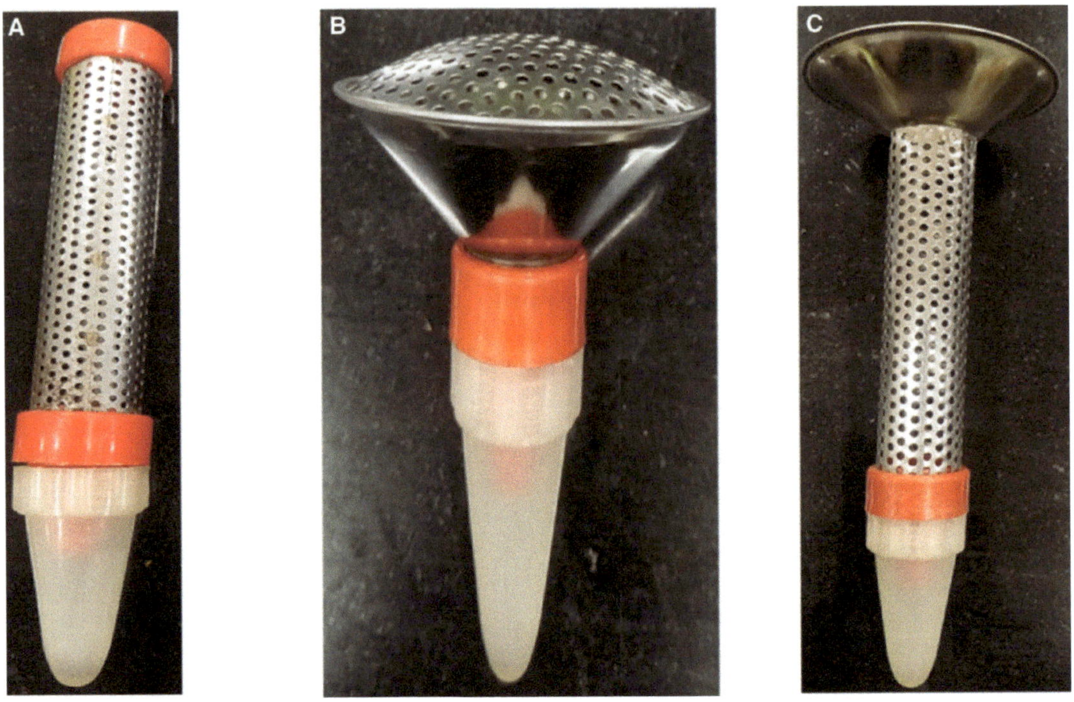

FIGURE 18.1 Images of various traps. (A) TNAU Insect Probe Trap, (B) TNAU Pitfall Trap, TNAU (C) Two-in-one Model Trap. (Photograph by the author.)

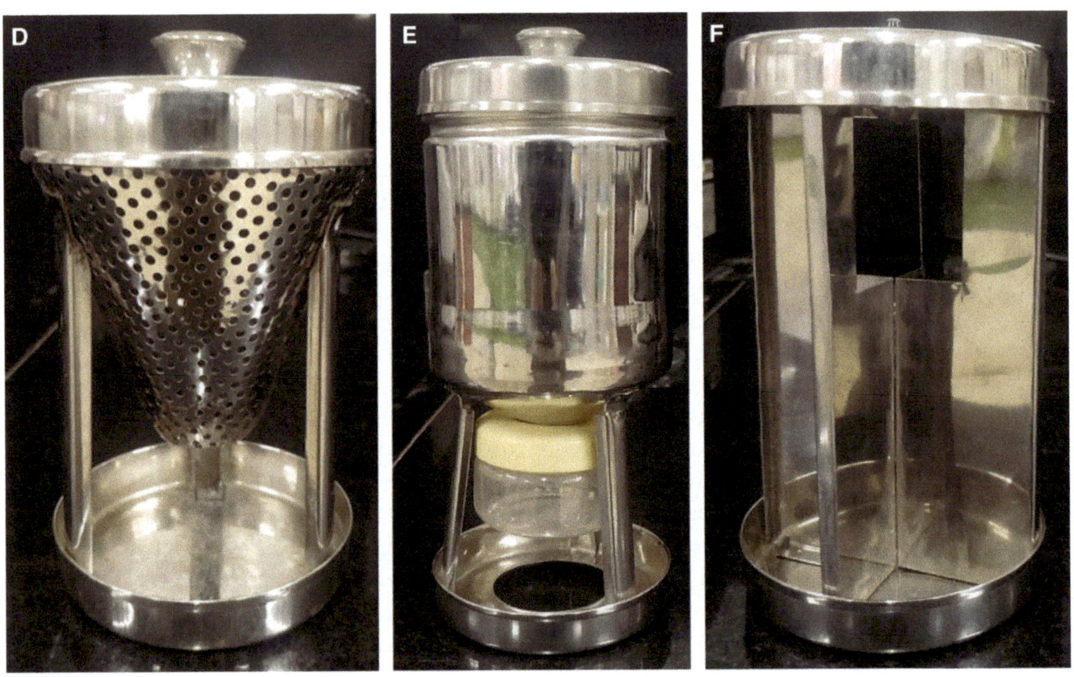

FIGURE 18.2 Images of various traps. (D) Automatic Insect Removal Bin, (E) UV Light Trap, (F) Indicator device. (Photograph by the author.)

catch is influenced by temperature, grain moisture content, and insect population density. The trap design, location, and duration of trapping also matter in their efficacy (Fargo et al. 1989). In a probe trap (Figure 18.1), if the holes are designed to slope upward instead of downward, it helps minimize the collection of grain debris inside the trap during the trapping period. Coating of "Fluon" around the neck of the collecting tube in the probe trap prevents the escape of trapped insects from the traps (Neethirajan et al. 2007).

18.2.2 Pheromone Traps

Pheromones are chemical compounds generated by insects used in communication. The efficiency of physical traps can be improved further by the incorporation of food attractants or pheromones in the trap (Ayertey 1989; Neethirajan et al. 2007). There are two types of pheromones: (1) aggregation pheromones and (2) sex pheromones. Sex pheromones are released by all moth pests as well as certain beetles, including *Trogoderma* spp. and *Stegobium paniceum*. Aggregation pheromones, released by the males of most beetle pests, attract both men and females of the same species. *Tribolium* species, *R. dominica*, and *Prostephanus truncates* are typical instances. Compared to aggregation pheromones, sex pheromones are more effective across longer distances (8–15 m). Insect behavioral reactions to pheromones are controlled by a number of variables, including temperature, photoperiod, and individual age. More than 30 stored product insects have had their pheromones identified; several of these have been synthesized and commercially produced for use in mass trapping, detection, monitoring, and control devices. To be used in the traps, these pheromones are impregnated into absorbent and slow-release membranes along with additional substances. When insects are drawn into the trap by the pheromone, they adhere to the replaceable adhesive sections. Pheromone lures are effective for 8–12 weeks, after which they need to be changed on a regular basis. To be used in the traps, these pheromones are incorporated into absorbent and slow-release membranes along with additional substances. When insects are drawn into the trap by the pheromone, they adhere to the replaceable adhesive sections. Pheromone lures are effective for 8–12 weeks, after which they need to be changed on a regular basis.

In preparation for use in monitoring, chemical attractants are usually coated or encased in a rubber or plastic lure that releases the active chemicals gradually over a few days or weeks. These lures are used in paper, plastic, and other types of traps. Most traps have an adhesive-coated surface or an entrance shaped like a funnel to capture the intended bug.

Certain pest traps have an adhesive covering them that also includes the chemical attractant. A range of traps can be used to find insects in stored products; some of them use artificial insect pheromones or food attractants. The LGB, *R. dominica*, the red flour beetle, *T. castaneum* (Herbst), which uses an aggregation pheromone, and the warehouse beetle, *Trogoderma variabile* (Ballion) (Banga et al. 2018), which uses a sex pheromone, are three stored-product insects that frequently coexist together and for which pheromones are available.

Pheromone traps can provide information about pest density; however, interpreting the results can be challenging due to a number of factors. The environment has an impact on trap catches. Insect flight and attractant release from lures are influenced by temperature, rainfall, wind direction, and speed. Many insects only fly and react to semiochemicals at specific times (dawn, lunchtime, dusk, or night) and temperatures (10–15°C), according to research.

For *Sitotroga cerealella*, commercial traps baited with pheromones include (Z,E)-7, 11-hexadecadien-1-ol acetate (or HAD). (Z,E)-9,12-tetradecadien-1-ol-acetate, also known as TDA, is used to treat pyralids such as *Ephestia* species and *Plodia interpunctella* (Hagstrum, Flinn, and Shuman 1996). When monitoring moth pests, the traps are set up at an ideal height of 2–3 m for every 14 m. Depending on the pests that need to be found and tracked, different trap densities are needed.

Chemical and optical cues or stimuli mediate the direction of insects to the traps. As a result, the effectiveness of the trap also depends on its shape and color scheme. Traps with darker stripes on a white background draw moths. Furthermore, they are more active in low light than in total darkness or during the day. As was previously indicated, several physical traps exhibit increased efficacy when used to capture insects when pheromones are present (Ayertey, 1989).

Pheromone traps have been used for crawling insects like *Tribolium* and *Trogoderma* species, as well as the active flyer *Lasioderma serricorne*, in the food sector and completed product warehouses. Only a 1.5–3 m radius surrounds the aggregation pheromones that are used in the traps. Pheromones have been identified for certain pests, including *Callosobruchus chinensis*, *O. surinamensis*, and *Cryptolestes* spp.; however, they have not been used in a commercial setting.

18.2.3 Electrical Conductance Method

Grain kernel weight, moisture content, diameter, and hardness are frequently measured using a single-kernel characterization system. The principles of electrical conductivity and compression force explain the operation of this system. In a circuit consisting of two resistors and a voltage divider, the kernel functions as a single resistor in the single-kernel characterization system. The voltage across the kernel is measured to track conductance. Low voltage readings are indicative of low kernel resistance, which is common in kernels with high moisture content. The conductance signal will most likely have a significant downward slope if a live insect is present inside a kernel. Infested kernels are distinguished from sound kernels based on the system's signal characteristics and the conductance signal's voltage level range.

Using electrical conductance, researchers found hidden interior insect infestations in wheat kernels. Their research revealed that there was not a single sound kernel that was incorrectly identified as infested and that the identification accuracy rates for all wheat samples were 88% for large-sized larvae and 87% for pupae.

Even with this low-cost technology, it takes time to examine individual grain kernels. Because of their low moisture content, infested kernels may go unnoticed, containing insect eggs and young larvae. Kernels containing dead internal insects cannot be detected by this method. When compared to soft X-ray inspection, this method's rates of insect detection are incredibly low. Since this approach only works with single kernels, the sample input poses additional challenges.

The electrically conductive roller mill technology uses compression force and electrical conductance to prevent infection in food grains that have been kept. One kernel serves as a resistor in the single-kernel characterization system with two resistors and a voltage-divider circuit. As the kernels are crushed between the rolls, the voltage is used to check the conductance of the kernels. When insects are present, the moisture content of the kernel rises, making it easier to distinguish between healthy and infected kernels (Pearson and Brabec 2007).

After testing 1 kg of wheat in roughly 2 min, Pearson and Brabec (2007) reported finding more than 70% of the kernels contaminated, along with *R. dominica* larvae and pupae. The two main insects that cause interior infestations were tested: the LGB (*R. dominica* F.) and the rice weevil (*S. oryzae* L.). It can be a helpful tool for classifying grains at receiving stations since it gives grain storage managers the information they need to decide whether to discard or fumigate a storage bin. To investigate the presence of LGB pieces in wheat flour, some work has been done with a conductive roller mill. They discovered that a conductance mill could test internally infested insects in low-density grains (up to three insects) in 1–2 kg of grain at a pace of 1 kg of grain per minute. In low moisture grains, this approach is not appropriate for identifying insect eggs, young larvae, or dead insects.

Pearson and Brabec (2007) employed a modified laboratory mill to identify immature LGB internal infestations in wheat and brown rice. Within 150 s, the modified conductance mill identified 97, 83, and 42% of large, medium, and small LGB larvae, respectively, in a 500-g sample of wheat and brown rice. When comparing wheat to thinner brown rice, the detection rate for wheat was greater. Using a conductive roller mill, the evidence of a maize weevil (*Sitophilus zeamais*) infestation in popcorn kernels at various stages of growth was found. After testing two mills, it was found that the faster mill recognized 75% of pupae and 80% of medium larvae, while the slower mill could only detect 81% of pupae, 91% of medium larvae, and 47% of small larvae.

18.2.4 Electromagnetic-Spectrum-Based Methods

18.2.4.1 Imaging Methods

18.2.4.1.1 Machine Vision

The inspection system for grain grading and insect infestation uses interpretative line prints and visual reference images (Banga et al., 2018). Grain kernels individually are compared in this system to the typical sample's photographic image. The drawbacks of this approach are its subjectivity, duration, and inability to detect interior infestations. It has been demonstrated that using computerized image analysis to find and identify different non-grain particles and insects in wheat has a lot of potential. A monochrome CCD (charge-coupled device) camera, a personal computer, and a high-speed integrated machine vision software package make up a machine vision system for spotting insects in grains. Using structural and chromatic data, researchers have identified *R. dominica* adults in wheat bulks with over 90% accuracy. Ridgway and Chambers (1996) developed a rapid machine vision method for the detection of adult beetles and determined that detection rates were 89% for commercial samples containing several insect species.

The slow rate of sample throughput is a major barrier to the development of commercially viable machine vision systems for insect detection in grain. Simple, fast, and reliable algorithms are needed to allow a statistically significant portion of a grain sample to be inspected in the short time available. Furthermore, device complexity, particularly in terms of camera type, computer specification, and sample delivery method, must be minimized to make this approach cost-efficient. Its limited ability to identify only exterior insects in grain bulks as opposed to inside insects that can be detected by near-infrared (NIR) spectroscopy and X-ray techniques is another drawback.

18.2.4.1.2 X-Ray Imaging

The only direct, non-destructive technique that can identify insect infestations in grain kernels is a soft X-ray (Neethirajan et al. 2007). Using soft X-ray imaging, Chelladurai et al. (2014) identified wheat kernels infested with *S. oryzae* (L) larvae and pupae-adults with over 97% accuracy. Additionally, they showed that a 99% accurate sound kernel identification system could examine 1 kg of grain in approximately 15 min.

Real-time images are produced using a fluoroscope operating at 65 A and 15 kV potential in a soft X-ray system. Singular kernels were scanned after being placed on plastic wrap on a platform that was positioned between the detecting system and the X-ray tube. A digital video maker digitized the images that developed on the detection screen after they were recorded by a CCD (charge coupled device) black and white camera. Insect-infested kernels were identified by processing the digital photos. Soft X-ray imaging is the only non-destructive, time-saving way for detecting insects in grains out of all the existing technologies. Grain intended for seed can be efficiently examined for mechanical damage using the soft X-ray method. It is challenging to discern weevil egg plugs and small larvae from heavier grain parts utilizing the soft X-ray approach (Chelladurai et al. 2014).

18.2.4.1.3 Thermal Imaging

Thermal imaging is a sort of night vision technology that improves object visibility in dimly light situations by detecting and translating the infrared radiation energy that an item releases into a viewable image. An object's heat signature, or the infrared radiation it emits as a function of temperature, is called a thermogram, and the image it produces. Thermal imaging cameras, detectors, lenses, and data-gathering instruments make up this imaging system. By identifying broken grains, foreign objects, and both internal and exterior infestation of grains, this technique helps manage stored food grains. When temperature variations are utilized to assess or quantify a process, it can be helpful. When it comes to determining material qualities and cutting costs, thermal imaging may be more advantageous than hyperspectral and fluorescence imaging.

Researchers have described using infrared thermal imaging with a camera, control panel, microwave applicator, conveyor, etc., to detect Rusty Grain Beetle (*Cryptolestes ferrugineus*) infestation in the wheat kernel. They employed the method to find *C. ferrugineus* within the seed coat of wheat kernels. To determine the insect's developmental stages, thermal pictures of the infected kernels were taken. The temperature distribution on the surface of the infested kernels showed a strong correlation (r = 0.83–0.91) with the respiration rate of the several stages of *C. ferrugineus*. For infected and sound kernels, the classification accuracy of the quadratic and linear functions was 83.5, 77.7, 77.6 and 83.0%, respectively. Chelladurai et al. (2014) employed thermal imaging to study *Callosobruchus maculatus*-infested moong beans. A total of 55.24–77.84% classification accuracy by LDA model and 75.45–91% classification accuracy by quadratic discriminant analysis (QDA) model were attained. Using the QDA model, it was possible to determine that more than 80% of moong beans were infested with *C. maculatus* in its early stages. A mid-infrared thermal camera was used to capture thermal pictures of the fungal-infected paddy at 0 (before heating), 180 (after heating), and 210 s (30 s after cooling). The moisture content was ascertained by using the average pixel of the image as a feature. When compared to non-fungal-infected paddy, fungal-infected paddy displayed higher average pixel values.

Chelladurai et al. (2014) obtained images of *C. maculatus*-infested soybeans as well as wholesome kernels using the soft X-ray and NIR hyperspectral imaging techniques. For data analysis, 48 features were retrieved from hyperspectral imaging and 33 characteristics from soft X-ray imaging. A total of 286 distinct phases of infestation were categorized using QDA and LDA models. For soft X-ray images, LDA was used to classify around 86% of non-infested grains and 83% of grains infected by *C. maculatus*. Principal component analysis (PCA) was used to classify hyperspectral data at 960, 1030, and 1440 nm in wavelengths. The accuracy of the classification of the egg and larvae stages was increased by combining hyperspectral and X-ray data.

18.2.4.2 Non-imaging Methods

18.2.4.2.1 Electronic Grain Probe Insect Counter (EGPIC)

The Electronic Grain Probe Insect Counter (EGPIC) is a type of automated passive grain probe that gathers data on insect pest infestation levels in stored food grains and displays it remotely. Perforated cylindrical tubes with insect collection receptacles pushed into the stored product mass are called insect probing traps, and the EGPIC design integrates these features.

It is designed to monitor and identify insect pests off-site. It is made up of a user interface, data recorder, system circuitry, and probe. Only low-voltage, high-impedance sensor leads are allowed to flow through the grains from the beam generation/detection circuitry to the sensor head due to the risk of grain dust explosions. Electric power and circuitry are kept outside of the storage structure. Sensor data is sent to a computer, which interprets the signals and creates time-stamped records of detection. The infrared diode generates a signal when an insect lands on the EGPIC sensor head across the infrared beam. The infrared phototransistor then receives this signal, causing a tiny declination in the light intensity that reaches the phototransistor. Insects escape the probe after the electronic circuitry system recognizes this minor declination and transforms it into a time-stamped insect count. It continuously monitors the insect's levels at any depth. During two years of storage, researchers have tested the commercial electronic grain probe trap in two bins containing 32.6 tons of wheat. They created a regression model and contrasted it with the EGPIC-estimated insect density. The anticipated insect density varied between 40 and 75% when compared to EGPIC. To estimate the densities of *C. ferrugineus* (Stephens), *R. dominica* (F.), and *T. castaneum* (Herbst) from trap capture counts, EGPIC were combined with an expert system called "Stored Grain Advisor Pro." Due to their similar sizes, the expert system's estimation of insect density was inefficient in distinguishing between *R. dominica* and *T. castaneum*.

18.2.4.2.2 NIR Spectroscopy

NIR spectroscopy has developed into a quick, dependable, precise, and affordable method for analyzing the composition of grains. Analyses that are both qualitative and quantitative can be conducted using this method. Information based on the reflectance characteristics of various compounds contained in a product is provided by the NIR method. The NIR is based on electromagnetic frequencies between 780 and 2500 nm being absorbed. Using traditional absorption spectroscopy, the amounts of components, including water, protein, fat, and carbohydrates, can be found.

The best way to identify individual wheat kernels containing live or dead internal rice weevils at different phases of their lives, according to Maghirang et al. (2002), is to use NIR technology. With an accuracy of 94, 92, 84, and 62%, respectively, they also classified sound kernels and kernels containing live pupae, large larvae, medium-sized larvae, and small larvae.

Several Coleopteran species have been identified using the NIR approach (Dowell et al. 1999); weevil infestation in wheat kernels (Perez-Mendoza et al. 2003); and both internal and exterior insect infestation in wheat have been detected (Ridgway and Chambers 1996; Dowell et al. 1999). One NIR system can scan a thousand kernels per second to find insects in them (Dowell et al. 1999). In order to identify insect pieces in wheat flour, Perez-Mendoza et al. (2003) contrasted an NIR system with the conventional flotation method. A NIR system, on the other hand, can process samples quickly (less than a minute per sample), doesn't require sample preparation, and may be readily automated to enable a more complex sampling procedure for big flour bulks.

The NIR approach is unable to distinguish between live and dead insects or identify low infestation levels in bulk samples (Perez-Mendoza et al. 2003). The device needs to be calibrated often because the NIR technique is highly sensitive to the moisture content of samples. Using a NIR vidicon camera, Ridgway and Chambers (1996) obtained NIR images of wheat samples that were infested. They came to the conclusion that NIR images can identify insects in grain samples by observing changes in kernel composition brought on by infestation after comparing the NIR images with the X-ray image of the sample. The movement of the larvae in the cavity caused by the heat produced by the lighting makes the NIR imaging method, which is an indirect method, unable to capture larvae.

Particle size of powders and other physical factors alter the spectra, obscuring the chemical information of the dietary items. As a result, NIR spectroscopy becomes a secondary technique that needs to be calibrated for the ingredient of interest against a reference method. Multivariate mathematics is typically used for calibration due to the intricacy of the spectra and the physics of diffuse transmittance and reflectance.

Chelladurai et al. (2014) identified the infested and sound soybean kernels using NIR hyperspectral imaging and soft X-ray imaging to identify *C. maculatus*. Using an automated NIRS system, researchers have found that live pupae and larvae of *S. oryzae* (L.) with 92–93% accuracy (400–1700 nm). Perez-Mendoza et al. (2003) compared the traditional flotation method with NIRS for wheat flour and discovered that NIRS was the faster method (less than 1 min/sample), allowing for bulk sample detection without sample preparation. Moldy maize kernels were detected with an 86.7% validation accuracy using Fourier transformation NIR (FT-NIR) technology, a quick detection tool, on characteristic wavelengths of 1466, 1530, 1926, 2321, and 2384 nm (Xia et al. 2015).

18.2.5 Ozone Fumigation

Large amounts of grain can be penetrated by ozone, which is also extremely unstable and breaks down quickly into oxygen without leaving any trace (Kells et al. 2001; McDonough et al. 2011b). Ozone is a desirable option for managing fungi and insects in products that have been preserved because of these qualities (Kells et al. 2001). Many researchers have carefully examined how ozone, in its gaseous state, kills insects in a variety of commodities, both in the lab and in the field, using a range of application procedures (Kells et al. 2001; McDonough et al. 2010, 2011a, b; Athanassiou and Arthur 2018).

McDonough et al. 2011(b) developed a model based on CT (concentration x exposure time) values, which are obtained by multiplying the needed exposure time (T) by the ozone concentration (C) to obtain the desired mortality level. The authors of that study determined the CT values for ozone treatments needed to achieve 100% death, and they discovered that *P. interpunctella* requires lower ozone concentrations, whereas *T. castaneum* requires longer exposure times (a mean exposure of 256,500 ppm-min and 183,000 ppm-min, respectively).

Additionally, Sousa et al. (2008) used adult *T. castaneum* insects in laboratory experiments. The insects were treated with 150-ppm ozone in plastic chambers to achieve a 95% mortality rate, with treatment times ranging from 23.35 to 31.98 h (CT ¼ 210,150 ppm-min and 287,820 ppm-min, respectively). Adult *T. castaneum* in cages buried in corn were treated with 50 parts per million ozone by Kells et al. (2001). The death rate of 92% for a CT value of 216,000 ppm-min after treatment is consistent with the findings of Sousa et al. 2008; McDonough, Campabadal, et al. (2011a).

Airspeed, temperature, and relative humidity (RH) all have a significant impact on how long ozone lasts. According to Mcclurkin and Maier (2010), ozonating grain works best in still air at low temperatures and low humidity. Similar to this, Donahaye, Navarro, and Leesch (2001) investigated the toxicity of gaseous ozone to adults of *Tribolium confusum* and various stages of *P. interpunctella*. He discovered that all stages of *P. interpunctella*, with the exception of eggs, were sensitive to ozone treatment in the lab for a 4-h exposure period at 300 ppm. Adults were the most easily killed life stage, followed by larvae and pupae; eggs were not impacted. According to Bonjour et al. (2011), *P. interpunctella* pupae were very responsive to ozone treatment; however, the species' eggs were not effectively controlled by it. The most ozone-tolerant stages of *T. castaneum* were the pupae and eggs, which needed 180 min at 1800 ppm ozone to achieve 100% death. Since the insect's respiratory system is most likely to be the primary factor in death following ozone exposure (Tiwari et al. 2010), ozone insecticidal activity that targets this system may be attributed to lower respiration rates in the eggs and pupae life stages (Hoback and Stanley 2001).

Mason, Woloshuk, and Maier (1991) observed that following exposure lengths of 3 and 5 days, respectively, 5 ppm of ozone caused the complete death of adult sawtoothed grain beetles, *O. surinamensis* (L.) (Coleoptera: Silvanidae) and *T. confusum*. The duration of exposure in these investigations was significantly greater than that examined by (Donahaye, Navarro, and Leesch 2001; Işikber and Öztekin 2009); however, the levels of ozone were significantly lower. As a result, variations in ozone concentration, exposure duration, or application technique, as well as variations in the strain of the insects, can be blamed for any variations in the effectiveness of ozone against stored product insect species. In this sense, applying ozone at short exposure times (24 h or less) is more crucial than using greater dosage rates for longer periods of time to truly establish it as a substitute for conventional fumigants.

A huge challenge in using ozone as a fumigant is getting the right dosage and maintaining continuous purging over the course of the treatment. Despite O_3's demonstrated efficacy as a fumigant, care must be given while selecting commodities because O_3 is a potent oxidant. Large-scale applications, pilot-scale research, and studies pertaining to the design and development of suitable storage structures with closed-loop components for O_3 treatment still needs to be established.

18.2.6 Low-Pressure Storage

Low-pressure storage refers to the planning and creation of a grain storage facility with a lower internal pressure. This represents one of the storage structure's innovations, which relies on changing the physical environment—typically accomplished by applying a vacuum and producing low-oxygen environments. Physiological functions are impacted by this reduced O_2 concentration. Under this approach, dehydration and hypoxia are the main causes of insect mortality (Navarro et al. 2001). Low-pressure storage refers to the planning and creation of a grain storage facility with a lower internal pressure. This represents one of the storage structure's innovations, which relies on changing the physical environment—typically accomplished by applying a vacuum and producing low-oxygen environments. Physiological functions are impacted by this reduced O_2 concentration.

Under this approach, dehydration and hypoxia are the main causes of insect mortality (Navarro et al. 2001). In a similar manner, Mbata et al. (2005a) studied the link between temperature and pressure in altered environmental settings and found that low pressure alone was less lethal to the various developmental stages of *T. castaneum*, *P. interpunctella*, and *R. dominica*. Furthermore, the longer a person was exposed to low pressure and the higher the temperature, the higher the fatality rate throughout all life stages. Mbata et al. (2005b) subjected *C. maculatus* to various low pressures and temperatures in a related investigation. According to their findings, adults had the highest susceptibility, with a fatality rate of 99% within 0.8 h at 32.5 mm Hg and 30°C. Navarro et al. (2001) discovered noteworthy outcomes for insect mortality when they employed a vacuum method to eradicate *T. castaneum*, *Ephestia cautella*, and *O. surinamensis*. The sensor-assisted CO_2 supply that was established more recently is a cause for concern for large-scale treatments (Banga et al. 2018). CO_2 can, however, be used in various ratios with other fumigants, depending on how effective it is. As an alternative to chemical fumigation, extensive field studies on modified structures with standard doses (a mixture of gases) may be given priority.

As a strategy for managing pests, low pressure offers a possible nonchemical substitute for fumigants like methyl bromide and phosphines, which are used to control bruchids (Mbata et al. 2005a). With a rise in temperature and exposure duration, the time-mortality response of red flour beetle eggs, early larvae, old larvae, pupae, and adult stages increased. More precisely, young larvae and eggs are more vulnerable to high temperatures than older larvae (Boina and Subramanyam 2004). At higher temperature gradients, rusticated grain beetle adults *C. ferrugineus* (Coleoptera: Laemophloeidae) elicited a faster response than at lower temperature gradients. The cowpea weevil, *C. maculatus*, is killed by low pressure, which produces a low oxygen-controlled atmosphere. In all flour-stored grain insects, high temperatures mixed with low pressures are more effective in killing more eggs than low temperatures and high pressures (Mbata et al. 2005b). When subjected to high temperatures and low pressure, the lethal time values of *T. castaneum* (Herbst), *P. interpunctella* (Hubner), and *R. dominica* (F) eggs, larvae, and pupae were decreased. However, the most resilient to low pressure were the eggs of each species (Mbata and Phillips 2001). When it comes to identifying insect pests and simulating their spread in grain bins, the movement and distribution of adults in grain provide crucial information.(Jian, Jayas, and White 2004).

Acoustic approaches are the most effective in detecting hidden infestations of stored-product insect larvae during their peak activity. When the larvae are chilled to a very low temperature, their periodic activity tends to reduce. Under unfavorable conditions, heat treatment increases larval activity and enhances the accuracy and speed of acoustic detection. It is possible to superheat food grains using infrared microwave radiation. High temperatures are produced via fluidized bed heating in order to rid the depth of infestations. If chilled grain samples are warmed and all samples are left undisturbed for 15–20 min before inspection, the detectability of larvae is improved, as given by Jian, Jayas, and White (2004).

18.3 CONTROL MEASURES OF INFESTATION IN THE STORAGE OF GRAINS

18.3.1 The Potential of Acoustic Detection for Stored Grain

Acoustic tracking is a limited technique used to identify and track the growth of concealed larvae that consume grains internally in items that have been stored. It allows for unobtrusive recording of the precise moment when the larvae cease devouring throughout the molting process from one developmental stage to the next phase (Lima et al. 2020). Monitoring the decrease and stoppage of feeding has been valuable in assessing the efficacy of entomopathogens, host plant resistance, hermetic storage, and pesticides (Azfar et al. 2018). With the use of impact acoustics, grain breakages have been identified among wheat, rice, and corn. Insect acoustic capturing has also been utilized for monitoring purposes. The use of acoustic technologies is very advantageous in detecting concealed larvae infestations in stored grain. Insect pests in both adult and juvenile stages are frequently found in grain samples. Distinguishing and extracting immature individuals concealed inside the grain kernels is a more challenging task compared to identifying and separating adults. To identify immature organisms, the conventional approach involves subjecting the grain samples to incubation. The incubation period might last for a maximum of two months (Fleurat-Lessard 2011). To minimize this delay, alternative approaches have been suggested that rely on identifying clear signs of insect presence or activity (primarily targeting more developed larval stages).

These methods include measuring respiratory gas release, using X-ray radiography, employing flotation techniques, utilizing Ninhydrin colorimetric methods (Fleurat-Lessard 2011), and employing acoustic indices for detection. The acoustic technique enables the identification of adult insects and their concealed young counterparts that are feeding on grain (Hagstrum, Flinn, and Gaffney 1998). It also allows for the non-invasive inspection of adult insects and immature ones in preserved grains. The provided information can be digitally transferred to both corporate and open databases. This data is intended for use by researchers, research data alliances, as well as regulatory authorities responsible for the protection of plant, animal, and human health. Further contributions to these databases will enhance the future computerization of investigations aimed at monitoring the global abundance, distribution, and diversity of insects (Høye et al. 2021). Furthermore, this report highlights the growing utilization of acoustic technology, along with other sensor-based technologies, for the detection and control of insect pests in stored grain. These technologies also hold significant potential for managing concealed pest insects in underground environments and trees, offering non-destructive alternatives to existing detection methods (Zhai et al. 2020). Acoustic sensing applications provide automated identification of insect pests, a task that will gain significance with the growth of human populations and the depletion of natural resources (Vincent et al. 2008). So far, commercialized pest detection technologies have not been able to match the variety, user-friendliness, and affordability of mobile phones that are now available. As a result, they are not quickly replacing commonly used methods such as pheromones or probing traps, screening, or other ways of detecting nuisance insects (Montgomery et al. 2021). Corporate manufacturers of pest acoustic monitoring technologies remain interested due to the possibility of enhanced portability and automation (Mankin et al. 2021).

18.3.1.1 Sensors Type Used in the Detection Process

18.3.1.1.1 Thickness-Shear Mode Sensors

Thickness-Shear Mode (TSM) resonator is a prevalent acoustic wave device utilized on sensors and analyzers. The operating concept of this device is based on the piezoelectric effect, which involves the generation of a voltage when mechanical stress is applied to certain materials, such as quartz/lithium niobate (Figure 18.3). A TSM resonator commonly consists of a slender piezoelectric crystal disc that is generally positioned sandwiched between two conducting electrodes (Thomas 2017). Applying an alternating voltage between these electrodes induces deformation in piezoelectric crystal, resulting in the generation of shear waves, which are mechanical vibrations. The shear waves in concern travel in a path transverse to the crystal's surface. The majority of the energy that is generated is focused in the thickness direction, which is why they are referred to as "Thickness-Shear Mode" (French et al. 2007).

The resonating frequency is dictated by the crystal plate's thickness and its elastic characteristics. Resonance occurs when the frequency of the applied voltage aligns with the intrinsic frequency of the quartz crystal, leading to the highest level of mechanical vibrations. The frequency at which resonance occurs is extremely responsive to changes in the mass loading or mechanical characteristics of the crystal surface (Bajaj et al. 2021).

18.3.1.1.2 Surface Acoustic Wave (SAW) Sensors

Surface Acoustic Wave (SAW) sensors are advanced devices used to detect variations in physical properties like pressure, temperature, and mass (Figure 18.3). They do this by translating these variations onto sound waves which propagate over the surface of a piezoelectric substrate (Liu, Lee, and Chahl 2017). These sensors function by sending and receiving sonic waves using interdigitated transducers that are designed on the surface of the substrate. When a wave meets a disturbance resulting from a specific factor, such as the attachment of a particular molecule in a

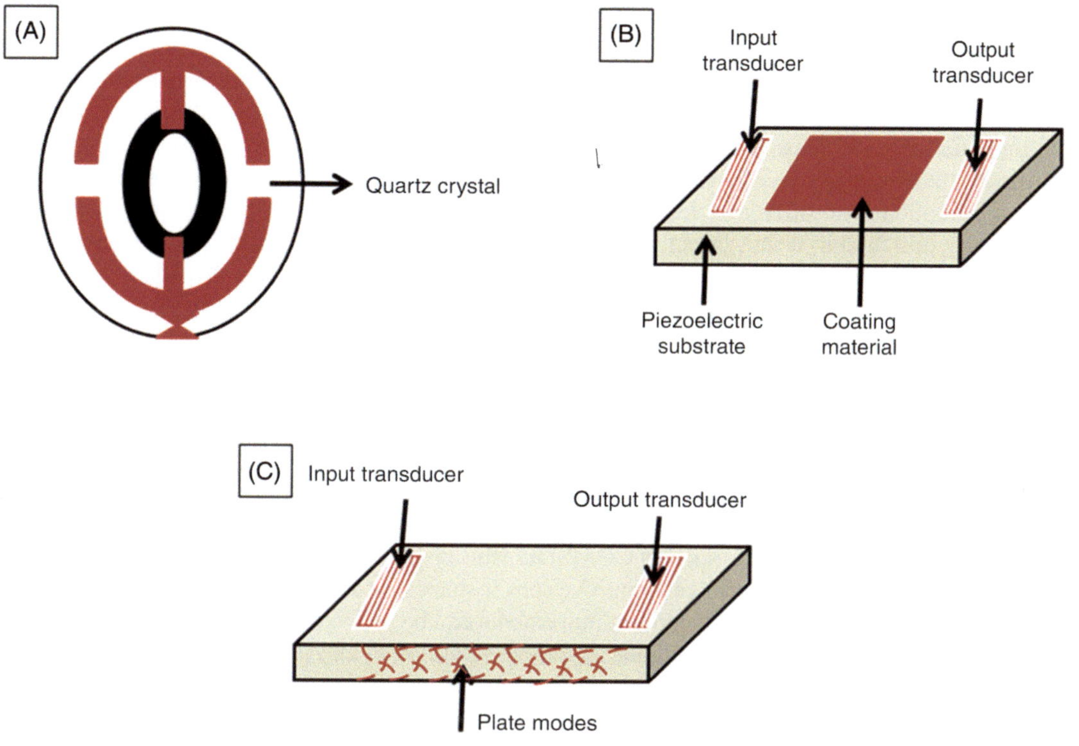

FIGURE 18.3 Schematics showcasing the distinct structures and mechanisms of (A) Thickness-Shear Mode, (B) Surface Acoustic Wave, and (C) Shear-Horizontal Acoustic Plate Mode sensors. (Illustration by the author.)

biochemical sensing scenario, it experiences a modification in its speed or intensity. This alteration is then identified and examined to deduce the existence or extent of the factor in question (Mankin et al. 2021).

18.3.1.1.3 Shear-horizontal Acoustic Plate Mode

The Shear-Horizontal Acoustic Plate Mode (SHAPM) sensor functions based on the propagation of shear-horizontal acoustic waves over the outer layer of a piezoelectric substrate using a complex methodology. The sensor generally comprises a slender plate made of piezoelectric material, such as lithium niobate or quartz, onto which interdigitated electrodes are applied (Thomas 2017). When a voltage is supplied, these electrodes produce an electric field that causes the shear-horizontal acoustic waves to get excited. SHAPM sensors differ from standard SAW sensors by generating and detecting shear-horizontal waves instead of longitudinal waves over the substrate surface. This provides special benefits in certain applications (Zida, Lin, and Khung 2021). The SHAPM sensor operates based on the relationship that exists that occurs between transmitted shear-horizontal vibrations and the object being measured analyte or the surroundings. When waves come across modifications in the mechanical characteristics or bulk deposition on the outermost layer of the substrate, such as caused by the existence of an analytical or biological analytical agent, they experience variations in velocity, phase, or attenuation (Figure 18.3). The sensor detects these modifications by analyzing the disturbances caused by the electrical signal produced by the interdigitated sensors. Through the examination of these changes in the wave features, the sensor can precisely ascertain the existence, amount, or attributes of the object being monitored component (Islam et al. 2018).

18.3.1.2 Features of the Acoustic System

To effectively detect and analyze the acoustic signals generated by insect movements within stored grains, it is necessary to use specialized equipment specifically designed for this purpose. Below are the essential equipment and tools required for this development:

***Acoustic sensors*:** Sensors of superior quality are necessary for collecting the noises produced by insects in stored grains. These sensors have to possess a broad spectrum of frequencies and exhibit exceptional sensitivity to detect subtle signals. Piezoelectric transmitters or microphones specifically designed to detect insect sounds are often used for this objective. Other sensors used are seismic sensors (Lima et al. 2020), microelectromechanical sensors (MEMS), and sonic tomography.

- ***Data acquisition system*:** A data collection system is necessary to gather and analyze the information obtained from the acoustic sensors. This system generally consists of converters that convert analog to digital (analog-to-digital converter [ADCs]), boosters, filters, and signal processing circuits to guarantee precise signal capture and conditioning.
- ***Computation*:** Computational technologies are essential for optimizing sensor performance by modeling and simulating the transmission of acoustic waves and the behavior of sensors. FEA software, such as COMSOL Multiphysics or ANSYS, allows engineers to model acoustic wave exchanges, improve sensor shapes, and forecast sensor effectiveness. The captured signals from the insect movements can be analyzed under some more different forms of computational methods as follows (Mankin et al. 2021).
- ***Preprocessing of signals*:** In general, the amplified analog recordings are subjected to spectrum-pass filtering, which involves concentrating on the frequency range of around 80–8000 Hz. This process of filtering occurs to catch the short impulses (3–30 m/s) that are generated by insect locomotion and consumption. An often-used classification technique involves calculating the frequency of signal segments that are above a certain threshold and then categorizing stored product specimens as infected if the frequency surpasses a minimum value defined via experimentation (Lima et al. 2020).
- ***Spectral and time-domain features*:** Certain characteristics of sound waves are very effective in accurately recognizing insect noises and distinguishing them from background noise. Methods that extract these characteristics encompass periodic recognition of patterns in the time domain, spectrum analysis, wavelet evaluation, formant assessment, and Independent Component Analysis(ICA).
- ***Classification models*:** The two techniques are the Gaussian Mixture Model (GMM) and Vector Quantization (VQ). These clustering methods are additionally employed to differentiate between characteristics of target signals and background sounds
- ***Classifiers based on neural networks*:** ANN are computational models that mimic the mechanisms of neuron connection in the nervous system. They take inputs that are weighted from data collected through observation, such as acoustic impulses, and generate one binary result. An ANN learns the right representation of the output by using reverse propagation or other techniques (Dong, Yan, and Wei 2018). Machine learning utilizes various neural networks such as CNN and (Probabilistic Neural Networks) PNN, as well as other methods like decision trees, HMM (Hidden Markov Model), SVM (Support Vector Machine), Bayesian classification algorithms, and (Perceptual Learning Prediction) PLP to enhance its predictive capabilities. This improvement is achieved through the analysis of information sets where the intended species of insects have been determined autonomously (Horvitz and Mulligan 2015).

Algorithms are used for categorizing or organizing data into different classes. Various techniques have been employed to differentiate target signals from various other signals, including, Linear Predictive Cepstral Coefficients (LPCCs), Sub-Band-Based Cepstral Coefficients (SBC),

(Santiago et al. 2017) wavelets (Mankin et al. 2021) (Mel Frequency Cepstral Coefficients) MFCCs, K Neighbors (Eliopoulos, Potamitis, and Kontodimas 2016) classifiers, and Medium and Complex tree classifications (Rathore et al. 2019).

18.3.1.3 Factors Influencing the Acoustic Signal

18.3.1.3.1 Size and Stages of Insects

The size and developmental stage of insects have a significant impact on the loudness and pace at which they produce sound. The sensors have the ability to detect *S. oryzae* larvae at a distance of 10–15 cm in grain (Mankin et al. 2011). In a study conducted by Hagstrum, Flinn, and Gaffney (1998), it was shown that adult *R. dominica* made 37 times more noises while they were exterior the grain kernels compared to larvae who were devouring within the grain. According to Hagstrum and Milliken (1991), adults can hear noises up to a distance of 18.5 cm. However, the rate of detection of noises from a little insect near a sensor may be comparable to those coming from a bigger insect that is farther away. Certain insects exhibit two different responses when perturbed: fleeing or feigning death. This behavior can have varying consequences on sound output, either good or negative. Agitating grain can temporarily decrease sound emission in the fourth-instar of *S. oryzae* for up to 20 min. Similarly, in another study by Hagstrum and Milliken (1991), it was seen that *T. castaneum* adults created 80 times more sounds than larvae. *S. oryzae* (Hickling and Wei 1995), and *C. maculatus* (E) larvae on cowpeas, *Vigna unguiculata* (L.) made more noises each instar (Njoroge 2017). Various research has examined the efficacy of raising the temperature of insects to enhance their recognition. Elevating the temperature inside the grain from 11°C to 17°C resulted in a significant boost in the sonic emission of *S. oryzae* larvae that were eating within, with a rise of up to 20–30 times.

The frequency range of insect sounds is positively correlated with temperature, particularly between 10 and 25–30°C, for both adult and juvenile individuals of insect species. This correlation is likely due to an upsurge in insect devouring activity (Mankin et al. 2021). The generation of sound by mature *T. castaneum* continued to rise until reaching a temperature of 40°C. The generation of sound by *C. maculatus* larvae diminished when the temperature exceeded 38°C in cowpeas.

Physical elements include background noise, the distance to the receiver, the receiver's spectral sensitivity, and the strength, duration, and spectral characteristic of the sound at the source. Unfavorable environmental conditions, insect behavior, and insect inactivity are examples of biological variables.

18.3.1.3.2 Time of Assessment

Insects emit noises with varying amplitudes at irregular intervals. Extending the evaluation period can enhance the probability of an insect within the sensor's active area generating an identifiable signal. Under circumstances involving little disruption and within the temperature range that facilitates the target species' hustle and bustle, listening periods of 30–180 s are appropriate for several detection purposes. The stored-product insects, specifically *S. cerealella*, *R. dominica*, and *S. oryzae*, make feeding noises in grain with frequencies of 61%, 71%, and 90%, respectively (Vick et al. 1988). Sampling periods of 180 s often capture at least one instance of activity. The infestation severity of stored grain beetles was accurately determined by utilizing eight microphones in a 1-kg sample bottle (Hagstrum, Flinn, and Gaffney 1998). Each sensor was examined for a duration of 10 s, repeated 72 times throughout a 12-min period. Insufficient knowledge about the behavior of the mysterious insect or the sound characteristics of the surface can result in a high number of incorrect results if the insects stop performing their tasks when perturbed. Therefore, it is recommended to conduct multiple tests over a 10–20 min time frame to figure out if the activity of the insects is enhanced periodically after inserting the waveguide through the surface. Extended sample periods can aid in the identification of distinctive sounds and rhythms in time that differentiate background noise from sounds emitted by the species of interest (Mankin et al. 2021).

18.3.1.4 Recent Application on Insect Control with Acoustic Sensors

Microphones with a high level of sensitivity were placed within a metal tube and injected into the grain mass. The tube had a dual purpose: safeguarding the delicate equipment from dust and external pressures, while also offering acoustical benefits by expanding the surface area for detecting beetle signals. Furthermore, the tube functioned as a beetle trap, capturing and documenting all noises emanating from a solitary beetle confined within the trap (Sutin et al. 2019). Some wavelet analysis was employed to differentiate between *S. oryzae* larvae and *S. zeamais* larvae in wheat, maize, and *Coix lacryma-jobi* (Poales Poaceae) millet (Geng, Zhang, and Zhao 2017).

Acoustic profiles were generated for *Prostephanus truncatus* (Horn), *S. zeamais* (Motschulsky), and *T. castaneum* in maize, as well as *Acanthoscelides obtectus* in beans. The frequency range with the highest spectral energy was determined to be between 3 and 8 kHz. The presence of banana weevil larvae (*Cosmopolites sordidus*) within the banana was identified using an acoustic method with a high accuracy rate of 90%. This method involved using a basic sensor (diaphragm stethoscope) coupled to a filter that effectively separated the signal produced by the larvae from any surrounding background noises (Vinatier and Vinatier 2013).

The integration of acoustic data, pit-fall trap-capture data, and background noise obtained in real-world situations within big grain storage facilities yields ample information for the creation of automated decision-support tools. Fürstenau (2023) worked on the development of an acoustic system named The "Beetle Sound Tube" technology which will be tailored to farmers and small storage facilities to detect and identify infestations in grain silos. This enables early treatment and minimal losses, including the administration of parasitoids via the tube system. Two tests were carried out in big wooden boxes filled with wheat and outfitted with data loggers and microphones to assess temperature, humidity, and acoustic properties. The studies intended to create an auditory early detection system for insect infestations. The recordings were analyzed by counting the number of insect signals at 15-s intervals. Monitoring grain quality in storage is critical for insect management and overall quality maintenance.

In order to control *P. truncates* and *S. zeamais* in stored food grains in Sub-Saharan Africa, Kiobia et al. (2015) created a sound and vibration-controlled device to identify the pests in maize storage. They discovered that within a 25 cm distance, the sensors could detect larval impulses. On the basis of variations in the spectrum and temporal pattern of insect noises connected with variations in physiological activity, software tools were also developed. Compared to low-energy movement sounds, feeding noises produced a broad, high-frequency spectrum. In comparison to a single impulse, trains of consecutive insect impulses (200 ms or less) established a more meaningful indicator of insect presence.

18.3.1.5 Limitations of Acoustic Sensing

When employing acoustic sensors to detect and monitor insects in a grain mass, it is important to consider many factors for precise results. These factors include determining the appropriate number of sensors needed, identifying the precise position inside the grain mass to install the sensors, and analyzing the data provided by the sensors (Mankin et al. 2021). In addition to these concerns, it is important to examine additional challenges.

- One such difficulty is that when detecting insects in silos, the acoustic signal weakens as the average spectrum of signals sent by the focused insects varies depending on the kind of grain and also changes with the distance from the insect's location (Preti, Verheggen, and Angeli 2021).
- The objective is to isolate and diminish the ambient noise originating from insects during the process of grain settling.
- Impact of ambient conditions on insect-generated sound.
- Differentiation of auditory signals emitted by adult and larval insects, as well as distinguishing signals from targeting bugs and non-target insects (Lima et al. 2020).

- An underlying drawback of the acoustic approach is its inability to identify the early developmental stages of insects, including eggs and pupae, as well as deceased insects present inside the grain. These undetected entities may have significant implications for human health. Insects' resonances may be disrupted by surrounding noises, which are often faint and susceptible to being overwhelmed by other sounds (Kaushik and Singhai 2018).
- Challenges encountered with utilizing sensors in a big grain mass include the proper deployment and design of the sensor to provide accurate readings. Monitoring factors related to the environment simultaneously necessitates a hardware design that is tailored for this purpose. Data security and privacy pose additional concerns. The testing and calibration of a device's sensor is a separate concern since the degree of sensitivity of the sensors may be influenced by several environmental factors. When employing a sensor system, it is important to consider data collecting, data processing, and data recording (Prince et al. 2019).
- One drawback of acoustic approaches is that they are not able to identify dead insects in grain or infestations by early stages of insect larvae.

18.3.2 Demystifying Irradiation for Stored Grain Protection

18.3.2.1 Unveiling the Molecular Processes of Irradiation in Insect Control

Irradiation affects biological creatures via two primary mechanisms. These effects may be categorized as non-thermal, meaning they are reversible, or thermal, which are irreversible. The point where both of these variables intersect is referred to as the threshold of microwave radiation, also known as the density of energy. Thermal impacts occur due to dielectric loss in the body. The body may experience localized overheating due to the movement of molecules and their mutual friction and collisions. This might result in the debilitation and death of the species (the elimination of living entities is predicated on this concept) (Novotny et al. 2013). Oxidative stress is the result of the presence of reactive oxygen species (ROS), which consist of superoxide, peroxyl radical, alkoxyl, the hydroxyl group, and nitric oxide (Gaikwad, Gaikwad, and Bhawane 2010). Antioxidants play a crucial role in all living organisms by preventing the oxidation of substances that might potentially cause oxidative harm (Halliwell, Zhao, and Whiteman 2000). Oxidative stress arises when the concentration of ROS surpasses the amount of antioxidants available in the host organism. The reduction of antioxidants may, therefore, function as a reliable biological indicator of oxidative stress.

Doses below 0.1-kGy or 100-Gy effectively prevent the growth of mature individuals and cause damage during the feeding stages. Insects exposed to radiation may survive for a while after treatment. The mid-gut of the insect is composed of undistinguished epithelium, rendering it very susceptible to radiation (Kabbashi et al. 2012). As a result, irradiated insects often stop feeding right after they are treated. The sensitivity of insect cells to radiation is influenced by their reproductive capacity and is indirectly correlated with their level of differentiation. Given that their tissue hasn't fully developed yet and is still dividing, eggs are their most vulnerable phase to ionizing radiation. Due to the sensitivity of gonadal cells, insects that can withstand the effects of radiation at permitted doses should become infertile. Insect larvae possess minimal cellular differentiation (Yamada et al. 2019).

Large doses are lethal to mature larvae, whereas lesser quantities inhibit larvae from advancing to the subsequent developmental phase, such as preventing pupae from emerging. The larval stage of insects exhibits resistance to radiation, but exposed larvae die after undergoing pupation (Yanagawa et al. 2020). Even though adults are an exceptionally resilient life stage, small amounts of irradiation can render them sterile, impairing their capacity to fly, locomote, and feed (Yanagawa et al. 2020).

18.3.2.2 Understanding the Importance of Dosage Levels for Stored Products

The insect developmental phase that has undergone the greatest evolutionary development is also the best capable of withstanding radiation. Female pests, including mites, are sometimes disinfected at a lower rate than males (Banga et al. 2018). Typically, irradiated insects have reduced lifespans compared to their non-irradiated counterparts. Low-oxygen conditions may boost chemical infestation and potentially increase radiation resistance. Accelerated dosing levels and the presence of oxygen were associated with increased mortality. Studies on stored-product insects revealed no correlation between decreased temperatures and insect sterility. In contrast to commonly held beliefs, there is no discernible disparity in radiation tolerance between pests in phases of diapause and non-diapause (Hallman et al. 2013). The merchant beetle, saw tooth beetle, and other cucujid grain beetles, along with the majority of flour beetles (tenebrionids), have a moderate level of sensitivity. The anobiids, cigarette beetles, spider beetles, skin beetles, and (ptinids) are z becoming more resistant. Stored products Due to their resistance to irradiation, Lepidoptera might require doses of 500 Gy to achieve sterilization of an entire population (Bakri, Mehta, and Lance 2005). Rice, wheat, and granary weevils (curculionids) have a similar level of sensitivity to bean or legume weevils (bruchids). The depressed flour beetle (*Palorus subdepressus* Wollaston) is a particularly resilient insect when it comes to infesting stored products. It underwent reproduction after being exposed to a radiation dose of 0.3 kGy (Krüger et al. 2018).

At least 33 nations authorize irradiation of stored commodities, while 14 allow it across the board (Hallman et al. 2013). After 300 Gy of radiation from gamma rays, Chiluwal et al. (2019) observed that female reproduction, the likelihood of hatching, and adult fecundity decreased in *C. chinensis* adults. Sileem, Mohamed, and Mehany (2019) tested gamma-irradiation and food-safe diatomaceous earth *upon T. castaneum*, *Sitophilus granarius*, and *R. dominica*. The amalgamation of gamma-radiation of 100 Gy and DE @ 1 g/kg caused total fatality with treatment conditions of 0, 64, 128, 192, and 256 Gy was evaluated against *P. interpunctella* (Hassan et al. 2019). As the usual fatal dosage, 256 Gy prevented all juvenile phases. For field usage, γ and x-ray irradiation, the interaction of additional suitable technologies, as well as more secure management of such radiations, must be explored. This information might be utilized as a viable alternative to current procedures.

A dose of 0.5 kGy is effective in controlling highly resistant beetles, juvenile moths, weevils, and grains and cereals. Certain robust adult moths could survive at this dose, but their progeny may not be able to reproduce. There are instances when it is unnecessary to establish conservative and demanding minimum doses. To effectively eliminate pests and their many life phases in nonfood agricultural products or animal feeds, it is advisable to provide a higher dose of irradiation. This approach helps minimize regulatory delays (Thistlewood and Judd 2019).

Exposure to this specific level of radiation produces a high-quality product without significant alterations to the functional characteristics of the grains or cereals. Increasing the dosage to 2–3 kGy may lead to the mortality of pests that are more susceptible during a 24-h period. However, this higher dosage may also cause alterations in starch which may have an impact on subsequent culinary uses. Wheat flour is very susceptible to exposure to radiation, yet it is often free from infestation except for eggs, which are probably the most vulnerable stage of life. To get optimal results, it is advised to use a minimum dosage of 0.25 kGy and a maximum dosage of 0.75 kGy (Yamada et al. 2019).

The minimal dosage required for controlling rice pests is typically between 0.40 and 0.50 kGy. (The smaller dosage is used when the targeted pest is the rice weevil.) The optimal dosage for rice preservation is 2–3 kGy, which effectively mitigates any quality issues (Jameel, Baker, and Nawar 2023).

18.3.2.3 Irradiation Equipment and Configurations

Irradiation may be achieved by three methods: using γ radiation emitted by a radioisotope (such as cobalt 60 in the case of commercially available irradiators), employing electron-beam accelerators, or generating X-rays by directing accelerated electrons toward an object anterior to their impact on the product. The schematic representation of irradiation setup for insect control shown in Figure 18.4. Each approach has practical and economic benefits as well as drawbacks.

X-rays are produced by directing electrons that are accelerated, emanating from an electron-beam accelerator, onto an intense metallic object, including tantalum. A significant amount of

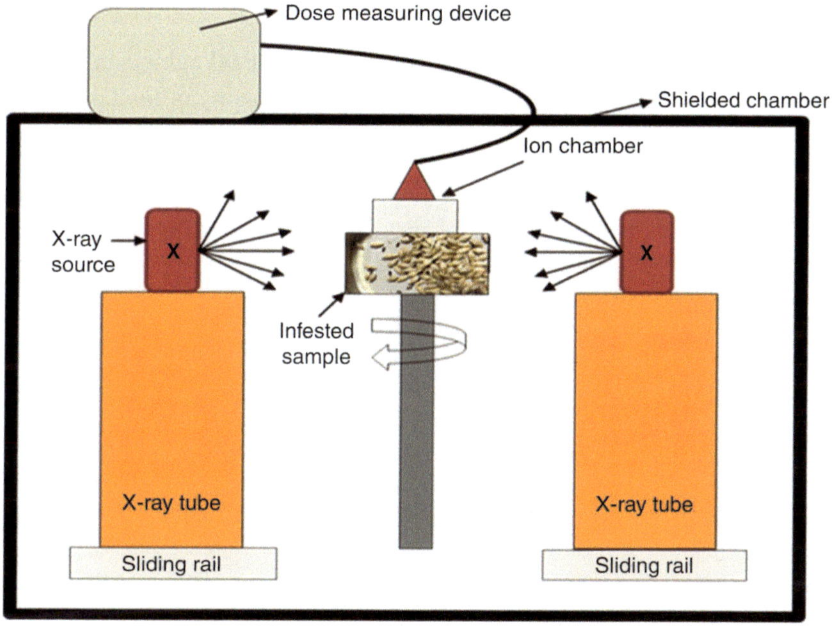

FIGURE 18.4 Schematic depiction of an insect control irradiation setup for stored grains, highlighting key components and their interactions. (Illustration by the author.)

energy is dissipated, but the resultant X-rays possess a high degree of penetrability. The ionization chamber used in this study has a maximum effective dose rate of 6.31 Gy/s, a maximal cumulative dosage of 17 kGy, and an energy dependency below 5% for the energy range between 45 keV and 1.33 MeV (Cheng et al. 2016). X-ray-based insect sterilization using a 5 MeV electron linear accelerator has been reported for Lepidoptera pests, including the sweet potato vine borer (*Omphisa anastomosalis*) (Follett and Neven 2006), *Amorbia emigratella* (Follett et al. 2007), and *Epiphyas postvittana* (Follett et al. 2013). Merely pupae of the sweet potato vine borer were examined, and it was shown that a dosage of 150 Gy effectively prevents the development of F1 adults (Follett et al. 2020). A dosage of 150 Gy effectively prevented the hatching of eggs in the offspring of irradiated neonates, larvae, and pupae of the *A. emigratella* and all preadult phases of life of the *E. postvittana*, except for pupae, that needed a maximum dose of 400 Gy for complete sterilization (Bond et al. 2019).

Electron-beam accelerator can produce X-rays that are significantly more intense. Bremsstrahlung is a kind of electromagnetic energy with a short wavelength. It is generated when electrons, which have undergone acceleration in a vacuum, collide with a metallic substrate such as tantalum/gold. Researchers performed a study on the rice weevil (*S. oryzae*) using an electron-beam accelerator of 250 kV. During the following experiment, the researchers employed 2 accelerators with varying energies, namely a 240-kV system and a 500-kV system, to study the effects on the LGB (*R. dominica*) with grain motions of 30 m/s (Kutsaev et al. 2021). Following the treatment, containers containing insects in wheat were placed in an environment with a temperature of 30°C and a relative humidity of 70%. Eggs, often located inside the pericarp of grains, were particularly susceptible to extermination. The adults, often located on the outer edge of the grains, would have received the highest level of radiation exposure from this technique. The presence of larvae and pupae within the kernel would provide the greatest difficulty for this approach (Chizh et al. 2018). The amount of radiation dosage varied between 200 and 1,600 Gy (0.2–1.6 kGy) (Cho et al. 2021).

18.3.3 Storage Structures to Outsmart Insects

18.3.3.1 Unveiling the Mechanisms of Insect Intrusion in Packaging Material

Food packaging plays a crucial role in ensuring the safety and quality of food products throughout their journey from producer to consumer. However, achieving this protection often requires striking a balance between duration and cost. This means understanding not only how long the packaging needs to keep food safe but also the economic implications of different packaging materials and designs. Unfortunately, food packaging isn't always an impenetrable barrier. Studies have shown that insects can be surprisingly adept at finding and exploiting small openings in packaging materials. Insects usually communicate by means of the Olfaction principle. Insects rely on their sense of smell to locate essential resources like food, mates, and egg-laying sites. This sensitivity to odors allows them to detect even faint traces of food emanating from packaging, leading them to search for weaknesses (Aulicky, Vendl, and Stejskal 2019).

Research by Mowery et al. (2002) highlighted the cunning of the saw-toothed grain beetle (*O. surinamensis*). Their findings revealed that females, attracted by food odors, lay eggs through tiny holes too small for their own bodies to fit through. This underscores the importance of meticulous sealing, even when employing odor barriers. The study also delved deeper into the factors influencing insect responses to food odors, including:

- *Species*: Different insect species have varying odor sensitivities.
- *Mating status*: Mating behavior and needs can influence odor attraction.
- *Sex*: Males and females often exhibit different responses to food odors.
- *Satiation*: Hungry insects are naturally more attracted to food smells.
- *Age*: Younger insects may be more susceptible to odor cues.

Recent research further explores the potential of bio-based repellents derived from natural sources like essential oils and plant extracts (Regnault-Roger, Vincent, and Arnason 2012; Kłyś, Malejky, and Nowak-Chmura 2017). These eco-friendly alternatives offer a sustainable solution, minimizing environmental impact while effectively deterring insect invaders. By accounting for both technical aspects and insect behavior, we can design food packaging that offers long-lasting protection without compromising affordability. This ensures the safe and enjoyable delivery of high-quality food from farm to table. Figure 18.5 elucidates the mechanism of insect intrusion within food packaging.

18.3.3.2 Hermetic Storage

A proper storage framework is crucial in insect management since it affects environmental factors, including humidity, warmth, and connectivity, which have implications for the behavior of insects and survival. Well-designed storage buildings may prevent or restrict infestations of insects through sustaining circumstances that are detrimental to insect reproduction, such as inadequate amounts of humidity and stable temperatures. Furthermore, including structural elements such as secured seals, screened openings, and sleek interior finishes may effectively hinder insect infiltration and restrict potential hiding spots. The implementation of efficient storage facilities decreases the reliance on chemical pesticides, limits the loss of products caused by taint or damage, and guarantees the integrity and nutritional value of food. This, in turn, supports economic effectiveness and long-term viability in the agriculture and food sectors.

The idea of segregating products and grain prompted hermetic sealing. As many traditional constructions are internally replenished with oxygen, hermetic storage uses de-oxygenation to lower the level of aerobic activity. This decrease in oxygen (O_2) raises carbon dioxide (CO_2), which impairs insect metabolism and kills them (Kumar et al. 2015). Tightly sealed or leak-proof container pest control increased grain humidity and desiccation and prevented volatiles from entering by enclosing them (Finkelman et al. 2017). Improved subterranean storage facilities for cereal grains,

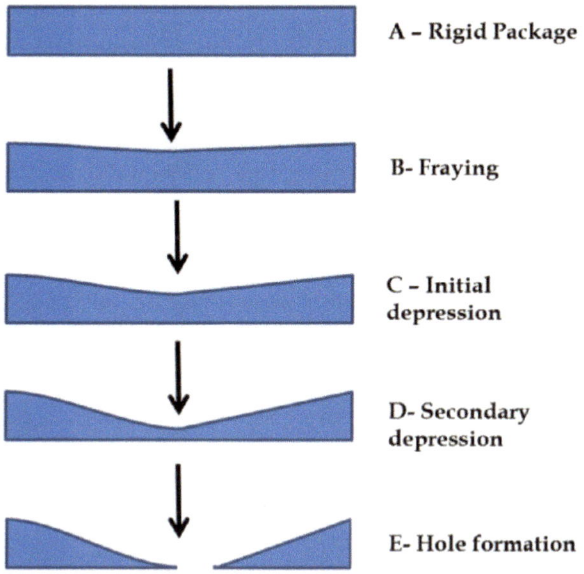

FIGURE 18.5 Visual representation of insect warfare on food packaging. (Illustration by the author.)

legumes, and oilseeds were still employed in Africa and Asia to eliminate insects with decreased O_2 (Somavat et al. 2017). Hermetic structures, including metal drums, silos (metal, cemented, and plastic), cocoons, bags, and polymeric bags and containers, are prevalent (Baributsa and Njoroge 2020). Regulated aeration, gas percentage changes, and insect-proof barriers improve hermetic structures for storage (Navarro 2012). Recent research evaluated the effectiveness of hermetic bags such as Eco tact bags R, Grain Pro R, PICS R, and Save Grain R, and against PP and jute woven bags (Aulicky et al. 2022). O_2 decrease to 6.4% eliminated insects within two months of storage. Until six months, grain quality remained preserved. In the lab and on farms, work was done to test triple-bagging with heavy-grade polyethylene against *C. maculatus* (Sanon, Dabiré-Binso, and Ba 2011). Due to O_2 deficiency in deliberately infected treatments, grains held for seven months were insect-free. Atta et al. (2020) examined *T. castaneum* damage in wheat preservation bags. *T. castaneum* died 100% in a hermetic bag after 30 days. Additionally, insect development rate (30%), wheat grain damage (10.16%), and wheat weight loss (0.5–1.0%) decreased. Baoua et al. (2014) compared PICS R bags for maize storage in West Africa, whereas work has been done to compare Bioxilo™ and sBag™ against stored corn insects in Central Mexico. Chigoverah and Mvumi (2016) demonstrated 100% mortality of insects in hermetically sealed bags and metal silos against stored-maize pests in modeled micro farmer locations. Many researchers have constructed HDPP (high density polypropylene) bags that control insects in smaller trials (De Groote et al. 2013). However, field and commercial hermetically storage structures are still deficient in efficiency and cost.

18.3.3.2.1 Balancing Benefits and Considerations: A Look at Hermetic Storage Challenges

Although hermetic storage is successful in managing insects in stored cereal grains, it does have certain drawbacks. A notable constraint is that it is unable to eliminate preexisting infestations of insects. Hermetic preservation might not completely halt an insect's reproductive phase, but it does provide an airtight atmosphere that starves insects of air and ultimately causes them to suffocate. For example, eggs or pupae that already exist in the grains that are stored may stay unharmed by the absence of oxygen, perhaps causing a reemergence of the population of pests once the container of storage is reopened (Covele et al. 2020). Instances of this constraint may be readily noticed in situations when airtight bags or containers are used to store grains that are contaminated with pests, such as the maize or the rice weevil. Even if the grains are stored in airtight containers, any viable eggs or pupae might still hatch when favorable circumstances return, requiring the implementation of further control techniques.

Hermetic storage is limited by its reliance on specific environmental circumstances to achieve maximum effectiveness (Adler et al. 2018). Although an airtight barrier effectively blocks oxygen, it does not account for other elements like temperature and humidity, which may adversely impact insect behavior and reproductive rates. In areas characterized by significant temperature variations or persistently high levels of humidity, airtight storage could prove less efficient in managing insect infestations (Mutambuki and Likhayo 2021). In tropical settings characterized by continuously humid conditions as well as elevated levels of moisture, relying only on airtight storage might not be enough to effectively manage the number of insects in grains that are stored. Additional measures such as temperature regulation or the use of desiccants may be necessary in certain situations to improve the effectiveness of hermetic storage (Sn, Vs, and Pr 2021).

18.3.4 Exploring the Potential of Sustainable Thermal Methods

18.3.4.1 Microwave Heating

Heating using microwaves utilizes electromagnetic radiation to generate thermal power that impacts the polar ions of compounds. Microwaves effectively eliminate pests from grains due to the insects' higher percentage of water, which causes them to heat up more rapidly than the grains. Microwave irradiation renders grain suitable for storing. Hence, the use of microwaves for heating might function as a nonchemical alternative approach for the eradication of pests in grains (Almaiman et al. 2021). Currently, the use of microwave treatment significantly prolongs the storage duration of grains in a manner that is both secure, cost-efficient, and highly efficient. Dielectric heating eradicates all life stages of insects with the use of volumetric heating (Wang et al. 2001). The alteration of insect DNA structure has a direct impact on their reproductive processes during volumetric heating (Lu et al. 2011). This phenomenon proves to be advantageous in the control of crop pests.

According to Walde et al. (2002), microwave radiation caused very little alterations in the protein composition of wheat (9.9%). However, when microwave irradiation was applied for more than 90 s, it did affect both the structure and content of the protein. Arnaud et al. (2005) used a 0.5th-order kinetic model to examine the impact of microwave heating on the heat-tolerant species *Amyelois transitella* (Walker navel orange worm) and *T. castaneum* (red flour beetle) within a temperature range of 48–52°C. According to the study findings, a death rate of 95% was seen when the temperature was between 50 and 52°C, with exposure times of 8–1.3 min. However, at a temperature of 48°C, it took 67 min to attain the same level of mortality. In their study, Vadivambal and Jayas (2010) discovered that exposing wheat kernels to 500 W microwave radiation for a duration of 56 s resulted in a reduction in germination. The research investigated the death rate of *L. serricorne*, *R. dominica*, and *T. confusum* on wheat grains exposed to temperatures ranging from 40 to 55°C for durations of 10–50 s. The mortality rate reached 100% after 50 s at a temperature of 50°C. Moreover, *T. castaneum* exhibited the highest resistance to microwave heating at a temperature of 48°C, while showed resistance at temperatures ranging from 50 to 52°C for *A. transitella*. Athanassiou and Arthur (2018) discovered that *T. castaneum* and *T. confusum* beetles exhibit heat tolerance up to 50°C, as indicated by the survival of adult individuals at this temperature. *T. castaneum* exhibits a higher tolerance to microwave radiation compared to other genera. Volumetric heating is more efficient than hot air exposure for heating grains since it requires a shorter period (Wang et al. 2001). While the impact on product culinary quality remained unaffected, these methods resulted in a reduction in the germination of seeds (Jiao et al. 2014). The effects of microwave treatment on the disinfestation of *S. granarius*, *S. oryzae*, and *R. dominica* for wheat samples have been studied and it found that all sorts of insects were completely eradicated after a 30-s treatment. The impact of a 25-s exposure on *S. oryzae* varied, resulting in a mortality rate of 100%, whereas the other two kinds had mortality rates of 93.25 and 87.00%, respectively. Exposures lasting 20 s led to death rates of 76, 70, and 60%, respectively. As the amount of exposure time reduced, the fatality rates also declined gradually.

18.3.5 Applications of Modified Atmosphere Technology in Grain Protection

Modified Atmospheric Storage (MAS) is a groundbreaking approach used to manage stored insects in rice and other cereals and pulses, providing a substitute for synthetic pesticides that might potentially harm humans as well as the surroundings (Constantin et al. 2020). This technique entails modifying the air composition around the stored grains to provide circumstances that hinder the proliferation of insect pests while maintaining the grain's value (Smit 2019).

18.3.5.1 Managing Pests: The Science behind MAS (Modified Atmospheric Storage)

The underlying premise of MAS (Modified Atmospheric Storage) for controlling insect pests are

- ***O_2 reduction*:** Insects demand oxygen for the process of breathing. Insects' metabolic functions may be impeded by lowering the oxygen levels in a preserved atmosphere, usually by replacing it with carbon dioxide or nitrogen. This leads to the inhibition of their development and growth (Likhayo et al. 2018).
- ***CO_2 accumulation*:** Increased concentrations of carbon dioxide may interfere with the respiratory mechanism of insects, resulting in asphyxiation and ultimate mortality (Riudavets et al. 2018).
- ***Humidity and temperature control*:** Humidity and temperature are essential for maintaining ideal conditions in the storage site. Decreased temperatures decelerate the activities of insects and diminish their rate of metabolism, but regulating humidity prevents the formation of mildew and preserves the nutritional value of grain (Ziegler, Paraginski, and Ferreira 2021).
- ***Hermetic sealing*:** To maintain the correct atmospheric conditions during storage spaces completely enclosed, preventing the transfer of air via the outside environment (Carvalho et al. 2019).

18.3.5.2 Recent Research Breakthroughs in MAS-based Insect Control

The concentrations of oxygen are often reduced to nearly 1–5% by replacing it with gases that are inert, like nitrogen or carbon dioxide. The decrease in oxygen levels results in an environment that is unsuitable for insect pests since they rely on oxygen for their respiratory process. In addition, the storage environment is exposed to increased quantities of carbon dioxide, with concentrations ranging from 30 to 70% (Tutuncu and Emekci 2019). Elevated levels of carbon dioxide interfere with the respiratory mechanism of insects, resulting in asphyxiation and ultimate fatalities (Wong-Corral, Castañé, and Riudavets 2013). Modified Atmospheric Storage (MAS) generators are crucial devices used in storage facilities and infrastructures to create and sustain the proper atmospheric conditions for pest management and grain sustainability. These generators often use technology, including CO_2 generators, gas mixing systems, and O_2 absorbers, to control the composition of the storage environment.

Gas blending devices merge gases such as carbon dioxide and nitrogen to get the specific quantities of carbon dioxide and oxygen required to impede insect activity and reproduction. Carbon-di-oxide generators emit carbon dioxide into the storage surroundings, often employing controlled burning of a fuel made from carbon (Shlomo Navarro 2012). Oxygen absorbers remove oxygen from the surroundings by chemical processes or absorption substances. These establishments are often equipped with control and monitoring systems to guarantee accurate management of the environment inside storage structures, such as warehouses or silos. Magnetic air separation generators are essential for improving the efficiency and long-term viability of grain storage operations (Riudavets et al. 2018). They provide a controlled environment that effectively eliminates insect pests and maintains the quality of the stored grain.

Utilizing Modified Atmosphere Packaging (MAP) with increased levels of CO_2 shows potential as an effective approach to manage stored product insect pests such as *Sitophilus* spp. and bruchids. *Sitophilus* insects in organic maize died completely within just 5 days when exposed to an atmosphere containing 40% CO_2. This is consistent with the conclusions drawn by Wong-Corral, Castañé, and Riudavets (2013), who obtained comparable outcomes in their studies on *R. dominica*, bruchids, and the rice weevil, respectively. The ability of MAP to target multiple life stages of organisms supports its effectiveness as a means of inhibition. Combining heat treatment with high levels of CO_2 and low levels of O_2 may effectively suppress the dormancy of codling moth *Cydia pomonella* Linnaeus, in walnuts. Specifically, temperatures over 44°C quickly eliminate the moths (Neven, Lehrman, and Hansen 2014). Therefore, using a combination of elevated temperatures, forced-air conditioning and a condition with reduced oxygen and increased carbon dioxide is an ecologically sound method for managing pests after harvest.

18.4 MECHANICAL VS. CHEMICAL: A WEIGHING OF INSECT CONTROL STRATEGIES

The major difference between the chemical methods and mechanical approaches for the detection and control of insect infestation in the stored grains has been discussed in Table 18.1.

TABLE 18.1
Difference between Mechanical and Chemical Approaches in the Detection and Control of Insect Pests in Stored Grains

Chemical Methods	Mechanical Approaches
Chemical Residue: Pesticides of a chemical nature have the potential to leave behind detrimental remnants on grains, hence presenting hazards to both human well-being and the natural surroundings (Bhunia et al. 2022).	*Non-Toxic Strategy*: Mechanical methods refrain from employing chemical pesticides, making them an environmentally friendly substitute for pest management, particularly significant for organic or ecologically concerned enterprises (Odjo et al. 2022).
Pesticide Tolerance: Insects may acquire immunity to synthetic pesticides over time, making them ineffectual and requiring the application of harsher chemicals.	*Physical Hurdle*: Implementing mechanical methods, such as hermetic sealing or employing physical traps, establishes a barrier that effectively obstructs insect pests from gaining possession of stored cereal grains, hence reducing the likelihood of infection (Athanassiou and Arthur 2018).
Health Risks: Chemical pesticides may provide health risks to personnel engaged in their application and to consumers owing to residues of pesticides on cereals.	*Minimal Possibility of Resistance*: Mechanical techniques generally do not aid in the emergence of resistance by insects simply because they are not reliant on chemical substances that pests can become accustomed to over time.
Expensive: Chemical pesticides may be costly to acquire, particularly for extensive grain storage facilities, and may need regular use (Odjo et al. 2022).	*Sustainability*: Utilizing mechanical methods, such as implementing efficient hygiene procedures and temperature regulation, fosters sustainability by decreasing dependence on external resources and minimizing the ecological footprint (Bhunia et al. 2022).
Accelerated Resistance: Insects can quickly acquire resistance to certain chemical pesticides, which requires the continuous creation of new chemical combinations.	*Minimal Residue Hazard*: Mechanical methods often do not result in the presence of chemical remnants on stored cereal grains, which is crucial for upholding food safety and complying with regulatory requirements.

(Continued)

TABLE 18.1 (Continued)

Difference between Mechanical and Chemical Approaches in the Detection and Control of Insect Pests in Stored Grains

Chemical Methods	Mechanical Approaches
Unintentional Consequences: Chemical pesticides might have negative impacts on non-target creatures, including pollinators and the natural enemies of insect pests (Athanassiou and Arthur 2018).	*Budget-Friendly*: Mechanical methods such as cleanliness, adequate ventilation, and the drying of grains are often affordable in the long run, necessitating low initial expenditure.
Immediate Efficacy: Pesticides made from chemicals may quickly suppress insect pests, but they generally need to be applied many times since pests can bounce back and develop resistance.	*Physical Extraction*: Utilizing mechanical techniques such as sifting or air separation may effectively eliminate insect pests and their eggs from stored grains, hence diminishing the population and averting further infestations (Bhunia et al. 2022).
Reliance: Excessive dependence on chemical pesticides might create a reliance that impedes the acceptance of alternative, environmentally friendly pest management approaches (Hamel, Rozman, and Liška 2020)	*Complementary Methodology*: Mechanical techniques may be used alongside chemical ways to augment the efficiency of pest control and decrease dependence on pesticides, hence promoting the implementation of IPM strategies (Odjo et al. 2022).
Challenges in Obtaining Organic Certification: The use of chemical pesticides may provide difficulties in obtaining organic certification as a result of the stringent regulations governing pesticide usage in organic farming (Bhunia et al. 2022).	*Environmental Effect Minimization*: Mechanical techniques have a negligible environmental effect since they do not transfer residual chemical compounds into the atmosphere or aid in the runoff of pesticides (Hamel, Rozman, and Liška 2020).
Environmental Cross-Contamination: Chemical pesticides have the potential to pollute the environment, including air, water, and soil, hence endangering biodiversity and the overall health of ecosystems.	*Organic Integration*: Mechanical techniques are in line with organic agricultural practices and may be employed in organic storage for grains to maintain accreditation and adhere to organic requirements (Athanassiou and Arthur 2018).

18.5 ADDRESSING CHALLENGES IN MECHANICAL INSECT PEST MANAGEMENT

The effectiveness of mechanical detection techniques in managing insect pests in stored grains is limited by their dependence on tangible obstacles or processes to hinder infestation.

18.5.1 Level of Sensitivity

- ***Threshold concerns*:** Several techniques, like as sifting and visual examination, exhibit insufficient sensitivity to identify incipient infestations at low levels, which may result in delayed intervention and exacerbated difficulties.
- ***Stage-specific constraints*:** The majority of techniques encounter difficulties in identifying eggs and pupae stages owing to their lack of movement or reduced dimensions. This has the potential to grossly overestimate the overall population and its capacity for expansion. Constraints regarding grain type: The efficacy of some mechanical techniques, such as sifting, may be compromised when dealing with grains that include stickiness or oiliness, hence impeding precise identification (Hamel, Rozman, and Liška 2020).
- ***Misidentification*:** Conventional methods such as visual examination are susceptible to human fallibility, which may result in the incorrect identification of non-insect particles as pests or the oversight of some species.

- ***Insufficient data*:** The majority of techniques just show if insects are present or not, without offering specific information about the species, age distribution, or size of the population. These facts are essential for making well-informed choices about pest treatment.
- ***Labor-intensive*:** Manual methods such as sifting and manual inspection need substantial time and workforce, particularly for substantial quantities of grain.
- ***Expenditure costs*:** The use of advanced technologies such as acoustic detection or electronic noses requires a greater initial expenditure and specialized technical knowledge for operation and maintenance (Odjo et al. 2022).
- ***Sieving*:** Sieving is a method that is very efficient in capturing adult insects, but it is not effective in capturing eggs, pupae, and smaller species. Has the potential to harm fragile grains and requires much time and effort (Hamel, Rozman, and Liška 2020).
- Visual examination is a subjective method that is susceptible to human error, making it prone to missing early infestations and perhaps underestimating the number of individuals in a population.

Insect traps are designed to selectively attract certain species, however, they might inadvertently collect unintended creatures, resulting in potential misinterpretations.

Acoustic detection may be costly and may have difficulties in distinguishing signals from background noise, hence reducing its accuracy in noisy settings.

In general, while mechanical detection systems have great benefits, their limits need a comprehensive strategy to efficiently control pests in stored grain (Banga et al. 2018). By integrating them with other techniques, such as thermal imaging, bioassays, or pheromone traps, and applying effective storage procedures like maintaining cleanliness and temperature control, a more holistic and successful approach to insect pest management may be achieved (Mankin et al. 2021).

Several mechanical methods for identification with potential are still in the research phase, missing defined techniques or economically viable solutions, which is impeding their wider adoption. Solely relying on mechanical detection fails to deal with existing infestations effectively. Further improvement is necessary to integrate it with mechanisms of control such as temperature modification or biological regulation (Cheng et al. 2016).

18.6 CONCLUSION

This chapter has delved into the world of mechanical approaches for managing insect pests in stored grain, examining their role in detection, prevention, and control. It is clear that these methods offer a powerful alternative to traditional chemical-based approaches, paving the way for a future focused on sustainability, safety, and long-term effectiveness.

For decades, chemical options have been the norm, but their environmental and health drawbacks are raising significant concerns. Issues like insect resistance, residue contamination, and harm to beneficial insects paint a troubling picture. Conversely, mechanical methods such as hermetic storage, physical barriers, and mechanical traps operate without the burden of harmful residues or the development of resistance. They effectively target specific pests while safeguarding beneficial insects and the broader ecosystem. Furthermore, their emphasis on prevention minimizes the need for reactive control measures, leading to reduced overall costs and a proactive approach to grain protection.

The future of managing insect pests in grain storage lies in harnessing the full potential of mechanical solutions. As research and development refine existing technologies and introduce innovative solutions, their effectiveness and user-friendliness will only improve. By transitioning toward these sustainable and responsible strategies, we can safeguard the safety and quality of our grain reserves while simultaneously protecting the environment and human health for future generations.

REFERENCES

Adler, C. S., G. Opit, B. Fürstenau, P. Kern, F. H. Arthur, C. G. Athanassiou, R. Bartosik, et al. 2018. "*Proceedings of the 12th International Working Conference on Stored Product Protection (IWCSPP)*" 46 3: 1130. doi:10.5073/jka.2018.463.121.

Almaiman, S. A., N. A. Albadr, S. Alsulaim, H. F. Alhuthayli, M. A. Osman, and A. B. Hassan. 2021. "Effects of Microwave Heat Treatment on Fungal Growth, Functional Properties, Total Phenolic Content, and Antioxidant Activity of Sorghum (*Sorghum bicolor* L.) Grain." *Food Chemistry* 348 (January). Elsevier Ltd: 128979. doi:10.1016/j.foodchem.2020.128979.

Arnaud, L., H. T. T. Lan, Y. Brostaux, and E. Haubruge. 2005. "Efficacy of Diatomaceous Earth Formulations Admixed with Grain against Populations of *Tribolium castaneum*." *Journal of Stored Products Research* 41 (2): 121–130. doi:10.1016/j.jspr.2003.09.004.

Athanassiou, C. G., and F. H. Arthur. 2018. "Recent Advances in Stored Product Protection." In *Recent Advances in Stored Product Protection*, June. Springer Berlin Heidelberg, 1–273. doi:10.1007/978-3-662-56125-6.

Atkinson, J. M. 1971. "An Improved Trap to Detect Beetles (Coleoptera) in Stored Grain." *The Canadian Entomologist* 539: 437–440. doi:10.4039/Ent105437-3.

Atta, B., M. Rizwan, A. M. Sabir, M. D. Gogi, and K. Ali. 2020. "Damage Potential of *Tribolium castaneum* (Herbst) (Coleoptera: Tenebrionidae) on Wheat Grains Stored in Hermetic and Non-Hermetic Storage Bags." *International Journal of Tropical Insect Science* 40 (1). International Journal of Tropical Insect Science: 27–37. doi:10.1007/s42690-019-00047-0.

Aulicky, R., J. A. Shah, V. Kolar, Z. Li, and V. Stejskal. 2022. "Control of Stored Agro-Commodity Pests *Sitophilus granarius* and *Callosobruchus chinensis* by Nitrogen Hypoxic Atmospheres: Laboratory and Field Validations." *Agronomy* 12 (11). doi:10.3390/agronomy12112748.

Aulicky, R., T. Vendl, and V. Stejskal. 2019. "Evaluation of Contamination of Packages Containing Cereal-Fruit Bars by Eggs of the Pest Indian Meal Moth (*Plodia interpunctella*, Lepidoptera) Due to Perforations in Their Polypropylene Foil Packaging." *Journal of Food Science and Technology* 56 (7). Springer India: 3293–3299. doi:10.1007/s13197-019-03799-2.

Ayertey, J. N. 1989. "Observations on the Efficiency of Two New Monitoring Devices for Storage Insect Pests." *Nigerian Journal of Entomology* 9 (1): 13–22. doi:10.36108/nje/8891/90.0130.

Azfar, S., A. Nadeem, A. B. Alkhodre, K. Ahsan, N. Mehmood, T. Alghmdi, and Y. Alsaawy. 2018. "Monitoring, Detection and Control Techniques of Agriculture Pests and Diseases Using Wireless Sensor Network: A Review." *International Journal of Advanced Computer Science and Applications* 9 (12): 424–433. doi:10.14569/IJACSA.2018.091260.

Bajaj, N., N. C. Giampietro, K. Mao, M. E. Rushton, N. A. Spomer, and J. F. Rhoads. 2021. "Searching for Bed Bugs: The Design, Development, and Evaluation of an Oscillator-Based Trans-2-Hexenal Sensor Array." *Sensors and Actuators, B: Chemical* 333 (September 2020). Elsevier B.V.: 129161. doi:10.1016/j.snb.2020.129161.

Bakri, A., K. Mehta, and D. R. Lance. 2005. "Sterilizing Insects with Ionizing Radiation." *Sterile Insect Technique: Principles and Practice in Area-Wide Integrated Pest Management*, 233–268. doi:10.1007/1-4020-4051-2_9.

Banga, K. S., N. Kotwaliwale, D. Mohapatra, and S. K. Giri. 2018. "Techniques for Insect Detection in Stored Food Grains: An Overview." *Food Control* 94 (December): 167–176. doi:10.1016/j.foodcont.2018.07.008.

Baoua, I. B., L. Amadou, B. Ousmane, D. Baributsa, and L. L. Murdock. 2014. "PICS Bags for Post-Harvest Storage of Maize Grain in West Africa." *Journal of Stored Products Research* 58. Elsevier Ltd: 20–28. doi:10.1016/j.jspr.2014.03.001.

Barak, A. V., and P. K. Harein. 1982. "Trap Detection of Stored-Grain Insects in Farm-Stored, Shelled Corn12." *Journal of Economic Entomology* 75 (1): 108–111. doi:10.1093/jee/75.1.108.

Baributsa, D., and A. W. Njoroge. 2020. "The Use and Profitability of Hermetic Technologies for Grain Storage among Smallholder Farmers in Eastern Kenya." *Journal of Stored Products Research* 87. Elsevier Ltd: 101618. doi:10.1016/j.jspr.2020.101618.

Bhunia, A. K., M. Shakeel, A. U. Paschapur, P. N. Guru, and A. S. Dukare. 2022. A Comprehensive Review on Advances in Storage Pest Management: Current Scenario and Future Prospects. *Frontiers in Sustainable Food Systems* 6: 993341.

Boina, D., and B. Subramanyam. 2004. "Relative Susceptibility of *Tribolium confusum* Life Stages Exposed to Elevated Temperatures." *Journal of Economic Entomology* 97.

Bond, J. G., A. R. Osorio, N. Avila, Y. Gómez-Simuta, C. F. Marina, I. Fernández-Salas, P. Liedo, et al. 2019. "Optimization of Irradiation Dose to Aedes Aegypti and Ae. Albopictus in a Sterile Insect Technique Program." *PLoS ONE* 14 (2). doi:10.1371/journal.pone.0212520.

Bonjour, E. L., G. P. Opit, J. Hardin, C. L. Jones, M. E. Payton, and R. L. Beeby. 2011. "Efficacy of Ozone Fumigation against the Major Grain Pests in Stored Wheat." *Journal of Economic Entomology* 104 (1): 308–316. doi:10.1603/EC10200.

Carvalho, M. O., P. Fradinho, M. J. Martins, A. Magro, A. Raymundo, and I. de Sousa. 2019. "Paddy Rice Stored under Hermetic Conditions: The Effect of Relative Humidity, Temperature and Storage Time in Suppressing *Sitophilus zeamais* and Impact on Rice Quality." *Journal of Stored Products Research* 80: 21–27. doi:10.1016/j.jspr.2018.11.002.

Chelladurai, V., K. Karuppiah, D. S. Jayas, P. G. Fields, and N. D. G. White. 2014. "Detection of *Callosobruchus maculatus* (F.) Infestation in Soybean Using Soft X-Ray and NIR Hyperspectral Imaging Techniques." *Journal of Stored Products Research* 57 (April): 43–48. doi:10.1016/j.jspr.2013.12.005.

Cheng, Y.-L., C.-Y. Lee, Y.-L. Huang, C. A. Buckner, R. M. Lafrenie, J. A. Dénommée, and J. M. Caswell. 2016. "We Are IntechOpen, the World's Leading Publisher of Open Access Books Built by Scientists, for Scientists TOP 1%." *Intech* 11 (tourism): 13.

Chigoverah, A. A., and B. M. Mvumi. 2016. "Efficacy of Metal Silos and Hermetic Bags against Stored-Maize Insect Pests under Simulated Smallholder Farmer Conditions." *Journal of Stored Products Research* 69. Elsevier Ltd: 179–189. doi:10.1016/j.jspr.2016.08.004.

Chiluwal, K., J. Kim, S. D. Bae, G. H. Roh, H. J. Park, and C. G. Park. 2019. "Effect of Gamma Irradiation on Fecundity, Sterility, and Female Sex Pheromone Production of *Callosobruchus chinensis* (Coleoptera: Bruchidae)." *Journal of Economic Entomology* 112 (1): 156–163. doi:10.1093/jee/toy317.

Chizh, T. V., N. N. Loy, A. N. Pavlov, M. S. Vorobyov, and S. Y. Doroshkevich. 2018. "Low-Energy Electron Beams for Protection of Grain Crops from Insect Pests and Diseases." *Journal of Physics: Conference Series* 1115 (2). doi:10.1088/1742-6596/1115/2/022025.

Cho, S. R., M. Kim, E. Shin, H. K. Kim, H. N. Koo, and G. H. Kim. 2021. "X-Ray Irradiation-Induced Abnormal Development and DNA Damage in *Phthorimaea operculella* (Lepidoptera: Gelechiidae)." *Applied Sciences (Switzerland)* 11 (11). doi:10.3390/app11115068.

Constantin, M., R. Jagadeesan, K. Chandra, P. Ebert, and M. K. Nayak. 2020. "Synergism between Phosphine (PH_3) and Carbon Dioxide (CO_2): Implications for Managing PH3Resistance in Rusty Grain Beetle (Laemophloeidae: Coleoptera)." *Journal of Economic Entomology* 113 (4): 1999–2006. doi:10.1093/jee/toaa081.

Covele, G., A. Gulube, L. Tivana, A. I. Ribeiro-Barros, M. O. Carvalho, A. Ndayiragije, and R. Nguenha. 2020. "Effectiveness of Hermetic Containers in Controlling Paddy Rice (*Oryza sativa* L.) Storage Insect Pests." *Journal of Stored Products Research* 89. doi:10.1016/j.jspr.2020.101710.

De Groote, H., S. C. Kimenju, P. Likhayo, F. Kanampiu, T. Tefera, and J. Hellin. 2013. "Effectiveness of Hermetic Systems in Controlling Maize Storage Pests in Kenya." *Journal of Stored Products Research* 53 (April): 27–36. doi:10.1016/j.jspr.2013.01.001.

Donahaye, E. J., S. Navarro, and J. Leesch. 2001. *Commercialization of the Electronic Grain Probe Insect Counter*. Executive Printing Services.

Dong, X., N. Yan, and Y. Wei. 2018. "Insect Sound Recognition Based on Convolutional Neural Network." *2018 3rd IEEE International Conference on Image, Vision and Computing, ICIVC 2018*. IEEE, 855–859. doi:10.1109/ICIVC.2018.8492871.

Dowell, F. E., J. E. Throne, D. Wang, and J. E. Baker. 1999. "Identifying Stored-Grain Insects Using Near-Infrared Spectroscopy." *Journal of Economic Entomology* 92 (1): 165–169.

Eliopoulos, P. A., I. Potamitis, and D. C. Kontodimas. 2016. "Estimation of Population Density of Stored Grain Pests via Bioacoustic Detection." *Crop Protection* 85. Elsevier Ltd: 71–78. doi:10.1016/j.cropro.2016.04.001.

Fargo, W. S., D. Epperly, G. W. Cuperus, B. C. Clary, and R. Noyes. 1989. "Effect of Temperature and Duration of Trapping on Four Stored Grain Insect Species." *Journal of Economic Entomology* 82 (3): 970–973. doi:10.1093/jee/82.3.970.

Finkelman, S., S. Navarro, Y. Lotan, T. Debruin, A. A. Isikber, M. Rindner, R. Dias, and A. Azrieli. 2017. "Insect Control of Cocoa Pests Using a Novel Vacuum Approach." *Advances in Stored Product Protection. Proceedings of the 8th International Working Conference on Stored Product Protection, York, UK, 22–26 July 2002*, no. June 2014: 579–582.

Fleurat-Lessard, F. 2011. "Monitoring Insect Pest Populations in Grain Storage: The European Context." *Stewart Postharvest Review* 7 (3). doi:10.2212/spr.2011.3.4.

Follett, P., G. Akepsimaidis, N. Meneses, M. Murdoch, and H. Kotilainen. 2020. "Advances in Insect Pest Management in Postharvest Storage of Cereals: Novel Techniques," no. August 2020: 319–338. doi:10.19103/as.2020.0072.13.

Follett, P. A., and L. G. Neven. 2006. "Current Trends in Quarantine Entomology." *Annual Review of Entomology* 51 (108): 359–385. doi:10.1146/annurev.ento.49.061802.123314.

Follett, P. A., K. Snook, A. Janson, B. Antonio, A. Haruki, M. Okamura, and J. Bisel. 2013. "Irradiation Quarantine Treatment for Control of *Sitophilus oryzae* (Coleoptera: Curculionidae) in Rice." *Journal of Stored Products Research* 52. Elsevier Ltd: 63–67. doi:10.1016/j.jspr.2012.09.004.

Follett, P. A., M.M. Yang, K.H. Lu, and T.W. Chen. 2007. "Irradiation for Postharvest Control of Quarantine Insects" 15 (May): 1–15.

French, L. A., D. F. McCann, M. Wark, S. Winters, and J. F. Vetelino. 2007. "A Lateral Field Excited Acoustic Wave Sensor." *TRANSDUCERS and EUROSENSORS '07 – 4th International Conference on Solid-State Sensors, Actuators and Microsystems* 00 (c): 1287–1290. doi:10.1109/SENSOR.2007.4300373.

Fürstenau, B. 2023. "Early Detection of Stored-Product Pest Insects in Grain Storage and in the Field in Germany." *Journal Fur Kulturpflanzen* 75 (9–10): 259–268. doi:10.5073/JfK.2023.09-10.04.

Gaikwad, Y. B., S. M. Gaikwad, and G. P. Bhawane. 2010. "Effect of Induced Oxidative Stress and Herbal Extracts on Acid Phosphatase Activity in Lysosomal and Microsomal Fractions of Midgut Tissue of the Silkworm, *Bombyx mori*." *Journal of Insect Science* 10 (113): 1–9. doi:10.1673/031.010.11301.

Geng, S., X. Zhang, and W. Zhao. 2017. "Detection and Analysis of the Stored Grain Insect Creeping Sound." *IOP Conference Series: Materials Science and Engineering* 242 (1). doi:10.1088/1757-899X/242/1/012093.

Hagstrum, D. W., F. W. Flinn, and J. J. Gaffney. 1998. "Temperature Gradient on *Tribolium castaneum* (Coleoptera: Tenebrionidae) Adult Dispersal in Stored Wheat." *Environmental Entomology* 27 (1): 123–129. doi:10.1093/ee/27.1.123.

Hagstrum, D. W., P. W. Flinn, and D. Shuman. 1996. "Automated Monitoring Using Acoustical Sensors for Insects in Farm-Stored Wheat." *Journal of Economic Entomology* 89.

Hagstrum, D. W., and G. A. Milliken. 1991. "Modeling Differences in Insect Developmental Times between Constant and Fluctuating Temperatures." *Annals of the Entomological Society of America* 84 (4): 369–379. doi:10.1093/aesa/84.4.369.

Halliwell, B., K. Zhao, and M. Whiteman. 2000. "The Gastrointestinal Tract: A Major Site of Antioxidant Action?." *Free Radical Research* 33 (6): 819–830. doi:10.1080/10715760000301341.

Hallman, G. J., V. Arthur, C. M. Blackburn, and A. G. Parker. 2013. "The Case for a Generic Phytosanitary Irradiation Dose of 250Gy for Lepidoptera Eggs and Larvae." *Radiation Physics and Chemistry* 89: 70–75. doi:10.1016/j.radphyschem.2013.03.046.

Hamel, D., V. Rozman, and A. Liška. 2020. "Storage of Cereals in Warehouses with or without Pesticides." *Insects* 11 (12): 1–21. doi:10.3390/insects11120846.

Hassan, R. S., E. A. Mahmoud, T. M. Sileem, and W. A. A. Sayed. 2019. "Evaluation of Fast Neutron Irradiation as a New Control Method against the Indian Meal Moth, *Plodia interpunctella* (Hübner)." *Journal of Radiation Research and Applied Sciences* 12 (1): 38–44. doi:10.1080/16878507.2019.1594133.

Hickling, R., and W. Wei. 1995. "Sound Transmission in Stored Grain." *Applied Acoustics* 45 (1): 1–8. doi:10.1016/0003-682X(94)00017-P.

Hoback, W. W., and D. W. Stanley. 2001. "Mini Review Insects in Hypoxia." *Journal of Insect Physiology* 47.

Horvitz, E., and D. Mulligan. 2015. "Machine Learning: Trends, Perspectives, and Prospects." *Science* 349 (6245): 253–255.

Høye, T. T., J. Ärje, K. Bjerge, O. L. P. Hansen, A. Iosifidis, F. Leese, H. M. R. Mann, K. Meissner, C. Melvad, and J. Raitoharju. 2021. "Deep Learning and Computer Vision Will Transform Entomology." *Proceedings of the National Academy of Sciences of the United States of America* 118 (2): 1–10. doi:10.1073/PNAS.2002545117.

Işikber, A. A., and S. Öztekin. 2009. "Comparison of Susceptibility of Two Stored-Product Insects, *Ephestia kuehniella* Zeller and *Tribolium confusum* Du Val to Gaseous Ozone." *Journal of Stored Products Research* 45 (3): 159–164. doi:10.1016/j.jspr.2008.12.003.

Islam, K., S. Damiati, J. Sethi, A. Suhail, and G. Pan. 2018. "Development of a Label-Free Immunosensor for Clusterin Detection as an Alzheimer's Biomarker." *Sensors (Switzerland)* 18 (1): 1–12. doi:10.3390/s18010308.

Jameel, M. A. R., S. Z. Baker, and M. H. Nawar. 2023. "Control of the Khapra Beetle *Trogoderma granarium* (Everts.) (Coleoptera : Dermastidae) Using Nanoparticles of Neem Leaf Extract." *IOP Conference Series: Earth and Environmental Science* 1225 (1). doi:10.1088/1755-1315/1225/1/012087.

Jian, F., D. S. Jayas, and N. D. G. White. 2004. Movement and Distribution of Adult Rusty Grain Beetle, *Cryptolestes ferrugineus* (Coleoptera: Laemophloeidae), in Stored Wheat in Response to Different Temperature Gradients and Insect Densities. *Journal of Economic Entomology* 97.

Jiao, J., Z. G. Li, Q. Y. Gai, X. J. Li, F. Y. Wei, Y. J. Fu, and W. Ma. 2014. "Microwave-Assisted Aqueous Enzymatic Extraction of Oil from Pumpkin Seeds and Evaluation of Its Physicochemical Properties, Fatty Acid Compositions and Antioxidant Activities." *Food Chemistry* 147. Elsevier Ltd: 17–24. doi:10.1016/j.foodchem.2013.09.079.

Kabbashi, E. E. B. M., E. G. H. Ahmed, S. A. Aljack, S. A. A. Hamad, M. E. Ahmed, and K. Elmamoun. 2012. "Use of Gamma Irradiation for the Control of the Red Flour Beetle (*Tribolium castaneum* Hrbst) in Bread Flour," *Proceeding of The Third International on Radiation Sciences and Applications* 44 (46).

Kaushik, R., and J. Singhai. 2018. "Sensing Technologies Used for Monitoring and Detecting Insect Infestation in Stored Grain." *International Journal of Engineering and Technology(UAE)* 7 (4.6 Special Issue 6): 169–173. doi:10.14419/ijet.v7i4.6.20456.

Kells, S. A., L. J. Mason, D. E. Maier, and C. P. Woloshuk. 2001. "Efficacy and Fumigation Characteristics of Ozone in Stored Maize." *Journal of Stored Products Research* 37.

Kiobia, D. O., S. D. Tumbo, J. Cantillo, B. B. Rohde, P. K. Mallikarjunan, and R. W. Mankin. 2015. "Characterization of Sounds in Maize Produced by Internally Feeding Insects: Investigations to Develop Inexpensive Devices for Detection of *Prostephanus truncatus* (Coleoptera: Bostrichidae) and *Sitophilus zeamais* (Coleoptera: Curculionidae) in Small-Scale." *Florida Entomologist* 98 (2): 405–409. doi:10.1653/024.098.0202.

Kłyś, M., N. Malejky, and M. Nowak-Chmura. 2017. "The Repellent Effect of Plants and Their Active Substances against the Beetle Storage Pests." *Journal of Stored Products Research* 74. Elsevier: 66–77.

Krüger, A. P., D. C. H. Schlesener, L. N. Martins, J. Wollmann, M. Deprá, and F. R. M. Garcia. 2018. "Effects of Irradiation Dose on Sterility Induction and Quality Parameters of *Drosophila suzukii* (Diptera: Drosophilidae)." *Journal of Economic Entomology* 111 (2): 741–746. doi:10.1093/jee/tox349.

Kumar, S., J. Park, E. Kim, J. Na, Y. S. Chun, H. Kwon, W. Kim, and Y. Kim. 2015. "Oxidative Stress Induced by Chlorine Dioxide as an Insecticidal Factor to the Indian Meal Moth, *Plodia interpunctella*." *Pesticide Biochemistry and Physiology* 124. Elsevier Inc.: 48–59. doi:10.1016/j.pestbp.2015.04.003.

Kutsaev, S. V., R. Agustsson, R. Berry, S. Boucher, and A. Y. Smirnov. 2021. "Electron Accelerator for Replacement of Radioactive Sources in Insect Sterilization Facilities." *Physics of Atomic Nuclei* 84 (10): 1743–1747. doi:10.1134/S1063778821100215.

Likhayo, P., A. Y. Bruce, T. Tefera, and J. Mueke. 2018. "Maize Grain Stored in Hermetic Bags: Effect of Moisture and Pest Infestation on Grain Quality." *Journal of Food Quality* 2018. doi:10.1155/2018/2515698.

Lima, M. C. F., M. E. Damascena de Almeida Leandro, C. Valero, L. C. P. Coronel, and C. O. G. Bazzo. 2020. "Automatic Detection and Monitoring of Insect Pests—A Review." *Agriculture (Switzerland)* 10 (5): 1–24. doi:10.3390/agriculture10050161.

Liu, H., S. H. Lee, and J. S. Chahl. 2017. "A Review of Recent Sensing Technologies to Detect Invertebrates on Crops." *Precision Agriculture.* 18. Springer US. doi:10.1007/s11119-016-9473-6.

Lu, P. J., H. C. Lin, W. T. Yu, and J. M. Chern. 2011. "Chemical Regeneration of Activated Carbon Used for Dye Adsorption." *Journal of the Taiwan Institute of Chemical Engineers* 42 (2). Taiwan Institute of Chemical Engineers: 305–311. doi:10.1016/j.jtice.2010.06.001.

Maghirang, E. B., F. E. Dowell, and J. E. Baker, and James E Throne. 2002. *Detecting Single Wheat Kernels Containing Live or Dead Insects Using Near-Infrared Reflectance Spectroscopy.* In 2002 ASAE Annual Meeting (p. 1). *American Society of Agricultural and Biological Engineers.*

Mankin, R., D. Hagstrum, M. Guo, P. Eliopoulos, and A. Njoroge. 2021. "Automated Applications of Acoustics for Stored Product Insect Detection, Monitoring, and Management." *Insects* 12 (3). doi:10.3390/insects12030259.

Mankin, R. W., D. W. Hagstrum, M. T. Smith, A. L. Roda, and M. T. K. Kairo. 2011. "Perspective and Promise: A Century of Insect Acoustic Detection and Monitoring." *American Entomologist* 57 (1): 30–44. doi:10.1093/ae/57.1.30.

Mason, L. J., C. P. Woloshuk, and D. E. Maier. 1991. "Efficacy of Ozone to Control Insects, Molds and Mycotoxins Linda." *International Conference on Controlled Atmosphere and Fumigation in Stored Products*, ed by Donahaye E. J., Navarro, S., Varnava, A., Printco, Ltd, Nicosia, Cyprus: 665–670.

Mbata, G. N., M. Johnson, T. W. Phillips, and M. Payton. 2005b. "Mortality of Life Stages of Cowpea Weevil (Coleoptera: Bruchidae) Exposed to Low Pressure at Different Temperatures." *Journal of Economic Entomology* 98.

Mbata, G. N., M. Johnson, T. W. Phillips, and M. Payton. 2005a. "Mortality of Life Stages of Cowpea Weevil (Coleoptera: Bruchidae) Exposed to Low Pressure at Different Temperatures." *Journal of Economic Entomology* 98 (3). Entomological Society of America: 107–0–1075. doi:10.1603/0022-0493-98.3.1070.

Mbata, G. N., and T. W. Phillips. 2001. "Effects of Temperature and Exposure Time on Mortality of Stored-Product Insects Exposed to Low Pressure." *Journal of Economic Entomology* 94 (5): 1302–1307.

Mcclurkin, J. D., and D. E. Maier. 2010. "Half-Life Time of Ozone as a Function of Air Conditions and Movement" 425. doi:10.5073/jka.2010.425.167.326.

McDonough, M. X., C. A. Campabadal, L. J. Mason, D. E. Maier, A. Denvir, and C. Woloshuk. 2011a. "Ozone Application in a Modified Screw Conveyor to Treat Grain for Insect Pests, Fungal Contaminants, and Mycotoxins." *Journal of Stored Products Research* 47 (3): 249–254. doi:10.1016/j.jspr.2011.04.001.

McDonough, M. X., L. J. Mason, and C. P. Woloshuk. 2011b. "Susceptibility of Stored Product Insects to High Concentrations of Ozone at Different Exposure Intervals." *Journal of Stored Products Research* 47 (4): 306–310. doi:10.1016/j.jspr.2011.04.003.

Mohan, S., M. Gopalan, P. C. Sundarababu, and V. V. Sreenarayanan. 1994. "Practical Studies on the Use of Light Traps and Bait Traps in the Management of *Rhyzopertha dominica* (F.) in Rice Warehouses." *International Journal of Pest Management* 40 (2): 148–152. doi:10.1080/09670879409371873.

Mohan, S., and A. Rajesh. 2016. "Tools for Stored Product Insects Management and Technology Transfer." *Indian Journal of Entomology* 78 (special): 59. doi:10.5958/0974-8172.2016.00026.2.

Montgomery, G. A., M. W. Belitz, R. P. Guralnick, and M. W. Tingley. 2021. "Standards and Best Practices for Monitoring and Benchmarking Insects." *Frontiers in Ecology and Evolution* 8 (January). doi:10.3389/fevo.2020.579193.

Mowery, S. V., M. A. Mullen, J. F. Campbell, and A. B. Broce. 2002. "Mechanisms Underlying Sawtoothed Grain Beetle (*Oryzaephilus surinamensis* [L.]) (Coleoptera: Silvanidae) Infestation of Consumer Food Packaging Materials." *Journal of Economic Entomology* 95 (6): 1333–1336. doi:10.1603/0022-0493-95.6.1333.

Mutambuki, K., and P. Likhayo. 2021. "Efficacy of Different Hermetic Bag Storage Technologies against Insect Pests and Aflatoxin Incidence in Stored Maize Grain." *Bulletin of Entomological Research* 111 (4): 499–510. doi:10.1017/S0007485321000213.

Navarro, S. 2012. "The Use of Modified and Controlled Atmospheres for the Disinfestation of Stored Products." *Journal of Pest Science* 85 (3). Springer: 301–322.

Navarro, S., J. Donahaye, R. Dias, A. Azrieli, M. Rindner, T. Phillips, and R. Noyes. 2001. *Application of Vacuum in a Transportable System for Insect Control*. Executive Printing Services.

Neethirajan, S., C. Karunakaran, D. S. Jayas, and N. D. G. White. 2007. "Detection Techniques for Stored-Product Insects in Grain." *Food Control* 18 (2): 157–162. doi:10.1016/j.foodcont.2005.09.008.

Neven, L. G., N. J. Lehrman, and L. D. Hansen. 2014. "Effects of Temperature and Modified Atmospheres on Diapausing 5th Instar Codling Moth Metabolism." *Journal of Thermal Biology* 42 (1). Elsevier: 9–14. doi:10.1016/j.jtherbio.2014.02.017.

Njoroge, A. W. 2017. "Acoustic Detection of Insect Pests of Stored Grains in Kenya." (Doctoral dissertation, University of Kassel, Department of Organic Agricultural Sciences, Kenya).

Novotny, M., J. Skramlik, K. Suhajda, and V. Tichomirov. 2013. "Sterilization of Biotic Pests by Microwave Radiation." *Procedia Engineering* 57. Elsevier B.V.: 1094–1099. doi:10.1016/j.proeng.2013.04.138.

Odjo, S., N. Bongianino, J. G. Regalado, M. L. C. Soto, N. Palacios-Rojas, J. Burgueño, and N. Verhulst. 2022. "Effect of Storage Technologies on Postharvest Insect Pest Control and Seed Germination in Mexican Maize Landraces." *Insects* 13 (10): 1–15. doi:10.3390/insects13100878.

Pearson, T., and D. L. Brabec. 2007. "Detection of Wheat Kernels with Hidden Insect Infestations with an Electrically Conductive Roller Mill." *Applied Engineering in Agriculture* 23 (5): 639–645. doi:10.13031/2013.23662.

Perez-Mendoza, J., J. E. Throne, F. E. Dowell, J. E. Baker, J. Throne, F. Dowell, and J. Baker. 2003. "Detection of Insect Fragments in Wheat Flour by Near-Infrared Spectroscopy". *Journal of Stored Products Research* 39 (3): 305–312.

Preti, M., F. Verheggen, and S. Angeli. 2021. "Insect Pest Monitoring with Camera-Equipped Traps: Strengths and Limitations." *Journal of Pest Science* 94 (2). Springer Berlin Heidelberg: 203–217. doi:10.1007/s10340-020-01309-4.

Prince, P., A. Hill, E. P. Covarrubias, P. Doncaster, J. L. Snaddon, and A. Rogers. 2019. "Deploying Acoustic Detection Algorithms on Low-Cost, Open-Source Acoustic Sensors for Environmental Monitoring." *Sensors (Switzerland)* 19 (3): 1–23. doi:10.3390/s19030553.

Rathore, D. S., B. Ram, B. L. Pal, and S. Malviya. 2019. "Analysis of Classification Algorithms for Insect Detection Using MATLAB." *SSRN Electronic Journal*, no. December. doi:10.2139/ssrn.3350283.

Reed, C. R., V. F. Wright, T. W. Mize, J. R. Pedersen, and B. J. Evans. 1991. "Pitfall Traps and Grain Samples as Indicators of Insects in Farm-Stored Wheat." *Journal of Economic Entomology* 84 (4): 1381–1387. doi:10.1093/jee/84.4.1381.

Regnault-Roger, C., C. Vincent, and J. T. Arnason. 2012. "Essential Oils in Insect Control: Low-Risk Products in a High-Stakes World." *Annual Review of Entomology* 57: 405–424. doi:10.1146/annurev-ento-120710-100554.

Ridgway, C., and J. Chambers. 1996. Detection of External and Internal Insect Infestation in Wheat by Near-Infrared Reflectance-Spectroscopy. *Journal of the Science of Food and Agriculture* 71.

Riudavets, J., M. J. Pons, J. Messeguer, and R. Gabarra. 2018. "Effect of CO2 Modified Atmosphere Packaging on Aflatoxin Production in Maize Infested with *Sitophilus zeamais*." *Journal of Stored Products Research* 77: 89–91. doi:10.1016/j.jspr.2018.03.005.

Sanon, A., L. C. Dabiré-Binso, and N. M. Ba. 2011. "Triple-Bagging of Cowpeas within High Density Polyethylene Bags to Control the Cowpea Beetle *Callosobruchus maculatus* F. (Coleoptera: Bruchidae)." *Journal of Stored Products Research* 47 (3): 210–215. doi:10.1016/j.jspr.2011.02.003.

Santiago, R. M. C., S. L. Rabano, R. K. D. Billones, E. J. Calilung, E. Sybingco, and E. P. Dadios. 2017. "Insect Detection and Monitoring in Stored Grains Using MFCCs and Artificial Neural Network." *IEEE Region 10 Annual International Conference, Proceedings/TENCON 2017-Decem* (November): 2542–2547. doi:10.1109/TENCON.2017.8228290.

Shendge, S. N., V. S. Pawar and P. R. Kale. 2021. "Novel Technique: Hermetic Storage and Its Application." *The Pharma Innovation Journal* 10 (8): 451–456.

Sileem, T., S. Mohamed, and A. Mehany. 2019. "First Trial to Assess Gamma-Irradiation and Diatomaceous Earth Combination on Mortality of Some Stored Product Insects." *Arab Journal of Nuclear Sciences and Applications* 52 (3): 115–120. doi:10.21608/ajnsa.2019.6159.1134.

Smit, R. 2019. "Postharvest Phytosanitary Disinfestation Strategies Using Thermal and Atmospheric Stress: Commodity and Insect Tolerances." (Doctoral dissertation, Stellenbosch: Stellenbosch University).

Somavat, P., H. Huang, S. Kumar, M. K. Garg, M. C. Danao, V. Singh, M. R. Paulsen, and K. D. Rausch. 2017. "Comparison of Hermetic Storage of Wheat with Traditional Storage Methods in India." *Applied Engineering in Agriculture* 33 (1): 121–130. doi:10.13031/aea.11792.

Sousa, A. H., L. R. D. A. Faroni, R. N. C. Guedes, M. R. Tótola, and W. I. Urruchi. 2008. "Ozone as a Management Alternative against Phosphine-Resistant Insect Pests of Stored Products." *Journal of Stored Products Research* 44 (4): 379–385. doi:10.1016/j.jspr.2008.06.003.

Sutin, A., A. Yakubovskiy, H. R. Salloum, T. J. Flynn, N. Sedunov, and H. Nadel. 2019. "Towards an Automated Acoustic Detection Algorithm for Wood-Boring Beetle Larvae (Coleoptera: Cerambycidae and Buprestidae)." *Journal of Economic Entomology* 112 (3): 1327–1336. doi:10.1093/jee/toz016.

Thistlewood, H. M. A., and G. J. R. Judd. 2019. "Twenty-Five Years of Research Experience with the Sterile Insect Technique and Area-Wide Management of Codling Moth, *Cydia pomonella* (L.), in Canada." *Insects* 10 (9). doi:10.3390/insects10090292.

Thomas, S. 2017. *Artificial Insect Olfaction School of Engineering*. The University of Warwick.

Tiwari, B. K., C. S. Brennan, T. Curran, E. Gallagher, P. J. Cullen, and C. P. O. Donnell. 2010. "Application of Ozone in Grain Processing." *Journal of Cereal Science* 51 (3): 248–255. doi:10.1016/j.jcs.2010.01.007.

Tutuncu, S., and M. Emekci. 2019. "Comparative Efficacy of Modified Atmospheres Enriched with Carbon Dioxide against *Cadra* (= *Ephestia*) *Cautella*." *Journal of the Science of Food and Agriculture* 99 (13): 5962–5968. doi:10.1002/jsfa.9871.

Vadivambal, R., and D. S. Jayas. 2010. "Non-Uniform Temperature Distribution during Microwave Heating of Food Materials-A Review." *Food and Bioprocess Technology* 3 (2): 161–171. doi:10.1007/s11947-008-0136-0.

Vick, K. W., J. C. Webb, B. A. Weaver, and C. Litzkow. 1988. "Sound Detection of Stored-Product Insects That Feed Inside Kernels of Grain1." *Journal of Economic Entomology* 81 (5): 1489–1493. doi:10.1093/jee/81.5.1489.

Vinatier, F., and C. Vinatier. 2013. "Acoustic Recording as a Non-Invasive Method to Detect Larval Infestation of *Cosmopolites sordidus*." *Entomologia Experimentalis et Applicata* 149 (1): 22–26. doi:10.1111/eea.12102.

Vincent, C., P. G. Weintraub, G. J. Hallman, and F. Fleurat-Lessard. 2008. "Insect Management with Physical Methods in Pre- and Post-Harvest Situations." In *Integrated Pest Management: Concepts, Tactics, Strategies and Case Studies*, no. January: 309–323. doi:10.1017/CBO9780511626463.025.

Walde, S. G., K. Balaswamy, V. Velu, and D. G. Rao. 2002. "Microwave Drying and Grinding Characteristics of Wheat (*Triticum aestivum*)." *Journal of Food Engineering* 55 (3): 271–276. doi:10.1016/S0260-8774(02)00101-2.

Wang, S., J. N. Ikediala, J. Tang, J. D. Hansen, E. Mitcham, R. Mao, and B. Swanson. 2001. "Radio Frequency Treatments to Control Codling Moth in In-Shell Walnuts." *Postharvest Biology and Technology* 22 (1): 29–38. doi:10.1016/S0925-5214(00)00187-3.

White, N. D. G., R. T. Arbogast, P. G. Fields, R. C. Hillmann, and S R Loschiavo. 1990. "The Development and Use of Pitfall and Probe Traps for Capturing Insects in Stored Grain". *Journal of the Kansas Entomological Society* 63 (4): 506–525. https://www.jstor.org/stable/25085217

Wong-Corral, F. J., C. Castañé, and J. Riudavets. 2013. "Lethal Effects of CO_2-Modified Atmospheres for the Control of Three Bruchidae Species." *Journal of Stored Products Research* 55 (October): 62–67. doi:10.1016/j.jspr.2013.08.005.

Xia, C., T. S. Chon, Z. Ren, and J. M. Lee. 2015. "Automatic Identification and Counting of Small Size Pests in Greenhouse Conditions with Low Computational Cost." *Ecological Informatics* 29 (P2). Elsevier B.V.: 139–146. doi:10.1016/j.ecoinf.2014.09.006.

Yamada, H., H. Maiga, J. Juarez, D. De Oliveira Carvalho, W. Mamai, A. Ali, N. S. Bimbile-Somda, A. G. Parker, D. Zhang, and J. Bouyer. 2019. "Identification of Critical Factors That Significantly Affect the Dose-Response in Mosquitoes Irradiated as Pupae." *Parasites & Vectors* 12 (1). BioMed Central: 435. doi:10.1186/s13071-019-3698-y.

Yanagawa, A., A. Kajiwara, H. Nakajima, E. Desmond-Le Quéméner, J. P. Steyer, V. Lewis, and T. Mitani. 2020. "Physical Assessments of Termites (Termitidae) under 2.45 GHz Microwave Irradiation." *Scientific Reports* 10 (1): 1–9. doi:10.1038/s41598-020-61902-6.

Yanagawa, A., M. Tomaru, A. Kajiwara, H. Nakajima, E. D. L. Quemener, J. P. Steyer, and T. Mitani. 2020. "Impact of 2.45 GHz Microwave Irradiation on the Fruit Fly, Drosophila Melanogaster." *Insects* 11 (9): 1–15. doi:10.3390/insects11090598.

Zhai, Z., J. F. Martínez, V. Beltran, and N. L. Martínez. 2020. "Decision Support Systems for Agriculture 4.0: Survey and Challenges." *Computers and Electronics in Agriculture* 170 (August 2019). Elsevier: 105256. doi:10.1016/j.compag.2020.105256.

Zida, S. I., Y. D. Lin, and Y. L. Khung. 2021. "Current Trends on Surface Acoustic Wave Biosensors." *Advanced Materials Technologies* 6 (6): 1–23. doi:10.1002/admt.202001018.

Ziegler, V., R. T. Paraginski, and C. D. Ferreira. 2021. "Grain Storage Systems and Effects of Moisture, Temperature and Time on Grain Quality – A Review." *Journal of Stored Products Research* 91 (March): 101770. doi:10.1016/j.jspr.2021.101770.

19 Nano-Based Disinfestation Methods for Stored Products

Aditi Negi and Jeyan A. Moses
Computational Modeling & Nanoscale Processing Unit, Department of Food Process Engineering, National Institute of Food Technology, Entrepreneurship and Management (NIFTEM-T), Thanjavur, Tamil Nadu, India

19.1 INTRODUCTION

Storage pests pose an equally significant threat as crop pests, causing severe damage to stored food products. Globally, food grain losses due to insect infestation were reported at 10%–40% per year. The major economic loss caused by storage insect pests is not always by consumption but also by the amount of contamination. Insect infestation affects the quantity and causes total quality loss, such as a decrease in nutritive value, seed germination, pH level and protein content, high moisture content, and free fatty acid content (Mohapatra et al. 2017). About 600 species of insects have been associated with stored grain products, and nearly 100 species of insect pests (mostly beetles or moths) of stored products cause economic losses. The estimated annual economic loss in India due to storage is approximately 14 million tons worth Rs. 7,000 crores in which insects alone account for nearly Rs. 1,300 crores (IGMRI 2019). These data are alarming, and thus, insect control is essential to save production.

There are several insect control alternatives that are accessible, including mechanical, physical, chemical, and biological practices, but fumigation is the most often used form of pest management. Among chemical disinfectants, malathion, chlorpyrifos-methyl, and deltamethrin are considered contact pesticides, while phosphine (PH_3) and methyl bromide (MeBr) are regarded as fumigants are most commonly used because of their broad applicability in a variety of storage environments, including bags, silos, warehouses, and mills. However, owing to negative effects on human health and the environment, the use of certain chemical disinfectants, such as MeBr, has been outlawed. Similarly, PH_3, chlorpyrifos-methyl, and fumigants have been demonstrated that insects acquired resistance to the chemical disinfectants and were not efficient in killing egg stages, particularly ones that are inside the grain kernels (Moirangthem and Baik 2021). Disinfestation using botanicals (volatiles, essential oils (EOs)) as a green alternative solution in insect management. The active ingredient present in botanicals works against pests, such as antifeedants, chemosterilants, repellents, toxicants, and growth regulators properties. Although the efficacy of many botanicals has been proven against stored pests, only 1% of botanical formulations are commercially available in the world market (Guru et al. 2022).

The recent invention significance and continuous advancements in nanotechnology have led to the development of novel methodologies and tactics aimed at the production or enhancement of toxicity toward the stored pest. In order to transform modern pest management methods, the implementation of nanomaterial-based products is being pursued. Nano-based insecticides are defined as any product that deliberately combines materials in the nm size range and/or claims specific qualities linked with these nano-size ranges. The materials possess a high surface area-to-volume ratio and exhibit unique physicochemical features, which make them capable of satisfying the growing demand as a result of their heightened reactivity. Furthermore, the augmentation of the surface area per unit mass of a substance allows for a greater extent of interaction between the nanomaterial and

 DOI: 10.1201/9781003609834-19

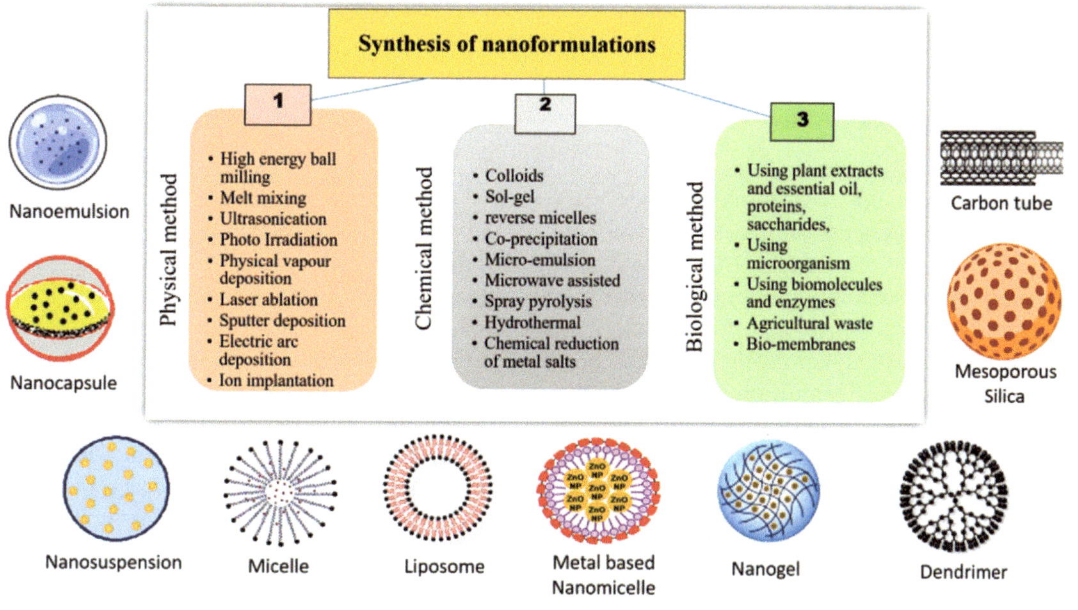

FIGURE 19.1 Classification of different nano formulates and synthesis methods.

its surrounding components, hence impacting its reactivity (Nile et al. 2020; Shelar et al. 2023). The unique characteristics of nanomaterials, such as changes in electrical conductivity, surface chemistry, and reactivity (Gamal 2018) and due to extremely small size, nanomaterials act as efficient agents in cellular penetration, which leads to toxic properties or lethal effects in the insects. According to previous research, nanoparticles have a much greater surface area compared to larger-scale materials of equivalent masses (Qiu et al. 2020).

Based on the above-mentioned characteristics, nanoformulations outperform traditional insecticides in stored pest control. Their ability to precisely target and penetrate pests means active chemicals are administered efficiently. The lower dose and sustained release of insecticide reduce residue concerns and extend the time of action in stored food facilities to control pests without reapplication. Their accurate targeting reduces environmental damage and supports biological balance in storage habitats by minimizing harm to non-target species. Additionally, nanoformulations may smoothly integrate into integrated pest management (IPM) methods, enabling comprehensive and sustainable pest control in stored settings. Different formulations, such as organic (such as polymers) and inorganic (such as metal oxides) nanoformulations, are similar to those of traditional formulations. Nano-based pesticides can be synthesized in a variety of forms, such as particles, emulsions, micelles, and nanocomposites, using different techniques, as shown in Figure 19.1. Detailed review of nanoparticle formulates and applications against stored product protection is discussed in consecutive sections.

19.2 NANO-PESTICIDE FORMULATES

Nanoformulations have shown great potential in the field of agriculture as biopesticides to protect crops as well as it has also been explored in stored pest control. According to their intended use, nanocides are created to increase solubility, limit the release of active components, stop degradation, etc. This offers advantages such as increased efficiency, reduced environmental impact, and targeted delivery. Modifications to the chemical nature carrier molecule have been made to achieve these goals and are categorized as organic, inorganic nano-sized metals and metal oxides, polymer, lipid, and clay-based nanomaterials are few to mention. The different nanomaterials that can be used as insecticides in stored pest management are discussed here.

19.2.1 Nanoemulsions

It is a biphasic dispersion where the active component (pesticide) of the formulation is distributed in the form of nano-sized droplets, either oil-in-water (O/W) or water-in-oil emulsion, where surfactant molecules act as an interfacial layer between the pesticide and water. They can encapsulate various biopesticides, such as EOs, botanical extracts, and microbial agents. Nanoemulsions provide enhanced solubility, prolonged release, and improved penetration of active ingredients, leading to better pest control. Based on the number and kind of surfactants, nanoemulsions are further divided into thermodynamically stable and kinetically stable categories. It is said to be a thermodynamically stable nanoemulsion if the pesticide is only partly soluble in the aqueous phase, whereas a stable emulsion spontaneously forms when all the components, such as pesticide, surfactant, and water, are mixed together. Initially, pesticide and surfactant form a two-phase system due to the insolubility of the active ingredient; however, through continuous shearing, pesticide droplets are well dispersed and mix in the nanoemulsion and stable for an extended period, and therefore it is considered as a kinetically stable nanoemulsion. Several research studies have shown the application of nanoemulsion against the disinfestation of stored grain pests. Nanoemulsion containing sweet orange (*Citrus sinensis*) EO formulation is acutely poisonous to confused flour beetle and flat grain beetle reported by (Giunti et al. 2019). The eucalyptus oil nanoemulsion with 8.57 nm particle size had better insecticidal efficacy against *Tribolium castaneum*. Nanoemulsions were subjected to centrifugation, thermodynamic stability, heating and cooling, and freezing and thawing tests and were shown to be extremely poisonous at 1:2 and 1:2.5 (LC_{50} 0.56 and 0.45 L/cm^2) of lethal concentrations against *T. castaneum* (Adak et al. 2020). Several other studies reported EO EO-based nanoemulsions were found toxic to the stored pest and also discussed in the section effect of nanoformulations on the ontogenic stage of insects.

19.2.2 Nano-Suspension

Nano-suspension or nano-dispersion consists of solid biopesticide particles dispersed in a liquid medium. They enhance the stability and bioavailability of poorly water-soluble active ingredients, leading to improved pest control. The surfactant molecules are constrained at the particle surface in nano-dispersions, with the non-polar sections interacting with the solid pesticide and the polar portions extending into the aqueous solution. In a study conducted by (Hazafa et al. 2022), nano-suspension was formulated with *Azadirachta indica* and *Chrysanthemum coronarium* using the anti-solvent precipitation procedure exhibiting unique pesticidal and antimicrobial properties. The red flour beetle (*T. castaneum*) and lesser grain borer (*Rhyzopertha dominica*) pests were used as test subjects for the pesticidal activity of nano-suspension, which demonstrated 100% mortality against both the insects with nano-suspensions of 100% concentration of *C. coronarium* and *A. indica*. Comparatively, nano-suspensions at 50% concentration with a 1:1 ratio of these two plant extracts demonstrated the highest 100% mortality after 72 h.

19.2.3 Polymer-Based Nanoparticles

Polymer-based formulations for pesticide, herbicide, and fungicide-controlled release are gaining popularity in pest management. These formulations protect light-sensitive active components and efficiently disperse them. Slow release is controlled by nanocarrier breakdown and bonding between chemicals, carriers, and environmental variables. Eco-friendly polymers protect EO and plant metabolites against degradation caused by light, water, air, and heat. In polymers, biodegradability is critical. Nanoformulations boost effectiveness while reducing environmental impact by using fewer solvents and surfactants. Compounds, effects, biocompatibility, and biodegradability are all accommodated via complex delivery networks. Polysaccharides such as chitosan, alginates, and starch are often used in nanopesticide formulations, as are polyesters such as polycaprolactone, polylactic acid, and polyethylene glycol (PEG) (Kashyap, Xiang, and Heiden 2015; Rani et al. 2017; Kumar et al. 2018). Encapsulation is accomplished by the use of various polymer forms, such as nanocapsules, nanospheres, nanogels, micelles, and nanofibers, particularly those based on chitosan. PEG-based formulations outperform other commercial solutions in the control of insect pests and nematodes (Loha et al. 2012; Pankaj et al. 2012). *T. castaneum* (Herbst) is efficiently combated by PEG-coated nanoparticles containing garlic EO (Yang et al. 2009).

19.2.3.1 Nanoencapsulation

Nanoencapsulation is a spherical structure composed of a polymer shell encompassing a core containing the biopesticide. This arrangement offers controlled release characteristics, safeguarding the active component from degradation and environmental influences. Nano-encapsulates can be designed to release the biopesticide in response to specific triggers, such as pH or temperature changes. A heterogeneous reservoir type structure, nanocapsules, or nano-encapsulation has an inner core chamber that restricts the hydrophobic or hydrophilic active component and is encircled by a polymer coating or membrane. In homogeneous vesicular formations called nanospheres, the bioactive component is evenly distributed throughout the polymer matrix. For instance, nanospheres of polymer-stabilized bifenthrin are being created. Negahban et al. (2012) reported that the repeated use of nanoencapsulated *Cuminum cyminum* oil in water emulsion (via poly (urea-formaldehyde) on red flour beetles attained LC_{50} value of 16.25 ppm over a seven-day period, surpassing the efficacy of oil tested alone. According to (Khoobdel, Ahsaei, and Farzaneh 2017), a nanocapsule formulation containing *Rosmarinus officinalis* EO showed significant efficacy in *T. castaneum*. In a study conducted by (Batool et al. 2021), it was shown that the administration of *Albizia procera's* (ApCP's) cysteine protease nanocapsule at concentrations of 7.0 and 3.5 mg/g led to 100% mortality of *Sitotroga cerealella*. Similarly, nanoencapsulated *Eucalyptus globulus*, and *Zataria multiflora* EOs demonstrated potency toward *Ephestia kuehniella* (Emamjomeh et al. 2023). The study conducted by Bayramzadeh et al. (2019), it showed that *Lavandula angustifolia* exhibited similar effectiveness to *C. cyminum* (cumin oil). Additionally, a combination of nanocapsulated EO and lowered sub-lethal phosphine concentrations resulted in a 50% reduction in *Sitophilus granarius* and *T. castaneum* population. The concentrations required for this control were measured at 42.51 and 78.99 l/l, respectively.

19.2.3.2 Nanogels

These are also known as nanoparticles made of hydrogel. These are produced by cross-linking water-absorbent polymeric particles, which increase their water absorption. Chitosan-myristic-based nanogel containing *Carum copticum* seed oil possesses more insecticidal and residual activity than EO alone and is found effective against *S. granarius*, *Tribolium confusum* (M Ziaee, Moharramipour, and Mohsenifar 2014). Further, the studies indicated that with an increase in exposure time, toxicity will increase.

19.2.3.3 Nano-Fibers

Thermally-induced phase separation and electrospinning are used to create nanofibers. For the control of a variety of lepidopteran insect pests, scientists have developed electrospun nanofibers that are loaded with the chemical (Z)-9-dodecenyl acetate, which is a component of pheromones and becomes entrenched in the polymer matrix. In another study conducted by Allahvaisi et al. (2017), a polylactic acid solution was mixed with *Mentha piperita* and *Salvia officinalis* and screened for contact toxicity against *Plodia interpunctella*. Because the polylactic acid nanofibers generate surface tension and slow release, they prolong the efficiency time. The product displayed greater contact toxicity than the pure EO, and *M. piperita* has shown greater toxicity than *S. officinalis*.

19.2.3.4 Carbon Nanotubes (CNTs) and Dendrimer

Carbon nanotubes (CNTs) offer high surface area, mechanical strength, and the ability to encapsulate both hydrophobic and hydrophilic compounds and can serve as carriers for biopesticides. CNTs can be functionalized to improve biopesticide loading and release characteristics. Dendrimers are highly branched, nanoscale polymers that can encapsulate biopesticides within their cavities. They offer a well-defined structure, tuneable properties, and controlled release kinetics, making them suitable for targeted and sustained pest management. Recently Edwards et al. (2020) developed polyamidoamine dendrimer-coated carbon nanotubes (PAMAM-CNT) as a delivery vehicle for dsRNA against *T. castaneum*. Results reported that PAMAM-CNT increases the efficacy of RNA interference in insect pest species.

19.2.3.5 Nanostructured Formulations of Microbial Biopesticides

Formulations of microbial biopesticides: Beneficial microorganisms, such as bacteria and fungi, can be formulated into nanoscale structures to improve their stability, viability, and efficacy. These formulations can enhance the colonization of pests and provide long-lasting pest control.

19.2.4 Lipid-Based Nanocarrier

Phospholipids are used to make lipid-based nanocarriers, which may self-assemble into numerous bilayers that restrict the aqueous phase (Sala et al. 2018). Because of its unique qualities, such as physiochemical storage stability, high loading capacity with target-oriented smart regulated active ingredient release system, and environmental safety. These nanocarriers, such as nanoliposomes and solid lipid nanoparticles (SLNs), are very efficient for release. These lipid-based nanomaterials have several advantages over other delivery systems, such as nanopolymers and nanoemulsions, such as slow rate of chemical degradation, incorporation of both hydrophobic and hydrophilic active ingredients, allowing wider ingredient application and make it a viable approach for commercially scalable products (Wei et al. 2017). Lipid-based nanocarriers can also be able to prevent photodegradation of active substances without the use of ultraviolet (UV) absorbers (Nguyen et al. 2012).

19.2.4.1 Nanoliposomes

These are nanoscale vesicular structures having an aqueous phase cavity surrounded by a phospholipid bilayer. The average nanoparticle size of these nanoliposomes is 70–350 nm, showing better physical stability, strength, and reduced polydispersity. The resistance of the nanocarriers to diffusion via the degrading agents has a significant impact on the stability of the active components. Because of the physiochemical instability of liposomes in aqueous dispersions, long-term storage cannot be suggested. The most frequent strategy for extending the shelf life of these nanoliposomes is lyophilization (Chen and Yada 2011). Faraji et al. (2020) developed nanoliposomes using different plants, EO *Mentha pulegium*, *Ferula gummosa*, using thin film ultrasonication and heating methods. The durability of the fumigant toxicity of *M. pulegium* was studied against *T. castaneum*. The study reported that nanoliposome-based EO possesses more fumigant toxicity than the EO alone.

19.2.4.2 Solid Lipid Nanoparticles

SLNs are innovative potential nanocarriers made of a high melting point lipid matrix and in a spherical form that is, disseminated in an aqueous solution. They also offer many benefits over conventional nanoformulations, including physical stability, superior biocompatibility, reduced toxicity, and improved distribution of lipophilic active components. Pardeshi et al. (2012) created SLN on a wide scale as a replacement for polymer-based nanoformulations for pesticide application. Nguyen et al. (2012) created second-generation SLN by integrating liquid lipids into the solid matrix to boost payload while avoiding quick UV degradation and active ingredient leakage. A study conducted by (Hosseinpour Jajarm et al. 2021) using *Ziziphora clinopodioides* Lam.-loaded SLNs were synthesized using concentrations (1%, 2%, and 2.5% w/v) using high shear homogenization and ultrasound. According to fumigant toxicity data, the LC_{50} values for SLN-EO (2.5%) and pure oil against *T. castaneum* were determined to be 30.602 and 68.303 μl/l air, respectively. The results of this study indicated that the impact of SLN-EO on red flour beetles was more detrimental. The combination of SLN-EO exhibited a prolonged persistence of toxicity for a duration of 14 days, while the pure oil showed a decline in its toxic effects after eight days.

19.2.5 Nanoparticles

Nanoparticles, including metal-based (e.g., silver, zinc, aluminum phosphide, copper) and polymeric nanoparticles, can be loaded with biopesticides. These nanoparticles possess unique properties that can disrupt insects through mechanisms, such as damaging their cuticles or affecting their nervous systems. These nanoparticles can be designed to release the active ingredients slowly over time, targeting action, reducing resistance development, increasing their efficacy, reducing the need for frequent applications, and lowering environmental impact (Jasrotia et al. 2022).

19.2.5.1 Silver Nanoparticles (AgNPs)

Silver nanoparticles (AgNPs) have emerged as a well-studied and beneficial class of nanomaterials for managing insect pest concerns. Their potential lies in their role as delivery vehicles for targeted agrochemical applications. AgNPs have strong insecticidal, bactericidal, antifungal, and antiviral activities, making them a potential insecticide. Additionally, their catalytic characteristics, antibacterial efficiency, increased chemical stability, and electrical conductivity have piqued the attention of pest control experts (Saratale et al. 2017).

The ability of silver-based pesticide formulations to deliver a greater dose of active components directly to target species outperforms that of traditional insecticides. These nano-based insecticides have an exceptional safety profile for non-target species, harmonizing with ecologically friendly methods (Athanassiou et al. 2018; Oliveira et al. 2019). According to reports, silver nanoparticles are efficient in controlling pests, such as *Sitophilus oryzae* (Pathipati and Kanuparthi 2021). Using green synthesis approaches (plant-mediated), AgNPs generated from plants such as *Euphorbia prostrata* and *Avicennia marina* outperformed the stored pest *S. oryzae*, *T. castaneum*, and *R. dominica* (Zahir et al. 2012; Sankar and Abideen 2015). (Vadlapudi and Amanchy 2017) used innovative synthesis techniques to create AgNPs from *Myriostachya wightiana* leaf extracts, which displayed toxic effects at a concentration of 50 g against *S. oryzae*, leading to a death rate of 29% after 24 h. AgNPs produced from rosemary extract showed 100% mortality in *S. granarius* adults after 120 h of exposure as reported by (Rashwan and Abu-Zaid 2018). Combining AgNPs with chemical pesticide malathion resulted in significant insecticidal activity toward red flour beetle (As and Thangapandiyan 2019). (Malathi et al. 2019) reported AgNPs derived from *Annona reticulata* leaf extract have been shown to have significant insecticidal effects against *S. oryzae*. Shahzadi et al. (2019) demonstrated the use of silver nanoparticles in conjunction with *Ricinus communis* plant oil demonstrated a repellency rate of 67.89% at a concentration of 15%. Similarly, when combined with Citrus paradise oil at the same dosage, the repellency rate was measured to be 28.31% against *T. castaneum*.

A study conducted by (Rani et al. 2019), it was shown that AgNPs synthesized using silver nitrate and *Moringa oleifera* leaf extract exhibited a complete mortality rate of *S. oryzae* following a 15-day period of exposure. These results highlight the versatility of silver nanoparticles as a tool in current insect pest control tactics.

19.2.5.2 Nanostructured Alumina (NSA)

Nanostructured alumina (NSA) dust or aluminum nanoparticles have been recognized as grain protectants against insect pest infestation (Stadler et al. 2017; Jasrotia et al. 2022). NSA disinfects stored grains through enhanced surface reactivity. It adsorbs contaminants like pest-attracting compounds and generates antimicrobial reactive oxygen species (ROS) upon light exposure. This dual action deters pests, inhibits microbial growth, and extends grain shelf life naturally, potentially reducing the need for traditional preservatives. Nano alumina may be a viable alternative to diatomaceous earth (DE)-based dust compositions. Stadler et al. (2012) demonstrated that the lethality of Nano alumina is superior to that of DE formulations against *S. oryzae*. M. M. Sabbour et al. (2012) produced two entomotoxins against *S. oryzae* using nanoparticles of Al_2O_3 and TiO_2. An aluminum-based toxin was shown to be more potent than nano TiO_2 against *S. oryzae*. The increased mortality was found when nano alumina dust of varying size and shape was used against pests *S. oryzae* and *R. dominica*. With all the findings, it was concluded that particle size reduction and increased surface area were not the sole important variables influencing pesticide efficiency.

Abo-Arab, Amal, and Hashem (2014) found that aluminum oxide nanoparticles at a dose of 1 g/kg caused 95.33% and 91.66% mortality of *S. oryzae* and *Sitophilus zeamais*, respectively, during 21 days of exposure. (Nasr 2015) demonstrated the antibacterial activity of Al_2O_3 NPs against *S. oryzae*. Nanoaluminum oxide exhibited a 100% death rate in rice weevil after 14 days when treated at a dose of 2 g/kg rice (El-Bendary, Abdel-Wahab, and El-Helaly 2016). López-García, Buteler, and Stadler (2018) studied the impact of NSA dust on *S. oryzae* in galvanized steel jars, and significant mortality rates were observed. Similarly, Das, Yadav, and Debnath (2019) confirmed that killing 90% of *S. oryzae* after four days of exposure to aluminum oxide at a 1 g/kg dose level. NSA works as seed protectant and insecticidal when sprayed at a rate of 400 mg/kg against three stored pests; 100% mortality was reported for *Stegobium paniceum*, 80.64% for *Oryzaephilus surinamensis*, and 79.41% for *T. confusum* by (Belhamel et al. 2020). Aluminum oxide nanoparticles caused complete mortality after seven days of exposure to 8,000 mg/kg wheat; additionally, after 60 days of exposure, all tested Al_2O_3-NPs concentrations (at different doses 1,000, 2,000, 4,000, and 8,000 mg/kg grain) significant reduction in *S. oryzae* offspring was observed in a dose-dependent manner (Ismail, Salama, and El-Ebiary 2021).

19.2.5.3 Titanium Dioxide Nanoparticles (TiO_2 NPs)

TiO_2 NPs have gained attention for their potential use in the disinfection and preservation of stored products the nanoparticles exhibit lower toxicity and negligible biological impacts. TiO_2 NPs possess photocatalytic properties that make them capable of generating ROS when exposed to UV light. This photocatalytic activity has applications in mitigating microbial contamination and extending the shelf life of stored items, particularly in agricultural and food industries (Jasrotia et al. 2022). In a study conducted by Abo-Arab, Amal, and Hashem 2014, it was shown that the application of TiO_2 NPs resulted in a mortality rate of 61.66% and 60.66% for *S. oryzae* and *S. zeamais*, respectively; when administered at a dose of 1 g/kg over a period of 21 days. In their study, Das, Yadav, and Debnath (2019) substantiated the efficacy of TiO_2 NPs in eradicating *S. oryzae*. The tested nanoparticles were administered at a dose of 2 g/kg, resulting in 90% mortality of *S. oryzae* during a 14-day period. In a study conducted by (Hilal et al. 2021), it was observed that a significant death rate of *T. castaneum* was recorded following exposure to a substantial concentration of TiO_2 nanoparticles. The cumulative mortality percentages for the treatments after one, three, and five days were found to be 15.30%, 23.57%, and 29.85%, respectively.

19.2.5.4 Zinc-Based Nanoparticles (ZnO NPs)

ZnO NPs present a promising way for disinfestation of stored pests due to their distinctive antimicrobial, optical, and physical properties. With a high surface area, ZnO NPs physically adhere to pests' cuticles, disrupting their protective layers. These nanoparticles generate ROS upon exposure to light or moisture, causing oxidative stress within pests' cells and subsequent damage. Moreover, ZnO NPs penetrate cell membranes, interfering with enzymes and disrupting metabolic pathways. This disruption extends to pests' reproductive and developmental processes, further reducing populations. Synergistic effects with other agents are possible. Various techniques may be used for the synthesis of these materials, including vapor transport and precipitation hydrothermal procedures (Raoufi 2013). Previous studies have shown the efficacy of nanoformulations containing ZnO in the management of *T. castaneum and S. oryzae*, resulting in increased mortality rates (Hamza 2012; Salem, Hamzah, and El-Taweelah 2015). A screening test was conducted by Nasr (2015) to evaluate and compare the impacts of ZnO NP and chemical insecticide pirimiphos methyl on *S. oryzae*. The findings demonstrated that micro ZnO may serve as a viable substitute for pirimiphos methyl, a substance known to have adverse impacts on human health. According to Malaikozhundan et al. (2017), the application of *Bacillus thuringiensis*-coated ZnO NP at a concentration of 25 μg/ml resulted in a significant decrease in the reproductive capacity and overall development time of the pulse beetle, *Callosobruchus maculatus*. Furthermore, this treatment led to the complete death of the beetles. According to (Haroun et al. 2020), the maximum concentration of ZnONPs shows a notable harmful impact on *S. oryzae* and *C. maculatus*. However, *T. castaneum* revealed a considerable level of tolerance. In a study conducted by Mohammed and Aswd (2019), it was observed that ZnO NP with 100 nm size and at 1,000 ppm concentration resulted in complete mortality of *C. maculatus*. The study conducted by Haider et al. (2020) demonstrated that the use of ZnO nanoparticles, together with leaf powder derived from *R. communis*, *Jatropha curcas*, and Citrus paradise, resulted in a death rate of 66.32% and 49.51% for *T. castaneum* and *Trogoderma granarium*, respectively.

The study conducted by Siddique et al. (2022) showed that the use of *E. globulus* extracts in the preparation of ZnO resulted in a death rate of 80.5% for *R. dominica* when exposed to a concentration of 600 ppm over a period of 15 days. In a similar vein, the study conducted by (Hilal et al. 2021) yielded findings that indicated a positive correlation between the concentration of ZnO nanoparticles and the cumulative death rate of *T. castaneum*. Specifically, following exposure periods of one, three, and five days, the treatments resulted in cumulative mortality rates of 20.42%, 27.08%, and 33.96%, respectively. (Dikbaş et al. 2021) demonstrated that the inclusion of chitinase derived from *Lactobacillus coryniformis* ZnO NPs at a concentration of 6 mg/l resulted in complete mortality of *S. zeamais* with an average mortality time of 2.4 days.

Nevertheless, it has been shown that ZnO NPs exhibit lower efficacy in managing the population of storage pests in comparison to nanoparticles like Ag, Al_2O_3, and TiO_2 (Raduw and Mohammed 2020; Mohammed and Nasr 2020; Wazid et al. 2020). The efficacy of aluminum oxide, titanium dioxide, and ZnO NPs against *S. oryzae*. The death rate of organisms exposed to NSA was found to be 90% at a concentration of 1 g/kg after a duration of four days. In contrast, nano ZnO and TiO_2 required a higher concentration of 2 g/kg and a longer duration of 14 days to attain the same level of mortality. Hence, this finding provides justification for the reduced efficacy of ZnO NPs in stored pest management (Das, Yadav, and Debnath 2019). However, owing to its antibacterial characteristics, it is advised for use in the management of agricultural pests (Jasrotia et al. 2022).

19.2.5.5 Copper (CuNPs) and Iron Oxide Nanoparticles (FeNPs)

The use of *Pseudomonas fluorescens* MAL2, a biotransformation agent for CuNPs, has a striking resemblance to its counterpart *P. fluorescens* DSM 12442T DSM. This strain exhibits remarkable efficacy in combatting *T. castaneum*, as shown by (El-Saadony et al. 2020). The mortality percentages of *S. granarius* and *R. dominica* caused by CuO-NPs, which were produced using metabolites derived from *Aspergillus niger* strain (G3-1), were reported to be 55%–94.4% and 70%–90%, respectively (Badawy et al. 2021). According to (Sivapriya, Azeez, and Deepa 2018), the use of a mixture of FeNPs and an aqueous extract derived from *Anthocephalus cadamba* has been shown to result in the complete death of *S. granarius*.

19.2.5.6 Silicon Dioxide (SiO_2)

These particles are often referred to as silica nanoparticles (SNPs) and possess enhanced thermal stability, exceptional biocompatibility with reduced toxic effects with various compounds and other polymers (Rahman and Padavettan 2012; Huang et al. 2019; Jasrotia et al. 2022). According to Rastogi et al. (2019), the mesoporous feature of these substances makes them highly suitable as effective nanocarriers for various agrochemicals. The broader range of applications for silica is attributed to its capacity to be precisely modified in terms of form, size, porosity, and crystallinity.

These substances have been identified as effective nanocarriers for the delivery of fungicides, biopesticides, pheromones, and growth stimulants (Barik, Sahu, and Swain 2008; Cáceres et al. 2019). They physically interact with pests, disrupting their protective barriers and respiratory systems. Their desiccant properties lead to dehydration and suffocation. SiO_2 NPs can disrupt cellular processes and generate ROS upon contact, causing pest mortality. The efficacy of a silica-based nanoformulation was evaluated against *T. castaneum*, resulting in successful pest control (Debnath et al. 2011). In a study conducted by (M Rouhani, Samih, and Kalantari 2013), it was shown that the application of SiO_2 at a dose of 2.06 g/kg resulted in 100% death of the cowpea weevil *C. maculatus*. According to (Satehi, Ziaee, and Ashrafi 2018), it was observed that the use of SNPs loaded with chlorpyrifos, referred to as Ch-SNPs, demonstrated efficient control against *R. dominica* and *T. confusum*. The study further revealed that the death rate of both pests rose proportionally with the concentration of Ch-SNPs applied.

According to Khalil (2019), the application of inorganic nano silica at a concentration of 0.90% resulted in a mortality rate of 50% among adult *T. castaneum* individuals during a treatment period of ten days. In a study conducted by (Rouhani et al. 2019), it was shown that SNPs have significant efficacy in combating *S. granarius*, resulting in complete death within a two-week period. According to (Abdelfattah and Zein 2019), the application of fumed SNPs, namely aerosil 200, at concentrations of 1.5 and 2 g/kg resulted in the complete death of *T. granarium* and *S. paniceum* after five and two days of treatment, respectively.

El-Naggar et al. (2020) demonstrated that the use of the sol-gel process in the production of SNPs resulted in successful control of *R. dominica*, *T. castaneum*, and *O. surinamensis*. The efficacy of fumigating maize grains using SiO_2-NPs was shown in effectively controlling four prevalent pests found in stored grain. In a study conducted by Wazid et al. (2020), it was found that green SNPs, with a concentration of 1,500 ppm, were successfully synthesized using spinach leaves, tulsi leaves, and paddy husks. These nanoparticles were shown to have potential as a substitute for malathion in combating storage pests. Three different kinds of silica, namely Aerosil 200, chemical silica, and bio-silica in the form of nanoparticles, were subjected to testing against three species of insects, namely *C. maculatus*, *R. dominica*, and *T. confusum*. The death of both adults and progeny of all three insects was seen across all tested materials. Furthermore, the mortality rate was shown to be positively correlated with both the concentration of the materials and the duration of exposure. According to Salem (2020), the susceptibility of *C. maculatus* adults was shown to be higher compared to *R. dominica* and *T. confusum* when exposed to the three investigated chemicals at different doses. The study conducted by Agrafioti et al. (2020) showed that the use of insect-resistant nets covered with SNPs resulted in the complete death of *S. oryzae*. According to Kar et al. (2021), when rice was treated with nano-silica particles measuring 30 nm in size at a dose level of 0.5 g/kg, the death rates of *S. oryzae* were observed to be 80% after 7 days and 97.4% after 14 days. In a study conducted by Saed et al. (2022), synthesized nano silica from sugarcane bagasse ash (SCBA) and employed as an additive to DE in order to enhance the insecticidal efficacy of Mamaghan DE. The results indicated that after a 14-day exposure period, the mortality rate of adult *T. confusum* was 86% and *R. dominica* insects 95%, respectively.

19.2.6 Nanocomposites

A nanocomposite is a material that consists of two or more distinct components, typically a matrix or host material and nanoparticles, combined at the nanoscale level to create a new material with enhanced properties. Nanocomposites are designed to take advantage of the unique properties and characteristics of nanoparticles, which are typically particles with dimensions in the nanometer range. Hossain et al. (2019) tested eight EOs (basil, cinnamon, eucalyptus, mandarin, oregano, peppermint, tea tree, and thyme) and one binary combination (thyme and oregano) for fumigant toxicity against the *S. oryzae*. Eucalyptus EO was the most hazardous, producing 100% death at 0.8 μl/ml after 24 h of exposure. The combination of oregano and thyme EO fumigated is more lethal than either alone. Oregano: thyme nanoemulsion (0.75%) in cellulose nanocrystal (CNC) composite films with chitosan/CNC, methyl cellulose/CNC, and polylactic acid/CNC caused prolonged diffusion matrices leads to 32%–51% mortality of *S. oryzae* after 14 days. The bioactive chitosan film containing the oregano: thyme nanoemulsion raised mortality from 79% to 100% after 200 gray irradiation.

19.3 METHOD OF PREPARATION

A nanoformulation is a complex colloid created by dispersing two immiscible liquids, mainly specific functionalized nanomaterials, into conventional solvents like water, oil, or surfactant. Researchers encounter various challenges when working with nanoformulations, and one of the foremost issues is achieving well-dispersed nanomaterials. This problem arises because neighboring nanoparticles tend to clump together due to van der Waals forces. Moreover, the difference in density between nanoparticles and the base solvent causes nanoparticles to settle at the bottom of the nanoformulations due to gravitational forces. To produce nanoformulations with stable dispersion, researchers have employed surfactants and various nanoparticle coating techniques. The manufacturing methods for nanoformulations can be categorized into two primary groups: single-step and two-step approaches. The single-step manufacturing method involves simultaneously synthesizing nanoparticles and dispersing them into the solvent, typically using techniques like physical vapor deposition or liquid chemical methods. This approach offers advantages such as reducing the likelihood of nanoparticle agglomeration or sedimentation by avoiding the intermediate steps of drying, transportation, storage, and dispersion. However, it's not suitable for large-scale or industrial nanoparticle production. One significant concern with the single-step method is the need to minimize residual chemical reagents in the nanoformulations, which can result from unstable chemical reactions and can compromise its purity.

In contrast, the two-step method is the most common approach for manufacturing nanoformulations in both industrial and academic research settings due to its cost-effectiveness and simplicity. This process begins with the dry powder production of nanomaterials like nanoparticles, nanofibers, or nanotubes. The synthesis of nanoparticles can be categorized into three main groups, such as biological, physical, and chemical approaches. Among these, biological methods, while intriguing, involve complex practices and thus are not commonly employed. Physical and chemical methods, on the other hand, are widely adopted and have been developed for both research and industrial applications. Physical methods often rely on large equipment that consumes significant energy, whereas chemical methods employ simpler techniques with lower energy consumption. Followed by nanoparticle synthesis, the next step in the two-step method is to mix the nanoparticles with the base liquid using various techniques, such as ultrasonication, intensive magnetic stirring, homogenizing, microwave homogenization, spray pyrolysis, co-precipitation (Figure 19.1). The challenge of sustaining dispersion stability is a fundamental disadvantage of the two-step approach. Therefore, it is common practice to use surfactants and coating agents to stabilize the nanoparticles within the base fluid. These additives find widespread use in a variety of applications to ensure the stability and functionality of nanoformulations (Boldoo et al. 2020).

19.4 METHOD OF APPLICATION

The successful utilization of nanoformulations in practical applications pivots to achieving efficient distribution. In the context of stored grain facilities, where the stakes are high, a diverse array of application methods harnessing nanotechnology has emerged to enhance pest control. These methods harness the unique properties of nanoparticles to optimize the delivery and effectiveness of pest control measures. Various techniques for the application of nanopesticides in stored grain facilities are available which are based on traditional spraying and dusting to more advanced approaches like nanocapsules and nanofiber deployment. Each method addresses specific challenges, providing comprehensive protection to safeguard stored grains and minimize post-harvest losses while considering regulatory compliance and safety. The choice of application method depends on various factors, such as the type of pest, storage conditions, and the specific nanopesticides formulation (Stejskal et al. 2021). Table 19.1 represents different application methods of nanopesticides with descriptions and examples.

TABLE 19.1
Application Methods of Nano-Based Pesticides in Stored Facilities

Application Method	Description	Formulations
Spraying	Apply nanopesticides as a spray using conventional equipment for better adhesion and penetration.	Nanostructured alumina (NSA) used against *Rhyzopertha dominica* (Stadler et al. 2012) Tween 80 (hydrophilic surfactant, 150 mg) (nanoencapsulation) with *Rosmarinus officinalis* essential oil (EO) against *Tribolium castaneum* (Khoobdel, Ahsaei, and Farzaneh 2017).
Dusting	Nanopesticides formulate as dustable powders to spread or dust over stored grains for enhanced coverage.	Glycine-nitrate-modified nano-alumina dust used against *Sitophilus oryzae* (Buteler et al. 2015) Biotransformed copper nanoparticles of *Pseudomonas fluorescens were* applied for *T. castaneum* disinfestation (El-Saadony et al. 2020).
Coating	Nanopesticides are mixed with a carrier material and applied as a thin film or coating on grain surfaces for long-lasting protection.	*Hazomalania voyronii* (Jum.) EO-based nanoemulsion (HvNE) is used as a wheat grain protectant against *Tribolium confusum*, *T. castaneum*, and *Tenebrio molitor* (Kavallieratos et al. 2021).
Fumigation	Nanopesticides are used alongside fumigation techniques to improve effectiveness in enclosed storage spaces.	Tween 80 emulsifier (Nanoemulsion) with *Citrus sinensis* peels extracts (cold pressed EO) *Cryptolestes ferrugineus*, *T. confusum* (Giunti et al. 2019). Chitosan and Tween 80 (nanoparticles) *with Peppermint* leaf extracts (EO) *against S. oryzae*, *T. castaneum* (Rajkumar et al. 2020). Tween 80 (nanoencapsulation) with *Coriandrum sativum*, Janesville EO used against *T. confusum* and *T. castaneum* (Sabbour 2020).
Nanocapsules	Encapsulated nanopesticides within tiny capsules for controlled release when suspended in a liquid or mixed with grains.	Polyurea-formaldehyde nanoencapsules with active ingredient *Cuminum cyminum* L. seed oil extract against *T. castaneum* (Negahban et al. 2012). Polyethersulfone nanocapsules with active ingredient *C. cyminum* and *Lavandula angustifolia* EOs against *Sitophilus granarius* and *T. castaneum* (Bayramzadeh et al. 2019).
Nanofibers	Nanofibers as carriers for nanopesticides, releasing active ingredients gradually when applied near stored grains.	Polylactic acid solution containing *Mentha piperita and Salvia officinalis* was added to test contact toxicity against *Plodia interpunctella* (Allahvaisi et al. 2017).
Packaging materials	Infuse nanopesticide particles into packaging materials to offer continuous protection while the packaging remains intact.	Citronella, oregano, and rosemary oil-based insect repellent packaging were reported (Licciardello et al. 2013). High-density polyethylene (HDPE) films infused with nanoparticles, ZnO, or Ag was developed. The HDPE films infused with 2% Ag and irradiated with a dosage of 75 kGy were shown as the most effective formulation, as adult *S. granarius* could not penetrate the loaded and irradiated film (Eyssa, Sawires, and Senna 2019).

(Continued)

TABLE 19.1 *(Continued)*
Application Methods of Nano-Based Pesticides in Stored Facilities

Application Method	Description	Formulations
Nanotechnology-infused storage structures/net	Design storage structures with nanotechnology features (coatings or materials) that release nanopesticides over time, providing ongoing protection.	Silica nanoparticles coated nets ED3, ED3-P (paraffin coated), ED5, and ED5-P (paraffin coated) against *S. oryzae* and alpha-cypermethrin-coated polyester nets (Carifend®), against *Lasioderma serricorne* (F.), *Ephestia elutella*, and *S. oryzae* (Agrafioti et al. 2020).
Bait Stations	Incorporate nanopesticides formulations into bait stations strategically placed within storage facilities.	Recently liquid baits formulation using liposome vesicles (so-called dsRNA lipoplexes) as carriers of dsRNA molecules. Results reported that the lipoplexes efficiently triggered lethal RNA interference in the cockroach digestive system (Wang et al. 2021).

19.5 MODE OF ACTION OF NANOFORMULATIONS

The development of new pesticides employing nanoparticles via various synthesis procedures has recently received substantial scientific interest. However, their precise mechanism of action against pests is still unresolved. Multiple researches have been led to evaluate their toxicity against a diverse array of insect pests (Athanassiou et al. 2018). Only a small number of studies have investigated the toxicokinetics and toxicodynamics of these agents against stored grain pests due to the largely unexplored region of nanopesticides. Toxicodynamics encompasses the study of the physiological, chemical, and molecular impact of pesticide compounds, along with the mechanisms underlying their actions, while toxicokinetics primarily pertains to the internal journey of an insecticide within an organism, encompassing absorption, distribution, metabolism, and excretion (Alzogaray and Zerba 2017; Jasrotia et al. 2022).

Several studies have delved into the mechanism of action of nanopesticides based on nanoparticles, such as silica, alumina, and silver against insects (Benelli 2018). Silver nanopesticides, through the modulation of antioxidants and detoxifying enzymes, induce a decline in acetylcholinesterase function, thereby triggering oxidative stress and cellular death. Furthermore, they have been shown to manipulate essential insect genes, leading to altered protein synthesis and suppressed gonadotropin secretion, culminating in developmental anomalies and reproductive deficiencies (Nair and Choi 2011). The reduced capacity of metal nanoparticles to bind with S and P accentuates membrane permeability, ultimately leading to disruption of cell organelle, membrane, and enzyme leading to cell death due to nucleic acid denaturation. Gold nanoparticles, beyond influencing growth and reproduction, have also demonstrated trypsin-inhibitory properties (Patil et al. 2016; Small et al. 2016).

Silicon dioxide and aluminum oxide nanopesticides function by adhering to the insect cuticle and subsequently absorbing waxes and lipids, resulting in insect desiccation as described by (Debnath et al. 2011; Arumugam et al. 2016). Another study by Ibrahim and Salem (2019) reported that nano zeolite formulations led to scratching and cuticle fragmentation upon contact with the body of *T. confusum*, eventually leading to insect dehydration and mortality. The toxic properties of NSA and its action mechanism against *S. oryzae* were elucidated by (Stadler et al. 2017). Stadler reported that NSA particles attached to the insect cuticle by surface effect and triboelectric forces induced dehydration by absorbing its protective wax layer, subsequently leading to the death of the insect (Figure 19.2).

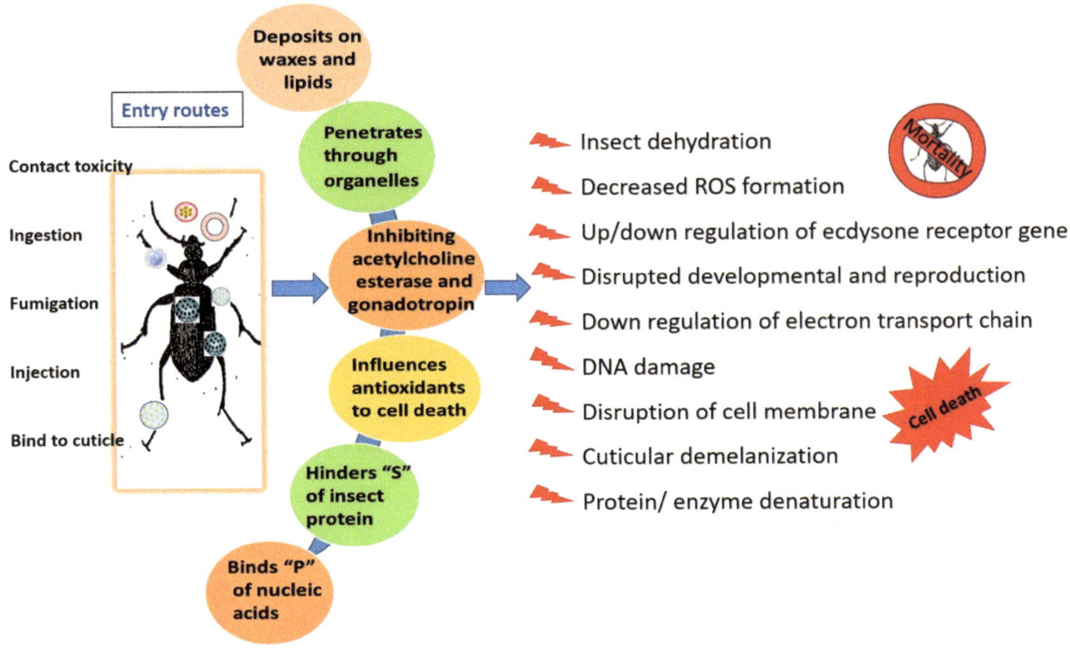

FIGURE 19.2 Mode of action of nano formulations.

19.6 EFFECT OF NANOFORMULATIONS ON ONTOGENIC STAGES OF INSECTS

The impact of pest control methods, whether through nanopesticides or nanoemulsions, varies significantly across different developmental stages of insects, encompassing the egg, larva, pupa, and adult phases. The influence of formulations can vary depending on factors, such as the type of nanomaterial, concentration, exposure duration, and insect species. Here, some of the nanoformulations and their effect on the developmental stages of insects have been summarized.

19.6.1 Egg Stage

Both nanopesticides and nanoemulsions exhibit a diverse range of effects on insect eggs. Nanopesticides, with their ability to breach egg shell barriers and interfere with embryonic development, have been linked to decreased hatching rates and the emergence of deformities in emerging larvae. Similarly, nanoemulsion formulations have demonstrated potential in controlling stored grain pests at the egg stage. These formulations, particularly those based on EOs like eucalyptus oil, have been associated with reduced hatching rates and the induction of deformities in emerging larvae (Pant et al. 2014). Nanoformulations containing a plant extract of *Mentha longifolia* with particle sizes ranging from 14 to 36 ppm formulation demonstrated enhanced fumigant toxicity against *C. maculatus*, and 50% of the egg population was effectively killed in 4.7 days at a concentration of 9 µl/l (Louni, Shakarami, and Negahban 2019). In a recent investigation by (Lakshmi et al. 2020), it was found that zinc oxide nanoparticles at a concentration of 200 ppm exhibited a 100% mortality rate and the lowest egg production when tested against *T. castaneum* and *C. maculatus* in green gram.

19.6.2 Larva Stage

Nano-based disinfectants exert notable influence during the larval stage of insects. Nanopesticides can disrupt larval growth and metamorphosis, affecting proper development. They may infiltrate respiratory systems, leading to respiratory distress and eventual larval mortality. Similarly, nanoemulsions have been observed to disrupt larval growth, impair feeding behavior, and induce respiratory distress, resulting in significant larval mortality. EO nanoformulations, such as *Pimpinella anisum* L. (Apiaceae) EO-based nanoemulsions, have demonstrated potent toxicity against larvae, impacting their growth and survival (Hashem et al. 2018). Nenaah (2014) demonstrated that nanoemulsions of *Achillea biebersteinii*, *Achillea santolina*, and *Achillea millefolium* EOs are effective against *T. castaneum* larvae and adults. *Hazomalania voyronii* EO-based nanoemulsion (HvNE) at 1,000 ppm concentration, resulted in 92.1%–97.4% mortality against *T. confusum*, *T. castaneum* larvae, and 100% *Tenebrio molitor* adults (Kavallieratos et al. 2021). *E. kuehniella* fifth instar larvae were destroyed using a nanoemulsion containing an extract of *M. longifolia* with particle sizes ranging from 14 to 36 nm. For the EO, LC_{50} was reported at 21,352 ppm, while nanoemulsion showed effectiveness toward larvae at 14,068 ppm (Louni, Shakarami, and Negahban 2018). Similarly, Isofuranodiene-based nanoemulsion 3% (w/w) derived from *Smyrnium olusatrum* EO at 1,000 ppm displayed higher activity at 98.6, 97.4, and 93.5% against *T. molitor* adults and larvae of *T. castaneum* and *T. confusum*, respectively, after seven days of exposure (Kavallieratos et al. 2021). *E. kuehniella* larvae were destroyed using a nanoemulsion containing a plant extract of *M. longifolia* with particles ranging from 14 to 36 ppm (Louni, Shakarami, and Negahban 2018). After four days of exposure, a nanoemulsion produced with *A. biebersteinii* EO at 10 l/l air induced 100% death of the second larvae of *T. castaneum* (Almadiy 2021). Citronella oil, moringa oil, and goatweed extract were recently turned into nano-oils, which demonstrated high death rates against *R. dominica* larvae (Sabbour 2020). Notably, the application of silver nanoparticles at a rate of 1.00 g/kg of seed resulted in an 83% mortality of *C. maculatus* adults and larvae (Rouhani, Samih, and Kalantari 2013). Similarly, AgNPs produced from *Nerium oleander* plant extract were shown to be effective against *T. castaneum* and *C. maculatus* larvae (Jafer and Annon 2018). The study conducted by Ziaee and Babamir (2021) included nano silica with three active components, namely deltamethrin, pyriproxyfen, and chlorpyrifos. The efficacy of this formulation was evaluated against *T. granarium* larvae, which are known to infest stored wheat. The experiments were conducted utilizing mosaic and galvanized steel surfaces. The study found that deltamethrin, followed by chlorpyrifos, had the highest efficacy in controlling *T. granarium* larvae. Additionally, the mortality rate was substantially greater in galvanized steel compared to mosaic surfaces.

19.6.3 Pupa Stage

The pupal stage, crucial for insect development, is affected by both nanopesticides and nanoemulsions. These nanoformulations have been found to disrupt pupation processes, potentially impeding the pupal-to-adult transition and resulting in the emergence of malformed pupae. EO nanoemulsions, such as those derived from *A. biebersteinii*, *A. santolina*, and *A. millefolium*, have shown efficacy toward *T. castaneum* pupae and adults (Nenaah 2014).

19.6.4 Adult Stage

The adult stage of insects experiences significant impacts from both nanopesticides and nanoemulsions. Nanopesticides disrupt critical physiological functions in adults upon contact, affecting neurotransmission, hormonal regulation, and behaviors, such as mating and egg-laying. Similarly, nanoemulsions disrupt essential physiological functions, leading to compromised behaviors and reduced longevity in adult insects. EO nanoformulations, like those based on *Mentha spicata* L. and *M. pulegium* L. EOs, have demonstrated higher mortality rates against adult insects. The LC_{50} values of both EOs and nanoformulations of *M. spicata* were 18.422 and 9.279 μl/ml, and for *M. pulegium*, 7.939 and 6.793 μl/ml were reported after 72 h (Heydarzadeh et al. 2019). Nanoemulsions derived from *A. indica* A. Juss (neem oil) have proven highly effective against adult stages stored pests, contributing to their management (Choupanian et al. 2017). Pant et al. (2014)

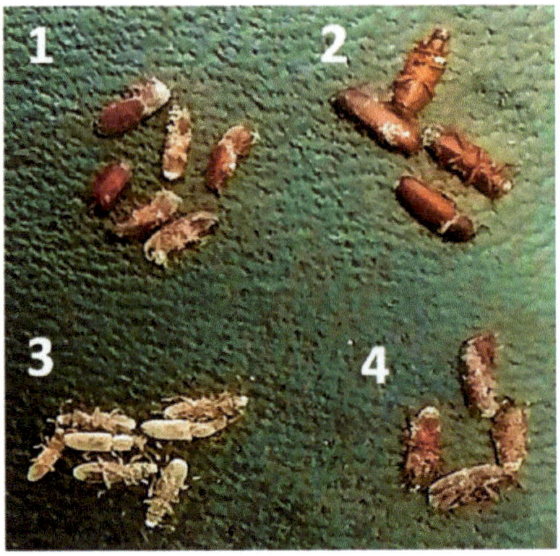

FIGURE 19.3 Effect of SiO_2-NPs on the morphology of stored products insects i.e., (1) *T. castaneum*, (2) *R. dominica*, (3) *O. surinamensis*, and (4) *S. oryzae* lead to death. (Image source El-Naggar et al. (2020) under CC BY.)

reported eucalyptus oil nanoemulsion containing 300 and 1,500 ppm concentrations of aqueous filtrate of Karanja and Jatropha plant caused 88%–100% mortality of adult *T. castaneum* within one day after the exposure. In another study, nanoemulsion containing *P. anisum* L. (Apiaceae) EO having (E)-anethole (81.2%) reported higher toxicity of LC_{50} at 9.3% v/v and a significant reduction in the progeny of *T. castaneum* was observed (Hashem et al. 2018). According to the study conducted by Choupanian et al. (2017), it was observed that nanoemulsions derived from neem oil (*A. indica* A. Juss) had a high level of effectiveness, ranging from 85% to 100% against *S. oryzae*. Similarly, the same nanoemulsions showed effectiveness ranging from 74% to 100% against *T. castaneum*. Anise (*P. anisum*), Artemisia (*Artemisia vulgaris*), fennel (*Foeniculum vulgare*), garlic (*Allium sativum*), lavender (*L. angustifolia*), mint (*M. piperita*), rosemary (*R. officinalis*), and sage (*S. officinalis*) nanoemulsions disclosed significant toxicity against *T. confusum*. However, the application of garlic oil nanoemulsion resulted in the maximum toxicity, while *P. anisum* was shown to be the best repellent (Palermo et al. 2021). Tasmanian blue gum (TBG) EO nanoformulations have shown LC_{50} values in contact and fumigant applications against *C. maculatus* ranging from 1.37 to 8.53 ppm and 0.05–0.44 ppm, respectively. A significant reduction and ovicidal effects were observed in the population (Ya-Ali, Yarahmadi, and Mehrnia 2020) (Figure 19.3).

Both nanopesticides and nanoemulsions, particularly those derived from EOs, have shown potential for habituation effects in stored grain pests. Long-term use of these formulations, especially continuous applications of EO Nanoemulsions, has raised concerns about habituation in certain pest species. This highlights the need for thoughtful application strategies and rotation to maintain the efficacy of these pest management methods (Giunti et al. 2021). Both nanopesticides and nanoemulsions demonstrate intricate and stage-specific effects on stored grain pests throughout their various developmental phases. Understanding these effects is crucial for the successful implementation of IPM strategies in stored grain facilities.

19.7 PROBABLE IMPACT ON GRAIN QUALITY, ADVANTAGES AND LIMITATIONS

Nanoformulations have gained interest in various applications, including stored grain disinfestation. They offer several advantages and limitations when used for this purpose: The impact of nano-based formulations on grain quality can be both beneficial and potentially concerning, depending on various factors such as the type of nanoformulations, the active ingredients used, and the application methods (Yadav et al. 2021). Here are some potential impacts to consider:

19.7.1 Advantages of Nanoformulations in Stored Grain Disinfestation

1. ***Enhanced efficacy***: Nanoformulations can improve the efficacy of insecticides and fumigants used for grain disinfestation. The increased surface area and improved penetration of nanoparticles can lead to better pest control, targeting hard-to-reach areas. The nanoscale delivery systems, such as nanocapsules and nano-suspensions, provide a more targeted and sustained release of active ingredients, ensuring that pests are exposed to the pesticide for longer durations. This controlled release allows for extended protection against pests and minimizes the need for frequent reapplication, which is common with conventional pesticides.
2. ***Reduced chemical residues***: Nanoformulations can allow for lower doses of active ingredients while maintaining high pest control efficacy. Reduced chemical exposure and the potential for lower doses contribute to safer working conditions in food storage areas. For instance, nanoemulsions can improve the uptake of active ingredients, making them more effective against pests. This heightened bioavailability ensures that a smaller quantity of pesticide is required for the same level of pest control, reducing the overall chemical load on the environment and agricultural produce.
3. ***Targeted delivery***: Nanoformulations can be engineered to release insecticides slowly over time, providing sustained protection against pests. This targeted delivery can help maintain grain quality during storage.
4. ***Improved formulation stability and extended shelf life***: Nanoformulations can protect sensitive active ingredients from degradation due to environmental factors, such as UV light or moisture, thereby maintaining their potency and extending their shelf life. This can lead to better-quality grains in terms of pest resistance and nutritional content.
5. ***Uniform coverage***: Nanoformulations can provide better coverage of grain surfaces due to their smaller droplet size and improved spreading properties. This uniform coverage can lead to consistent pest control and reduced variability in grain quality.
6. ***Reduced environmental impact***: Conventional pesticides can lead to soil and water contamination, affecting non-target organisms and posing ecological risks. In contrast, nanoformulated pesticides, with their protective nanocarrier, mitigate the environmental impact. The use of nanoformulations can lead to reduced pesticide runoff and environmental contamination due to their enhanced efficiency and lower application rates. These carriers shield the active ingredients from premature degradation and ensure that they reach the intended target with minimal dispersion, thus reducing the risk of collateral damage.
7. ***Less resistance development***: The novel mechanisms of action enabled by nanoformulations can potentially slow down the development of resistance in insect populations compared to traditional insecticides.

19.7.2 Concerns and Considerations

1. ***Residue concerns***: While nanoformulations can reduce chemical residues, there might be concerns about the potential accumulation of nanoparticles or other components on grain surfaces. Regulatory agencies typically set maximum residue limits, and careful assessment of potential health and environmental risks associated with nanomaterials is necessary.
2. ***Nutritional impact***: The application of nanoformulations could potentially affect the nutritional composition of grains, especially if the emulsions alter the absorption or

bioavailability of nutrients. Thorough studies are needed to ensure that grain quality and nutritional value are not compromised.

3. ***Unintended effects***: The interaction of nanoformulations with grains, pests, and other components of the storage environment could lead to unintended consequences, affecting grain quality or safety. It may also potentially affect their texture, taste, or cooking properties. Research is required to assess any unintended consequences on grain quality. Nanomaterials are generally safe when they are maintained at their optimal size, but when they approach nanoscale dimensions, particularly non-biodegradable materials like metals, they can exhibit toxic properties. These nanomaterials have the potential to accumulate and provoke immune responses, leading to toxicity in mammals, plants, and soils. Additionally, they may disrupt the functioning of detoxifying enzymes (Deka et al. 2021).
4. ***Environmental impact***: The use of nanoformulations in agriculture might have broader environmental implications, such as the potential accumulation of nanoparticles in soil or water systems. Proper risk assessments and environmental monitoring are essential to minimize negative effects.
5. ***Regulatory considerations***: Nanoformulations may require additional regulatory scrutiny and approval before widespread use in stored grain disinfestation due to concerns about potential environmental and human health impacts. Compliance with regulations is crucial to mitigate potential negative impacts on grain quality and food safety. Nanomaterials possess unique properties that demand specialized testing beyond standard guidelines. Different regions, like India, the US, Europe, and Australia, have developed specific regulations to address these concerns. For example, India introduced guidelines for nano-agricultural products, while the US recommends assessing the health and environmental risks of nanopesticides. Europe regulates plant protection products, nanomaterial residues in food, and labor safety through various regulations.
6. ***Cost***: Nano-based pesticides tend to be more expensive than traditional pesticides due to their complex development and production processes. Research and development costs, the need for precision in targeting pests, expensive nanomaterials, rigorous regulatory compliance, and limited-scale production contribute to their higher cost. However, these pesticides often offer improved environmental and health benefits, enhanced efficacy, and more precise application. Over time, as technology advances and adoption increases, economies of scale may help mitigate the cost difference between nano-based and traditional pesticides.
7. ***Scale-up challenges***: Scaling up the production and use of nanopesticides for stored pest control poses challenges in achieving uniform distribution, preventing agglomeration, and ensuring precise dosing. Maintaining manufacturing consistency, balancing costs, and addressing environmental and regulatory concerns are also key issues.
8. ***Limited research***: While nanoformulations show promise, there is still a need for more research to fully understand their effects on stored grain quality, pest populations, and the environment.
9. ***Application techniques***: Developing suitable application methods for nanoformulations in large storage facilities may require adjustments to existing practices.

19.8 PRESENT RESEARCH STATUS AND COMMERCIALIZATION LEVELS

Nanoformulations have the potential to revolutionize stored grain disinfestation by offering enhanced pest control and reduced environmental impact. These formulations are carefully designed to maximize effectiveness, stability, and active component release. Advances in dispersion, adhesion, and penetration of these pesticides inside stored grains provide a potential solution to pest resistance, and the ability to reduce chemical treatment's environmental effect has also prompted research. Nanoformulations' continuous slow-release capabilities are worth investigating to safeguard grain quality and keep pests at bay. Due to their unique modes of action, nanoformulations may prevent stored pests from developing resistance. Pilot-scale and storage conditions have been conducted for stored pest disinfestation efficacy before commercialization, and preliminary studies show promising insect control and grain quality maintenance. However, the commercialization of nanoformulations as disinfectants will face challenges, such as scalability, cost-effectiveness, regulatory clearances, customer acceptability, and storing conditions that need to be addressed carefully. Thus, researchers and companies need to ensure that their products meet safety and environmental standards. Further adoption of nanoformulations by stakeholders, public perception, market demand, and economic feasibility possibly change the pest control approaches. For this research institutes, governments, and private companies need to collaborate to help finance, share, and transfer cutting-edge innovations as commercialized nanoformulations.

19.9 CONCLUSION

Nano-based disinfestation methods present a promising solution for improving stored grain pest control. Advantages of nanopesticides over conventional agrochemicals, such as lower dosage requirements, increased solubility, and targeted delivery, thereby minimizing environmental impact. These formulations improve grain quality and ensure uniform pest control through better coverage and stability, reducing chemical residues and pesticide runoff. However, concerns about nanoparticle toxicity, biodegradability, resistance development, and broader environmental implications necessitate careful evaluation and regulation. Current research work focuses on developing environmentally sustainable and cost-effective nanopesticides alongside assessing their eco-toxicological impact and establishing regulatory frameworks. Nano-biopesticides offer eco-friendly alternatives with high efficacy, inhibiting pathogens and promoting biodegradation. Despite their promise, commercialization faces hurdles such as higher costs, scale-up challenges, and public acceptance. Collaboration among research institutes, governments, and private companies is essential to ensure safety, environmental standards, and successful market integration, ultimately enhancing food security and sustainability in the agriculture and food sector.

REFERENCES

Web Reference

Abdelfattah, N. A., and D. M. Zein. 2019. "Biological Studies and Toxicity Experiments of AEROSIL 200 Nanoparticles on Adults and Larvae of Some Stored Grain Insects." *International Journal of Entomology Research* 4 (1): 103–108.

Abo-Arab, R. B., M. H. Amal, and A. S. Hashem. 2014. "Comparative Bioactivity of Aluminum Oxide (Al_2O_3), Titanium Dioxide (TiO_2) Nanoparticles and Malathion on *Sitophilus oryzae* L. and *Sitophilus zeamais* (Motsch.)." *Global Journal of Agriculture and Food Safety Sciences* 1 (2), 25–37.

Adak, T., N. Barik, N. B. Patil, B. G. Gadratagi, M. Annamalai, A. K. Mukherjee, and P. C. Rath. 2020. "Nanoemulsion of Eucalyptus Oil: An Alternative to Synthetic Pesticides against Two Major Storage Insects (*Sitophilus oryzae* (L.) and *Tribolium castaneum* (Herbst)) of Rice." *Industrial Crops and Products* 143: 111849. doi:10.1016/j.indcrop.2019.111849.

Agrafioti, P., S. Faliagka, E. Lampiri, M. Orth, M. Pätzel, N. Katsoulas, and C. G. Athanassiou. 2020. "Evaluation of Silica-Coated Insect Proof Nets for the Control of *Aphis fabae*, *Sitophilus oryzae*, and *Tribolium confusum*." *Nanomaterials* 10(9): 1658. doi:10.3390/nano10091658.

Allahvaisi, S., K. T. Jahromi, S. Imani, and M. Khanjani. 2017. "Efficacy of Electrospun Bionanofibers as Fumigant Pesticides in Foodstuff Storage." *Journal of Plant Protection Research* 57 (1): 72–80.

Almadiy, A. A. 2021. "Insecticidal and Acetylcholinesterase Inhibitory Activities of *Achillea biebersteinii* Essential Oil and Its Nanoemulsion and Major Monoterpenes against *Tribolium castaneum*." *Journal of Asia-Pacific Entomology* 24 (4): 1170–1178. doi:10.1016/j.aspen.2021.11.002.

Alzogaray, R. A., and E. N. Zerba. 2017. "*Rhodnius prolixus* Intoxicated." *Journal of Insect Physiology* 97: 93–113. doi:10.1016/j.jinsphys.2016.04.004.

Arumugam, G., V. Velayutham, S. Shanmugavel, and J. Sundaram. 2016. "Efficacy of Nanostructured Silica as a Stored Pulse Protector against the Infestation of Bruchid Beetle, *Callosobruchus maculatus* (Coleoptera: Bruchidae)." *Applied Nanoscience* 6: 445–450. doi:10.1007/s13204-015-0446-2.

As, A. A., and S. Thangapandiyan. 2019. "Comparative Bioassay of Silver Nanoparticles and Malathion on Infestation of Red Flour Beetle, *Tribolium castaneum*." *The Journal of Basic & Applied Zoology* 80 (1): 1–10. doi:10.1186/s41936-019-0124-0.

Athanassiou, C. G., N. G. Kavallieratos, G. Benelli, D. Losic, P. U. Rani, and N. Desneux. 2018. "Nanoparticles for Pest Control: Current Status and Future Perspectives." *Journal of Pest Science* 91: 1–15. doi:10.1007/s10340-017-0898-0.

Badawy, A. A., N. A. H. Abdelfattah, S. S. Salem, M. F. Awad, and A. Fouda. 2021. "Efficacy Assessment of Biosynthesized Copper Oxide Nanoparticles (CuO-NPs) on Stored Grain Insects and Their Impacts on Morphological and Physiological Traits of Wheat (*Triticum aestivum* L.) Plant." *Biology* 10 (3): 233. doi:10.3390/biology10030233.

Barik, T. K., B. Sahu, and V. Swain. 2008. "Nanosilica—From Medicine to Pest Control." *Parasitology Research* 103: 253–258. doi:10.1007/s00436-008-0975-7.

Batool, M., D. Hussain, A. Akrem, M. N. Haq, S. Saeed, and Q. Saeed. 2021. "Nanoencaspsulation of Cysteine Protease for the Management of Stored Grain Pest, *Sitotroga cerealella* (Olivier)." *Journal of King Saud University – Science* 33 (4): 101404. doi:10.1016/j.jksus.2021.101404.

Bayramzadeh, N., F. Mehrkhou, A. A. Pourmirza, and M. Mahmoudian. 2019. "Fumigant Toxicity of Two Nano-Capsulated Essential Oils with Sublethal Rate of Phosphine against Three Stored-Product Pests." *Journal of Agricultural Science and Technology* 21 (4): 857–872.

Belhamel, C., L. Boulekbache–Makhlouf, S. Bedini, C. Tani, T. Lombardi, P. Giannotti, K. Madani, K. Belhamel, and B. Conti. 2020. "Nanostructured Alumina as Seed Protectant against Three Stored-Product Insect Pests." *Journal of Stored Products Research* 87: 101607. doi:10.1016/j.jspr.2020.101607.

Benelli, G. 2018. "Mode of Action of Nanoparticles against Insects." *Environmental Science and Pollution Research* 25 (13): 12329–12341. doi:10.1007/s11356-018-1850-4.

Boldoo, T., J. Ham, E. Kim, and H. Cho. 2020. "Review of the Photothermal Energy Conversion Performance of Nanofluids, Their Applications, and Recent Advances." *Energies* 13 (21): 5748.

Buteler, M., S. W. Sofie, D. K. Weaver, D. Driscoll, J. Muretta, and T. Stadler. 2015. "Development of Nanoalumina Dust as Insecticide against *Sitophilus oryzae* and *Rhyzopertha dominica*." *International Journal of Pest Management* 61 (1): 80–89. doi:10.1080/09670874.2014.1001008.

Cáceres, M., C. V. Vassena, M. D. Garcerá, and P. L. Santo-Orihuela. 2019. "Silica Nanoparticles for Insect Pest Control." *Current Pharmaceutical Design* 25 (37): 4030–4038.

Chen, H., and R. Yada. 2011. "Nanotechnologies in Agriculture: New Tools for Sustainable Development." *Trends in Food Science & Technology* 22 (11): 585–594. doi:10.1016/j.tifs.2011.09.004.

Choupanian, M., D. Omar, M. Basri, and N. Asib. 2017. "Preparation and Characterization of Neem Oil Nanoemulsion Formulations against *Sitophilus oryzae* and *Tribolium castaneum* Adults." *Journal of Pesticide Science* 42 (4): 158–165. doi:10.1584/jpestics.D17-032.

Das, S., A. Yadav, and N. Debnath. 2019. "Entomotoxic Efficacy of Aluminium Oxide, Titanium Dioxide and Zinc Oxide Nanoparticles against *Sitophilus oryzae* (L.): A Comparative Analysis." *Journal of Stored Products Research* 83: 92–96. doi:10.1016/j.jspr.2019.06.003.

Debnath, N., S. Das, D. Seth, R. Chandra, S. C. Bhattacharya, and A. Goswami. 2011. "Entomotoxic Effect of Silica Nanoparticles against *Sitophilus oryzae* (L.)." *Journal of Pest Science* 84: 99–105.

Deka, B., A. Babu, C. Baruah, and M. Barthakur. 2021. "Nanopesticides: A Systematic Review of Their Prospects with Special Reference to Tea Pest Management." *Frontiers in Nutrition* 8: 686131. doi:10.3389/fnut.2021.686131.

Dikbaş, N., S. Uçar, G. Tozlu, T. Özer, and R. Kotan. 2021. "Bacterial Chitinase Biochemical Properties, Immobilization on Zinc Oxide (ZnO) Nanoparticle and Its Effect on *Sitophilus zeamais* as a Potential Insecticide." *World Journal of Microbiology and Biotechnology* 37: 1–14.

Edwards, C. H., C. R. Christie, A. Masotti, A. Celluzzi, A. Caporali, and E. M. Campbell. 2020. "Dendrimer-Coated Carbon Nanotubes Deliver dsRNA and Increase the Efficacy of Gene Knockdown in the Red Flour Beetle *Tribolium castaneum*." *Scientific Reports* 10 (1): 12422. doi:10.1038/s41598-020-69068-x.

El-Bendary, H. M., A. S. Abdel-Wahab, and A. A. El-Helaly. 2016. "Entomo-Toxic Assay of Nano-Aluminum Oxide and Nano-Zinc Oxide against *Sitophilus oryzae* as a Promising Insecticide." *Egyptian Journal of Zoology* 174 (4083): 1–8. doi:10.12816/0034708.

El-Naggar, M. E., N. R. Abdelsalam, M. M. G. Fouda, M. I. Mackled, M. A. M. Al-Jaddadi, H. M. Ali, M. H. Siddiqui, and E. E. Kandil. 2020. "Soil Application of Nano Silica on Maize Yield and Its Insecticidal Activity against Some Stored Insects after the Post-Harvest." *Nanomaterials* 10 (4): 739. doi:10.3390/nano10040739.

El-Saadony, M. T., M. E. Abd El-Hack, A. E. Taha, M. M. G. Fouda, J. S. Ajarem, S. N. Maodaa, A. A. Allam, and N. Elshaer. 2020. "Ecofriendly Synthesis and Insecticidal Application of Copper Nanoparticles against the Storage Pest *Tribolium castaneum*." *Nanomaterials* 10 (3): 587. doi:10.3390/nano100 30587.

Emamjomeh, L., S. Imani, K. T. Jahromi, and S. Moharramipour. 2023. "Nanoencapsulation Enhances the Contact Toxicity of *Eucalyptus globulus* Labill and *Zataria multiflora* Boiss Essential Oils against the Third Instar Larvae of *Ephestia kuehniella* (Lepidoptera: Pyralidae)." *International Journal of Pest Management* 69 (3): 207–214. doi:10.1080/09670874.2020.1871529.

Eyssa, H. M., S. G. Sawires, and M. M. Senna. 2019. "Gamma Irradiation of Polyethylene Nanocomposites for Food Packaging Applications against Stored-product Insect Pests." *Journal of Vinyl and Additive Technology* 25 (S1): E120–E129.

Faraji, Z., J. Shakarami, J. Varshosaz, and S. Jafari. 2020. "Encapsulation of Essential Oils of *Mentha pulegium* and *Ferula gummosa* Using Nanoliposome Technology as a Safe Botanical Pesticide." *Journal of Applied Biotechnology Reports* 7 (4): 237–242.

Gamal, M. M. Z. 2018. "Nano-Particles: A Recent Approach for Controlling Stored Grain Insect Pests." Academia Journal of Agricultural Research 6 (5): 88–94.

Giunti, G., O. Campolo, F. Laudani, L. Zappalà, and V. Palmeri. 2021. "Bioactivity of Essential Oil-Based Nano-Biopesticides toward *Rhyzopertha dominica* (Coleoptera: Bostrichidae)." *Industrial Crops and Products* 162 (April): 113257. doi:10.1016/j.indcrop.2021.113257.

Giunti, G., D. Palermo, F. Laudani, G. M. Algeri, O. Campolo, and V. Palmeri. 2019. "Repellence and Acute Toxicity of a Nano-Emulsion of Sweet Orange Essential Oil toward Two Major Stored Grain Insect Pests." *Industrial Crops and Products* 142: 111869. doi:10.1016/j.indcrop.2019.111869.

Guru, P. N., D. Mridula, A. S. Dukare, B. M. Ghodki, A. U. Paschapur, I. Samal, M. N. Raj, V. K. Padala, M. Rajashekhar, and A. R. N. S. Subbanna. 2022. "A Comprehensive Review on Advances in Storage Pest Management: Current Scenario and Future Prospects." *Frontiers in Sustainable Food Systems* 6 (November). doi:10.3389/fsufs.2022.993341.

Haider, Z., A. R. Ayub, M. U. Tahir, S. A. Sajid, and M. Khalid. 2020. "Efficacy and Characterization of Synthesized Zinc Oxide Nanoparticles against *Tribolium castaneum* and *Trogoderma granarium*." *International Journal of Scientific and Research Publications* 10 (3): 131–151. doi: 10.29322/IJSRP.10.03.2020.p9912.

Hamza, A. M. 2012. "Efficacy and Safety of Non-Traditional Methods as Alternatives for Control of *Sitophilus oryzae* (L.)(Coleoptera: Curculionidae) in Rice Grains." *Egyptian Journal of Biological Pest Control* 22 (2): 103–108.

Haroun, S., M. E. Ali, Elnaggar, and M. Z. Doaa, and Rehab Ibrahim Gad. 2020. "Insecticidal Efficiency and Safety of Zinc Oxide and Hydrophilic Silica Nanoparticles against Some Stored Seed Insects." *Journal of Plant Protection Research* 60 (1): 77–85.

Hashem, A. S., S. S. Awadalla, G. M. Zayed, F. Maggi, and G. Benelli. 2018. "*Pimpinella anisum* Essential Oil Nanoemulsions against *Tribolium castaneum*—Insecticidal Activity and Mode of Action." *Environmental Science and Pollution Research* 25: 18802–18812. doi:10.1007/s11356-018-2068-1.

Hazafa, A., N. Jahan, M. A. Zia, K-U. Rahman, M. Sagheer, and M. Naeem. 2022. "Evaluation and Optimization of Nanosuspensions of *Chrysanthemum coronarium* and *Azadirachta indica* Using Response Surface Methodology for Pest Management." *Chemosphere* 292 (April): 133411. doi:10.1016/j.chemosphere.2021.133411.

Heydarzadeh, A., O. Valizadegan, M. Negahban, and F. Mehrkhou. 2019. "Efficacy of *Mentha spicata* and *Mentha pulegium* Essential Oil Nanoformulation On Mortality and Physiology of *Tribolium castaneum* (Col.: Tenebrionidae)." *Journal of Crop Protection* 8 (4): 501–520.

Hilal, S. M., A. S. Mohmed, N. M. Barry, and M. H. Ibrahim. 2021. "Entomotoxicity of TiO2 and ZnO Nanoparticles against Adults *Tribolium castaneum* (Herbest) (Coleoptera: Tenebrionidae)." *IOP Conference Series: Earth and Environmental Science* 910: 12088. doi:10.1088/1755-1315/910/1/012088.

Hossain, F., P. Follett, S. Salmieri, K. D. Vu, M. Harich, and M. Lacroix. 2019. "Synergistic Effects of Nanocomposite Films Containing Essential Oil Nanoemulsions in Combination with Ionizing Radiation for Control of Rice Weevil *Sitophilus oryzae* in Stored Grains." *Journal of Food Science* 84 (6): 1439–1446. doi:10.1111/1750-3841.14603.

Hosseinpour Jajarm, F., G. Moravvej, M. M. Awal, and S. Golmohammadzadeh. 2021. "Insecticidal Activity of Solid Lipid Nanoparticle Loaded by *Ziziphora clinopodioides* Lam. against *Tribolium castaneum* (Herbst, 1797) (Coleoptera: Tenebrionidae)." *International Journal of Pest Management* 67 (2): 147–154. doi:10.1080/09670874.2020.1713420.

Huang, L., S. Yang, J. Chen, J. Tian, Q. Huang, H. Huang, Y. Wen, F. Deng, X. Zhang, and Y. Wei. 2019. "A Facile Surface Modification Strategy for Fabrication of Fluorescent Silica Nanoparticles with the Aggregation-Induced Emission Dye through Surface-Initiated Cationic Ring Opening Polymerization." *Materials Science and Engineering: C* 94: 270–278. doi:10.1016/j.msec.2018.09.042.

Ibrahim, S. S., and N. Y. Salem. 2019. "Insecticidal Efficacy of Nano Zeolite against *Tribolium confusum* (Col., Tenebrionidae) and *Callosobruchus maculatus* (Col., Bruchidae)." *Bulletin of the National Research Centre* 43: 1–8. doi:10.1186/s42269-019-0128-4.

Indian Grain Storage Management & Research Institute (IGMRI) (2019) https://www.igmri.dfpd.gov.in/igmri/insect-pests

Ismail, T., M. A. Salama, and M. El-Ebiary. 2021. "Entomotoxic Effects of Synthesized Aluminum Oxide Nanoparticles against *Sitophilus oryzae* and Their Toxicological Effects on Albino Rats." *Toxicology and Industrial Health* 37 (10): 594–602. doi:10.1177/07482337211035000.

Jafer, F. S., and M. R. Annon. 2018. "Larvicidal Effect of Pure and Green-Synthesized Silver Nanoparticles against *Tribolium castaneum* (Herb.) and *Callosobruchus maculatus* (Fab.)." *Journal of Global Pharma Technology* 10 (3): 448–454.

Jasrotia, P., M. Nagpal, C. N. Mishra, A. K. Sharma, S. Kumar, U. Kamble, A. K. Bhardwaj, P. L. Kashyap, S. Kumar, and G. P. Singh. 2022. "Nanomaterials for Postharvest Management of Insect Pests: Current State and Future Perspectives." *Frontiers in Nanotechnology* 3. doi:10.3389/fnano.2021.811056.

Kar, S., R. N. Nayak, N. R. Sahoo, C. K. Bakhara, M. K. Panda, U. S. Pal, and L. M. Bal. 2021. "Rice Weevil Management through Application of Silica Nano Particle and Physico-Chemical and Cooking Characterization of the Treated Rice." *Journal of Stored Products Research* 94: 101892. doi:10.1016/j.jspr.2021.101892.

Kashyap, P. L., X. Xiang, and P. Heiden. 2015. "Chitosan Nanoparticle Based Delivery Systems for Sustainable Agriculture." *International Journal of Biological Macromolecules* 77: 36–51. doi:10.1016/j.ijbiomac.2015.02.039.

Kavallieratos, N. G., E. P. Nika, A. Skourti, N. Ntalli, M. C. Boukouvala, C. T. Ntalaka, F. Maggi, R. Rakotosaona, M. Cespi, and D. R. Perinelli. 2021. "Developing a *Hazomalania voyronii* Essential Oil Nanoemulsion for the Eco-Friendly Management of *Tribolium confusum*, *Tribolium castaneum* and *Tenebrio molitor* Larvae and Adults on Stored Wheat." *Molecules* 26 (6): 1812. doi:10.3390/molecules26061812.

Kavallieratos, N. G., A. Skourti, E. P. Nika, C. T. Ntalaka, M. C. Boukouvala, G. Bonacucina, M. Cespi, R. Petrelli, L. Cappellacci, and F. Maggi. 2021. "Isofuranodiene-Based Nanoemulsion: Larvicidal and Adulticidal Activity against Tenebrionid Beetles Attacking Stored Wheat." *Journal of Stored Products Research* 93: 101859.

Khalil, M. S. 2019. "Efficacy of Some Nanoparticles against the Adults of Red Flour Beetle Triboliumcastaneum (Herbst) under Laboratory Conditions." *Annals of Agricultural Science, Moshtohor* 57 (2): 535–540. doi:10.21608/assjm.2019.44940.

Khoobdel, M., S. M. Ahsaei, and M. Farzaneh. 2017. "Insecticidal Activity of Polycaprolactone Nanocapsules Loaded with *Rosmarinus officinalis* Essential Oil in *Tribolium castaneum* (Herbst)." *Entomological Research* 47 (3): 175–184. doi:10.1111/1748-5967.12212.

Kumar, S., H. Meena, S. Chakraborty, and B. C. Meikap. 2018. "International Journal of Mining Science and Technology Application of Response Surface Methodology (RSM) for Optimization of Leaching Parameters for Ash Reduction from Low-Grade Coal." *International Journal of Mining Science and Technology* 28 (4): 621–629. doi:10.1016/j.ijmst.2018.04.014.

Lakshmi, S. J., R. S. Roopa Bai, H. C. T. R. Sharanagouda, C. T. Ramachandra, and S. Nadagouda. 2020. "Effect of Zinc Oxide Nanoparticles on Pulse Beetle (*Callosobruchus maculatus*) (Col.: Chrysomelidae) in Greengram." *Journal of Entomological Zoology Studies* 8 (4): 297–300.

Licciardello, F., G. Muratore, P. Suma, A. Russo, and C. Nerín. 2013. "Effectiveness of a Novel Insect-Repellent Food Packaging Incorporating Essential Oils against the Red Flour Beetle (*Tribolium castaneum*)." *Innovative Food Science & Emerging Technologies* 19 (July): 173–180. doi:10.1016/J.IFSET.2013.05.002.

Loha, K. M., N. A. Shakil, J. Kumar, M. K. Singh, and C. Srivastava. 2012. "Bio-Efficacy Evaluation of Nanoformulations of β-Cyfluthrin against *Callosobruchus maculatus* (Coleoptera: Bruchidae)." *Journal of Environmental Science and Health, Part B* 47 (7): 687–691. doi:10.1080/03601234.2012.669254.

López-García, G. P., M. Buteler, and T. Stadler. 2018. "Testing the Insecticidal Activity of Nanostructured Alumina on *Sitophilus oryzae* (L.) (Coleoptera: Curculionidae) under Laboratory Conditions Using Galvanized Steel Containers." *Insects* 9 (3): 87.

Louni, M., J. Shakarami, and M. Negahban. 2018. "Insecticidal Efficacy of Nanoemulsion Containing *Mentha longifolia* Essential Oil against *Ephestia kuehniella* (Lepidoptera: Pyralidae)." *Journal of Crop Protection* 7 (2): 171–182.

Louni, M. O. J. D. E. H., J. Shakarami, and M. Negahban. 2019. "Study on Insecticidal Properties of Nanoemulsion *Mentha longifolia* L. (Lamiaceae) Essential Oil against *Callosobruchus maculatus* (Fabricius)(Coleoptera: Chrysomelidae)." *Journal of Entomological Society of Iran* 39 (2): 151–163.

Malaikozhundan, B., B. Vaseeharan, S. V. kumar, and M. P. Thangaraj. 2017. "*Bacillus thuringiensis* Coated Zinc Oxide Nanoparticle and Its Biopesticidal Effects on the Pulse Beetle, *Callosobruchus maculatus*." *Journal of Photochemistry and Photobiology B: Biology* 174: 306–314. doi:10.1016/j.jphotobiol.2017.08.014.

Malathi, S., G. Rameshkumar, R. L. Rengarajan, T. Rajagopal, S. Muniasamy, and P. Ponmanickam. 2019. "Phytofabrication of Silver Nanoparticles Using *Annona reticulata* and Assessment of Insecticidal and Bactericidal Activities." *Journal of Environmental Biology* 40 (4): 626–633. doi:10.22438/jeb/40/4/mrn-934.

Mohammed, A. M., and S. A. Aswd. 2019. "Effect of Some Nanoparticles on the Stages Biology of the Southern Cowpea Beetle *Callosobruchus maculatus* (Fab.) (Coleoptera: Bruchidae)." *Journal of Education and Science* 28 (3): 188–199.

Mohammed, T. G., and M. E. H. Nasr. 2020. "Preparation, Characterization and Biological Efficacy of Eucalyptus Oil Nanoemulsion against the Stored Grain Insects." *Asian Journal of Advances in Agricultural Research* 13 (2): 41–51. doi:10.9734/ajaar/2020/v13i230102.

Mohapatra, D., S. Kumar, N. Kotwaliwale, and K. K. Singh. 2017. "Critical Factors Responsible for Fungi Growth in Stored Food Grains and Non-Chemical Approaches for Their Control." *Industrial Crops and Products* 108: 162–182.

Moirangthem, T. T., and O.-D. Baik. 2021. "Disinfestation of Stored Grains Using Non-Chemical Technologies – A Review." *Trends in Food Science & Technology* 107: 299–308.

Nair, P. M. G., and J. Choi. 2011. "Identification, Characterization and Expression Profiles of Chironomus Riparius Glutathione S-Transferase (GST) Genes in Response to Cadmium and Silver Nanoparticles Exposure." *Aquatic Toxicology* 101 (3–4): 550–560. doi:10.1016/j.aquatox.2010.12.006.

Nasr, M. 2015. "Impact of Nanoparticle Zinc Oxide and Aluminum Oxide against Rice Weevil *Sitophilus oryzae* (Coleoptera: Curculionidae) under Laboratory Conditions." *Egyptian Journal of Plant Protection Research* 3 (3): 30–38.

Negahban, M., S. Moharramipour, M. Zandi, S. A. Hashemi, and F. Ziayee. 2012. "Nano-Insecticidal Activity of Essential Oil from *Cuminum cyminum* on *Tribolium castaneum*," no. October: 6368.

Nenaah, G. E. 2014. "Chemical Composition, Toxicity and Growth Inhibitory Activities of Essential Oils of Three Achillea Species and Their Nano-Emulsions against *Tribolium castaneum* (Herbst)." *Industrial Crops and Products* 53: 252–260. doi:10.1016/j.indcrop.2013.12.042.

Nguyen, H. M., I. C. Hwang, J. W. Park, and H. J. Park. 2012. "Photoprotection for Deltamethrin Using Chitosan-coated Beeswax Solid Lipid Nanoparticles." *Pest Management Science* 68 (7): 1062–1068. doi:10.1002/ps.3268.

Nile, S. H., V. Baskar, D. Selvaraj, A. Nile, J. Xiao, and G. Kai. 2020. "Nanotechnologies in Food Science: Applications, Recent Trends, and Future Perspectives." *Nano-Micro Letters* 12 (1): 45. doi:10.1007/s40820-020-0383-9.

Oliveira, C. R., C. E. C. Domingues, N. F. S. de Melo, T. C. Roat, O. Malaspina, M. Jones-Costa, E. C. M. Silva-Zacarin, and L. F. Fraceto. 2019. "Nanopesticide Based on Botanical Insecticide Pyrethrum and Its Potential Effects on Honeybees." *Chemosphere* 236: 124282. doi:10.1016/j.chemosphere.2019.07.013.

Palermo, D., G. Giunti, F. Laudani, V. Palmeri, and O. Campolo. 2021. "Essential Oil-Based Nano-Biopesticides: Formulation and Bioactivity against the Confused Flour Beetle *Tribolium confusum*." *Sustainability* 13 (17): 9746. doi:10.3390/su13179746.

Shakil, N. A., J. Kumar, M. K Singh, and K. Singh. 2012. "Bioefficacy Evaluation of Controlled Release Formulations Based on Amphiphilic Nano-Polymer of Carbofuran against Meloidogyne Incognita Infecting Tomato." *Journal of Environmental Science and Health, Part B,* Pesticides, Food Contaminants, and Agricultural Wastes 47 (6): 520–528. doi:10.1080/03601234.2012.665667.

Pant, M., S. Dubey, P. K. Patanjali, S. N. Naik, and S. Sharma. 2014. "Insecticidal Activity of Eucalyptus Oil Nanoemulsion with Karanja and Jatropha Aqueous Filtrates." *International Biodeterioration & Biodegradation* 91. Elsevier: 119–127. doi:10.1016/j.ibiod.2013.11.019.

Pardeshi, C., P. Rajput, V. Belgamwar, A. Tekade, G. Patil, K. Chaudhary, and A. Sonje. 2012. "Solid Lipid Based Nanocarriers: An Overview/Nanonosači Na Bazi Čvrstih Lipida: Pregled." *Acta Pharmaceutica* 62 (4): 433–472.

Pathipati, U. R., and P. L. Kanuparthi. 2021. "Silver Nanoparticles for Insect Control: Bioassays and Mechanisms." In *Silver Nanomaterials for Agri-Food Applications*, 471–494. Elsevier.

Patil, C. D., H. P. Borase, R. K. Suryawanshi, and S. V. Patil. 2016. "Trypsin Inactivation by Latex Fabricated Gold Nanoparticles: A New Strategy towards Insect Control." *Enzyme and Microbial Technology* 92: 18–25. doi:10.1016/j.enzmictec.2016.06.005.

Qiu, L., N. Zhu, Y. Feng, E. E. Michaelides, G. Żyła, D. Jing, X. Zhang, P. M. Norris, C. N. Markides, and O. Mahian. 2020. "A Review of Recent Advances in Thermophysical Properties at the Nanoscale: From Solid State to Colloids." *Physics Reports* 843: 1–81. doi:10.1016/j.physrep.2019.12.001.

Raduw, G. G., and A. A. Mohammed. 2020. "Insecticidal Efficacy of Three Nanoparticles for the Control of Khapra Beetle (*Trogoderma granarium*) on Different Grains." *Journal of Agricultural and Urban Entomology* 36 (1): 90–100. doi:10.3954/1523-5475-36.1.90.

Rahman, I. A., and V. Padavettan. 2012. "Synthesis of Silica Nanoparticles by Sol-gel: Size-dependent Properties, Surface Modification, and Applications in Silica-polymer Nanocomposites—A Review." *Journal of Nanomaterials* 2012 (1): 132424. doi:10.1177/07482337211035000.

Rajkumar, V., C. Gunasekaran, C. A. Paul, and J. Dharmaraj. 2020. "Development of Encapsulated Peppermint Essential Oil in Chitosan Nanoparticles: Characterization and Biological Efficacy against Stored-Grain Pest Control." *Pesticide Biochemistry and Physiology* 170 (June): 104679. doi:10.1016/j.pestbp.2020.104679.

Rani, A., M. Bisht, C. Pande, G. Tewari, S. Bhatt, and M. Matiyani. 2017. "Effect of Drying on the Volatiles of Leaves of *Murraya koenigii* (L.) Spreng." *Journal of Essential Oil Bearing Plants* 20 (2): 552–558. doi:10.1080/0972060X.2017.1317606.

Rani, S. S., C. G. L. Justin, K. Gunasekaran, and S. S. J. Roseleen. 2019. "Efficacy of Green Synthesized Silver Nanoparticle, Plant Powders and Oil against Rice Weevil *Sitophilus oryzae* L.(Coleoptera: Curculionidae) on Sorghum Seeds." *Journal of Pharmacognosy and Phytochemistry* 8 (5): 38–42.

Raoufi, D. 2013. "Synthesis and Microstructural Properties of ZnO Nanoparticles Prepared by Precipitation Method." *Renewable Energy* 50: 932–937. doi:10.1016/j.renene.2012.08.076.

Rashwan, R. S., and A. A. Abu-Zaid. 2018. "Evaluation of Nutritive Values and Microbial Content of Wheat Grains Treated with Some Types of Nano Particles to Control *Sitophilus granaries* (L.)." *Arab Universities Journal of Agricultural Sciences* 26 (1): 293–301. doi:10.21608/ajs.2018.14006.

Rastogi, A., D. K. Tripathi, S. Yadav, D. K. Chauhan, M. Živčák, M. Ghorbanpour, N. I. El-Sheery, and M. Brestic. 2019. "Application of Silicon Nanoparticles in Agriculture." *3 Biotech* 9: 1–11. doi:10.1007/s13205-019-1626-7.

Rouhani, M., M. A. Samih, and S. Kalantari. 2013. "Insecticidal Effect of Silica and Silver Nanoparticles on the Cowpea Seed Beetle, *Callosobruchus maculatus* F. (Col.: Bruchidae)." *Journal of Entomological Research* 4 (4): 297–307.

Rouhani, M., M. A. Samih, M. Zarabi, K. Beiki, M. Gorji, and M. R. Aminizadeh. 2019. "Synthesis and Entomotoxicity Assay of Zinc and Silica Nanoparticles against *Sitophilus granarius* (Coleoptera: Curculionidae)." *Journal of Plant Protection Research* 59 (1): 26–31.

Sabbour, M. M. A. 2020. "Efficacy of Nano-Formulated Certain Essential Oils on the Red Flour Beetle *Tribolium castaneum* and Confused Flour Beetle, *Tribolium confusum* (Coleoptera: Tenebrionidae) under Laboratory and Storage Conditions." *Bulletin of the National Research Centre* 44: 1–7. doi:10.1186/s42269-020-00336-6.

Sabbour, M. M., H. R. Alipour, S. Abbasi, S. Ghavidel, H. F. Farahani, M. Sabaghi, and M. K. Rahman. 2012. "Entomotoxicity Assay of Two Nanoparticle Materials 1-(Al2O3 and TiO2) against *Sitophilus oryzae* under Laboratory and Store Conditions in Egypt." *Journal of Applied Sciences* 1 (4): 103–108.

Saed, B., M. Ziaee, A. R. Kiasat, and M. Jafari nasab. 2022. "Preparation of Nanosilica from Sugarcane Bagasse Ash for Enhanced Insecticidal Activity of Diatomaceous Earth against Two Stored-Products Insect Pests." *Toxin Reviews* 41 (2): 516–522. doi:10.1080/15569543.2021.1903038.

Sala, M., R. Diab, A. Elaissari, and H. Fessi. 2018. "Lipid Nanocarriers as Skin Drug Delivery Systems: Properties, Mechanisms of Skin Interactions and Medical Applications." *International Journal of Pharmaceutics* 535 (1–2): 1–17.

Salem, A. A. 2020. "Comparative Insecticidal Activity of Three Forms of Silica Nanoparticles on Some Main Stored Product Insects." *Journal of Plant Protection and Pathology* 11 (4): 225–230. doi:10.21608/jppp.2020.96009.

Salem, A. A., A. M. Hamzah, and N. M. El-Taweelah. 2015. "Aluminum and Zinc Oxides Nanoparticles as a New Methods for Controlling the Red Flour Beetles, *Tribolium castaneum* (Herbest) Compared to Malathion Insecticide." *Journal of Plant Protection and Pathology* 6 (1): 129–137. doi:10.21608/jppp.2015.53186.

Sankar, M. V., and S. Abideen. 2015. "Pesticidal Effect of Green Synthesized Silver and Lead Nanoparticles Using *Avicennia marina* against Grain Storage Pest *Sitophilus oryzae*." *International Journal of Nanomaterials and Biostructures* 5 (3): 32–39.

Saratale, G. D., R. G. Saratale, G. Benelli, G. Kumar, A. Pugazhendhi, D.-S. Kim, and H.-S. Shin. 2017. "Anti-Diabetic Potential of Silver Nanoparticles Synthesized with Argyreia Nervosa Leaf Extract High Synergistic Antibacterial Activity With Standard Antibiotics against Foodborne Bacteria." *Journal of Cluster Science* 28: 1709–1727. doi:10.1007/s10876-017-1179-z.

Satehi, A. B., M. Ziaee, and A. Ashrafi. 2018. "Silica Nanoparticles: A Potential Carrier of Chlorpyrifos in Slurries to Control Two Insect Pests of Stored Products." *Entomologia Generalis* 37 (1). doi:10.1127/entomologia/2017/0406.

Shahzadi, K., F. Bashir, K. Nazir, S. Seemab, M. A. A. Shaker, I. Waheed, A. Shehzaib, and K. Fatima. 2019. "Impacts of Green Synthesized Silver-Nanoparticles Against *Tribolium castaneum* (Coleoptera: Tenebrionidae)." *International Journal of Entomology Research* 4 (6): 114–117.

Shelar, A., S. H. Nile, A. V. Singh, D. Rothenstein, J. Bill, J. Xiao, M. Chaskar, G. Kai, and R. Patil. 2023. "Recent Advances in Nano-Enabled Seed Treatment Strategies for Sustainable Agriculture: Challenges, Risk Assessment, and Future Perspectives." *Nano-Micro Letters* 15 (1): 54. doi:10.1007/s40820-023-01025-5.

Siddique, M. A., M. Hasan, M. Sagheer, and S. T. Sahi. 2022. "Comparative Toxic Effects of *Eucalyptus globulus* L. (Myrtales: Myrtaceae) and Its Green Synthesized Zinc Oxide Nanoparticles (ZnONPs) against *Rhyzopertha dominica* (F.) (Coleoptera: Bostrichidae)." *International Journal of Tropical Insect Science*, 1–10. doi:10.1007/s42690-021-00691-5.

Sivapriya, V., A. N. Azeez, and S. V. Deepa. 2018. "Phyto Synthesis of Iron Oxide Nano Particles Using the Agro Waste of *Anthocephalus cadamba* for Pesticidal Activity against Sitophilus Granaries." *Journal of Entomology and Zoology Studies* 6: 1050–1057.

Small, T., M. A. Ochoa-Zapater, G. Gallello, A. Ribera, F. M. Romero, A. Torreblanca, and M. D. Garcerá. 2016. "Gold-Nanoparticles Ingestion Disrupts Reproduction and Development in the German Cockroach." *Science of the Total Environment* 565: 882–888. doi:10.1016/j.scitotenv.2016.02.032.

Stadler, T., M. Buteler, D. K. Weaver, and S. Sofie. 2012. "Comparative Toxicity of Nanostructured Alumina and a Commercial Inert Dust for *Sitophilus oryzae* (L.) and *Rhyzopertha dominica* (F.) at Varying Ambient Humidity Levels." *Journal of Stored Products Research* 48: 81–90. doi:10.1016/j.jspr.2011.09.004.

Stadler, T., G. P. L. García, J. G. Gitto, and M. Buteler. 2017. "Nanostructured Alumina: Biocidal Properties and Mechanism of Action of a Novel Insecticide Powder." Alma Mater Studiorum, Univ Bologna.

Stejskal, V., T. Vendl, R. Aulicky, and C. Athanassiou. 2021. "Synthetic and Natural Insecticides: Gas, Liquid, Gel and Solid Formulations for Stored-Product and Food-Industry Pest Control." *Insects* 12 (7): 590.

Vadlapudi, V., and R. Amanchy. 2017. "Phytofabrication of Silver Nanoparticles Using *Myriostachya wightiana* as a Novel Bioresource, and Evaluation of Their Biological Activities." *Brazilian Archives of Biology and Technology* 60: e17160329. doi:10.1590/1678-4324-2017160329.

Wang, S., S. Miao, B. Yang, Z. Wang, Q. Liu, R. Wang, X. Du, Y. Ren, and Y. Lu. 2021. "Initial Characterization of the Vitellogenin Receptor from a Psocoptera Insect: Function Analysis and RNA Interference in Liposcelis Entomophila (Enderlein)." *Journal of Stored Products Research* 92: 101803.

Wazid, S. N., A. Prabhuraj, R. H. Naik, N. M. Shakuntala, and H. Sharanagouda. 2020. "The Persistence of Residual Toxicity of Zinc, Copper and Silica Green Nanoparticles against Important Storage Pests." *Journal of Entomology and Zoology Studies* 8: 1207–1211.

Wei, W., X. Lu, Z. Wang, B. Pérez, J. Liu, C. Wu, M. Dong, F. Feng, H. Mu, and Z. Guo. 2017. "Single-Component Solid Lipid Nanocarriers Prepared with Ultra-Long Chain Amphiphilic Lipids." *Journal of Colloid and Interface Science* 505: 392–401.

Ya-Ali, P., F. Yarahmadi, and M. A. Mehrnia. 2020. "Efficacies of Two Nano-Formulations of Tasmanian Blue Gum Essential Oil to Control *Callosobruchus maculatus*." *Journal of Economic Entomology* 113 (3): 1555–1562. doi:10.1093/jee/toaa069.

Yadav, J., P. Jasrotia, P. L. Kashyap, A. K. Bhardwaj, S. Kumar, M. Singh, and G. P. Singh. 2021. "Nanopesticides: Current Status and Scope for Their Application in Agriculture." *Plant Protection Science* 58 (1): 1–17.

Yang, F.-L., X.-G. Li, F. Zhu, and C.-L. Lei. 2009. "Structural Characterization of Nanoparticles Loaded with Garlic Essential Oil and Their Insecticidal Activity against *Tribolium castaneum* (Herbst) (Coleoptera: Tenebrionidae)." *Journal of Agricultural and Food Chemistry* 57 (21): 10156–10162. doi:10.1021/jf9023118.

Zahir, A. A., A. Bagavan, C. Kamaraj, G. Elango, and A. A. Rahuman. 2012. "Efficacy of Plant-Mediated Synthesized Silver Nanoparticles against *Sitophilus oryzae*." *Journal of Biopesticides* 5: 95.

Ziaee, M., and A. Babamir. 2021. "Insecticidal Efficacy of Silica Nanoparticles Loaded with Several Insecticides in Controlling Khapra Beetle Larvae, *Trogoderma granarium* on Mosaic and Galvanized Steel Surfaces" 43 (January): 35–47. doi:10.22055/PPR.2020.15975.

Ziaee, M., S. Moharramipour, and A. Mohsenifar. 2014. "Toxicity of *Carum copticum* Essential Oil-loaded Nanogel against *Sitophilus granarius* and *Tribolium Confusum*." *Journal of Applied Entomology* 138 (10): 763–771. doi:10.1111/jen.12133.

20 Non-Chemical Methods for Grain Protection

Challenges and Future Directions

P. Santhoshkumar, S. Ashwin Kumar, and Jeyan A. Moses
Computational Modeling and Nanoscale Processing Unit, Department of Food Process Engineering, National Institute of Food Technology Entrepreneurship and Management Thanjavur (NIFTEM-T), Tamil Nadu, India

20.1 INTRODUCTION

Stored products refer to commodities, such as grains, seeds, nuts, and dried fruits kept in storage facilities for later use or sale. These storage facilities, ranging from silos to warehouses, play a crucial role in preserving the quality and quantity of these products. However, they are also vulnerable to infestations by pests, such as insects, rodents, and molds. These pests can cause significant damage by contaminating the stored products, leading to spoilage, loss of nutritional value, and economic losses for farmers and distributors. To prevent infestations, proper storage techniques such as temperature control, sanitation, and use of insecticides are essential. Regular monitoring and inspection of stored products are also necessary to detect and address any pest issues promptly, ensuring the integrity and safety of the stored goods (Afrose Subaitha et al. 2024).

20.2 LOSSES DURING STORAGE

Losses during the storage of food commodities can occur due to various factors, including physical, chemical, and biological degradation. Physical factors, such as moisture, temperature fluctuations, and mechanical damage, can lead to spoilage and loss of quality. Chemical reactions such as oxidation and enzymatic activity can cause changes in flavor, color, and nutritional value, rendering the food unsuitable for consumption. Biological factors, such as microbial growth, insect infestations, and mold, can also contaminate stored food, leading to spoilage and deterioration. Improper storage conditions exacerbate these issues, as inadequate ventilation, poor hygiene, and improper packaging provide favorable environments for spoilage organisms to thrive (Kumar and Kalita 2017). Additionally, inadequate pest control measures can result in significant losses due to insect and rodent infestations. Losses during storage not only impact food quality and safety but also result in economic losses for producers, distributors, and consumers. To mitigate these losses, proper storage facilities equipped with temperature and humidity control, adequate ventilation, and pest management systems are essential. Furthermore, implementing good agricultural practices, proper handling techniques, and regular monitoring of stored food commodities can help minimize losses and ensure the availability of safe and high-quality food for consumers.

Insect infestation poses a significant threat to food commodities during storage, leading to substantial losses. Beetles, weevils, moths, and mites are common pests attracted to stored grains, seeds, and other food products, where they breed and feed, contaminating the food with feces and secretions (Yaseen et al. 2019). Factors such as poor sanitation, inadequate storage conditions, and lack of pest control measures exacerbate infestation problems. Insects can enter storage facilities through openings or contaminated shipments, accelerating spoilage and rendering products unfit

DOI: 10.1201/9781003609834-20

for consumption (Mohapatra, Kar, and Giri 2015). Besides economic losses, insect-infested food products may pose health risks due to potential pathogens and allergens (Abdelfattah, Hasan, and Qari 2022). Preventive measures, such as sanitation, proper storage practices, and integrated pest management (IPM) strategies, are crucial for mitigating losses. Regular monitoring, timely intervention, and judicious use of insecticides help control insect populations and preserve the quality and safety of stored food commodities.

20.3 CONVENTIONAL APPROACHES FOR STORAGE OF GRAINS

The conventional storage system offers a wide range of storage options, from basic storerooms and barns to larger, more technologically advanced warehouses, including chemical methods-adapted systems (Ziegler, Paraginski, and Ferreira 2021). Conventional chemical storage methods involve diverse effects on stored commodities; for instance, fumigants (methyl bromide, sulfuryl fluoride, phosphine) kill pests (insects, mites, and rodents); desiccants (alumina, silica gel) are used to absorb moisture from the air in storage containers, preventing mold growth and extending the shelf life of products; antimicrobials (acetic acid, benzoic acid, propionic acid) act as preservatives to inhibit the growth of bacteria, yeasts, molds, and insecticides (organophosphates, carbamates, pyrethroids) control insect infestations in stored commodities. They can be applied as surface treatments, aerosols, or incorporated into packaging materials to prevent insect damage (De Corato 2020). While chemical methods can be effective in preserving food quality and preventing losses during storage, their use requires careful application to ensure safety for consumers and minimize environmental impact. Additionally, regulatory guidelines and safety protocols must be followed to prevent contamination and ensure compliance with food safety standards. Challenges associated with insect infestation in food storage include rapid reproduction rates, resistance to pesticides, and difficulty in detecting hidden infestations. Controlling infestations requires constant vigilance, effective pest management strategies, and proper sanitation practices to minimize damage to stored commodities and ensure food safety (Anukiruthika, Jian, and Jayas 2021).

The need for alternatives to chemical methods for insect infestation in food storage arises due to concerns about environmental impact, pesticide residues, and the development of resistance to pests. IPM approaches, biological controls like natural predators, and physical methods such as heat treatment offer sustainable alternatives that reduce reliance on chemicals while maintaining effective pest control and ensuring food safety (Morrison, Scully, and Campbell 2021).

20.4 CHALLENGES AND FUTURE DIRECTIONS

Insect infestations, rodent activity, and mold growth pose significant challenges in maintaining the quality and safety of stored commodities. Temperature fluctuations, humidity levels, and exposure to light can affect the shelf life and quality of stored products, leading to spoilage and deterioration. Mechanical damage during handling, improper stacking, and inadequate packaging can result in product loss and compromise the integrity of stored commodities. Bacterial, fungal, and yeast growth can contaminate stored food, leading to foodborne illnesses and product spoilage. The presence of chemical residues from pesticides, cleaning agents, or packaging materials can pose health risks and affect the safety and quality of stored commodities. Theft, vandalism, and unauthorized access to storage facilities can result in loss of inventory and compromise food safety protocols. Despite the range of non-chemical pesticides proven to be effective against stored grain pests, they still lack a strong reputation in the market. The major challenges are categorized as follows:

20.4.1 Organic Alternatives

Regarding organic-based pesticides, ensuring the availability of raw materials for large-scale industrial use requires substantial arable land for cultivating botanicals. Additionally, variations in the quality and efficacy of botanicals due to seasonal and regional differences must be considered. Natural-material-sourced formulations, such as those based on spices and herbs, may experience changes in their activity during storage or between production and application, affecting their real-world performance. Moreover, there is a lack of knowledge regarding the bioactive components of organic pesticides and their mechanisms of action against storage pests. To refine the process, studies should focus on deeply characterizing bio-based extracts to identify specific components and additional active principles that could be effective in pest control applications.

20.4.2 Scalability

The novel approaches in recently developed areas, such as nanotechnology and organic-based pesticides, have been well-received. However, the large-scale production and application methods for these innovations in stored products remain underexplored. Due to their relatively high production costs and limited commercial exploration, these novel methods are still quite expensive. Nanotechnology in agriculture shows promising results but also requires effective techniques for detecting and quantifying nanoparticles across various matrices. Non-chemical methods, such as ultraviolet, microwave, radiofrequency, or ozone treatment, are also relatively costly and complex to maintain on a commercial scale. Consequently, industrialists are more focused on achieving immediate profitability and meeting consumer demands rather than prioritizing ecological sustainability.

20.4.3 Cost-Benefit Ratio

Modeling-based approaches can help complement assessment studies to determine the appropriate dosages for effective applications. It required manpower along with the additional technical requirements. Proper processing facilities are also essential for the extraction and formulation of these insecticides. However, the limited availability of materials and specific processes can increase the cost of botanical pesticides compared to conventional synthetic pesticides. While some reports claim these formulations are high-cost, others suggest they are cost-effective. Generally, many bio-based formulations tend to be more expensive than synthetic ones, although no comprehensive comparative assessment has been recorded.

20.4.4 Impact on Environment, Plants, Animals, and Humans

Pesticide majorly affects the health of humans, animals, and plants. For instance, pesticides will affect the nervous system, energy system, circulatory system, reproduction system, and water imbalance in the human body. Additionally, it also destroys the environment, including the air, water, and soil quality. For instance, fumigants are often highly toxic chemicals that pose significant health and safety risks to workers and bystanders if not handled properly. Inhalation or skin exposure to fumigants can cause serious health problems, including respiratory issues, neurological effects, and even death in extreme cases. Many fumigants are potent greenhouse gases or ozone-depleting substances, contributing to environmental degradation and climate change. Improper application or disposal of fumigants can lead to soil, air, and water pollution, harming ecosystems and wildlife. Residues from fumigants can persist in treated materials or environments, posing long-term risks to human health and the environment. Proper residue management practices, such as ventilation and aeration of treated spaces, are essential to ensure residual fumigants dissipate safely and do not contaminate food products or living areas. Pests and pathogens can develop resistance to fumigants over time, reducing their efficacy and necessitating the use of alternative chemicals or methods. Continuous reliance on fumigants can accelerate resistance development, requiring IPM strategies to maintain effectiveness while minimizing resistance risks.

20.4.5 Hybrid Approaches

To store and process the grains, diverse hybrid approaches were recently minted. For instance, the non-thermal methods were used in the process of denaturing the insects. Applying natural or synthetic pesticides to stored grains showed that synergistic effects produced better outcomes than using individual pesticides alone. Recent trends related to biological-based nanoparticle synthesis are tedious; consequently, scalability majorly affects the industrial scale application. However, these kinds of hybrid approaches, syntheses and their processing on stored grains were still unexplored by the researchers. For example, Figure 20.1 illustrates the complex procedure for the synthesis of nano-based biopesticides. While the radiation approach is effective in controlling pests, it can raise consumer concerns regarding food irradiation and potential long-term health effects, even though the process is generally considered safe and approved by various health authorities.

20.4.6 Underlying Mechanisms for Inactivation of Each Life Stage

The concentration and mode of action of pesticides in stored grains depend on the specific pesticides and their sources. These factors affect the insects' neurological and cell membrane systems (Figure 20.2). The lethality effects vary across different stages of the insect life cycle and species based on the pesticide and its concentration. High doses of non-thermal radiation negatively impact egg development, with irradiated larvae often dying before reaching the pupal stage. Non-thermal radiation can cause surface distortion during pupae development. While adults are highly susceptible to death from various non-chemical methods, larvae and eggs show more resistance to these disinfestation techniques. A major drawback of non-chemical and non-thermal disinfestation

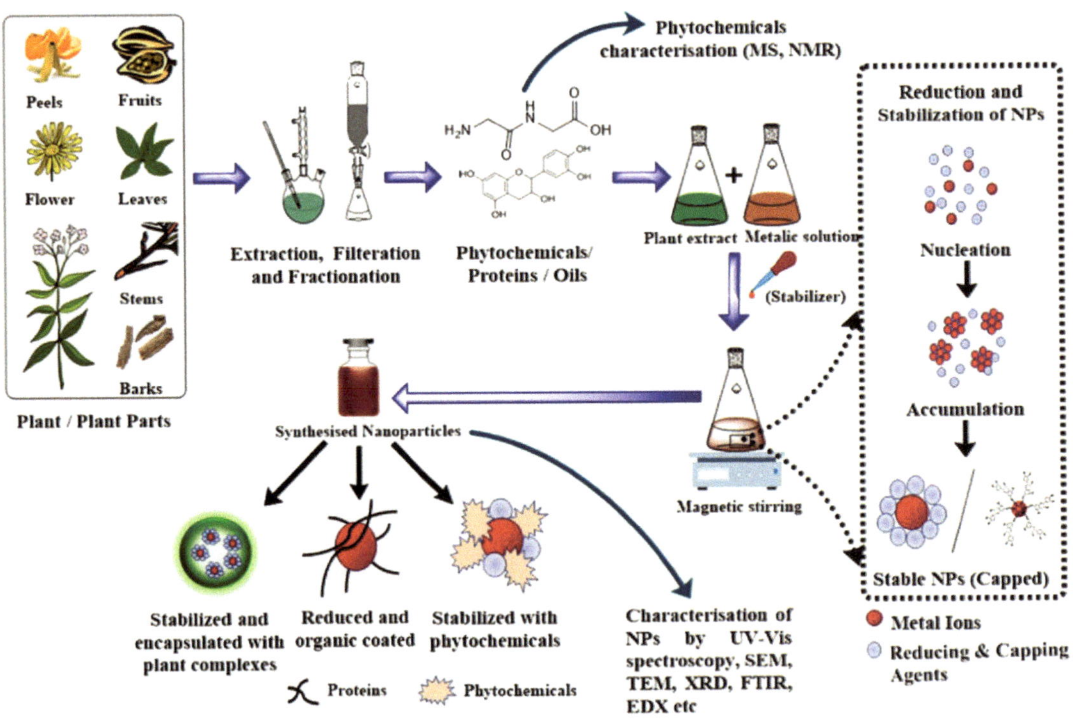

FIGURE 20.1 Synthesis procedure for bio-nano pesticides. (Baliyarsingh and Pradhan 2023.)

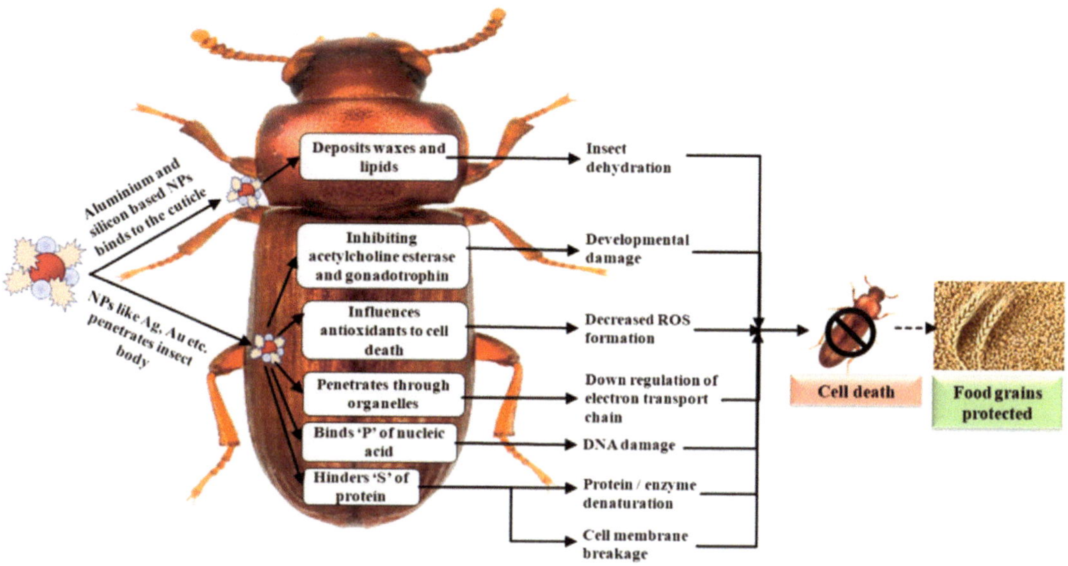

FIGURE 20.2 Underlying mechanisms of nano pesticides effects on stored grain insects. (Baliyarsingh and Pradhan 2023.)

methods is their potential to affect the inherent quality parameters of grains at higher concentrations and longer exposure times. Such treatments can negatively impact properties like color, protein digestibility, water absorption capacity, free fatty acids, amino acid profile, and grains' acid and peroxide values.

20.4.7 Impact on Food Quality

The use of non-chemical pesticides on stored products can have various impacts on food quality. Temperature treatments, such as freezing or heating, can affect the nutritional content and alter the texture, flavor, and color of foods. For example, high temperatures might degrade certain vitamins (e.g., Vitamin C), while freezing might preserve most nutrients effectively. In contrast, in the modified atmosphere case, it can maintain the nutritional quality by slowing down the respiration rate of the produce, thus reducing the degradation of nutrients. In the case of botanical pesticides, it might impart a certain odor or taste to the stored products, which can be desirable or undesirable depending on the concentration and type of botanical used. Physical and modified atmosphere treatments can effectively preserve nutritional content and extend shelf life, though they might sometimes alter sensory qualities. Botanical pesticides and biological controls require careful management to avoid negative impacts on taste and safety.

20.4.8 Policies and Regulation Framework

The commercialization of non-chemical pesticides faces significant challenges, particularly in the regulatory frameworks that necessitate safe evaluation through toxicological research. For instance, to promote botanical pesticides in the market, it is crucial to educate farmers on application techniques and production methods to enhance the practicality of these pesticides. Collaboration among government agencies, industries, and entrepreneurs, and raising awareness about the environmental hazards of synthetic pesticides, is necessary to position botanical insecticides as a sustainable pest management alternative. Further research is required to address the limitations of botanical pesticide commercialization, including application concentration, stability, volatility under environmental conditions, and their effects on grain quality and shelf life. In the aspect of fumigants, compliance with regulations governing the use of protective fumigants can be complex and stringent, requiring thorough training, permits, and documentation. Regulatory agencies impose strict guidelines to minimize risks to human health and the environment, which can increase the administrative burden and cost of fumigation operations. Nevertheless, the novel-based solution to control the storage insects in a research and development base, when it starts the commercialization, needs to establish new regulatory policies, or it has to meet the regulatory frameworks.

Addressing these challenges requires comprehensive storage management strategies, including proper sanitation practices, pest control measures, temperature and humidity control, suitable packaging materials, regular monitoring, and adherence to food safety regulations. By effectively managing these aspects, stakeholders can minimize losses, ensure product quality, and uphold food safety standards throughout the storage process. However, irrespective of previous chapters, the challenges are addressed in Table 20.1.

TABLE 20.1
Summarization of Challenges in Non-Chemical Methods for Grain Protection

S. No.	Chapter Titles	Challenges and Future Outlook
1	Grain storage ecosystem	• Insects, rodents, and fungi can infest stored grains, leading to spoilage and significant economic losses • Inconsistent temperatures, high moisture levels, and improper ventilation can lead to condensation, mold growth, and changes in grain quality • Regular cleaning of storage facilities to remove debris, old grain, and other potential contaminants • Monitoring and controlling moisture levels through proper ventilation, aeration, and use of desiccants or drying equipment • Implementing temperature monitoring systems and utilizing climate control technologies to maintain optimal storage conditions • Employing IPM techniques, such as sanitation, sealing entry points, and using insecticides or traps • Routine inspections to detect any signs of spoilage, infestation, or structural issues in storage facilities • Sampling and testing grains for quality parameters, such as moisture content, temperature, and presence of contaminants
2	Pests of stored products	• Stored product pests comprise a diverse array of insect species, each with unique biological characteristics and vulnerabilities, necessitating tailored control approaches for effective management • The complex life cycles of stored product pests, involving multiple developmental stages, require precise timing and targeting of control measures to interrupt their reproductive cycles and prevent infestations • Pests often exhibit cryptic behavior, hiding within stored commodities and infrastructure, making detection and eradication challenging and requiring thorough inspection and monitoring techniques • Prolonged exposure to chemical pesticides has led to the development of resistance in many stored product pests, undermining the efficacy of conventional control methods and necessitating the exploration of alternative strategies

(Continued)

TABLE 20.1 *(Continued)*

Summarization of Challenges in Non-Chemical Methods for Grain Protection

S. No.	Chapter Titles	Challenges and Future Outlook
3	Detection of stored product pest	• Emerging technologies for pest detection, such as remote sensing drones or DNA barcoding equipment, may be costly to implement and require specialized training, limiting their accessibility to some storage facilities • Ensuring the sensitivity and accuracy of detection methods, particularly non-destructive techniques, can be challenging, as factors like the density and composition of stored commodities may affect their efficacy • Integrating emerging techniques with existing pest management systems and protocols can pose logistical challenges, requiring adjustments to workflows and staff training • Both conventional and emerging detection methods may be prone to false positives (incorrectly identifying pests) or false negatives (missing infestations), highlighting the importance of validation and calibration to minimize errors • Adhering to regulatory standards and guidelines regarding pest detection methods and reporting can present challenges for storage facilities, particularly when implementing new technologies or non-traditional approaches
4	Chemical methods of disinfestation	• Pests and pathogens can develop resistance to chemical agents over time, rendering them ineffective. Continuous use of the same chemicals can accelerate this process, leading to the emergence of resistant strains • Many chemical disinfectants can have adverse effects on the environment, including soil, water, and non-target organisms. Runoff from treated areas can contaminate water sources and harm aquatic life, while persistent chemicals can accumulate in the soil and disrupt ecosystems • Some chemical disinfectants may pose risks to human health, especially if used improperly or in high concentrations. Exposure to these chemicals can cause skin irritation, respiratory problems, or more serious health issues in extreme cases • Compliance with regulations and guidelines regarding the use of chemical disinfectants can be challenging for industries and agricultural practices. Strict regulations may limit the types and concentrations of chemicals that can be used, adding complexity and cost to disinfection processes • Residues from chemical disinfectants can remain on treated surfaces or in treated products, requiring careful management to ensure they do not pose risks to human health or contaminate food products. Residue management is particularly important in industries such as agriculture and food processing, where consumer safety is paramount
5	Conventional approaches	• Incorporating disinfestation principles into solar-based systems requires careful engineering to ensure effective pest control while maintaining grain quality. Balancing factors, such as temperature, humidity, and exposure time, is crucial to achieving optimal results without compromising the integrity of the grain • Successful disinfestation relies on a deep understanding of the life cycles and behaviors of target insect species. Different insects may have varying susceptibilities to heat, making it essential to tailor disinfestation methods to specific pest types encountered in grain storage facilities • While solar-based disinfestation methods offer environmentally friendly pest control solutions, their effects on grain quality must be thoroughly evaluated. Factors, such as moisture content, germination rates, and nutritional value, may be influenced by the application of heat, requiring careful monitoring and mitigation strategies to minimize any negative impacts

(Continued)

TABLE 20.1 *(Continued)*
Summarization of Challenges in Non-Chemical Methods for Grain Protection

S. No.	Chapter Titles	Challenges and Future Outlook
		• Designing solar tunnels and other solar-based systems to effectively disinfest grain presents unique engineering challenges. Ensuring uniform heating throughout the storage environment, optimizing airflow to distribute heat evenly, and implementing safety measures to prevent overheating or combustion are critical considerations in system design • Despite promising advancements in solar-based disinfestation technologies, widespread adoption faces barriers related to scalability, cost-effectiveness, and regulatory approval. Bridging the gap between research findings and practical implementation requires further investment in pilot projects, demonstration trials, and collaborative efforts among academia, industry, and regulatory bodies
6	Bio-based approaches	• Achieving consistent efficacy across different batches of bio-based products can be challenging due to variations in raw material quality, extraction methods, and microbial strains. Developing standardized protocols for product formulation, application, and quality control is essential to ensure reliable pest management outcomes • Prolonged use of bio-based approaches may lead to the development of resistance in target pest populations, reducing treatment effectiveness over time. Implementing IPM strategies that combine multiple control tactics, rotating products with different modes of action, and monitoring for resistance emergence are critical for preserving the long-term viability of bio-based solutions • Bio-based products often face regulatory hurdles related to safety, efficacy, and labeling requirements. Obtaining regulatory approval for new formulations or active ingredients can be time-consuming and expensive, delaying their commercialization and market entry. Additionally, consumer perception and acceptance of bio-based pest management solutions may influence market demand and adoption rates • The production, formulation, and application of bio-based products may entail higher costs compared to conventional chemical pesticides. Ensuring cost-effectiveness and affordability for end-users, particularly small-scale farmers or grain storage facilities in developing countries, is crucial for promoting widespread adoption and sustainable pest management practices • Promoting the adoption of bio-based approaches requires knowledge transfer and capacity-building initiatives to educate farmers, extension agents, and other stakeholders about their benefits, proper application techniques, and IPM principles. Investing in training programs, demonstration trials, and outreach activities can facilitate technology transfer and foster local expertise in bio-based pest management strategies
7	Biological control methods	• Ensuring the effectiveness of biological control agents relies on their specificity to target pests and their ability to achieve sufficient pest suppression levels. Matching the right natural enemies or entomopathogens to the target pest species and optimizing application protocols are critical for achieving desired outcomes • This can be influenced by environmental factors, such as temperature, humidity, and habitat suitability for natural enemies. Adverse weather conditions or habitat disturbances may disrupt the balance between pests and their natural enemies, compromising the efficacy of biological control strategies • Obtaining regulatory approval for the release or use of biological control agents may require extensive safety and efficacy testing, as well as compliance with biosecurity regulations. Market acceptance and adoption may also be influenced by factors, such as consumer perceptions, labeling requirements, and availability of trained personnel for implementation • This is most effective when integrated with other pest management tactics as part of an IPM approach. Coordinating the timing and application of biological control agents with cultural, mechanical, and chemical control measures requires careful planning and collaboration among stakeholders

(Continued)

TABLE 20.1 *(Continued)*
Summarization of Challenges in Non-Chemical Methods for Grain Protection

S. No.	Chapter Titles	Challenges and Future Outlook
8	Use of inert materials	• Achieving optimal coverage and distribution of inert materials within storage facilities can be challenging, especially in large-scale grain storage operations. Developing efficient application methods and equipment that ensure uniform dispersal of materials while minimizing dust generation and worker exposure is essential for maximizing effectiveness and safety • Continuous monitoring of pest populations and their susceptibility to inert materials is necessary to detect signs of resistance development. Implementing IPM strategies that combine inert materials with other control tactics, such as sanitation or temperature management, can help mitigate the risk of resistance and prolong the effectiveness of these treatments • Inert materials may be subject to regulatory oversight and labeling requirements, particularly in food storage and processing facilities. Ensuring compliance with safety regulations and product standards, as well as providing adequate training and documentation for workers handling these materials, is essential for regulatory compliance and consumer safety • Promoting awareness and understanding of inert materials and their proper application methods among grain producers, storage facility operators, and pest management professionals is crucial for their effective adoption and implementation. Providing educational resources, training programs, and technical support can help build capacity and confidence in utilizing inert materials as part of IPM strategies • Further research is needed to explore new formulations, application techniques, and synergistic combinations of inert materials with other pest management tools. Investing in research and development efforts to address knowledge gaps, optimize efficacy, and overcome practical challenges can drive innovation and improve the effectiveness of inert materials in grain pest management
9	Aeration	• Achieving uniform air distribution throughout the grain mass is challenging, especially in large storage facilities with complex geometries. Variations in airflow can lead to uneven moisture and temperature distribution, increasing the risk of spoilage • Determining the most effective aeration strategy for a particular grain type and storage environment can be challenging. Factors, such as grain moisture content, ambient conditions, and storage structure design, must be considered to develop an optimal aeration plan • Predicting airflow patterns within grain storage structures is complex due to the influence of various factors, such as grain compaction, storage geometry, and airflow equipment. Accurately modeling airflow behavior is essential for designing efficient aeration systems • Balancing the need for adequate aeration with energy efficiency is a significant challenge. Aeration systems can consume substantial amounts of energy, particularly in large-scale storage facilities. Developing energy-efficient aeration systems while ensuring optimal grain quality is a key challenge • Continuous monitoring and control of moisture content, temperature, and airflow are essential for grain quality preservation. However, implementing effective monitoring and control systems, especially in remote or large-scale facilities, presents challenges, such as sensor placement, data interpretation, and automation complexity. Developing robust monitoring and control solutions is critical for mitigating spoilage risks

(Continued)

TABLE 20.1 *(Continued)*
Summarization of Challenges in Non-Chemical Methods for Grain Protection

S. No.	Chapter Titles	Challenges and Future Outlook
10	Extreme temperature-based approaches	• One of the primary challenges is minimizing energy consumption while implementing extreme temperature-based approaches. High energy demands can lead to increased operational costs and environmental concerns • Ensuring uniform treatment of grains throughout the storage facility poses a challenge. Variations in temperature or moisture levels may result in uneven treatment, leading to potential spoilage or pest infestation • The initial investment and ongoing maintenance costs associated with specialized equipment for extreme temperature-based approaches can be substantial, posing financial challenges for grain storage operators • Maintaining grain quality while implementing extreme temperature-based approaches is crucial. Excessive heat or moisture levels can lead to grain damage, loss of nutritional value, or changes in texture and taste, necessitating careful management to avoid quality degradation • Continuous monitoring and precise control of temperature and moisture levels are essential but challenging tasks. Implementing robust monitoring systems and automated control mechanisms is necessary to ensure grain safety and quality, particularly in large-scale storage facilities where manual monitoring is impractical • Different grain types and storage conditions may require tailored approaches for effective temperature-based treatment. Developing flexible strategies that can adapt to various grain types and storage environments is a key challenge • Meeting regulatory requirements and food safety standards adds another layer of complexity to implementing extreme temperature-based approaches. Ensuring compliance with relevant regulations while maintaining operational efficiency is essential but challenging
11	Dielectric heating and advanced thermal approaches	• Achieving uniform heating throughout the grain mass is crucial to ensure effective treatment and prevent localized hot spots or cold spots. However, the dielectric properties of grains can vary, leading to uneven heating and potential quality issues. Developing methods to ensure uniform heating across the entire grain bulk is a tedious task • Implementing microwave or radiofrequency heating systems in grain storage facilities requires specialized equipment design and integration. Ensuring compatibility with existing infrastructure and optimizing system performance while minimizing interference with other operations poses technical challenges • Dielectric heating methods are sensitive to moisture content, and excessive moisture can lead to inefficient heating or even equipment malfunction. Achieving precise control over moisture content throughout the grain bulk, especially in large storage facilities, presents a significant challenge • Optimizing energy efficiency is essential to minimize operational costs and environmental impact when using advanced thermal approaches. Dielectric heating systems can consume significant amounts of energy, and improving efficiency without compromising treatment effectiveness • Maintaining grain quality during dielectric heating or advanced thermal treatments is critical. Excessive heating can also lead to grain damage, loss of nutritional value, or changes in sensory • Ensuring the safety of personnel and grain storage facilities during dielectric heating or advanced thermal treatments is paramount. Addressing potential hazards, such as fire risks, equipment malfunctions, or electromagnetic interference, requires careful planning and implementation of safety measures

(Continued)

TABLE 20.1 (Continued)
Summarization of Challenges in Non-Chemical Methods for Grain Protection

S. No.	Chapter Titles	Challenges and Future Outlook
		• Meeting regulatory requirements and safety standards for food processing and grain storage adds complexity to implementing advanced thermal approaches. Ensuring compliance with relevant regulations while maintaining operational efficiency is a challenge that requires close attention to detail
12	Non-thermal technologies	• Ensuring uniform treatment of grains throughout the storage facility is crucial for effective pest control and microbial reduction. Achieving consistent exposure to cold plasma or ultraviolet (UV) light across the entire grain bulk, including hard-to-reach areas, is a challenge • Penetration depth of non-thermal technologies like cold plasma and UV light into the grain mass can be limited, particularly in denser or larger grain volumes. Overcoming this challenge to ensure adequate treatment depth while maintaining treatment efficacy is essential • Determining the optimal parameters for cold plasma or UV light treatment, such as exposure time, intensity, and wavelength, requires thorough experimentation and optimization. Balancing treatment effectiveness with energy efficiency and grain quality preservation presents a challenge • Non-thermal treatments should not compromise grain quality or nutritional value. However, prolonged exposure to cold plasma or UV light may lead to undesired effects, such as color changes, texture alterations, or nutrient degradation. Developing treatment protocols that minimize adverse effects on grain quality while ensuring effective treatment • Designing and integrating cold plasma or UV light systems into existing grain storage facilities can be complex. Ensuring compatibility with storage infrastructure, scalability, and reliability of equipment while minimizing operational disruptions • Ensuring the safety of personnel and grains during non-thermal treatments is paramount. Addressing potential hazards such as ozone generation in cold plasma systems or UV light exposure requires careful implementation of safety measures and monitoring protocols • Meeting regulatory requirements and food safety standards for non-thermal treatments in grain storage systems adds complexity. Ensuring compliance with relevant regulations while maintaining treatment effectiveness and operational efficiency is essential but challenging
13	Ionizing radiation	• Ionizing radiation technologies are subject to strict regulations and guidelines regarding their use in food processing and storage. Ensuring compliance with regulatory requirements, such as radiation dose limits and labeling regulations, is essential but can be complex and costly to implement • There may be public concerns or misconceptions regarding the safety of irradiated food products. Overcoming negative perceptions and building consumer trust in the safety and benefits of irradiation poses a significant challenge for widespread adoption in grain storage systems • Implementing ionizing radiation facilities in grain storage systems requires substantial initial investment in specialized equipment such as irradiators or electron beam generators. Additionally, infrastructure modifications may be necessary to accommodate radiation facilities, adding to the overall cost and complexity • Ensuring uniform irradiation of grains throughout the storage facility is critical for effective pest control and microbial reduction. However, achieving consistent exposure to ionizing radiation across the entire grain bulk, including densely packed areas or grain layers, can be challenging. The penetration depth of radiation into the grain mass also needs to be considered to ensure thorough treatment

(Continued)

TABLE 20.1 *(Continued)*
Summarization of Challenges in Non-Chemical Methods for Grain Protection

S. No.	Chapter Titles	Challenges and Future Outlook
		• Ionizing radiation can induce chemical and physical changes in grains, affecting their quality attributes such as texture, flavor, and nutritional content. Minimizing adverse effects on grain quality while achieving desired treatment outcomes presents a significant challenge in the application of ionizing radiation in grain storage systems • Ensuring the safety of personnel, consumers, and the environment during irradiation processes is paramount. Addressing potential hazards such as radiation leakage, worker exposure, and radioactive waste disposal requires stringent safety protocols and monitoring mechanisms • While ionizing radiation is effective for pest control, there is growing interest in alternative non-chemical pest management methods due to concerns over environmental impact and resistance development. Exploring IPM approaches that combine ionizing radiation with other non-chemical methods presents a challenge for optimizing pest control efficacy while minimizing environmental impact
14	Use of other gases non-chemical fumigation	• Ensuring that the chosen gas effectively controls pests and molds while preserving grain quality is a primary challenge. Balancing treatment efficacy with minimal impact on grain viability, nutritional quality, and sensory attributes requires careful optimization and understanding of gas-grain interactions • Achieving uniform distribution and penetration of the fumigant gas throughout the grain mass is essential for effective pest control. However, grain bulk density, storage configuration, and airflow patterns can affect gas distribution, leading to uneven treatment and potential pest survival • Some fumigant gases, such as ozone, can pose safety risks to personnel and the environment if not handled properly. Addressing potential hazards, such as ozone exposure or carbon dioxide buildup, requires implementing appropriate safety measures and monitoring protocols to ensure worker and environmental safety • While non-chemical fumigation methods may not leave behind chemical residues, managing residual gas levels and ensuring safe ventilation post-treatment are essential. Residue management involves minimizing gas concentrations to safe levels and addressing potential environmental impacts, such as ozone depletion or greenhouse gas emissions • Compliance with regulatory requirements and guidelines governing gas usage, exposure limits, and residue management is critical. Meeting regulatory standards for food safety, worker safety, and environmental protection poses challenges in implementing non-chemical fumigation methods in grain storage systems • The efficacy of non-chemical fumigation methods may vary depending on the type of grain and storage conditions. Factors, such as grain moisture content, temperature, and aeration methods, can influence gas interactions and treatment effectiveness, posing challenges for consistent pest control across different grain types and storage environments • Continuous monitoring and control of gas concentrations, environmental conditions, and treatment parameters are necessary to ensure effective fumigation and prevent overexposure or underexposure. Implementing robust monitoring systems and automated control mechanisms to achieve precise treatment delivery is challenging, particularly in large-scale grain storage facilities

(Continued)

TABLE 20.1 *(Continued)*
Summarization of Challenges in Non-Chemical Methods for Grain Protection

S. No.	Chapter Titles	Challenges and Future Outlook
15	Modifying gas concentration in storage environments	• Determining the optimal gas composition and levels for storage conditions requires careful consideration of factors, such as grain type, moisture content, temperature, and desired storage duration. Achieving the right balance of gases to inhibit pests, molds, and microbial growth while preserving grain quality poses a challenge • Maintaining the desired gas composition within storage environments can be challenging due to gas leakage and permeation through packaging materials or storage structures. Addressing factors, such as gas diffusion rates, sealing integrity, and storage facility design, is essential to minimize gas losses and maintain effective gas concentrations • Continuous monitoring and control of gas concentrations, environmental conditions, and storage parameters are critical for ensuring storage efficacy and grain quality preservation. Implementing robust monitoring systems and automated control mechanisms to maintain desired gas levels and respond to fluctuations presents technical challenges, especially in large-scale storage facilities • Implementing gas modification techniques, such as modified atmosphere packaging (MAP), controlled atmosphere storage (CAS), hermetic storage, or vacuum/low-pressure storage, may require specialized equipment and infrastructure modifications. Ensuring compatibility with existing storage facilities, scalability, and reliability of equipment while minimizing operational disruptions poses technical and financial challenges • Manipulating gas concentrations in storage environments can pose safety risks to personnel and the environment. Addressing potential hazards, such as oxygen depletion or gas leakage, requires implementing appropriate safety measures, training protocols, and emergency response plans to ensure worker and environmental safety • Compliance with regulatory requirements and guidelines governing gas modification techniques, food safety standards, and environmental regulations is essential. Ensuring adherence to relevant regulations for gas usage, storage conditions, and safety protocols poses challenges in implementing these techniques in grain storage systems • Maintaining grain quality during storage under modified gas environments is crucial. Factors, such as moisture content, temperature, and gas exposure levels, can affect grain quality attributes, such as texture, flavor, and nutritional value. Balancing pest control efficacy with grain quality preservation poses challenges in optimizing storage conditions
16	Insect-resistant/repellent packaging	• Developing packaging concepts that effectively repel or inhibit insect infestation while maintaining product integrity and usability is challenging. Balancing the need for protective barriers against pests with practical considerations, such as ease of handling, storage efficiency, and cost-effectiveness, requires innovative design approaches • Identifying suitable materials with inherent insect-repellent properties or incorporating insect-repellent additives into packaging materials presents challenges. Factors, such as material compatibility with food products, durability, recyclability, and regulatory compliance, must be considered when selecting packaging materials for insect-resistant applications • Ensuring the durability and long-term effectiveness of insect-resistant packaging materials under various storage conditions, including temperature fluctuations, humidity levels, and mechanical stresses, is essential. Testing for durability, shelf-life stability, and performance against different insect species poses technical challenges

(Continued)

TABLE 20.1 (Continued)
Summarization of Challenges in Non-Chemical Methods for Grain Protection

S. No.	Chapter Titles	Challenges and Future Outlook
		• Compliance with regulatory requirements and standards governing food contact materials, insecticides, and packaging additives is critical. Ensuring that insect-resistant packaging materials meet safety, quality, and environmental regulations presents challenges in formulation, labeling, and certification • Consumer acceptance of insect-resistant packaging may vary due to perceptions of safety, effectiveness, and environmental impact. Overcoming consumer skepticism and building trust in the efficacy and safety of insect-resistant packaging materials pose challenges in marketing and communication efforts • Insects may develop resistance to insect-repellent or growth regulator compounds over time, reducing the effectiveness of insect-resistant packaging solutions. Implementing strategies to minimize resistance development, such as rotating insecticides or utilizing multiple modes of action, presents challenges in pest management • The cost-effectiveness of insect-resistant packaging solutions compared to traditional packaging methods is a key consideration for adoption. Balancing the benefits of pest prevention and product preservation with the additional costs associated with insect-resistant materials or additives poses economic challenges for grain storage operators • Integrating insect-resistant packaging with other pest management practices, such as sanitation, monitoring, and chemical treatments, is essential for comprehensive pest control. Developing IPM strategies that optimize the synergistic effects of packaging solutions with other control methods presents logistical challenges
17	Insect traps	• Ensuring that insect traps effectively capture target pests while minimizing non-target captures is crucial. Optimizing trap design, placement, baiting strategies, and trap maintenance to maximize pest capture rates presents challenges, especially considering variations in insect behavior and species-specific preferences • Continuous monitoring and detection of insect populations using traps are essential for early pest detection and intervention. However, ensuring timely inspection, trap maintenance, and data collection in large-scale storage facilities can be logistically challenging, requiring efficient monitoring protocols and resources • Selecting trapping methods and baits that are specific to target pest species while minimizing attraction to beneficial insects or non-target species is challenging. Developing trap designs and bait formulations that selectively attract and capture specific pest species poses technical challenges in pest management • Determining optimal trap placement and coverage within storage facilities to maximize pest detection and capture efficacy presents challenges. Factors, such as storage layout, airflow patterns, pest behavior, and environmental conditions, must be considered to ensure comprehensive coverage and effectiveness of trapping efforts • Environmental conditions, such as temperature, humidity, and lighting, can influence trap efficacy and pest behavior. Addressing fluctuations in environmental conditions and their impact on trap performance requires adaptive management strategies and robust trap designs • Regular maintenance and refurbishment of traps are essential to ensure continued efficacy and reliability. Challenges associated with trap maintenance include trap cleaning, bait replenishment, equipment repair, and disposal of captured pests, especially in remote or large-scale storage facilities • Implementing and maintaining insect traps in grain storage systems can incur significant costs in terms of equipment, materials, labor, and ongoing monitoring efforts. Balancing the benefits of trap-based pest management with associated costs and resource constraints poses economic challenges for grain storage operators

(Continued)

TABLE 20.1 *(Continued)*
Summarization of Challenges in Non-Chemical Methods for Grain Protection

S. No.	Chapter Titles	Challenges and Future Outlook
		• Integrating trap-based pest monitoring and control with other pest management practices, such as sanitation, cultural controls, and chemical treatments, is essential for comprehensive pest management. Developing IPM strategies that optimize the synergistic effects of trapping with other control methods presents logistical and coordination challenges
18	Mechanical approaches	• Ensuring that mechanical methods effectively remove pests while minimizing damage to grain and preserving grain quality is crucial. Achieving high efficiency in pest removal, especially for small insects or larvae, poses challenges due to variations in pest density, grain characteristics, and equipment design • Mechanical methods of pest removal may require handling large volumes of grain, which can be labor-intensive and time-consuming. Minimizing grain losses, dust generation, and operational disruptions during the removal process presents logistical challenges in grain storage facilities • Developing and optimizing equipment for sieving, impaction, and insect removal requires careful consideration of factors such as grain size, moisture content, pest morphology, and processing capacity. Balancing equipment complexity, cost, and performance to meet operational needs poses engineering challenges • Integrating mechanical pest removal equipment with existing storage infrastructure, such as grain bins, silos, or conveyor systems, presents technical challenges. Ensuring compatibility, scalability, and efficient operation within the storage environment while minimizing installation and retrofitting costs requires careful planning and design • Ensuring the reliability and consistent performance of mechanical pest removal equipment over time requires regular maintenance, calibration, and repair. Challenges associated with equipment downtime, component wear, and replacement parts availability can impact operational efficiency and grain storage management • Mechanical pest removal methods may require significant energy consumption, particularly for large-scale operations. Addressing energy efficiency and environmental sustainability concerns, such as minimizing carbon footprint and waste generation, presents challenges in equipment design and operation • Integrating mechanical pest removal methods with other pest management practices, such as monitoring, sanitation, and chemical treatments, is essential for comprehensive pest control. Developing IPM strategies that optimize the synergistic effects of mechanical methods with other control measures presents coordination and implementation challenges • Ensuring worker safety and ergonomic considerations during the operation and maintenance of mechanical pest removal equipment is paramount. Addressing potential hazards, such as pinch points, entanglement risks, and ergonomic strain, requires implementing appropriate safety protocols and training programs
19	Nano-based techniques	• Ensuring the stability and uniform dispersion of nanoparticles in grain storage environments is crucial for effective pest control. Challenges related to nanoparticle aggregation, sedimentation, and interactions with grain surfaces must be addressed to maintain consistent efficacy over time • Understanding the potential toxicity and environmental impact of nanoparticles on non-target organisms, humans, and ecosystems is essential. Addressing safety concerns, regulatory requirements, and risk assessment protocols for nanoparticle use in grain storage presents challenges in ensuring safe and sustainable pest management practices

(Continued)

TABLE 20.1 *(Continued)*
Summarization of Challenges in Non-Chemical Methods for Grain Protection

S. No.	Chapter Titles	Challenges and Future Outlook
		• Investigating the interactions between nanoparticles and grain components, such as proteins, lipids, and carbohydrates, is critical for understanding their efficacy and potential effects on grain quality. Challenges related to nanoparticle adsorption, penetration, and stability within the grain matrix must be addressed to optimize pest control outcomes while minimizing adverse effects on grain properties • Managing nanoparticle release and residues in grain storage environments is important for ensuring food safety and environmental sustainability. Challenges associated with nanoparticle persistence, mobility, and degradation require the development of effective residue management strategies to minimize potential risks and impacts • Developing efficient delivery systems and application methods for nanoparticles in grain storage presents technical challenges. Optimizing nanoparticle formulations, encapsulation techniques, and application protocols to ensure uniform coverage and targeted delivery to pest-infested areas requires innovative approaches and technology development • Monitoring and mitigating the risk of pest resistance development to nanoparticle-based treatments is essential for long-term pest management success. Understanding the mechanisms of resistance and implementing strategies to minimize selection pressure on pest populations pose challenges in maintaining the efficacy and sustainability of nanoparticle-based pest control • Scaling up nanoparticle-based pest management technologies from laboratory-scale research to commercial application presents logistical and economic challenges. Addressing issues related to production scalability, cost-effectiveness, and market acceptance requires collaboration among researchers, industry stakeholders, and regulatory agencies • Integrating nanoparticle-based pest management strategies with other pest control methods, such as monitoring, sanitation, and biological control, is important for comprehensive pest management. Challenges related to optimizing synergistic effects, minimizing conflicts, and ensuring compatibility with existing practices require interdisciplinary collaboration and strategic planning

20.5 CONCLUDING REMARKS

In summary, preventing insect infestation and reducing chemical residues in food requires innovative, sustainable storage solutions. Methods include hermetic storage, modified atmosphere packaging, natural preservatives, bio-based packaging, and controlled atmosphere storage, which have the potential to address these challenges while maintaining food safety and quality. Implementing these alternatives enhances environmental sustainability and meets consumer demand for healthier, natural food choices. Further, by exploring and implementing innovative storage techniques, it may be possible to move closer to a resilient, sustainable food system for the future. Overall, when proper guidance is applied, non-chemical methods can maintain or even enhance food quality while ensuring safety and reducing environmental impact.

REFERENCES

Abdelfattah, N. A. H., E. A. Hasan, and S. H. Qari. 2022. "Cytotoxicity and Genotoxicity Evaluation of Some Stored Grain Insects and Their Infested Flour Using the BHK-21 Cell Line in an In Vitro Experimental Model." *Journal of Food Quality* 2022 (December): 1–12. doi:10.1155/2022/6415310.

Afrose Subaitha, Z., P. Santhoshkumar, J. A. Moses, and M. Loganathan. 2024. "Nonchemical Strategies for Stored Product Pest Management: Exploring the Potential of Spices, Herbs, and Their Formulations." *Food Control* 158 (October 2023). Elsevier Ltd: 110212. doi:10.1016/j.foodcont.2023.110212.

Anukiruthika, T., F. Jian, and D. S. Jayas. 2021. "Movement and Behavioral Response of Stored Product Insects under Stored Grain Environments – A Review." *Journal of Stored Products Research* 90 (January): 101752. doi:10.1016/j.jspr.2020.101752.

Baliyarsingh B. Pradhan CK. 2023. "Prospects of plant-derived metallic nanopesticides against storage pests-A review." *Journal of Agriculture and Food Research.* 14, 100687. doi:https://dx.doi.org/10.1016/j.jafr.2023.100687.

De Corato, U. 2020. "Improving the Shelf-Life and Quality of Fresh and Minimally-Processed Fruits and Vegetables for a Modern Food Industry: A Comprehensive Critical Review from the Traditional Technologies into the Most Promising Advancements." *Critical Reviews in Food Science and Nutrition* 60 (6): 940–975. doi:10.1080/10408398.2018.1553025.

Kumar, D., and P. Kalita. 2017. "Reducing Postharvest Losses during Storage of Grain Crops to Strengthen Food Security in Developing Countries." *Foods* 6 (1): 8. doi:10.3390/foods6010008.

Mohapatra, D., A. Kar, and S. K. Giri. 2015. "Insect Pest Management in Stored Pulses: An Overview." *Food and Bioprocess Technology* 8 (2): 239–265. doi:10.1007/s11947-014-1399-2.

Morrison, W. R., E. D. Scully, and J. F. Campbell. 2021. "Towards Developing Areawide Semiochemical-Mediated, Behaviorally-Based Integrated Pest Management Programs for Stored Product Insects." *Pest Management Science* 77 (6): 2667–2682. doi:10.1002/ps.6289.

Yaseen, M., T. Kausar, B. Praween, S. J. Shah, Y. Jan, S. S. Shekhawat, M. Malik, and Z. R. A. Ahmad Azad. 2019. "Insect Pest Infestation during Storage of Cereal Grains, Pulses and Oilseeds." In *Health and Safety Aspects of Food Processing Technologies*, 209–234. Cham: Springer International Publishing. doi:10.1007/978-3-030-24903-8_8.

Ziegler, V., R. T. Paraginski, and C. D. Ferreira. 2021. "Grain Storage Systems and Effects of Moisture, Temperature and Time on Grain Quality – A Review." *Journal of Stored Products Research* 91 (March): 101770. doi:10.1016/j.jspr.2021.101770.

Index

Note: Locators in *italics* represent figures and **bold** indicate tables in the text.

D

K

L

M

Q

R

S

T

U

V

W

X

Z